ENERGY AND VARIATIONAL METHODS IN APPLIED MECHANICS

ENERGY AND VARIATIONAL METHODS IN APPLIED MECHANICS

With an Introduction to the Finite Element Method

J. N. REDDY

Department of Engineering Science and Mechanics
Virginia Polytechnic Institute and State University
Blacksburg, Virginia

A WILEY-INTERSCIENCE PUBLICATION

JOHN WILEY & SONS

New York • Chichester • Brisbane • Toronto • Singapore

Library of Congress Cataloging in Publication Data:

Reddy, J. N. (Junuthula Narasimha), 1945–
    Energy and variational methods in applied mechanics.

    "A Wiley-Interscience publication."
    Bibliography: p.
    Includes index.
    1. Mechanics, Applied—Mathematics. 2. Calculus of
variations. 3. Finite element method. 4. Force and
energy. I. Title.
TA350.R39 1984   620'.001'515353   84-3605
ISBN 0-471-89673-X

To

The Red, White and Blue
The Land of Opportunity

ఎందరో మహానుభావులు
అందరికి నా వందనములు

PREFACE

The increasing use of finite element methods in engineering and applied science has shed new light on the importance of energy and variational methods. The number of engineering courses and research papers that make use of variational and energy methods has also grown very rapidly in recent years. In view of the increased use of the variational methods (including the finite element method), there is a need to introduce the concepts of energy and variational methods and their use in the formulation and solution of equations of mechanics to both undergraduate and beginning graduate students. This book is intended for senior undergraduate students and beginning graduate students in aerospace, civil and mechanical engineering, and applied mechanics, who have had a course in ordinary and partial differential equations.

The text is organized into four chapters. Chapter 1 is essentially a review, especially for graduate students, of the equations of applied mechanics. Much of the chapter can be assigned as reading material to the student. The equations of bars, beams, torsion, and plane elasticity presented in Section 1.7 are used to illustrate concepts from energy and variational methods.

Chapter 2 deals with the study of the basic topics from variational calculus, virtual work and energy principles, and energy methods of mechanics. The instructor can omit Section 2.4 on stationary principles and Section 2.5 on Hamilton's principle if he or she wants to cover all of Chapter 4. Classical variational methods of approximation (e.g., the methods of Ritz, Galerkin, Kantorovich, etc.) and the finite element method are introduced and illustrated in Chapter 3 via linear problems of science and engineering, especially solid mechanics. A unified approach, more general than that found in most solid mechanics books, is used to introduce the variational methods. As a result, the student can readily extend the methods to other subject areas of solid mechanics as well as to other branches of engineering.

The classical variational methods and the finite element method are put to work in Chapter 4 in the derivation and approximate solution of the governing equations of elastic plates and shells. In the interest of completeness, and for use as a reference for approximate solutions, exact solutions of plates and shells are also included. Keeping the current developments in composite-material structures in mind, a brief but reasonably complete, discussion of laminated plates and shells is included in Sections 4.3 and 4.4.

The book contains many example and exercise problems that illustrate, test, and broaden the understanding of the topics covered. A long list of references, by no means complete or up-to-date, is provided in the Bibliography at the end of the book.

The author wishes to acknowledge with great pleasure and appreciation the encouragement and support by Professor Daniel Frederick (ESM Department at Virginia Tech.) during the course of this writing and the skillful typing of the manuscript by Mrs. Vanessa McCoy. The author is also thankful to the many students who, by their comments, contributed to the improvement of this book. Special thanks to K. Chandrashekhara, Glenn Creamer, C. F. Liu and Paul Heyliger for their help in proofreading the galleys and pages, and to Dr. Ozden Ochoa for constructive comments on a preliminary draft of the manuscript. It is a pleasure to acknowledge, with many thanks, the cooperation of the technical staff at Wiley, New York (Frank Cerra, Christina Mikulak, and Lisa Morano).

J. N. Reddy

Blacksburg, Virginia
June 1984

CONTENTS

1

A REVIEW OF THE
EQUATIONS OF MECHANICS

1.1 INTRODUCTION

The phrase "energy methods" in the present study refers to methods that make use of the energy of a system to obtain values of the unknown at a specific point. These include Castigliano's theorems, unit-dummy-load and displacement methods, and Betti's and Maxwell's theorems. These methods are limited to the (exact) determination of generalized displacements or generalized forces at fixed points in the structure, and they cannot be used to determine the solution (i.e., displacements and/or forces) as a function of position in the structure. The phrase "variational methods" refers to methods that make use of the variational principles, such as the principle of virtual displacements, to determine approximate displacements as continuous functions of position in a body. In the classical sense, *variational principle* has to do with the minimization of a functional, which includes all the intrinsic features of the problem, such as the governing equations, boundary and/or initial conditions, and constraint conditions.

Variational formulations can be useful in three related ways. First, many problems of mechanics are posed in terms of finding the extremum (i.e., minima or maxima) and thus, by their nature, can be formulated in terms of variational statements. Second, there are problems that can be formulated by other means, such as by vector mechanics (e.g., Newton's laws), but these can also be formulated by means of variational principles. Third, variational formulations form a powerful basis for obtaining approximate solutions to practical problems, many of which are intractable otherwise. The last two applications provide the motivation for the present study. Below we consider an example of each usage of variational formulations.

An example of problems posed in terms of finding the minimum is provided by the Brachistochrone problem: determine the geometry of the centerline of a chute between two points $A:(0,0)$ and $B:(a, b)$ in a vertical plane such that a material particle of mass m, sliding without friction under its own weight, travels from point A to point B in the shortest (i.e., minimum) time. The variational formulation of the problem begins with the application of the principle of the conservation of energy. If $y = u(x)$ is the equation of the curve joining point A to point B, at some instant of time t, the velocity of the particle can be obtained by equating the energies at time $t = 0$ and time t. At time $t = 0$, the energy is only due to the potential energy. Therefore, we have

$$mgb = mg(b - u) + \tfrac{1}{2}mv^2$$

or

$$mgu = \tfrac{1}{2}mv^2 \quad \text{or} \quad v = \sqrt{2gu}$$

where g is the acceleration owing to gravity and v is the velocity of the particle. If s represents the distance along $y = u(x)$, measured from A, we have

$$v = \frac{ds}{dt} = \frac{\sqrt{(dx)^2 + (du)^2}}{dt} = \sqrt{1 + (u')^2}\,\frac{dx}{dt}$$

where u' denotes du/dx. The time taken by the particle in traveling a distance ds is

$$dt = \sqrt{\frac{1 + (u')^2}{v^2}}\ dx = \sqrt{\frac{1 + (u')^2}{2gu}}\ dx$$

The total time taken by a particle in going from point A to point B is given by

$$T(u) = \int_0^a \sqrt{\frac{1 + (u')^2}{2gu}}\ dx \tag{a}$$

Thus, the problem is reduced to the following variational problem: find a function $u(x)$ within the class of functions with continuous derivatives [so that T in Eq. (a) can be evaluated] that satisfies the *end conditions*

$$u(0) = 0, \qquad u(a) = b \tag{b}$$

and minimizes the integral expression in Eq. (a). One should note that this problem cannot be formulated directly from vector mechanics.

As an illustration of problems that can be formulated by vector mechanics, consider the situation of a simply supported beam bent by a uniformly

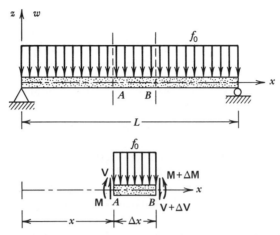

Figure 1.1 Equilibrium of a simply supported beam under uniform loading.

distributed load f_0 (see Fig. 1.1). Consider an element of length Δx in the beam and label the shear forces (V), bending moments (M), and externally applied force (f_0) on the element. Application of the Newton's second law to the free-body-diagram of the element gives

$$\Sigma F_y = 0: V - f_0 \Delta x - (V + \Delta V) = 0$$

$$(+)\Sigma M_B = 0: -V\Delta x - M + (M + \Delta M) + (f_0 \Delta x)\frac{\Delta x}{2} = 0$$

Dividing by Δx and taking the limit $\Delta x \to 0$, one obtains

$$\frac{dV}{dx} = -f_0, \qquad \frac{dM}{dx} = V$$

Eliminating V from the equations, we obtain

$$\frac{d^2 M}{dx^2} + f_0 = 0 \tag{c}$$

One can recall from a course on the mechanics of deformable bodies that the bending moment is related to the transverse deflection $w(x)$ by the flexure equation

$$M = EI\frac{d^2 w}{dx^2} \tag{d}$$

where E is Young's modulus and I is the moment of inertia of the beam (both of which could be functions of x). Combining Eq. (d) with Eq. (c), one obtains

$$\frac{d^2}{dx^2}\left(EI\frac{d^2w}{dx^2}\right) + f_0 = 0 \qquad (0 < x < L) \tag{e}$$

Equation (e) can be solved for w with the boundary conditions (of the simply supported beam)

$$w(0) = w(L) = 0, \qquad M(0) = M(L) = 0 \tag{f}$$

The problem of bending of beams can also be formulated by means of a variational principle. As shown later in this book, the variational problem associated with the simply-supported beam involves determining a function $w(x)$ that satisfies the geometric boundary conditions

$$w(0) = w(L) = 0 \tag{g}$$

and minimizes the integral expression

$$\Pi(w) = \int_0^L \left[\frac{EI}{2}\left(\frac{d^2w}{dx^2}\right)^2 + f_0 w\right] dx \tag{h}$$

The expression in Eq. (h) is known as the *total potential energy* of the simply-supported beam. The first term represents the strain energy and the second term represents the work done by the distributed load. In other words, the solution for Eqs. (e) and (f) is also the function that satisfies the boundary conditions in Eq. (g) and minimizes the total potential energy.

The third use of variational formulations lies in the determination of approximate solutions of problems. As an example, consider the beam problem discussed above. As an approximation for w, one can choose

$$w_1(x) = c_1 x(L - x) \qquad (c_1 = \text{constant})$$

which has a continuous second-order derivative and which satisfies the boundary conditions in Eq. (g). The function w_1 represents an approximation to the true solution if it minimizes the functional in Eq. (h). Substituting w_1 for w into (h), one obtains

$$\Pi(c_1) = \int_0^L \left[\frac{EI}{2}(-2c_1)^2 + c_1(Lx - x^2)f_0\right] dx$$

$$= 2EILc_1^2 + \frac{L^3}{6}f_0 c_1$$

Now w_1 is an approximate solution to a simply-supported beam under uniform loading if c_1 minimizes $\Pi = \Pi(c_1)$. From the calculus of ordinary functions, a necessary condition for a function to attain its minimum (or maximum) is that its first derivative with respect to its argument be zero. In the present case, we have

$$\frac{d\Pi}{dc_1} = 0 = 4EILc_1 + \frac{f_0 L^3}{6}$$

from which we get $c_1 = -f_0 L^2/24EI$, and the solution w_1 becomes

$$w_1(x) = -\frac{f_0 L^2}{24EI} x(L - x) \tag{i}$$

The exact solution of Eqs. (e) and (f) is given by

$$w(x) = -\frac{f_0}{24EI} x(L - x)(L^2 + Lx - x^2) \tag{j}$$

The approximate solution is 50% in error in predicting the maximum deflection. The accuracy can be improved by adding higher-order terms in x to w_1. For example, the three-parameter (i.e., three undetermined constants) approximation that satisfies the boundary conditions in Eq. (g) is given by

$$w_3(x) = c_1 x(L - x) + c_2 x^2(L - x) + c_3 x^3(L - x)$$

The parameters c_1, c_2, and c_3 can be determined by requiring that $\Pi = \Pi(c_1, c_2, c_3)$ is a minimum. From the calculus of several variables, one finds that the necessary condition for Π to attain a minimum is

$$\frac{\partial \Pi}{\partial c_1} = 0, \qquad \frac{\partial \Pi}{\partial c_2} = 0, \qquad \frac{\partial \Pi}{\partial c_3} = 0$$

Thus, there are three equations for the three unknowns. Solving these equations yields

$$c_1 = -\frac{f_0 L^2}{24EI}, \qquad c_2 = -\frac{f_0 L}{24EI}, \qquad c_3 = \frac{f_0}{24EI}$$

which when substituted into w_3 yields a three-parameter solution that coincides with the exact solution. The method just outlined is known as the Ritz method.

The objective of the present study is to introduce energy and variational principles of mechanics and to illustrate their use in the derivation and solution of the equations of applied mechanics, including plane elasticity, plates, and shells. To keep the scope of the book within reasonable limits, only linear problems are considered. Although stability and vibration problems are introduced via examples and exercises, a detailed study of these topics is omitted. Of course, interested readers can readily extend the ideas presented herein to stability, vibration, dynamics, and nonlinear problems.

1.1.1 Configuration and Coordinates

The major objective of this chapter is to review the field equations governing a deformable body. All deformable bodies considered are assumed to contain no gaps or empty spaces. This assumption permits us to describe the macroscopic motion and deformation of a solid continuum and to define stress and strain at a point.

A solid body consists of infinitely many particles, called *material points*. The simultaneous position of all material points of the body is called the *configuration* of the body. An analytical description of the configuration of a body requires a fixed coordinate system. The mechanics of deformable bodies deals with the description of all configurations that a body can assume under the action of external specified loads. The set of all configurations a body can assume is called the *configuration space*, and each configuration is called an *element* of the configuration space. For example, the sum as well as each term in the series

$$\sum_{i=1}^{\infty} c_i \sin \frac{i\pi x}{L} \tag{1.1.1}$$

represents a possible configuration of a beam of length L, simply supported at both ends, and subjected to transverse loads. The set $\{\sin(i\pi x/L)\}_{i=1}^{\infty}$ spans a configuration space for the beam, and any linear combination of elements from the set specifies a configuration. The values of the constants c_i depend on the loading and governing differential equations of the beam.

A *rigid body* is one for which the distance between any two material points remains the same at all times. The configuration of a rigid body can be specified by six generalized coordinates: three translational coordinates and three rotational coordinates. A deformable body requires, in general, infinitely many coordinates to represent its configuration. For example, the configuration of a simply supported beam is specified by an infinite set $\{\sin(i\pi x/L)\}$,

$$w = \sum_{i=1}^{\infty} c_i \sin \frac{i\pi x}{L} \tag{1.1.2}$$

If the series is approximated by a finite number of terms, the beam configuration is effectively approximated by a finite number of generalized coordinates:

$$w_n = \sum_{i=1}^{n} c_i \sin \frac{i\pi x}{L} \tag{1.1.3}$$

We discuss such approximations in Chapter 3.

1.1.2 Notation

In the present study, a vector quantity is denoted by a boldface letter. If $\mathbf{A}$ is a vector, its magnitude is denoted by italic letter A. The unit vectors along the Cartesian rectangular axes (x_1, x_2, x_3) are denoted by $(\hat{\mathbf{e}}_1, \hat{\mathbf{e}}_2, \hat{\mathbf{e}}_3)$.

It is useful to represent a plane area as a vector. The direction is denoted by a unit vector drawn normal to that plane. The direction of the normal is that in which a right-handed screw advances as it is rotated counterclockwise around the boundary of the plane area. The unit normal vector is denoted by $\hat{\mathbf{n}}$. Then a plane area of magnitude A is denoted by $\mathbf{A} = A\hat{\mathbf{n}}$.

The Cartesian components of a vector $\mathbf{A}$ are denoted by (A_1, A_2, A_3). In component form, $\mathbf{A}$ is represented by

$$\mathbf{A} = A_1\hat{\mathbf{e}}_1 + A_2\hat{\mathbf{e}}_2 + A_3\hat{\mathbf{e}}_3 \tag{1.1.4}$$

The basis vectors $\hat{\mathbf{e}}_i$ in an orthogonal coordinate system satisfy the dot-product relations

$$\hat{\mathbf{e}}_1 \cdot \hat{\mathbf{e}}_2 = \hat{\mathbf{e}}_1 \cdot \hat{\mathbf{e}}_3 = \hat{\mathbf{e}}_2 \cdot \hat{\mathbf{e}}_3 = 0$$

and in general

$$\hat{\mathbf{e}}_i \cdot \hat{\mathbf{e}}_j = \delta_{ij} = \begin{cases} 1, & i = j \\ 0, & i \neq j \end{cases} \tag{1.1.5}$$

where δ_{ij} is called the *Kronecker delta*. Further, $\hat{\mathbf{e}}_i$ satisfy the following cross-product relations:

$$\hat{\mathbf{e}}_1 \times \hat{\mathbf{e}}_2 = \hat{\mathbf{e}}_3, \qquad \hat{\mathbf{e}}_2 \times \hat{\mathbf{e}}_3 = \hat{\mathbf{e}}_1, \qquad \hat{\mathbf{e}}_3 \times \hat{\mathbf{e}}_1 = \hat{\mathbf{e}}_2$$

or in short

$$\hat{\mathbf{e}}_i \times \hat{\mathbf{e}}_j = \varepsilon_{ijk}\hat{\mathbf{e}}_k \tag{1.1.6}$$

where ε_{ijk} is the *permutation symbol*

$$\varepsilon_{ijk} = \begin{cases} 1 & \text{if } ijk \text{ are in cyclic order} \\ & \text{and are not repeated } (i \neq j \neq k) \\ -1 & \text{if } ijk \text{ are not in cyclic order} \\ & \text{and are not repeated } (i \neq j \neq k) \\ 0 & \text{if any of } ijk \text{ are repeated} \end{cases} \tag{1.1.7}$$

It is customary in (solid) mechanics literature to abbreviate a summation of terms by repeating an index, which indicates summation over all values of that index. For example, the summation

$$\mathbf{A} = A_1\hat{\mathbf{e}}_1 + A_2\hat{\mathbf{e}}_2 + A_3\hat{\mathbf{e}}_3$$

$$= \sum_{i=1}^{3} A_i\hat{\mathbf{e}}_i$$

can be shortened to

$$\mathbf{A} = A_i\hat{\mathbf{e}}_i \qquad (1.1.8)$$

The repeated index i is called a *dummy index* and, therefore, can be replaced by *any other symbol that has not already been used in that expression*. Thus, we can write

$$F_i \equiv \hat{\mathbf{e}}_j \frac{\partial A_i}{\partial x_j}$$

$$= \hat{\mathbf{e}}_m \frac{\partial A_i}{\partial x_m} \qquad (1.1.9)$$

The Kronecker delta and permutation symbols can be used to write dot products and cross products of arbitrary vectors in a rectangular Cartesian coordinate system:

$$\mathbf{A} \cdot \mathbf{B} = (A_i\hat{\mathbf{e}}_i) \cdot (B_j\hat{\mathbf{e}}_j)$$

$$= A_i B_j \delta_{ij}$$

$$= A_i B_i \qquad (1.1.10)$$

$$\mathbf{A} \times \mathbf{B} = (A_i\hat{\mathbf{e}}_i) \times (B_j\hat{\mathbf{e}}_j)$$

$$= A_i B_j \varepsilon_{ijk}\hat{\mathbf{e}}_k \qquad (1.1.11)$$

The Kronecker delta and permutation symbols are related by

$$\varepsilon_{ijk}\varepsilon_{imn} = \delta_{jm}\delta_{kn} - \delta_{jn}\delta_{km} \qquad (1.1.12)$$

This relation is referred to as ε-δ *identity*.

If $(\hat{\mathbf{e}}_1, \hat{\mathbf{e}}_2, \hat{\mathbf{e}}_3)$ and $(\hat{\bar{\mathbf{e}}}_1, \hat{\bar{\mathbf{e}}}_2, \hat{\bar{\mathbf{e}}}_3)$ are sets of basis vectors of two Cartesian rectangular coordinate systems (x_1, x_2, x_3) and $(\bar{x}_1, \bar{x}_2, \bar{x}_3)$, respectively, the transformation law between the two systems can be expressed as

$$\hat{\bar{\mathbf{e}}}_i = a_{ij}\hat{\mathbf{e}}_j \qquad (1.1.13)$$

The coefficients a_{ij} denote the direction cosines of the $(\bar{x}_1, \bar{x}_2, \bar{x}_3)$ system with respect to (x_1, x_2, x_3) system:

$$a_{ij} = \text{cosine of the angle between } \hat{\bar{\mathbf{e}}}_i \text{ and } \hat{\mathbf{e}}_j \qquad (1.1.14)$$

1.1.3 Integral Relations

In the forthcoming sections, we encounter the need for the conversion of a volume integral to a surface integral and a surface integral to a line integral. Such integral relations are known as the gradient, divergence, and curl theorems and we review them here. Let V denote a three-dimensional region in space surrounded by surface S. Let dS be a differential element of the surface S with $\hat{n}$ the unit outward normal to S and dV a differential volume element. The following integral relations, taken from a course in advanced calculus, hold for any closed surface S:

Gradient Theorem (Green's Theorem)

$$\int_V \operatorname{grad} \phi \, dV = \oint_S \hat{n}\phi \, dS \qquad (1.1.15)$$

Divergence Theorem (Gauss' Theorem)

$$\int_V \operatorname{div} \mathbf{A} \, dV = \oint_S \hat{n} \cdot \mathbf{A} \, dS \qquad (1.1.16)$$

Curl Theorem (Stokes' Theorem)

$$\int_V \operatorname{curl} \mathbf{A} \, dV = \oint_S \hat{n} \times \mathbf{A} \, dS \qquad (1.1.17)$$

where $\phi = \phi(x_1, x_2, x_3)$, $\mathbf{A} = \mathbf{A}(x_1, x_2, x_3)$ are scalar and vector functions, respectively, and a circle around the surface integral denotes the integration over the *entire* surface S. All three theorems can be combined into a single relation,

$$\int_V \nabla * \mathbf{F} \, dV = \oint_S \hat{n} * \mathbf{F} \, dS \qquad (1.1.18)$$

where ∇ is the del operator and $\nabla *$ denotes one of the three operations: gradient (F can be a scalar), divergence, or curl. Equations (1.1.15)–(1.1.18) are also valid for two dimensional domains (S becomes a curve). Throughout this book we use symbol R to denote a region, whether two-dimensional or three-dimensional, and S to denote its boundary, whether a surface or a curve. Also, we use $d\mathbf{x}$ to denote either the volume $dV = dx_1 \, dx_2 \, dx_3$ or area $dA = dx_1 \, dx_2$, depending on whether R is a volume or an area.

For $\mathbf{A} = \operatorname{grad} \phi$ in Eq. (1.1.16), we get

$$\int_V \operatorname{div}(\operatorname{grad} \phi) \, dV \equiv \int_V \nabla^2 \phi \, dV = \oint_S \hat{n} \cdot \operatorname{grad} \phi \, dS \qquad (1.1.19)$$

The quantity $\hat{n} \cdot \nabla \phi$ is called the *normal derivative* of ϕ on the surface S, and is denoted by

$$\frac{\partial \phi}{\partial n} \equiv \hat{\mathbf{n}} \cdot \nabla \phi = n_i \frac{\partial \phi}{\partial x_i} \qquad (1.1.20)$$

where n_i are the direction cosines of the unit normal

$$\hat{\mathbf{n}} = n_i \hat{\mathbf{e}}_i \qquad (1.1.21)$$

For additional study of vectors and tensors and their applications to problems in mechanics, the reader is asked to consult the text by Reddy and Rasmussen (1982)[†].

1.1.4 Classification of Equations

In much of this study, we are concerned with the motion or equilibrium of elastic bodies under the action of externally applied mechanical and/or thermal loads. The equations governing the motion of a solid body can be classified into four basic categories:

1. Kinetics (conservation of linear momentum).
2. Kinematics (strain–displacement equations).
3. Thermodynamics (first and second laws of thermodynamics).
4. Constitutive equations (stress–strain relations).

Kinetics is the study of the static or dynamic equilibrium of forces acting on a body. Kinematics is a study of the geometry of motion and deformation without consideration of the forces causing the motion. The thermodynamic principles are concerned with the relations among heat, work, and thermodynamic properties of the body. The constitutive equations describe the constitutive behavior of the body and relate the dependent variables introduced in the kinetic description to those in the kinematic description. These equations are supplemented by appropriate boundary and initial conditions. In the following sections, we review the notions of stress and strain at a point and derive the field equations of solid mechanics.

Exercises 1.1

1. Let X denote the configuration space of a mechanical system and $\mathbf{x}$ and $\mathbf{y}$ be any arbitrary elements in X. The *distance* (or *metric*) between $\mathbf{x}$ and $\mathbf{y}$, denoted $d(\mathbf{x}, \mathbf{y})$, should satisfy the following properties (see Reddy and Rasmussen, 1982).
   (i) $d(\mathbf{x}, \mathbf{y}) = 0$ if and only if $\mathbf{x} = \mathbf{y}$ (nonnegative).
   (ii) $d(\mathbf{x}, \mathbf{y}) = d(\mathbf{y}, \mathbf{x}) > 0$ for $\mathbf{x} \neq \mathbf{y}$ (symmetry).
   (iii) $d(\mathbf{x}, \mathbf{y}) + d(\mathbf{y}, \mathbf{z}) \geq d(\mathbf{x}, \mathbf{z})$ for any $\mathbf{x}, \mathbf{y}$, and $\mathbf{z}$ in X (triangle inequality).

[†]See the Bibliography at the end of the book.

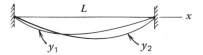

Figure E1

Note that, as a special case, one has the n-dimensional euclidean space R^n by setting $X = R^n$, where $\mathbf{x}$ denotes a point in R^n and $d(\mathbf{x}, \mathbf{y})$ denotes the *euclidean distance* $|\mathbf{x} - \mathbf{y}|$,

$$|\mathbf{x} - \mathbf{y}| = \left(\sum_{i=1}^{n} |x_i - y_i|^2 \right)^{1/2}$$

Now consider the transverse motion of an elastic cable of length L, fixed at its ends (see Fig. E1). Let Y denote the configuration space of the cable, which consists of all real-valued continuous functions $y(x, t)$ defined on the interval $0 \le x \le L$ and vanishing at $x = 0$ and $x = L$. For any time t, define the distance in Y by

$$d(y_1, y_2) \equiv \max_{0 \le x \le L} \left[|y_1(x, t) - y_2(x, t)|, \quad t = \text{fixed} \right]$$

Show that $d(y_1, y_2)$ defined above satisfies the properties (i)–(iii). Here, y denotes the transverse deflection of the cable.

2. Consider a clamped beam of length L. Let H denote the configuration space that consists of all real-valued continuous functions $w(x)$ defined on the interval $0 \le x \le L$ and satisfy the end conditions

$$w(0) = \frac{dw}{dx}(0) = w(L) = \frac{dw}{dx}(L) = 0$$

Define the distance between w_1 and w_2 by

$$d(w_1, w_2) \equiv \left[\int_0^L |w_1(x) - w_2(x)|^2 \, dx \right]^{1/2}$$

Show that $d(w_1, w_2)$ satisfies the properties of a distance. *Hint:* To establish the triangle inequality, use the *Minkowski inequality* for integrals (see Reddy and Rasmussen, 1982):

$$\left[\int_R |x(t) \pm y(t)|^p \, dt \right]^{1/p} \le \left[\int_R |x(t)|^p \, dt \right]^{1/p} + \left[\int_R |y(t)|^p \, dt \right]^{1/p}$$

for $1 \le p < \infty$, where R denotes the interval (or domain) over which x and y are defined.

3. In Exercise 2, determine the length, $d(w, 0)$, of the element

$$w = c_1 \left(1 - \cos \frac{\pi x}{L} \right) + c_2 \left(1 - \cos \frac{2\pi x}{L} \right)$$

where c_1 and c_2 are constants.

4. If a mechanical system requires m arbitrarily chosen coordinates to
    define its configuration, the system is said to have m *degrees of
    freedom*. If m degrees of freedom defining the configuration are related
    by $p(< m)$ constraints, the number of *independent* degrees of freedom,
    called *generalized coordinates*, is equal to $n = m - p$. For the follow-
    ing systems with discrete coordinates, identify m, p, and n (see Fig.
    E4):
    (a) A billiard table with two balls.
    (b) Motion of a pendulum in a plane.

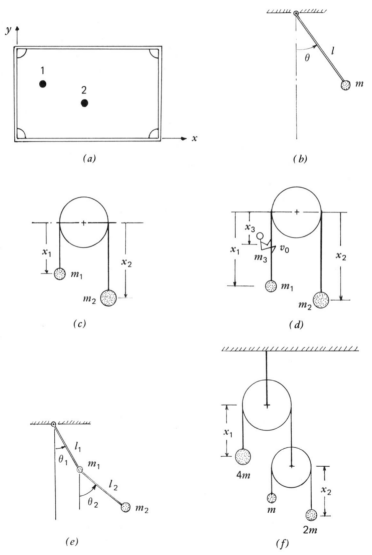

Figure E4

    (c) Two masses (m_1 and m_2) attached to an inextensible cord that is suspended over a frictionless stationary pulley.

    (d) Same as in (c), with a monkey of mass m_3 climbing up the cord above mass m_1 with a speed v_0 relative to mass m_1.

    (e) A double pendulum of lengths l_1 and l_2 in a plane.

    (f) The two-pulley system shown in the figure.

5. For continuous systems, such as beams, the configuration can only be described by discrete coordinates in combination with *coordinate functions*. For example, the configuration of a simply supported beam can be described by

$$w = \sum_{n=1}^{\infty} c_n \phi_n(x)$$

where c_n denote the discrete coordinates and ϕ_n denote the coordinate functions (Fig. E5). An example of the trigonometric form of ϕ_n is given by $\sin(n\pi x/L)$. In general, the number of degrees of freedom for continuous systems is not finite. Give an *algebraic* form of ϕ_1 that can be used to represent the configuration.

6. Verify the following identities:

    (a) $\delta_{ij}\delta_{ij} = \delta_{ii} = 3$.

    (b) $\varepsilon_{ijk}\varepsilon_{ijk} = 6$.

    (c) $A_i A_j \varepsilon_{ijk} = 0$.

    (d) $B_{ij}\varepsilon_{ijk} = 0$ if and only if $B_{ij} = B_{ji}$ (or $[B] = [B]^T$).

    (e) determinant of a 3×3 matrix $[A]$ can be expressed by

$$|A| = \varepsilon_{ijk}a_{1i}a_{2j}a_{3k} = \tfrac{1}{6}\varepsilon_{ijk}\varepsilon_{rst}a_{ir}a_{js}a_{kt}$$

where a_{ij} denotes the element in the ith row and jth column of the matrix $[A]$.

7. Using the index notation and the ε-δ identity, prove the following vector identities:

    (a) $(\mathbf{A} \times \mathbf{B})\cdot(\mathbf{C} \times \mathbf{D}) = (\mathbf{A}\cdot\mathbf{C})(\mathbf{B}\cdot\mathbf{D}) - (\mathbf{A}\cdot\mathbf{D})(\mathbf{B}\cdot\mathbf{C})$.

    (b) $\operatorname{div}(\mathbf{A} \times \mathbf{B}) = \operatorname{curl}\mathbf{A}\cdot\mathbf{B} - \operatorname{curl}\mathbf{B}\cdot\mathbf{A}$.

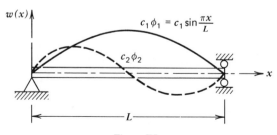

Figure E5

(c) $\text{grad}(r^n) = nr^{n-2}\mathbf{r}$, $r^2 = \mathbf{r} \cdot \mathbf{r}$.

(d) $\text{curl}[f(r)\mathbf{r}] = \mathbf{0}$, $f(r)$ is a continuous function of r.

8. Determine the transformation matrix relating the orthonormal basis vectors $(\hat{\mathbf{e}}_1, \hat{\mathbf{e}}_2, \hat{\mathbf{e}}_3)$ and the orthonormal basis vectors $(\hat{\mathbf{e}}_1', \hat{\mathbf{e}}_2', \hat{\mathbf{e}}_3')$, when $\hat{\mathbf{e}}_i'$ are given by

   (a) $\hat{\mathbf{e}}_1'$ is along the vector $\hat{\mathbf{e}}_1 - \hat{\mathbf{e}}_2 + \hat{\mathbf{e}}_3$, and $\hat{\mathbf{e}}_2'$ is perpendicular to the plane $2x_1 + 3x_2 + x_3 - 5 = 0$.

   (b) $\hat{\mathbf{e}}_1'$ is along the line segment connecting point $(1, -1, 3)$ to $(2, -2, 4)$, and $\hat{\mathbf{e}}_3' = (-\hat{\mathbf{e}}_1 + \hat{\mathbf{e}}_2 + 2\hat{\mathbf{e}}_3)/\sqrt{6}$.

   (c) $\hat{\mathbf{e}}_3' = \hat{\mathbf{e}}_3$, and the angle between $x_1' - $ axis and $x_1 - $ axis is $30°$.

9. Show that the vector area of a closed surface is zero:

$$\oint_S \hat{\mathbf{n}}\, dS = 0$$

and that the volume V enclosed by a surface S is given by

$$V = \tfrac{1}{3}\oint_S \mathbf{r} \cdot \hat{\mathbf{n}}\, dS$$

where $\mathbf{r}$ is the position vector.

10. Using the integral identities, establish the following relations:

    (a) $\int_V v\nabla^2 u\, d\mathbf{x} = \int_V - \text{grad}\, v \cdot \text{grad}\, u\, d\mathbf{x} + \oint_S v(\partial u/\partial n)\, dS$.

    (b) $\int_V v\nabla^4 u\, d\mathbf{x} = \int_V \nabla^2 v\nabla^2 u\, d\mathbf{x} + \oint_S [v(\partial/\partial n)(\nabla^2 u) -$
    $(\partial v/\partial n)\nabla^2 u]\, dS$

    (c) $\int_A v\{(\partial/\partial x)[k_1(\partial u/\partial x)] + (\partial/\partial y)[k_2(\partial u/\partial y)]\}\, dx\, dy$
    $= \oint_S v[k_1(\partial u/\partial x)n_x + k_2(\partial u/\partial y)n_y]\, dS$
    $- \int_A [k_1(\partial v/\partial x)(\partial u/\partial x) + k_2(\partial v/\partial y)(\partial u/\partial y)]\, dx\, dy$

11. Write the one-dimensional form of the identities in Exercise 10.

1.2 KINETICS

1.2.1 The Stress at a Point

Forces acting on a body can be classified as *internal* and *external*. The internal forces resist the tendency of one part of the body to be separated from another part. The external forces are those transmitted by the body. The external forces can be classified as *body forces* and *surface forces*. Body forces act on the distribution of mass inside the body. Examples of body forces are provided by the gravitational and magnetic forces. Body forces are usually measured per unit mass or unit volume of the body. They are reckoned per unit volume in

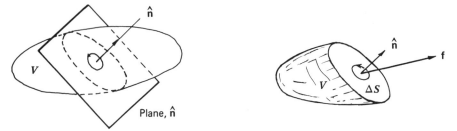

Figure 1.2 Surface force **f** on area Δs of the cross section obtained by plane $\hat{n}$.

the present study. Surface forces are contact forces acting on the boundary surface of the body. Examples of surface forces are provided by applied forces on the surface of the body. Surface forces are reckoned per unit area.

Consider a body occupying the volume V and bounded by surface S. The surface force/unit area acting on an elemental area dS is called the *traction* or *stress vector* acting on the element. The concept also applies to a surface created by slicing the body with a plane whose normal is $\hat{n}$ (see Fig. 1.2). Consider the surface force **f** acting on a small portion ΔS of the surface area S of the body. The stress vector on ΔS is the vector $\mathbf{f}/\Delta S$, which has the same direction as **f**, but with magnitude equal to $|\mathbf{f}|/\Delta S$. The stress vector at a point P on ΔS is defined by

$$\mathbf{t} = \lim_{\Delta S \to 0} \frac{\mathbf{f}}{\Delta S} \tag{1.2.1}$$

Since the magnitude and direction depend on the orientation of the plane that created the surface, we denote it by $\mathbf{t}^{(\hat{n})}$, where $\hat{n}$ denotes the normal to the plane. In other words, $\mathbf{t}^{(\hat{n})}$ is the stress vector at point P on the plane $\hat{n}$ passing through the point. The component of **t** that is in the direction of $\hat{n}$ is called the *normal* stress. The component of **t** that is normal to $\hat{n}$ is called *shear* stress.

It is useful to express the stress vector at a point, on a given plane, in terms of its components along three mutually perpendicular axes. Let $\mathbf{t}^{(i)}$ denote the stress vector at point P on a plane perpendicular to x_i − axis (see Fig. 1.3). Each vector $\mathbf{t}^{(i)}$ can be resolved into components along the coordinate lines

$$\mathbf{t}^{(i)} = \sigma_{ij}\hat{\mathbf{e}}_j \qquad (i, j = 1, 2, 3) \tag{1.2.2}$$

where σ_{ij} denotes the component of stress vector $\mathbf{t}^{(i)}$ along the x_j-direction. In Eq. (1.2.2), summation on repeated indices is implied. The stress components σ_{ij} are shown on three perpendicular planes in Fig. 1.3. It should be noted that the cube used to indicate the stress components has no dimensions; it is a point cube so that all nine components are acting at the point P. Thus, the state of stress at a point is characterized, in general, by nine components of stress.

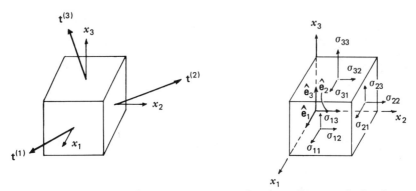

Figure 1.3   Traction vectors and stress components on three mutually perpendicular planes at a point.

1.2.2   Stress Tensor and the Cauchy Formula

As noted, the surface force acting on a small element of area in a continuous body depends not only on the magnitude of the area but also the orientation of the area. If we choose to represent a plane area by a vector whose magnitude is equal to that of the area and whose direction is equal to that of the normal to the surface, it is clear that the stress vector $\mathbf{t}$ is a point function of the unit normal $\hat{\mathbf{n}}$, which denotes the orientation of the surface ΔS:

$$\mathbf{t}(\hat{\mathbf{n}}) = \lim_{\Delta S \to 0} \frac{\mathbf{f}}{\Delta S} \qquad (1.2.3a)$$

Because of Newton's third law of action and reaction, we have

$$\mathbf{t}(-\hat{\mathbf{n}}) = -\mathbf{t}(\hat{\mathbf{n}}) \qquad (1.2.3b)$$

We now establish a relationship between the stress vector acting on a given plane and the stress vectors on three mutually perpendicular planes, all stress vectors being defined at the same point. To do this, we set up an infinitesimal tetrahedron in orthogonal cartesian coordinates (see Fig. 1.4).

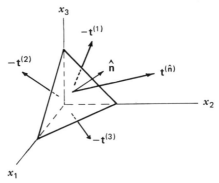

Figure 1.4   Stresses on a tetrahedral element.

If $-\mathbf{t}^{(1)}$, $-\mathbf{t}^{(2)}$, $-\mathbf{t}^{(3)}$, and $\mathbf{t}^{(\hat{n})}$ denote the stress vectors in the outward directions on the faces of the infinitesimal tetrahedron whose areas are ΔS_1, ΔS_2, ΔS_3, and ΔS, we have by Newton's second law ($\mathbf{F} = m\mathbf{a}$),

$$\mathbf{t}^{(\hat{n})}\,\Delta S - \mathbf{t}^{(1)}\,\Delta S_1 - \mathbf{t}^{(2)}\,\Delta S_2 - \mathbf{t}^{(3)}\,\Delta S_3 + \Delta V\mathbf{f} = \rho\,\Delta V\mathbf{a} \qquad (1.2.4)$$

where ΔV is the volume of the tetrahedron, ρ is the density, $\mathbf{f}$ is the body force/unit volume, and $\mathbf{a}$ is the acceleration. To simplify the equation in (1.2.4), we note that the areas ΔS_i are related to ΔS. By the gradient theorem, Eq. (1.1.15), we have

$$\int_V \nabla \phi \, dx_1 \, dx_2 \, dx_3 = \oint_S \hat{n}\phi \, dS \qquad (1.2.5)$$

where ϕ is a function of position in V. For $\phi = 1$, Eq. (1.2.5) implies that the vector sum of the total area of a closed surface is zero (see Exercise 1.1.9). Thus, for the tetrahedron we have

$$0 = \oint_S \hat{n} \, dS \equiv \Delta S\,\hat{n} - \Delta S_1\,\hat{e}_1 - \Delta S_2\,\hat{e}_2 - \Delta S_3\,\hat{e}_3 \qquad (1.2.6)$$

where $\hat{e}_i$ denotes the unit vector along the x_i-coordinate. By taking the dot-product of Eq. (1.2.6) with $\hat{e}_i$, we obtain

$$\Delta S_i = \Delta S\,(\hat{n}\cdot\hat{e}_i)$$

$$= \Delta S\,n_i \qquad (i = 1,2,3) \qquad (1.2.7)$$

The volume of the tetrahedron can be expressed as

$$\Delta V = \frac{h}{3}\,\Delta S \qquad (1.2.8)$$

where h is the perpendicular distance from the origin of the coordinate system to the slant surface. Substituting for ΔS_i and ΔV from Eqs. (1.2.7) and (1.2.8) into Eq. (1.2.4), we obtain

$$\mathbf{t}^{(\hat{n})} = (\hat{n}\cdot\hat{e}_1)\mathbf{t}^{(1)} + (\hat{n}\cdot\hat{e}_2)\mathbf{t}^{(2)} + (\hat{n}\cdot\hat{e}_3)\mathbf{t}^{(3)} + \frac{h}{3}(\rho\mathbf{a} - \mathbf{f}) \qquad (1.2.9)$$

In the limit when the tetrahedron shrinks to a point, $h \to 0$, we obtain

$$\mathbf{t}^{(\hat{n})} = (\hat{n}\cdot\hat{e}_1)\mathbf{t}^{(1)} + (\hat{n}\cdot\hat{e}_2)\mathbf{t}^{(2)} + (\hat{n}\cdot\hat{e}_3)\mathbf{t}^{(3)}$$

$$= \hat{n}\cdot\left(\hat{e}_1\mathbf{t}^{(1)} + \hat{e}_2\mathbf{t}^{(2)} + \hat{e}_3\mathbf{t}^{(3)}\right)$$

$$= \hat{n}\cdot\boldsymbol{\sigma} = \boldsymbol{\sigma}'\cdot\hat{n} \qquad (1.2.10)$$

where σ is the stress dyadic (or tensor of second order),

$$\sigma = \hat{e}_1 t^{(1)} + \hat{e}_2 t^{(2)} + \hat{e}_3 t^{(3)}$$

$$= \hat{e}_i t^{(i)}$$

$$= \hat{e}_i \sigma_{ij} \hat{e}_j \tag{1.2.11}$$

where in Eq. (1.2.2) is used to express $t^{(i)}$ in terms of the stress components σ_{ij}. The last expression in Eq. (1.2.11) is called the *nonion* form of the stress dyadic. In view of Eq. (1.2.11), Eq. (1.2.10) takes the form

$$t^{(\hat{n})} = n_i \sigma_{ij} \hat{e}_j \quad \text{or} \quad t_j^{(\hat{n})} = n_i \sigma_{ij} \tag{1.2.12a}$$

which is known as the *Cauchy stress formula*. In matrix form, Eq. (1.2.12a) can be expressed as

$$\begin{Bmatrix} t_1^{(\hat{n})} \\ t_2^{(\hat{n})} \\ t_3^{(\hat{n})} \end{Bmatrix} = \begin{bmatrix} \sigma_{11} & \sigma_{12} & \sigma_{13} \\ \sigma_{21} & \sigma_{22} & \sigma_{23} \\ \sigma_{31} & \sigma_{32} & \sigma_{33} \end{bmatrix}^T \begin{Bmatrix} n_1 \\ n_2 \\ n_3 \end{Bmatrix} \tag{1.2.12b}$$

where $[\cdot]^T$ denotes the transpose of the enclosed matrix.

The normal and shearing components of the stress vector $t^{(\hat{n})}$ are given by

$$t_N = t^{(\hat{n})} \cdot \hat{n}$$

$$= \sigma_{ij} n_i n_j \tag{1.2.12c}$$

$$t_S = \sqrt{|t^{(\hat{n})}|^2 - t_N^2} \tag{1.2.12d}$$

Example 1.1

The state of stress at a point in a body is

$$[\sigma] = \begin{bmatrix} 1 & -1 & 0 \\ -1 & 2 & -1 \\ 0 & -1 & 2 \end{bmatrix} \text{psi}$$

We wish to find the stress vector on the plane defined by the equation $2x_1 + x_2 - x_3 = 1$.

The unit normal to the plane is given by

$$\hat{n} = \frac{\nabla(2x_1 + x_2 - x_3 - 1)}{|\nabla(2x_1 + x_2 - x_3 - 1)|}$$

$$= \frac{2\hat{e}_1 + \hat{e}_2 - \hat{e}_3}{\sqrt{6}}$$

The components of the stress vector are given by

$$
\begin{Bmatrix} t_1 \\ t_2 \\ t_3 \end{Bmatrix} = \begin{bmatrix} 1 & -1 & 0 \\ -1 & 2 & -1 \\ 0 & -1 & 2 \end{bmatrix} \begin{Bmatrix} 2 \\ 1 \\ -1 \end{Bmatrix} \frac{1}{\sqrt{6}}
$$

$$
= \frac{1}{\sqrt{6}} \begin{Bmatrix} 1 \\ 1 \\ -3 \end{Bmatrix}
$$

or

$$
\mathbf{t}^{(\hat{n})} = \frac{1}{\sqrt{6}} (\hat{e}_1 + \hat{e}_2 - 3\hat{e}_3)
$$

The normal and shearing components are given by

$$
t_N = \hat{\mathbf{n}} \cdot \mathbf{t}^{(\hat{n})} = \tfrac{1}{6}(2 + 1 + 3) = 1 \text{ psi}
$$

$$
t_S = \sqrt{|\mathbf{t}^{(n)}|^2 - t_N^2}
$$

$$
= \sqrt{\tfrac{1}{6}(1 + 1 + 9) - 1} = \sqrt{\tfrac{5}{6}} \text{ psi}
$$

1.2.3 Transformations of Stress Components

In Sections 1.2.1 and 1.2.2, the stress tensor was represented in terms of Cartesian components in an arbitrarily selected coordinate system (x_1, x_2, x_3). When the components of stress are known in one coordinate system, it is of interest to determine the components of stress in another coordinate system. The relationship between the components of stress tensor in one coordinate system and the components of the same stress tensor in another coordinate system is known as the *transformation of stress*. Such transformations are useful in determining the maximum values of normal and shear stresses and the planes on which they act. Here we establish the transformation of stress.

Consider two sets of orthogonal coordinate systems: (x_1, x_2, x_3) and (x_1', x_2', x_3'). Let $(\hat{e}_1, \hat{e}_2, \hat{e}_3)$ and $(\hat{e}_1', \hat{e}_2', \hat{e}_3')$ denote the sets of basis vectors in the two coordinate systems. Suppose that the two coordinate systems are related by the transformation equation (1.1.13):

$$
\hat{e}_i' = a_{ij}\hat{e}_j \tag{1.2.13}
$$

where a_{ij} denote the direction cosines

$$a_{ij} = \hat{\mathbf{e}}'_i \cdot \hat{\mathbf{e}}_j \tag{1.2.14}$$

We note that the stress dyadic at a point is invariant; that is, it does not depend on the coordinate system. However, the components do depend on the coordinate system. Now we wish to relate the stress components in the x_i-system to those in the x'_i-system. We have

$$\boldsymbol{\sigma} = \sigma_{ij}\hat{\mathbf{e}}_i\hat{\mathbf{e}}_j \quad \text{(in the unprimed system)}$$

$$= \sigma'_{ij}\hat{\mathbf{e}}'_i\hat{\mathbf{e}}'_j \quad \text{(in the primed system)} \tag{1.2.15}$$

and

$$\sigma_{ij}\hat{\mathbf{e}}_i\hat{\mathbf{e}}_j = \sigma'_{kl}\hat{\mathbf{e}}'_k\hat{\mathbf{e}}'_l$$

$$= \sigma'_{kl}a_{ki}a_{lj}\hat{\mathbf{e}}_i\hat{\mathbf{e}}_j$$

or

$$\left(\sigma_{ij} - \sigma'_{kl}a_{ki}a_{lj}\right)\hat{\mathbf{e}}_i\hat{\mathbf{e}}_j = 0$$

$$\sigma_{ij} = \sigma'_{kl}a_{ki}a_{lj} \quad \text{or} \quad [\sigma] = [A]^T[\sigma'][A] \tag{1.2.16a}$$

Similarly, one can show that

$$\sigma'_{ij} = a_{ik}a_{jl}\sigma_{kl} \quad \text{or} \quad [\sigma'] = [A][\sigma][A]^T \tag{1.2.16b}$$

As a special case, one can obtain the normal stress along a given direction from Eq. (1.2.16b):

$$\sigma'_{nn} = a_{nk}a_{nl}\sigma_{kl} \quad \text{(no sum on } n) \tag{1.2.17}$$

Example 1.2

Consider the transformation from the rectangular cartesian system (x, y, z) to the cylindrical coordinate system (r, θ, z). We have $(x'_1 = r, x'_2 = \theta, x'_3 = z)$

$$\hat{\mathbf{e}}_r = \cos\theta\,\hat{\mathbf{e}}_x + \sin\theta\,\hat{\mathbf{e}}_y$$

$$\hat{\mathbf{e}}_\theta = -\sin\theta\,\hat{\mathbf{e}}_x + \cos\theta\,\hat{\mathbf{e}}_y$$

$$\hat{\mathbf{e}}_z = \hat{\mathbf{e}}_z \tag{1.2.18}$$

and

$$\begin{bmatrix} \sigma_{rr} & \sigma_{r\theta} & \sigma_{rz} \\ \sigma_{\theta r} & \sigma_{\theta\theta} & \sigma_{\theta z} \\ \sigma_{zr} & \sigma_{z\theta} & \sigma_{zz} \end{bmatrix}$$

$$= \begin{bmatrix} \cos\theta & \sin\theta & 0 \\ -\sin\theta & \cos\theta & 0 \\ 0 & 0 & 1 \end{bmatrix} \begin{bmatrix} \sigma_{xx} & \sigma_{xy} & \sigma_{xz} \\ \sigma_{yx} & \sigma_{yy} & \sigma_{yz} \\ \sigma_{zx} & \sigma_{zy} & \sigma_{zz} \end{bmatrix} \begin{bmatrix} \cos\theta & -\sin\theta & 0 \\ \sin\theta & \cos\theta & 0 \\ 0 & 0 & 1 \end{bmatrix}$$

$$= \begin{bmatrix} \cos\theta & \sin\theta & 0 \\ -\sin\theta & \cos\theta & 0 \\ 0 & 0 & 1 \end{bmatrix} \begin{bmatrix} \sigma_{xx}\cos\theta + \sigma_{xy}\sin\theta & -\sigma_{xx}\sin\theta + \sigma_{xy}\cos\theta & \sigma_{xz} \\ \sigma_{yx}\cos\theta + \sigma_{yy}\sin\theta & -\sigma_{yx}\sin\theta + \sigma_{yy}\cos\theta & \sigma_{yz} \\ \sigma_{zx}\cos\theta + \sigma_{zy}\sin\theta & -\sigma_{zx}\sin\theta + \sigma_{zy}\cos\theta & \sigma_{zz} \end{bmatrix}$$

Performing the matrix multiplication, we obtain

$$\sigma_{rr} = \sigma_{xx}\cos^2\theta + (\sigma_{xy} - \sigma_{yx})\cos\theta\sin\theta + \sigma_{yy}\sin^2\theta$$

$$\sigma_{r\theta} = (\sigma_{yy} - \sigma_{xx})\cos\theta\sin\theta + \sigma_{xy}\cos^2\theta - \sigma_{yx}\sin^2\theta$$

$$\sigma_{\theta r} = (\sigma_{yy} - \sigma_{xx})\cos\theta\sin\theta - \sigma_{xy}\sin^2\theta + \sigma_{yx}\cos^2\theta$$

$$\sigma_{\theta\theta} = \sigma_{xx}\sin^2\theta + \sigma_{yy}\cos^2\theta - (\sigma_{xy} + \sigma_{yx})\cos\theta\sin\theta$$

$$\sigma_{rz} = \sigma_{xz}\cos\theta + \sigma_{yz}\sin\theta$$

$$\sigma_{\theta z} = -\sigma_{xz}\sin\theta + \sigma_{yz}\cos\theta$$

$$\sigma_{zr} = \sigma_{zx}\cos\theta + \sigma_{zy}\sin\theta$$

$$\sigma_{z\theta} = -\sigma_{zx}\sin\theta + \sigma_{zy}\cos\theta$$

$$\sigma_{zz} = \sigma_{zz} \tag{1.2.19}$$

Note that we have not assumed symmetry of the stress tensor. When the stress components are symmetric (i.e., $\sigma_{xy} = \sigma_{yx}, \sigma_{xz} = \sigma_{zx}, \sigma_{yz} = \sigma_{zy}$), then only six of the nine equations in Eq. (1.2.19) are linearly independent.

The following expressions involving the stress components remain unchanged (i.e., invariant) under coordinate transformations:

$$I_1 \equiv \sigma_{ii} = \sigma_{11} + \sigma_{22} + \sigma_{33}$$

$$I_2 \equiv \tfrac{1}{2}\left(\sigma_{ii}\sigma_{jj} - \sigma_{ij}\sigma_{ji}\right) = \sigma_{11}\sigma_{22} + \sigma_{22}\sigma_{33} + \sigma_{33}\sigma_{11} - \left(\sigma_{12}^2 + \sigma_{13}^2 + \sigma_{23}^2\right)$$

$$I_3 \equiv \det[\sigma] = |\sigma_{ij}| \qquad (1.2.20)$$

The quantities I_1, I_2, and I_3 are called the *stress invariants*.

1.2.4   Principal Stresses and Principal Planes

For a given state of stress, the determination of maximum normal and shear stresses at a point is of considerable interest in the design of structures. Here we consider the question of finding the maximum (and minimum) normal stresses, called *principal stresses*, at a point. The planes associated with the principal stresses are called *principal planes*. It is clear from Eq. (1.2.12d) that $t_N = |\mathbf{t}^{(\hat{n})}|$ if and only if the shear stress t_S is zero. In other words, the maximum normal stress at a point acts on a plane on which the shear stress is zero. In that case, we have

$$\mathbf{t}^{(\hat{n})} = \lambda\hat{\mathbf{n}} \qquad (1.2.21)$$

where λ denotes the magnitude of the normal stress and $\hat{\mathbf{n}}$ denotes the unit normal to the plane on which the maximum stress is acting. We also have from the general formula in Eq. (1.2.10)

$$\mathbf{t}^{(\hat{n})} = \boldsymbol{\sigma}^t \cdot \hat{\mathbf{n}} \qquad (1.2.22)$$

Equating these two equations, and assuming that $\boldsymbol{\sigma}^t \equiv \boldsymbol{\sigma}$, we obtain

$$\boldsymbol{\sigma} \cdot \hat{\mathbf{n}} - \lambda\hat{\mathbf{n}} = 0$$

or in component form,

$$\left(\sigma_{ij} - \lambda\delta_{ij}\right)n_j = 0, \qquad ([\sigma] - \lambda[I])\{n\} = \{0\} \qquad (1.2.23)$$

Here $[I]$ denotes the identity matrix. Equation (1.2.23) has a nontrivial solution ($n_i \neq 0$, $i = 1, 2, 3$) only if the determinant of the coefficient matrix is zero:

$$\begin{vmatrix} \sigma_{11} - \lambda & \sigma_{12} & \sigma_{13} \\ \sigma_{12} & \sigma_{22} - \lambda & \sigma_{23} \\ \sigma_{13} & \sigma_{23} & \sigma_{33} - \lambda \end{vmatrix} = 0 \qquad (1.2.24)$$

The determinant, for a given state of stress, is a cubic polynomial in λ. Therefore, three planes exist on which the shearing stresses are zero. In general, the polynomial is called the *characteristic polynomial*, the three values of λ are called *characteristic values* or *eigenvalues*, and the associated (unit) normal vectors are called *characteristic vectors* or *eigenvectors*. In the context of stress analysis, the eigenvalues are called the *principal stresses* and the planes defined by the eigenvectors are called the *principal planes*.

The characteristic polynomial is of the form

$$-\lambda^3 + I_1\lambda^2 - I_2\lambda + I_3 = 0 \tag{1.2.25}$$

where I_1, I_2, and I_3 are stress invariants defined in Eq. (1.2.20). For a symmetric matrix $[\sigma] = [\sigma]'$, the eigenvalues can be shown to be real. Furthermore, when the eigenvalues are distinct ($\lambda_i \neq \lambda_j$, for $i \neq j$), the associated eigenvectors are mutually orthogonal:

$$\hat{\mathbf{n}}^{(i)} \cdot \hat{\mathbf{n}}^{(j)} = \delta_{ij} \tag{1.2.26}$$

where $\hat{\mathbf{n}}^{(i)}$ denotes the normalized eigenvector associated with the eigenvalue λ_i.

The eigenvectors $\hat{\mathbf{n}}^{(i)}$ associated with the eigenvalue λ_i can be determined from Eq. (1.2.23). For any eigenvalue λ_i, Eq. (1.2.23) gives only two linearly independent equations for three components $n_j^{(i)}$ ($j = 1, 2, 3$) of $\hat{\mathbf{n}}^{(i)}$. The third equation is provided by the relation

$$\left(n_1^{(i)}\right)^2 + \left(n_2^{(i)}\right)^2 + \left(n_3^{(i)}\right)^2 = 1 \tag{1.2.27}$$

Since we are interested only in the direction of the vector perpendicular to the plane of maximum stress, we can set one of the components of the vector $\hat{\mathbf{n}}^{(i)}$ arbitrarily to a fixed value and determine the remaining two components from Eq. (1.2.23) [and not make use of Eq. (1.2.27)].

Example 1.3

Consider the state of stress at a point given in Example 1.1. We wish to determine the principal stresses at the point. We have

$$0 = \begin{vmatrix} 1 - \lambda & -1 & 0 \\ -1 & 2 - \lambda & -1 \\ 0 & -1 & 2 - \lambda \end{vmatrix}$$

$$= -\lambda^3 + I_1\lambda^2 - I_2\lambda + I_3 \tag{1.2.28}$$

where

$$I_1 = 1 + 2 + 2 = 5$$

$$I_2 = 1 \times 2 + 2 \times 2 + 2 \times 1 - \left[(-1)^2 + (0)^2 + (-1)^2\right] = 6$$

$$I_3 = (4 - 1) + (-2) = 1$$

To find the roots of the polynomial, we introduce the transformation

$$\bar{\lambda} = \lambda - \tfrac{1}{3}I_1 \tag{1.2.29}$$

Substituting Eq. (1.2.29) into Eq. (1.2.28), we get

$$-(\bar{\lambda})^3 + \bar{I}_2\bar{\lambda} - \bar{I}_3 = 0 \tag{1.2.30}$$

where

$$\bar{I}_2 = \tfrac{1}{2}(\bar{\sigma}_{ii}\bar{\sigma}_{jj} - \bar{\sigma}_{ij}\bar{\sigma}_{ij})$$

$$\bar{I}_3 = \det[\bar{\sigma}]$$

$$\bar{\sigma}_{ij} = \sigma_{ij} - \tfrac{1}{3}\sigma_{kk}\delta_{ij} \tag{1.2.31}$$

The stress components $\bar{\sigma}_{ij}$ are called *deviatoric components* of the stress tensor $[\sigma]$. The cubic equation in (1.2.30) is of a special form that allows a direct computation of its roots. The three roots are given by

$$\bar{\lambda}_i = 2\cos\alpha_i\left(-\frac{\bar{I}_2}{3}\right)^{1/2} \tag{1.2.32a}$$

$$\alpha_1 = \frac{1}{3}\left\{\cos^{-1}\left[\frac{\bar{I}_3}{2}\left(-\frac{3}{\bar{I}_2}\right)^{3/2}\right]\right\}, \qquad \alpha_2 = \alpha_1 + \frac{2\pi}{3}, \qquad \alpha_3 = \alpha_1 - \frac{2\pi}{3}$$

$$\tag{1.2.32b}$$

For the example problem at hand, we have

$$[\bar{\sigma}] = \begin{bmatrix} 1 - \frac{5}{3} & -1 & 0 \\ -1 & 2 - \frac{5}{3} & -1 \\ 0 & -1 & 2 - \frac{5}{3} \end{bmatrix}$$

$$\bar{I}_2 = (-\tfrac{2}{3})(\tfrac{1}{3}) + (\tfrac{1}{3})(\tfrac{1}{3}) + (\tfrac{1}{3})(-\tfrac{2}{3}) - [(-1)^2 + (0)^2 + (-1)^2] = -\tfrac{7}{3}$$

$$\bar{I}_3 = (-\tfrac{2}{3})[\tfrac{1}{3} \times \tfrac{1}{3} - 1] + [-\tfrac{1}{3}] = \tfrac{7}{27}$$

$$\alpha_1 = \tfrac{1}{3}\left\{\cos^{-1}\left[\frac{7}{54}\left(\frac{9}{7}\right)^{3/2}\right]\right\} = \frac{1}{3}\left[\cos^{-1}\left(\frac{1}{2\sqrt{7}}\right)\right] = 26.37^0$$

$$\alpha_2 = 146.37^0, \qquad \alpha_3 = -93.63^0$$

$$\bar{\lambda}_1 = 2\cos 26.37^0(\tfrac{7}{9})^{1/2} = 1.5803$$

$$\bar{\lambda}_2 = 2\cos 146.37^0(\tfrac{7}{9})^{1/2} = -1.4687$$

$$\bar{\lambda}_3 = 2\cos(-93.63^0)(\tfrac{7}{9})^{1/2} = -0.1117$$

Finally, the principal stresses are given by

$$\lambda_1 = 3.247, \qquad \lambda_2 = 0.198, \qquad \lambda_3 = 1.555$$

The plane associated with $\lambda_1 = 3.247$ is given by solving the equations

$$-2.247n_1^{(1)} - n_2^{(1)} = 0$$

$$-n_1^{(1)} - 1.247n_2^{(1)} - n_3^{(1)} = 0$$

$$-n_2^{(1)} - 1.247n_3^{(1)} = 0$$

which yield

$$n_2^{(1)} = -2.247n_1^{(1)}, \qquad n_3^{(1)} = -1.802n_1^{(1)}$$

The plane is defined by its normal

$$\mathbf{N}^{(1)} = (1, -2.247, -1.802)$$

or the unit normal

$$\hat{\mathbf{n}}^{(1)} = (0.328, -0.737, -0.591)$$

1.2.5 Equations of Motion

Equations of motion for solids can be derived using either Newton's second law of motion or the principle of conservation of linear momentum. First, we shall derive the equations using Newton's second law and then employing the principle of conservation of linear momentum.

Consider the stresses and body forces on an infinitesimal element of an elastic body. In general, the stresses in a body vary with respect to position. Figure 1.5 shows the stresses acting on the various faces of an infinitesimal parallelepiped. Note that the state of stress shown in Fig. 1.3 is at a *point*, whereas that shown in Fig. 1.5 is on an infinitesimal parallelepiped. The sum of the forces in the x_1-direction is given by

$$\left(\sigma_{11} + \frac{\partial \sigma_{11}}{\partial x_1} \Delta x_1 \right) \Delta x_2 \Delta x_3 + \left(\sigma_{21} + \frac{\partial \sigma_{21}}{\partial x_2} \Delta x_2 \right) \Delta x_1 \Delta x_3$$

$$+ \left(\sigma_{31} + \frac{\partial \sigma_{31}}{\partial x_3} \Delta x_3 \right) \Delta x_1 \Delta x_2 - \sigma_{11} \Delta x_2 \Delta x_3$$

$$- \sigma_{21} \Delta x_1 \Delta x_3 - \sigma_{31} \Delta x_1 \Delta x_2 + f_1 \Delta x_1 \Delta x_2 \Delta x_3$$

$$= \left(\frac{\partial \sigma_{11}}{\partial x_1} + \frac{\partial \sigma_{21}}{\partial x_2} + \frac{\partial \sigma_{31}}{\partial x_3} + f_1 \right) \Delta x_1 \Delta x_2 \Delta x_3$$

By Newton's second law of motion, the sum of the forces is equal to the

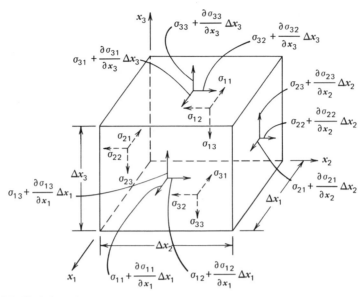

Figure 1.5 Variation of stress components on a parallelepiped element of dimensions Δx_1, Δx_2, and Δx_3.

product of mass and acceleration in the x_1-direction

$$\left(\rho\,\Delta x_1\,\Delta x_2\,\Delta x_3\right)\frac{\partial^2 u_1}{\partial t^2}$$

where ρ is the density. Thus we have, upon dividing throughout by $\Delta x_1\,\Delta x_2\,\Delta x_3$,

$$\frac{\partial \sigma_{11}}{\partial x_1} + \frac{\partial \sigma_{21}}{\partial x_2} + \frac{\partial \sigma_{31}}{\partial x_3} + f_1 = \rho\frac{\partial^2 u_1}{\partial t^2} \qquad (1.2.33a)$$

Similarly, the application of Newton's second law to forces along the x_2- and x_3-directions gives respectively,

$$\frac{\partial \sigma_{12}}{\partial x_1} + \frac{\partial \sigma_{22}}{\partial x_2} + \frac{\partial \sigma_{32}}{\partial x_3} + f_2 = \rho\frac{\partial^2 u_2}{\partial t^2} \qquad (1.2.33b)$$

$$\frac{\partial \sigma_{13}}{\partial x_1} + \frac{\partial \sigma_{23}}{\partial x_2} + \frac{\partial \sigma_{33}}{\partial x_3} + f_3 = \rho\frac{\partial^2 u_3}{\partial t^2} \qquad (1.2.33c)$$

or, in index notation,

$$\frac{\partial \sigma_{ji}}{\partial x_j} + f_i = \rho\frac{\partial^2 u_i}{\partial t^2} \qquad (i = 1, 2, 3) \qquad (1.2.34)$$

Now we derive the equations of motion from the principle of conservation of linear momentum. This principle states that *the time rate of change of the total momentum of a given body equals the vector sum of all the external forces acting on the body*, provided that Newton's third law of action and reaction governs the internal forces. The momentum of the elemental volume dV is (mass $\times$ velocity) $= (\rho \, dV)\partial \mathbf{u}/\partial t$. The total momentum is given by

$$\int_V \rho \frac{\partial \mathbf{u}}{\partial t} \, dV$$

The time rate of the momentum is

$$\frac{d}{dt} \int_V \rho \frac{\partial \mathbf{u}}{\partial t} \, dV = \int_V \frac{d}{dt}(\rho \, dV)\frac{\partial \mathbf{u}}{\partial t} + \int_V \rho \frac{\partial^2 \mathbf{u}}{\partial t^2} \, dV \qquad (1.2.35)$$

where d/dt denotes the total (material) time derivative. The first integral on the right-hand side is equal to zero because of the principle of conservation of mass of a given material

$$\frac{d}{dt}(\rho \, dV) = 0 \qquad (1.2.36)$$

The momentum principle can now be expressed as

$$\int_V \rho \frac{\partial^2 \mathbf{u}}{\partial t^2} \, dV = \int_V \mathbf{f} \, dV + \oint_S \mathbf{t} \, dS$$

Using Eq. (1.2.10), the stress vector $\mathbf{t}$ can be expressed in terms of the stress tensor $\boldsymbol{\sigma}$ and the unit normal $\hat{\mathbf{n}}$. We obtain

$$\int_V \rho \frac{\partial^2 \mathbf{u}}{\partial t^2} \, dV - \int_V \mathbf{f} \, dV - \oint_S \hat{\mathbf{n}} \cdot \boldsymbol{\sigma} \, dS = \mathbf{0} \qquad (1.2.37a)$$

Using the divergence theorem, Eq. (1.1.16), the surface integral in (1.2.37a) can be converted to a volume integral:

$$-\int_V \left(\nabla \cdot \boldsymbol{\sigma} + \mathbf{f} - \rho \frac{\partial^2 \mathbf{u}}{\partial t^2} \right) dV = \mathbf{0} \qquad (1.2.37b)$$

The equation should hold for arbitrary region V. This implies that the integrand of the left-hand side of Eq. (1.2.37b) should be identically equal to zero. This gives the vector form of the equations of motion:

$$\nabla \cdot \boldsymbol{\sigma} + \mathbf{f} = \rho \frac{\partial^2 \mathbf{u}}{\partial t^2} \qquad (1.2.38)$$

In a rectangular cartesian coordinate system, Eq. (1.2.38) takes the form of Eq. (1.2.34):

$$\frac{\partial \sigma_{ji}}{\partial x_j} + f_i = \rho \frac{\partial^2 u_i}{\partial t^2}$$

In unabridged form, Eq. (1.2.34) is given by

$$\frac{\partial \sigma_{11}}{\partial x_1} + \frac{\partial \sigma_{21}}{\partial x_2} + \frac{\partial \sigma_{31}}{\partial x_3} + f_1 = \rho \frac{\partial^2 u_1}{\partial t^2}$$

$$\frac{\partial \sigma_{12}}{\partial x_1} + \frac{\partial \sigma_{22}}{\partial x_2} + \frac{\partial \sigma_{32}}{\partial x_3} + f_2 = \rho \frac{\partial^2 u_2}{\partial t^2} \qquad (1.2.39)$$

$$\frac{\partial \sigma_{13}}{\partial x_1} + \frac{\partial \sigma_{23}}{\partial x_2} + \frac{\partial \sigma_{33}}{\partial x_3} + f_3 = \rho \frac{\partial^2 u_3}{\partial t^2}$$

The equations of equilibrium are obtained by setting the right-hand side of Eq. (1.2.38) equal to zero:

$$\nabla \cdot \sigma + f = 0 \qquad (1.2.40)$$

Equation (1.2.40) contains three equations [see Eq. (1.2.39)] relating nine stress components. In the following paragraphs, we show that the stress tensor is symmetric ($\sigma_{ij} = \sigma_{ji}$) when no body moments and couples exist. Even when the stress tensor is symmetric, the equations of equilibrium are not sufficient for the determination of the six stress components ($\sigma_{11}, \sigma_{12} = \sigma_{21}, \sigma_{13} = \sigma_{31}, \sigma_{22}, \sigma_{23} = \sigma_{32}, \sigma_{33}$). Additional equations are required, and these include the strain–displacement equations (see Section 1.3) and constitutive equations (see Section 1.5).

1.2.6 Symmetry of the Stress Tensor

When a body is not subjected to distributed couples (i.e., volume-dependent couples are not present), one can establish the symmetry of the stress tensor. Here we prove the symmetry first using Newton's second law for moments and then using the principle of moment of momentum.

Consider the moment of all forces acting on the parallelepiped shown in Fig. 1.5. Taking the moment of the forces about the x_3-axis (use the right-hand screw rule for positive moment), we obtain

$$-\left(\sigma_{11} + \frac{\partial \sigma_{11}}{\partial x_1}\Delta x_1\right)\Delta x_2 \Delta x_3 \cdot \frac{\Delta x_2}{2} - \left(\sigma_{21} + \frac{\partial \sigma_{21}}{\partial x_2}\Delta x_2\right)\Delta x_1 \Delta x_3 \cdot \Delta x_2$$

$$+\left(\sigma_{12} + \frac{\partial \sigma_{12}}{\partial x_1}\Delta x_1\right)\Delta x_2 \Delta x_3 \cdot \Delta x_1 + \left(\sigma_{22} + \frac{\partial \sigma_{22}}{\partial x_2}\Delta x_2\right)\Delta x_1 \Delta x_3 \cdot \frac{\Delta x_1}{2}$$

$$-\left(\sigma_{31} + \frac{\partial \sigma_{31}}{\partial x_3}\Delta x_3\right)\Delta x_1 \Delta x_2 \cdot \frac{\Delta x_2}{2} + \left(\sigma_{32} + \frac{\partial \sigma_{32}}{\partial x_3}\Delta x_3\right)\Delta x_1 \Delta x_2 \cdot \frac{\Delta x_1}{2}$$

$$+\sigma_{11}\Delta x_2 \Delta x_3 \cdot \frac{\Delta x_2}{2} + \sigma_{31}\Delta x_1 \Delta x_2 \cdot \frac{\Delta x_2}{2} - \sigma_{22}\Delta x_1 \Delta x_3 \cdot \frac{\Delta x_1}{2}$$

$$-\sigma_{32}\Delta x_1 \Delta x_2 \cdot \frac{\Delta x_1}{2} - f_1 \Delta x_1 \Delta x_2 \Delta x_3 \cdot \frac{\Delta x_2}{2} + f_2 \Delta x_1 \Delta x_2 \Delta x_3 \cdot \frac{\Delta x_1}{2} = 0$$

Dividing throughout by $\Delta x_1 \Delta x_2 \Delta x_3$ and taking the limit $\Delta x_1 \to 0$, $\Delta x_2 \to 0$, and $\Delta x_3 \to 0$, we obtain

$$\sigma_{12} - \sigma_{21} = 0 \qquad (1.2.41a)$$

Similar considerations of moments about the x_1-axis and x_2-axis give, respectively, the relations

$$\sigma_{23} - \sigma_{32} = 0 \qquad (1.2.41b)$$

$$\sigma_{13} - \sigma_{31} = 0 \qquad (1.2.41c)$$

Next, we prove the symmetry of the stress tensor using the principle of moment of momentum. The principle states that *the time rate of change of moment of momentum for a given material body is equal to the vector sum of the moment of the external forces acting on the body*:

$$\frac{d}{dt}\int_V \mathbf{r} \times \left(\rho \frac{\partial \mathbf{u}}{\partial t}\right) dV = \int_V \mathbf{r} \times \mathbf{f} \, dV + \oint_S \mathbf{r} \times \mathbf{t} \, dS \qquad (1.2.42)$$

Using Eq. (1.2.36) and (1.2.38), Eq. (1.2.42) can be expressed as

$$\int_V \mathbf{r} \times (\nabla \cdot \boldsymbol{\sigma}) \, dV - \oint_S \mathbf{r} \times \mathbf{t} \, dS = \mathbf{0}$$

To further simplify the equation, we express it in rectangular component form:

$$\oint_S \varepsilon_{ijk} x_i t_j \hat{\mathbf{e}}_k \, dS = \int_V \varepsilon_{ijk} x_i \frac{\partial \sigma_{mj}}{\partial x_m} \hat{\mathbf{e}}_k \, dV \qquad (1.2.43)$$

Using Eq. (1.2.12a) to write the stress vector components in the form $t_j = n_m \sigma_{mj}$, we have

$$\oint_S \varepsilon_{ijk} x_i n_m \sigma_{mj} \, dS = \int_V \varepsilon_{ijk} \frac{\partial}{\partial x_m} (x_i \sigma_{mj}) \, dV$$

$$= \int_V \left(\varepsilon_{ijk} \delta_{im} \sigma_{mj} + \varepsilon_{ijk} x_i \frac{\partial \sigma_{mj}}{\partial x_m} \right) dV$$

Substituting the above identity into Eq. (1.2.43), we obtain

$$\int_V \varepsilon_{ijk} \sigma_{ij} \, dV = 0$$

or

$$\varepsilon_{ijk} \sigma_{ij} = 0 \quad \text{for } k = 1, 2, 3 \tag{1.2.44}$$

This implies that the components of the stress tensor are symmetric:

$$k = 1: \qquad \sigma_{23} - \sigma_{32} = 0$$

$$k = 2: \qquad \sigma_{31} - \sigma_{13} = 0$$

$$k = 3: \qquad \sigma_{12} - \sigma_{21} = 0$$

Thus, there are only six stress components that are independent.

Exercises 1.2

1. Derive the Cauchy formula in two dimensions by considering the equilibrium of the right-angle wedge shown in Fig. E1.
2. Derive the equations of equilibrium for a two-dimensional body by summing the forces in the x_1- and x_2-directions.
3. Using the wedge in Exercise 1, derive the transformation relations relating the normal and shear stresses, σ_N and σ_S, on the slant edge to

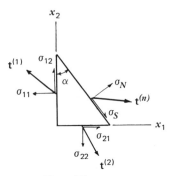

Figure E1

stresses σ_{11}, σ_{22}, and $\sigma_{12} = \sigma_{21}$:

$$\sigma_N = \sigma_{11}\cos^2\alpha + \sigma_{22}\sin^2\alpha - 2\sigma_{12}\sin\alpha\cos\alpha$$

$$\sigma_S = (\sigma_{11} - \sigma_{22})\sin\alpha\cos\alpha + \sigma_{12}(\cos^2\alpha - \sin^2\alpha)$$

4. Show that the principal stresses at a point in a two-dimensional body are given by

$$\sigma_{max} = \frac{\sigma_{11} + \sigma_{22}}{2} + \sqrt{\left(\frac{\sigma_{11} - \sigma_{22}}{2}\right)^2 + \sigma_{12}^2}$$

$$\sigma_{min} = \frac{\sigma_{11} + \sigma_{22}}{2} - \sqrt{\left(\frac{\sigma_{11} - \sigma_{22}}{2}\right)^2 + \sigma_{12}^2}$$

and the orientation of the principal planes is given by

$$\alpha = \tfrac{1}{2}\tan^{-1}\left[\frac{-\sigma_{12}}{(\sigma_{11} - \sigma_{22})/2}\right]$$

5. Determine whether the following stress fields are possible in a structural member free of body forces:

(a) $\sigma_{11} = c_1 x_1 + c_2 x_2 + c_3 x_1 x_2$, $\sigma_{12} = \sigma_{21} = -c_3(x_2^2/2) - c_1 x_2$, $\sigma_{22} = c_4 x_1 + c_1 x_2$

(b) $\sigma_{11} = x_1^2 - 2x_1 x_2 + 3x_3$, $\sigma_{12} = \sigma_{21} = -x_1 x_2 + x_2^2$, $\sigma_{13} = \sigma_{31} = -x_1 x_3$, $\sigma_{22} = x_2^2$, $\sigma_{23} = -x_2 x_3$, $\sigma_{33} = (x_1 + x_2)x_3$

(c) $\sigma_{11} = 3x_1 + 5x_2$, $\sigma_{12} = 4x_1 - 3x_2$, $\sigma_{22} = 2x_1 - 4x_2$

6. Given the following state of stress ($\sigma_{ij} = \sigma_{ji}$),

$$\sigma_{11} = -2x_1^2, \qquad \sigma_{12} = -7 + 4x_1 x_2 + x_3, \qquad \sigma_{13} = 1 + x_1 - 3x_2,$$

$$\sigma_{22} = 3x_1^2 - 2x_2^2 - 5x_3, \qquad \sigma_{23} = 0, \qquad \sigma_{33} = -5 + x_1 + 3x_2 + 3x_3,$$

determine the body force components for which the stress field describes a state of equilibrium.

7. For the cantilever beam bent by a point load at the free end (see Fig. E7), the bending moment M_3 about the x_3-axis is given by $M_3 = Px_1$. The bending stress σ_{11} is given by

$$\sigma_{11} = -\frac{M_3 x_2}{I_3} = -\frac{Px_1 x_2}{I_3}$$

where I_3 is the moment of inertia of the cross section about the

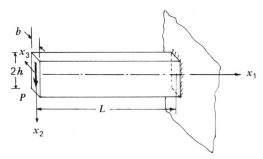

Figure E7

x_3-axis. Starting with this equation, use the two-dimensional equilibrium equations to determine the stresses σ_{22} and σ_{12} as functions of x_1 and x_2.

8. Repeat Exercise 7 for the case in which the cantilever beam is bent by uniformly distributed load f_0/unit length.

9. For the state of stress given in Exercise 6, determine the stress vector at point (x_1, x_2, x_3) on the plane $x_1 + x_2 + x_3 = $ constant. What are the normal and shearing components of the stress vector at point $(1, 1, 3)$?

10. Find the principal stresses and their orientation at point $(1, 2, 1)$ for the state of stress given in Exercise 6.

11. Repeat Exercise 10 for point $(1, 1, 3)$.

12. Find the maximum principal stress and its orientation for the state of stress

$$[\sigma] = \begin{bmatrix} 3 & 5 & 8 \\ 5 & 1 & 0 \\ 8 & 0 & 2 \end{bmatrix} \text{psi}$$

13. Show that the characteristic polynomial for $[\sigma]$ can be expressed in terms of the invariants I_1, I_2, and I_3 as

$$\lambda^3 - I_1\lambda^2 - I_2\lambda - I_3 = 0$$

where

$$I_1 = \sigma_{ii}, \qquad I_2 = \tfrac{1}{2}(\sigma_{ij}\sigma_{ji} - \sigma_{ii}\sigma_{jj}), \qquad I_3 = \det[\sigma]$$

1.3 KINEMATICS

1.3.1 Strain at a Point

Kinematics is a study of the geometry of motion and deformation without consideration of the forces causing them. To describe the motion of a deformable body, we establish a fixed reference frame in three-dimensional space and denote the Cartesian coordinates of a point by $\mathbf{x} = (x_1, x_2, x_3)$.

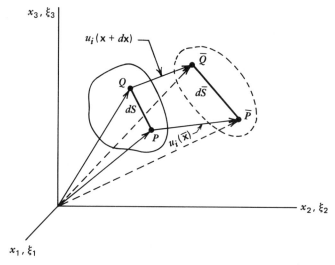

Figure 1.6 Relative displacements between two neighboring points in a body. P: (x_1, x_2, x_3). Q: $(x_1 + dx_1, x_2 + dx_2, x_3 + dx_3)$. $\bar{P}$: (ξ_1, ξ_2, ξ_3). $\bar{Q}$: $(\xi_1 + d\xi_1, \xi_2 + d\xi_2, \xi_3 + d\xi_3)$.

Consider a body in its unstressed state, the configuration to which the body will return after the removal of the forces causing the deformation. A body is said to undergo *rigid body motion* if the distance between any two points in the body remains unchanged. Thus, a measure of the deformation (i.e., change in geometry) in a body is provided by the change in the distance between points in the body. Consider two neighboring points P: (x_1, x_2, x_3) and Q: $(x_1 + dx_1, x_2 + dx_2, x_3 + dx_3)$ in the reference configuration (see Fig. 1.6). Under the action of externally applied forces, the body deforms, and the points P and Q move to new places: $\bar{P}$: (ξ_1, ξ_2, ξ_3) and $\bar{Q}$: $(\xi_1 + d\xi_1, \xi_2 + d\xi_2, \xi_3 + d\xi_3)$, respectively. Since points P and Q are arbitrary, the discussion applies to any point in the body. The motion of a particle occupying position $\mathbf{x}$ in the undeformed body to point $\boldsymbol{\xi}$ in the deformed body can be expressed by the transformation

$$\xi_i = \xi_i(x_1, x_2, x_3, t) \qquad (1.3.1)$$

where t denotes time. By assumption, a continuous medium cannot have gaps or overlaps. Therefore, a one-to-one correspondence exists between points in the undeformed body and points in the deformed body. Consequently, a unique inverse to (1.3.1) holds:

$$x_i = x_i(\xi_1, \xi_2, \xi_3, t) \qquad (1.3.2)$$

The distances between points P and Q and points $\bar{P}$ and $\bar{Q}$ are given, respectively, by

$$(dS)^2 = dx_i\, dx_i, \qquad (d\bar{S})^2 = d\xi_i\, d\xi_i \qquad (1.3.3)$$

where summation on repeated subscripts is implied. The change in the distance between points P and Q is given by

$$(d\bar{S})^2 - (dS)^2 = d\xi_i \, d\xi_i - dx_i \, dx_i$$

$$= \frac{\partial \xi_i}{\partial x_m} \frac{\partial \xi_i}{\partial x_n} dx_m \, dx_n - dx_i \, dx_i$$

$$= \left(\frac{\partial \xi_m}{\partial x_i} \frac{\partial \xi_m}{\partial x_j} - \delta_{ij} \right) dx_i \, dx_j$$

$$= 2\varepsilon_{ij} \, dx_i \, dx_j \qquad (1.3.4)$$

where ε_{ij} are the components of the Green strain tensor at point P:

$$\varepsilon_{ij} = \frac{1}{2} \left(\frac{\partial \xi_m}{\partial x_i} \frac{\partial \xi_m}{\partial x_j} - \delta_{ij} \right) \qquad (1.3.5)$$

Note that the change in a line element is zero if and only if $\varepsilon_{ij} = 0$. The strains can be expressed in terms of the displacement components at point P:

$$u_m = \xi_m - x_m \qquad (1.3.6)$$

We have

$$\frac{\partial \xi_m}{\partial x_i} = \frac{\partial u_m}{\partial x_i} + \delta_{mi} \qquad (1.3.7)$$

$$\varepsilon_{ij} = \frac{1}{2} \left[\left(\frac{\partial u_m}{\partial x_i} + \delta_{mi} \right) \left(\frac{\partial u_m}{\partial x_j} + \delta_{mj} \right) - \delta_{ij} \right]$$

$$= \frac{1}{2} \left(\frac{\partial u_i}{\partial x_j} + \frac{\partial u_j}{\partial x_i} + \frac{\partial u_m}{\partial x_i} \frac{\partial u_m}{\partial x_j} \right) \qquad (1.3.8)$$

In unabridged notation, typical strain components are given by

$$\varepsilon_{11} = \frac{\partial u_1}{\partial x_1} + \frac{1}{2} \left[\left(\frac{\partial u_1}{\partial x_1} \right)^2 + \left(\frac{\partial u_2}{\partial x_1} \right)^2 + \left(\frac{\partial u_3}{\partial x_1} \right)^2 \right]$$

$$\varepsilon_{12} = \frac{1}{2} \left(\frac{\partial u_1}{\partial x_2} + \frac{\partial u_2}{\partial x_1} + \frac{\partial u_1}{\partial x_1} \frac{\partial u_1}{\partial x_2} + \frac{\partial u_2}{\partial x_1} \frac{\partial u_2}{\partial x_2} + \frac{\partial u_3}{\partial x_1} \frac{\partial u_3}{\partial x_2} \right) \quad (1.3.9)$$

When the derivatives of the displacement components are small compared to unity, their products can be neglected:

$$\frac{\partial u_i}{\partial x_j} \ll 1, \qquad \left(\frac{\partial u_i}{\partial x_j}\right)^2 \approx 0 \tag{1.3.10}$$

In this case, the strain components ε_{ij} become the *infinitesimal strain* components e_{ij},

$$e_{ij} = \frac{1}{2}\left(\frac{\partial u_i}{\partial x_j} + \frac{\partial u_j}{\partial x_i}\right) \tag{1.3.11}$$

1.3.2   Transformation of Strain Components

Next we establish a relation between the components of strain at a point in one coordinate system and those at the same point in another coordinate system. Consider two coordinate systems (x_1, x_2, x_3) and (x_1', x_2', x_3'), and the strain components ε_{ij} and ε_{ij}' in the respective coordinate systems. The two coordinate systems are related by the transformation

$$x_i' = a_{ij}x_j, \qquad x_i = a_{ji}x_j' \tag{1.3.12a}$$

or

$$\{x'\} = [A]\{x\}, \qquad \{x\} = [A]^T\{x'\} \tag{1.3.12b}$$

where a_{ij} are the direction cosines

$$a_{ij} = \hat{e}_i' \cdot \hat{e}_j \tag{1.3.13}$$

Here $\hat{e}_i'$ and $\hat{e}_i$ denote the basis vectors in the primed and unprimed coordinates, respectively. From Eq. (1.3.12), it follows that

$$a_{ij}a_{mj} = \delta_{im} \quad \text{or} \quad [A][A]^T = [I] \tag{1.3.14}$$

In view of Eq. (1.3.12a), the displacement components in the two coordinate systems are related by

$$u_i' = a_{ij}u_j \tag{1.3.15}$$

From Eqs. (1.3.12) and (1.3.15), we have

$$\frac{\partial u_i'}{\partial x_j'} = a_{im}\frac{\partial u_m}{\partial x_j'}$$

$$= a_{im}a_{jn}\frac{\partial u_m}{\partial x_n} \tag{1.3.16}$$

and

$$\varepsilon'_{ij} = \frac{1}{2}\left(\frac{\partial u'_i}{\partial x'_j} + \frac{\partial u'_j}{\partial x'_i} + \frac{\partial u'_k}{\partial x'_i}\frac{\partial u'_k}{\partial x'_j}\right)$$

$$= \frac{1}{2}\left(a_{im}a_{jn}\frac{\partial u_m}{\partial x_n} + a_{in}a_{jm}\frac{\partial u_m}{\partial x_n} + a_{kp}a_{iq}\frac{\partial u_p}{\partial x_q}a_{kr}a_{js}\frac{\partial u_r}{\partial x_s}\right)$$

$$= a_{im}a_{jn}\frac{1}{2}\left(\frac{\partial u_m}{\partial x_n} + \frac{\partial u_n}{\partial x_m} + \frac{\partial u_r}{\partial x_m}\frac{\partial u_r}{\partial x_n}\right)$$

$$= a_{im}a_{jn}\varepsilon_{mn} \qquad\qquad (1.3.17a)$$

where, in arriving at the last step, we made several changes in dummy subscripts and made use of Eq. (1.3.14). In matrix form, Eq. (1.3.17a) can be written as

$$[\varepsilon'] = [A][\varepsilon][A]^T \qquad\qquad (1.3.17b)$$

which is analogous to Eq. (1.2.16b) for a stress tensor.

Example 1.4

Consider a rectangular block of dimensions $a \times b \times h$, where h is very small compared to a and b. Suppose that the block is deformed into the diamond shape shown in Fig. 1.7. By inspection, the geometry of the deformed body can be described as follows: let (ξ_1, ξ_2, ξ_3) denote the coordinates of a material point in the undeformed configuration. In the deformed configuration, the material point has the coordinates,

$$x_1 = \xi_1 + \frac{e_0}{b}\xi_2$$

$$x_2 = \xi_2 + \frac{e_0}{a}\xi_1$$

$$x_3 \approx \xi_3 \qquad \text{(because } h \ll a \text{ or } b\text{)} \qquad (1.3.18)$$

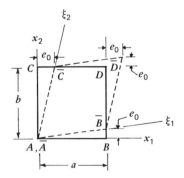

Figure 1.7 Strains in a rectangular block.

Thus, the displacement components of the material point are

$$u_1 = x_1 - \xi_1 = \frac{e_0}{b}\xi_2$$

$$u_2 = x_2 - \xi_2 = \frac{e_0}{a}\xi_1$$

$$u_3 = x_3 - \xi_3 = 0 \tag{1.3.19}$$

Note that $\xi_i = x_i$ in the undeformed body. Therefore, we can write the displacements in terms of the original position of the material point

$$u_1 = \frac{e_0}{b}x_2, \qquad u_2 = \frac{e_0}{a}x_1, \qquad u_3 = 0$$

The strains are given by

$$\varepsilon_{11} = \frac{1}{2}\left(\frac{e_0}{a}\right)^2, \qquad \varepsilon_{12} = \frac{e_0}{2b} + \frac{e_0}{2a}, \qquad \varepsilon_{22} = \frac{1}{2}\left(\frac{e_0}{b}\right)^2$$

$$\varepsilon_{13} = \varepsilon_{23} = \varepsilon_{33} = 0 \tag{1.3.20}$$

The same result can be obtained using the elementary mechanics of materials approach. For example, a line element AB in the undeformed body moves to position $\overline{AB}$. Then the strain in the line AB is given by

$$\varepsilon_{11} = \varepsilon_{AB} = \frac{\overline{AB} - AB}{AB}$$

$$= \frac{1}{a}\sqrt{a^2 + e_0^2} - 1$$

$$= \sqrt{1 + \left(\frac{e_0}{a}\right)^2} - 1$$

$$= \left[1 + \frac{1}{2}\left(\frac{e_0}{a}\right)^2 + \cdots\right] - 1$$

$$\approx \frac{1}{2}\left(\frac{e_0}{a}\right)^2$$

When e_0/b and e_0/a are small compared to unity, their squares can be neglected, and the infinitesimal strain components are given by

$$e_{11} \approx 0, \qquad e_{22} \approx 0, \qquad e_{12} = \frac{e_0}{2ab}(a + b) \tag{1.3.21}$$

Now consider a line element oriented at an angle from the x_1-axis. For example, the diagonal line AC deforms and becomes $\overline{AC}$ in the deformed body. The normal strain in AC is given by

$$\varepsilon_\theta = \varepsilon_{AC} = \frac{\overline{AC} - AC}{AC}$$

$$= \frac{\sqrt{(a + e_0)^2 + (b + e_0)^2} - \sqrt{a^2 + b^2}}{\sqrt{a^2 + b^2}}$$

$$= \sqrt{1 + 2\left[\frac{e_0^2 + e_0(a + b)}{a^2 + b^2}\right]} - 1$$

$$\approx \frac{e_0^2 + e_0(a + b)}{a^2 + b^2} \tag{1.3.22}$$

The same result can be obtained from the transformation equation (1.3.17a). We have

$$\varepsilon_{11}' = a_{1i}a_{1j}\varepsilon_{ij}$$

$$= a_{11}a_{11}\varepsilon_{11} + a_{12}a_{12}\varepsilon_{22} + a_{11}a_{12}\varepsilon_{12} + a_{12}a_{11}\varepsilon_{21}$$

where

$$a_{11} = \cos\theta, \qquad a_{12} = \sin\theta, \qquad a_{22} = \cos\theta, \qquad a_{21} = -\sin\theta$$

$$\varepsilon_{11}' = \cos^2\theta\,\varepsilon_{11} + \sin^2\theta\,\varepsilon_{22} + 2\sin\theta\cos\theta\,\varepsilon_{12} \equiv \varepsilon_\theta$$

where $\theta = \tan^{-1}(b/a)$. Substituting for $\cos\theta$, $\sin\theta$, and ε_{ij} into ε_{ij}', we get

$$\varepsilon_{11}' = \frac{a^2}{a^2 + b^2}\frac{1}{2}\left(\frac{e_0}{a}\right)^2 + \frac{b^2}{a^2 + b^2}\frac{1}{2}\left(\frac{e_0}{b}\right)^2 + 2\frac{ab}{a^2 + b^2}\frac{e_0(a + b)}{2ab}$$

$$= \frac{e_0^2 + e_0(a + b)}{a^2 + b^2}$$

This completes the example.

1.3.3 Strain Compatibility Equations

When a sufficiently differentiable displacement field is given, the computation of the strains is straightforward: one can make use of the equations in (1.3.8) to compute the strains. However, when the strain components are given, the determination of the displacements is not always possible, because there are six strain components related to three displacement components. Stated in other words, there are six differential equations involving three unknowns. Thus, the six equations should be compatible with each other in the sense that any three equations should give the same displacement field.

To further understand the situation, let us consider the two-dimensional case. The strain–displacement relations of the infinitesimal theory are given by

$$\frac{\partial u_1}{\partial x_1} = e_{11}$$

$$\frac{\partial u_2}{\partial x_2} = e_{22}$$

$$\frac{\partial u_1}{\partial x_2} + \frac{\partial u_2}{\partial x_1} = 2e_{12} \tag{1.3.23}$$

If e_{ij} are given as functions of x_1 and x_2, they cannot be arbitrary; e_{11}, e_{22}, and e_{12} should have a relationship that the three equations are compatible. This relation can be derived as follows: differentiate the third equation with respect to x_1 and x_2 to obtain

$$\frac{\partial^3 u_1}{\partial x_1 \partial x_2^2} + \frac{\partial^3 u_2}{\partial x_1^2 \partial x_2} = 2\frac{\partial^2 e_{12}}{\partial x_1 \partial x_2} \tag{1.3.24}$$

The third derivatives of u_1 and u_2 needed in Eq. (1.3.24) can be computed from the first two equations in Eq. (1.3.23):

$$\frac{\partial^3 u_1}{\partial x_1 \partial x_2^2} = \frac{\partial^2 e_{11}}{\partial x_2^2}, \qquad \frac{\partial^2 u_2}{\partial x_1^2 \partial x_2} = \frac{\partial^2 e_{22}}{\partial x_1^2}$$

Substituting these relations into Eq. (1.3.24), we obtain a relationship between the derivatives of the strain components:

$$\frac{\partial^2 e_{11}}{\partial x_2^2} + \frac{\partial^2 e_{22}}{\partial x_1^2} = 2\frac{\partial^2 e_{12}}{\partial x_1 \partial x_2} \tag{1.3.25}$$

Equation (1.3.25) is called the compatibility equation for a two-dimensional strain field.

The discussion given above can be generalized to three-dimensional elasticity. First, we form the following derivatives of the strains defined in (1.3.11):

$$e_{ij,kl} = \tfrac{1}{2}\left(u_{i,jkl} + u_{j,ikl}\right)$$

$$e_{kl,ij} = \tfrac{1}{2}\left(u_{k,lij} + u_{l,kij}\right)$$

$$e_{lj,ki} = \tfrac{1}{2}\left(u_{l,jki} + u_{j,lki}\right)$$

$$e_{ki,lj} = \tfrac{1}{2}\left(u_{k,ilj} + u_{i,klj}\right) \tag{1.3.26}$$

Next, we add the first two equations and the last two equations separately:

$$e_{ij,kl} + e_{kl,ij} = \tfrac{1}{2}\left(u_{i,jkl} + u_{j,ikl} + u_{k,lij} + u_{l,kij}\right)$$

$$e_{lj,ki} + e_{ki,lj} = \tfrac{1}{2}\left(u_{l,jki} + u_{j,lki} + u_{k,ilj} + u_{i,lkj}\right)$$

We immediately note that the right sides of the two expressions are the same (note that the order of the subscripts following the comma is interchangeable). Therefore, we have

$$e_{ij,kl} + e_{kl,ij} = e_{lj,ki} + e_{ki,lj} \tag{1.3.27}$$

Equation (1.3.27) forms the necessary and sufficient conditions for the existence of a single-valued displacement field (when the strains are given). Although Eq. (1.3.27) yields 81 equations, only six of them are different from each other, and the remaining are either trivial or linear combinations of the six equations. These six equations are given by

$$\frac{\partial^2 e_{11}}{\partial x_2^2} + \frac{\partial^2 e_{22}}{\partial x_1^2} = 2\frac{\partial^2 e_{12}}{\partial x_1 \, \partial x_2}$$

$$\frac{\partial^2 e_{22}}{\partial x_3^2} + \frac{\partial^2 e_{33}}{\partial x_2^2} = 2\frac{\partial^2 e_{23}}{\partial x_2 \, \partial x_3}$$

$$\frac{\partial^2 e_{11}}{\partial x_3^2} + \frac{\partial^2 e_{33}}{\partial x_1^2} = 2\frac{\partial^2 e_{13}}{\partial x_1 \, \partial x_3}$$

$$\frac{\partial}{\partial x_1}\left(-\frac{\partial e_{23}}{\partial x_1} + \frac{\partial e_{13}}{\partial x_2} + \frac{\partial e_{12}}{\partial x_3}\right) = \frac{\partial^2 e_{11}}{\partial x_2 \, \partial x_3}$$

$$\frac{\partial}{\partial x_2}\left(-\frac{\partial e_{13}}{\partial x_2} + \frac{\partial e_{12}}{\partial x_3} + \frac{\partial e_{23}}{\partial x_1}\right) = \frac{\partial^2 e_{22}}{\partial x_1 \, \partial x_3}$$

$$\frac{\partial}{\partial x_3}\left(-\frac{\partial e_{12}}{\partial x_3} + \frac{\partial e_{23}}{\partial x_1} + \frac{\partial e_{13}}{\partial x_2}\right) = \frac{\partial^2 e_{33}}{\partial x_1 \, \partial x_2} \tag{1.3.28}$$

Example 1.5

Consider the problem of an isotropic cantilever beam bent by a load P at the free end (see Fig. 1.8). From the elementary beam theory (see Section 1.7 for a review), we can calculate the strains. They are given by

$$e_{11} = -\frac{Px_1 x_2}{EI}, \qquad e_{22} = v\frac{Px_1 x_2}{EI}, \qquad e_{12} = -\frac{(1+v)P}{2EI}\left(h^2 - x_2^2\right)$$

$$\tag{1.3.29}$$

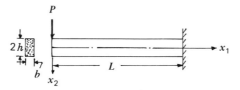

Figure 1.8 Bending of a cantilever beam.

Now let us see if these strains are compatible. Substituting e_{ij} into Eq. (1.3.25), we get

$$0 + 0 = 0$$

Although the two-dimensional strains are compatible, the three-dimensional strains are not compatible. For example, using the additional strains

$$e_{33} = \frac{\nu P x_1 x_2}{EI}, \qquad e_{13} = e_{23} = 0 \qquad (1.3.30)$$

one can verify that all of the equations except the last one in Eq. (1.3.28) are satisfied.

Exercises 1.3

1. The two-dimensional displacement field in a member is given by

$$u_1 = x_1\left[x_1^2 x_2 + c_1\left(2c_2^3 + 3c_2^2 x_2 - x_2^3\right)\right]$$

$$u_2 = -x_2\left(2c_2^3 + \tfrac{3}{2}c_2^2 x_2 - \tfrac{1}{4}x_2^3 + \tfrac{3}{2}c_1 x_1^2 x_2\right)$$

where c_1 and c_2 are constants. Find the linear strains.

2. Find the linear strains associated with the displacements

$$u_1 = \left[x_1 x_2(2 - x_1) - c_1 x_2 + c_2 x_2^3\right]$$

$$u_2 = -\left[c_3 x_2^2(1 - x_1) + (3 - x_1)\frac{x_1^2}{3} + c_1 x_1\right]$$

3-4. Determine the displacements and strains in the (x_1, x_2) system for the bodies shown in Figs. E3 and E4.

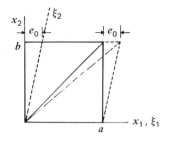

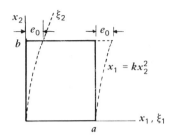

Figure E3   Figure E4

5. Find the linear strains associated with the displacement field,

$$u_1 = u_1^0(x_1, x_2) + x_3\psi_1(x_1, x_2)$$

$$u_2 = u_2^0(x_1, x_2) + x_3\psi_2(x_1, x_2)$$

$$u_3 = u_3^0(x_1, x_2)$$

6. Determine whether the following strain fields are possible in a continuous body

(a) $[e] = \begin{bmatrix} (x_1^2 + x_2^2) & x_1 x_2 \\ x_1 x_2 & x_2^2 \end{bmatrix}$

(b) $[e] = \begin{bmatrix} x_3(x_1^2 + x_2^2) & 2x_1 x_2 x_3 & x_3 \\ 2x_1 x_2 x_3 & x_2^2 & x_1 \\ x_3 & x_1 & x_3^2 \end{bmatrix}$

7. Find the normal and shear strains in the diagonal element of the rectangular block in Exercise 3.

8. The biaxial state of strain at a point is given by $\varepsilon_{11} = 800 \times 10^{-6}$ in./in., $\varepsilon_{22} = 200 \times 10^{-6}$ in./in., $\varepsilon_{12} = 400 \times 10^{-6}$ in./in. Find the principal strains and their directions.

9. Find the axial strain in the diagonal element of Exercise 3 (a) using the basic definition of normal strain, and (b) using the strain transformation equations.

1.4 THERMODYNAMIC PRINCIPLES

1.4.1 Introduction

The first law of thermodynamics is commonly known as the principle of balance of energy, and it can be regarded as a statement of the interconvertibility of heat and work. The law does not place any restrictions on the direction of the process. For instance, in the study of mechanics of particles and rigid bodies, the kinetic energy and potential energy can be fully transformed from one to the other in the absence of friction and other dissipative mechanisms. From our experience, we know that mechanical energy that is converted into heat cannot all be converted back into mechanical energy. For example, the motion (i.e., kinetic energy) of a flywheel can all be converted into heat (i.e., internal energy) by means of a friction brake; if the whole system is insulated, the internal energy causes the temperature of the system to rise. Although the first law does not restrict the reversal process, namely the conversion of heat to internal energy and internal energy to motion (the flywheel), such a reversal cannot occur because the frictional dissipation is an *irreversible process*. The second law of thermodynamics places restrictions on such processes. To restrict the scope of the present coverage to reasonable limits, we give a brief discussion of the two laws of thermodynamics. A reader who is eager to skim this chapter fast can omit Sections 1.4.2 and 1.4.3.

1.4.2 The First Law of Thermodynamics: Energy Equation

The first law of thermodynamics states that the time-rate of change of the total energy (i.e., sum of the kinetic energy and the internal energy) is equal to the sum of the rate of work done by the external forces and the change of heat content per unit time:

$$\frac{d}{dt}(K + U) = W + H \tag{1.4.1}$$

Here, K denotes the kinetic energy, U is the internal energy, W is the power input, and H is the heat input to the system.

The kinetic energy of the system is given by

$$K = \frac{1}{2}\int_V \rho \frac{\partial \mathbf{u}}{\partial t} \cdot \frac{\partial \mathbf{u}}{\partial t}\, dV \tag{1.4.2}$$

If $\hat{U}_0$ is the energy/unit mass (or *specific internal energy*), the total internal energy of the system is given by

$$U = \int_V \rho \hat{U}_0\, dV \tag{1.4.3}$$

The kinetic energy (K) of a system is the energy associated with the macroscopically observable velocity of the continuum. The kinetic energy associated with the (microscopic) motions of molecules of the continuum is a part of the internal energy; the elastic strain energy and other forms of energy are also parts of internal energy.

The power input, in the nonpolar case, consists of the rate at which the external surface tractions $\mathbf{t}$/unit area and body forces $\mathbf{f}$/unit volume are doing work on the mass system instantaneously occupying the volume V bounded by S:

$$W = \oint_S \mathbf{t} \cdot \frac{\partial \mathbf{u}}{\partial t}\, dS + \int_V \mathbf{f} \cdot \frac{\partial \mathbf{u}}{\partial t}\, dV$$

$$= \oint_S (\hat{\mathbf{n}} \cdot \boldsymbol{\sigma}) \cdot \frac{\partial \mathbf{u}}{\partial t}\, dS + \int_V \mathbf{f} \cdot \frac{\partial \mathbf{u}}{\partial t}\, dV$$

$$= \int_V \left[\nabla \cdot \left(\boldsymbol{\sigma} \cdot \frac{\partial \mathbf{u}}{\partial t} \right) + \mathbf{f} \cdot \frac{\partial \mathbf{u}}{\partial t} \right] dV$$

$$= \int_V \left[(\nabla \cdot \boldsymbol{\sigma} + \mathbf{f}) \cdot \frac{\partial \mathbf{u}}{\partial t} + \boldsymbol{\sigma} : \nabla \frac{\partial \mathbf{u}}{\partial t} \right] dV \qquad (1.4.4)$$

where : denotes the "double-dot" product $\boldsymbol{\sigma}^1 : \boldsymbol{\sigma}^2 = \sigma_{ij}^1 \sigma_{ji}^2$. The expression in parentheses equals $\rho(\partial^2 \mathbf{u}/\partial t^2)$ by the equations of motion, Eq. (1.2.38). Also, by virtue of the symmetry of the stress tensor, the last expression in the integrand of Eq. (1.4.4) can be replaced by $\boldsymbol{\sigma} : \dot{\boldsymbol{\varepsilon}}$, where $\dot{\boldsymbol{\varepsilon}} = (\partial \boldsymbol{\varepsilon}/\partial t)$. Hence,

$$W = \frac{d}{dt} \int_V \frac{\rho}{2} \frac{\partial u_i}{\partial t} \frac{\partial u_i}{\partial t}\, dV + \int_V \sigma_{ij} \dot{\varepsilon}_{ij}\, dV \qquad (1.4.5)$$

The rate of heat input consists of conduction through the surface S and internal heat generation in the body (possibly from a radiation field):

$$H = -\oint_S \mathbf{q} \cdot \hat{\mathbf{n}}\, dS + \int_V \rho Q\, dV \qquad (1.4.6)$$

where $\mathbf{q}$ is the heat flux vector normal to the surface, and Q is the internal heat source/unit mass.

Substituting Eqs. (1.4.2)–(1.4.6) into Eq. (1.4.1), we obtain

$$\frac{d}{dt} \int_V \left(\frac{\rho}{2} \frac{\partial u_i}{\partial t} \frac{\partial u_i}{\partial t} + \rho \hat{U}_0 \right) dV = \frac{d}{dt} \int_V \frac{1}{2} \rho \frac{\partial u_i}{\partial t} \frac{\partial u_i}{\partial t}\, dV + \int_V \sigma_{ij} \dot{\varepsilon}_{ij}\, dV$$

$$- \oint_S \mathbf{q} \cdot \hat{\mathbf{n}}\, dS + \int_V \rho Q\, dV \qquad (1.4.7)$$

or [see Eqs. (1.2.35) and (1.2.36)]

$$\int_V \left(\rho \frac{\partial \hat{U}_0}{\partial t} - \sigma_{ij} \dot{\varepsilon}_{ij} + \frac{\partial q_i}{\partial x_i} - \rho Q \right) dV = 0 \qquad (1.4.8)$$

for arbitrary volume V. Thus, the conservation of energy implied by the first law of thermodynamics is given by

$$\rho \frac{\partial \hat{U}_0}{\partial t} = \sigma_{ij} \dot{\varepsilon}_{ij} + \rho Q - \frac{\partial q_i}{\partial x_i} \qquad (1.4.9)$$

1.4.3 The Second Law of Thermodynamics

The second law of thermodynamics for a reversible process states that there exists a function $\eta = \eta(\varepsilon_{ij}, T)$, called the *specific entropy*, such that

$$d\eta = \left[\frac{-(1/\rho)q_{i,i} + Q}{T} \right] dt \qquad (1.4.10)$$

is a perfect differential. Equation (1.4.10) is called the *entropy equation of state*. Combining Eqs. (1.4.9) and (1.4.10) we arrive at the equation

$$\sigma_{ij} \dot{e}_{ij} + \rho T \dot{\eta} = \dot{U}_0 \qquad (1.4.11)$$

where the superposed dot indicates differentiation with respect to time.

For irreversible processes, the second law of thermodynamics states that

$$\frac{d\eta}{dt} \geqq \frac{Q}{T} - \frac{1}{\rho} \mathrm{div}\left(\frac{\mathbf{q}}{T} \right) \qquad (1.4.12)$$

which is known as the *Clausius–Duhem inequality*.

Exercises 1.4

1. Express the energy equation in the form $(Q = 0)$

$$\frac{\partial}{\partial t}\left(\rho \hat{U}_0 + \frac{\rho}{2} \dot{u}^2 \right) = \mathrm{div}(\boldsymbol{\sigma} \cdot \dot{\mathbf{u}}) + \mathbf{f} \cdot \dot{\mathbf{u}} - \mathrm{div}\,\mathbf{q}$$

2. Assuming that the specific internal energy is a function of the strains and temperature, $\hat{U}_0 = \hat{U}_0(e_{ij}, T)$, and that $d\eta = (1/T)[d\hat{U}_0 - (1/\rho)\sigma_{ij}\,de_{ij}]$, $\hat{\sigma}_{ij} \equiv \partial \hat{U}_0 / \partial e_{ij}$, and $\hat{c} \equiv \partial \hat{U}_0 / \partial T$, show that

$$d\eta = \frac{\hat{\sigma}_{ij} - (\sigma_{ij}/\rho)}{T} de_{ij} + \frac{\hat{e}}{T} dT$$

and, because $d\eta$ is a perfect differential, that

$$\hat{\sigma}_{ij} = \frac{1}{\rho}\left(\sigma_{ij} - T \frac{\partial \sigma_{ij}}{\partial T} \right), \qquad \frac{\partial \hat{c}}{\partial e_{ij}} = 0$$

Hint: Note that if dU is a perfect differential one has

$$\frac{\partial \hat{\sigma}_{ij}}{\partial e_{rs}} = \frac{\partial \hat{\sigma}_{rs}}{\partial e_{ij}}, \qquad \frac{\partial \hat{\sigma}_{ij}}{\partial T} = \frac{\partial \hat{c}}{\partial e_{ij}}, \qquad \text{etc.}$$

3. Using the results of Exercise 2 and the energy equation (1.4.9), show that

$$\rho d\Psi = \rho \frac{\partial \Psi}{\partial e_{ij}} de_{ij} + \rho \frac{\partial \Psi}{\partial T} dT$$

$$= \left(\sigma_{ij} - T \frac{\partial \sigma_{ij}}{\partial T} \right) de_{ij}$$

where

$$\Psi = \hat{U}_0(e_{ij}, T) - T\eta(e_{ij}, T)$$

1.5   CONSTITUTIVE EQUATIONS

1.5.1   Generalized Hooke's Law

A material body is said to be *ideally elastic* when the body recovers (under isothermal conditions) its original form completely upon removal of the forces causing deformation, and there is a one-to-one relationship between the state of stress and state of strain. The generalized Hooke's law relates the nine components of stress to the nine components of strain by the linear relation,

$$\sigma_{ij} = c_{ijkl} e_{kl} \tag{1.5.1}$$

where e_{kl} are the infinitesimal strain components, σ_{ij} are the Cauchy stress components, and c_{ijkl} are the material parameters. A material is said to be *homogeneous* if the parameters c_{ijkl} do not vary from point to point in the body. The nine equations in Eq. (1.5.1) contain 81 parameters. However, owing to the symmetry of both σ_{ij} and e_{kl}, it follows that

$$c_{ijkl} = c_{jikl}, \qquad c_{ijkl} = c_{ijlk} \tag{1.5.2}$$

and there are only 36 constants. Equation (1.5.1) can also be expressed in the matrix form

$$\begin{Bmatrix} \sigma_{11} \\ \sigma_{22} \\ \sigma_{33} \\ \sigma_{23} \\ \sigma_{31} \\ \sigma_{12} \end{Bmatrix} = \begin{bmatrix} c_{11} & c_{12} & c_{13} & c_{14} & c_{15} & c_{16} \\ c_{21} & c_{22} & c_{23} & c_{24} & c_{25} & c_{26} \\ c_{31} & c_{32} & c_{33} & c_{34} & c_{35} & c_{36} \\ c_{41} & c_{42} & c_{43} & c_{44} & c_{45} & c_{46} \\ c_{51} & c_{52} & c_{53} & c_{54} & c_{55} & c_{56} \\ c_{61} & c_{62} & c_{63} & c_{64} & c_{65} & c_{66} \end{bmatrix} \begin{Bmatrix} e_{11} \\ e_{22} \\ e_{33} \\ 2e_{23} \\ 2e_{31} \\ 2e_{12} \end{Bmatrix} \tag{1.5.3a}$$

or

$$\sigma_i = c_{ij} e_j \qquad (1.5.3b)$$

where

$$\sigma_1 = \sigma_{11},\ \sigma_2 = \sigma_{22},\ \sigma_3 = \sigma_{33},\ \sigma_4 = \sigma_{23},\ \sigma_5 = \sigma_{13},\ \sigma_6 = \sigma_{12}$$

$$e_1 = e_{11},\ e_2 = e_{22},\ e_3 = e_{33},\ e_4 = 2e_{23},\ e_5 = 2e_{13},\ e_6 = 2e_{12}$$

$$c_{11} = c_{1111},\ c_{12} = c_{1122},\ c_{14} = c_{1123} = c_{1132},$$

$$c_{44} = c_{2323} = c_{2332} = c_{3223} = c_{3232}, \quad \text{etc.} \qquad (1.5.4)$$

Note that σ_j and e_j do not constitute the components of a tensor. Therefore, one cannot use Eqs. (1.2.16) and (1.3.17) to transform σ_i and e_i.

1.5.2 Strain Energy Density Function

In ideal elasticity, one assumes that all of the input work is converted into internal energy in the form of recoverable stored elastic strain energy. We denote the strain energy per unit volume by $U_0 = U_0(e_{ij})$. Therefore,

$$\frac{\partial U_0}{\partial t} = \frac{\partial U_0}{\partial e_{ij}} \frac{\partial e_{ij}}{\partial t} \qquad (1.5.5)$$

The specific internal energy $\hat{U}_0$ is related to the strain energy density U_0 by $U_0 = \rho \hat{U}_0$. Substituting Eq. (1.5.5) into the energy equation (1.4.9) and observing that $Q = q_i = 0$ for the isothermal case, we get

$$\frac{\partial U_0}{\partial e_{ij}} \dot{e}_{ij} = \sigma_{ij} \dot{e}_{ij}$$

$$\sigma_{ij} = \frac{\partial U_0}{\partial e_{ij}} \qquad (1.5.6)$$

When temperature effects are involved, the strain energy density is a function of strains and temperature:

$$U_0 = U_0(e_{ij}, T) \qquad (1.5.7)$$

We introduce the *free-energy function*, $\Psi(e_{ij}, T)$, by

$$\rho \Psi = U_0(e_{ij}, T) - \rho T \eta(e_{ij}, T) \qquad (1.5.8)$$

Since Ψ is a function of the strains and temperature, we have

$$\dot{\Psi} = \frac{\partial \Psi}{\partial e_{ij}} \dot{e}_{ij} + \frac{\partial \Psi}{\partial T} \dot{T}$$

and

$$\dot{U}_0 = \rho\dot{\Psi} + \rho\dot{T}\eta + \rho T\dot{\eta}$$

$$= \rho\frac{\partial \Psi}{\partial e_{ij}} \dot{e}_{ij} + \left(\rho\frac{\partial \Psi}{\partial T} + \rho\eta\right)\dot{T} + \rho T\dot{\eta} \qquad (1.5.9)$$

Substituting Eq. (1.4.11) into Eq. (1.5.9) yields

$$0 = \left(\rho\frac{\partial \Psi}{\partial e_{ij}} - \sigma_{ij}\right)\dot{e}_{ij} + \rho\left(\frac{\partial \Psi}{\partial T} + \eta\right)\dot{T}$$

Since e_{ij} and T are independent of each other, it follows that

$$\rho\frac{\partial \Psi}{\partial e_{ij}} = \sigma_{ij}, \qquad \frac{\partial \Psi}{\partial T} = -\eta \qquad (1.5.10)$$

Thus, there exists a free-energy function Ψ such that the stress components and entropy can be derived from it according to Eq. (1.5.10) and the energy equation is identically satisfied.

If the deformation process is isothermal and reversible, we have $\Psi = \Psi(e_{ij})$ and, hence, $\eta = 0$. Then from Eqs. (1.5.6) and (1.5.10) we get

$$\rho d\Psi = dU_0 = \sigma_{ij} de_{ij} \qquad (1.5.11)$$

Equation (1.5.11) can be used to reduce the number of independent elastic constants to 21 [note that Eq. (1.5.11) is a result obtained from thermodynamics considerations]. The substitution of the generalized Hooke's law, Eq. (1.5.1), into Eq. (1.5.10) gives

$$\rho\frac{\partial \Psi}{\partial e_{ij}} = c_{ijkl}e_{kl} \qquad (1.5.12)$$

The partial differentiation of Eq. (1.5.12) with respect to e_{kl} gives

$$\rho\frac{\partial}{\partial e_{kl}}\left(\frac{\partial \Psi}{\partial e_{ij}}\right) = c_{ijkl} \qquad (1.5.13a)$$

Interchanging the indices kl with ij, we obtain

$$\rho\frac{\partial}{\partial e_{ij}}\left(\frac{\partial \Psi}{\partial e_{kl}}\right) = c_{klij} \qquad (1.5.13b)$$

Since the order of partial differentiation is unimportant (i.e., immaterial), it follows that

$$c_{ijkl} = c_{klij} \qquad (1.5.14)$$

Because of the symmetry condition in Eq. (1.5.14), the array c_{ij} in Eq. (1.5.3) is symmetric and the number of independent constants is reduced from 36 to 21. Thus, for an anisotropic elastic body for which the strain energy density function exists, there are 21 material constants.

1.5.3 Elastic Symmetry

Generally speaking, the elastic moduli c_{ij} relating the cartesian components of stress and strain depend on the orientation of the coordinate system. When c_{ij} do not depend on the orientation, the material is *isotropic*; otherwise, the material is *anisotropic*.

Plane of elastic symmetry. When the elastic moduli c_{ij} at a point remain invariant for every pair of coordinate systems that are mirror images of each other in a plane, the plane is called a plane of elastic symmetry for the material at the point. Whenever one plane of elastic symmetry exists, the number of constants reduces from 21 to 13. For example, if $x_1 x_2$-plane is the plane of elastic symmetry, the coefficient matrix takes the form ($c_{ij} = c_{ji}$):

$$\begin{bmatrix} c_{11} & c_{12} & c_{13} & 0 & 0 & c_{16} \\ c_{12} & c_{22} & c_{23} & 0 & 0 & c_{26} \\ c_{13} & c_{23} & c_{33} & 0 & 0 & c_{36} \\ 0 & 0 & 0 & c_{44} & c_{45} & 0 \\ 0 & 0 & 0 & c_{45} & c_{55} & 0 \\ c_{16} & c_{26} & c_{36} & 0 & 0 & c_{66} \end{bmatrix} \qquad (1.5.15)$$

Note that, in this case, a state of pure shear strain can give rise to normal stress, for example,

$$\sigma_{11} = c_{16}(2e_{12})$$

Whenever three mutually orthogonal planes of elastic symmetry for a material exist, the material is said to be *orthotropic*. In this case, the coefficient matrix in Eq. (1.5.15) is further simplified by setting

$$c_{16} = c_{26} = c_{36} = c_{45} = 0$$

Hence, for an orthotropic material, the number of independent elastic constants is nine. Thus, in orthotropic materials, normal stresses do not cause shearing strains and shearing stresses do not cause extensional strains. Aragonite crystal, an example of an orthotropic material, has the following values for its elastic constants (in million psi)

$$c_{11} = 23.2, \qquad c_{12} = 5.41, \qquad c_{13} = 0.25$$

$$c_{22} = 12.6, \qquad c_{23} = 2.28, \qquad c_{33} = 12.3$$

$$c_{44} = 6.9, \qquad c_{55} = 3.71, \qquad c_{66} = 6.10 \qquad (1.5.16)$$

The compliance form of the Hooke's law that relates strains to stresses is given by

$$e_{ij} = S_{ijkl}\sigma_{kl} \quad \text{or} \quad e_i = S_{ij}\sigma_j \qquad (1.5.17)$$

where S_{ijkl} are the components of the compliance tensor S, and S_{ij} are the coefficients of the compliance matrix, defined analogous to Eqs. (1.5.3) and (1.5.4). Since the stiffness and compliance matrices are mutually inverse, we can relate S_{ij} to c_{ij} for orthotropic materials by

$$S_{11} = \frac{c_{22}c_{33} - c_{23}^2}{c}, \qquad S_{12} = \frac{c_{13}c_{23} - c_{12}c_{33}}{c}$$

$$S_{22} = \frac{c_{33}c_{11} - c_{13}^2}{c}, \qquad S_{13} = \frac{c_{12}c_{23} - c_{13}c_{22}}{c}$$

$$S_{33} = \frac{c_{11}c_{22} - c_{12}^2}{c}, \qquad S_{23} = \frac{c_{12}c_{13} - c_{23}c_{11}}{c}$$

$$S_{44} = \frac{1}{c_{44}}, \qquad S_{55} = \frac{1}{c_{55}}, \qquad S_{66} = \frac{1}{c_{66}}$$

$$c = c_{11}c_{22}c_{33} - c_{11}c_{23}^2 - c_{22}c_{13}^2 - c_{33}c_{12}^2 + 2c_{12}c_{23}c_{13} \qquad (1.5.18a)$$

The converse relationship (i.e., c_{ij} in terms of S_{ij}) is provided by interchanging the symbols S and c in Eq. (1.5.8a).

$$[S] = \begin{bmatrix} \dfrac{1}{E_1} & -\dfrac{\nu_{12}}{E_1} & -\dfrac{\nu_{13}}{E_1} & 0 & 0 & 0 \\[2mm] -\dfrac{\nu_{21}}{E_2} & \dfrac{1}{E_2} & -\dfrac{\nu_{23}}{E_2} & 0 & 0 & 0 \\[2mm] -\dfrac{\nu_{31}}{E_3} & -\dfrac{\nu_{32}}{E_3} & \dfrac{1}{E_3} & 0 & 0 & 0 \\[2mm] 0 & 0 & 0 & \dfrac{1}{G_{23}} & 0 & 0 \\[2mm] 0 & 0 & 0 & 0 & \dfrac{1}{G_{13}} & 0 \\[2mm] 0 & 0 & 0 & 0 & 0 & \dfrac{1}{G_{12}} \end{bmatrix} \qquad (1.5.18b)$$

In practice, the components of the compliance matrix (S_{ij}) are determined more directly than those of the stiffness matrix (c_{ij}). For orthotropic material, the compliance matrix, in terms of the engineering constants, is given by Eq.

(1.5.18b), wherein

E_1, E_2, E_3 = Young's moduli in 1, 2, and 3 directions, respectively

ν_{ij} = Poisson's ratio of transverse strain in the j-direction to the axial strain in the i-direction, when stressed in the i-direction

$= -\varepsilon_j/\varepsilon_i$ (for $\sigma_i = \sigma$ and all other stresses are zero)

G_{23}, G_{13}, G_{12} = shear moduli in the 2-3, 1-3, and 1-2 planes, respectively

$$(1.5.19)$$

For an orthotropic material, the following reciprocal relationship holds:

$$\frac{\nu_{ij}}{E_i} = \frac{\nu_{ji}}{E_j} \qquad i, j = 1, 2, 3 \qquad (1.5.20)$$

The relationship between the stiffness constants and the engineering constants is given by

$$c_{11} = E_1 \frac{1 - \nu_{23}\nu_{32}}{\Delta}, \qquad c_{12} = E_1 \frac{\nu_{21} + \nu_{31}\nu_{23}}{\Delta} = E_2 \frac{\nu_{12} + \nu_{32}\nu_{13}}{\Delta}$$

$$c_{13} = E_1 \frac{\nu_{31} + \nu_{21}\nu_{32}}{\Delta} = E_3 \frac{\nu_{13} + \nu_{12}\nu_{23}}{\Delta}, \qquad c_{22} = E_2 \frac{1 - \nu_{13}\nu_{31}}{\Delta}$$

$$c_{23} = E_2 \frac{\nu_{32} + \nu_{12}\nu_{31}}{\Delta} = E_3 \frac{\nu_{23} + \nu_{21}\nu_{13}}{\Delta}, \qquad c_{33} = E_3 \frac{1 - \nu_{12}\nu_{21}}{\Delta}$$

$$c_{44} = G_{23}, \qquad c_{55} = G_{31}, \qquad c_{66} = G_{12}$$

$$\Delta = 1 - \nu_{12}\nu_{21} - \nu_{23}\nu_{32} - \nu_{31}\nu_{13} - 2\nu_{21}\nu_{32}\nu_{13} \qquad (1.5.21)$$

An isotropic material has an infinite number of planes of elastic symmetry. In other words, in an isotropic material, the elastic moduli are independent of the orientation of the coordinate system (but can depend on the position of the point). In this case, the stress–strain relationship is of the form

$$\sigma_{ij} = 2\mu e_{ij} + \lambda \delta_{ij} e_{kk} \qquad (1.5.22)$$

where μ and λ are called the Lamé constants. Thus, the constitution of an isotropic material at a point is completely described by two elastic parameters.

The Lamé constants are related to the shear modulus G, Young's modulus E, and Poisson's ratio ν by

$$\mu = G = \frac{E}{2(1+\nu)}, \qquad \lambda = \frac{\nu E}{(1+\nu)(1-2\nu)} \qquad (1.5.23)$$

The coefficients c_{ij} for isotropic materials have the following meaning:

$$c_{11} = c_{22} = c_{33} = \lambda + 2\mu, \qquad c_{12} = c_{13} = c_{23} = \lambda, \qquad c_{44} = c_{55} = c_{66} = \mu$$

$$(1.5.24)$$

and all other c_{ij}'s are zero.

Example 1.6

Consider the cantilever beam of Example 1.5. Overlooking the fact that the strains are not compatible (only one of the six equations is violated), we can compute the displacements u_1 and u_2. We have

$$\frac{\partial u_1}{\partial x_1} = -\frac{P x_1 x_2}{EI} \quad \text{or} \quad u_1 = -\frac{P x_1^2 x_2}{2EI} + f(x_2)$$

$$\frac{\partial u_2}{\partial x_2} = \frac{\nu P x_1 x_2}{EI} \quad \text{or} \quad u_2 = \frac{\nu P x_1 x_2^2}{2EI} + g(x_1)$$

where f and g are functions of the integration. Substituting u_1 and u_2 into the shear strain $2e_{12}$, we obtain

$$-\frac{P x_1^2}{2EI} + \frac{df}{dx_2} + \frac{\nu P x_2^2}{2EI} + \frac{dg}{dx_1} = -\frac{(1+\nu)}{EI} P(h^2 - x_2^2)$$

Separating the x_1 and x_2 terms, we have

$$-\frac{dg}{dx_1} + \frac{P x_1^2}{2EI} - \frac{(1+\nu) P h^2}{EI} = \frac{df}{dx_2} - \frac{(2+\nu)P}{2EI} x_2^2$$

Since the left side depends only on x_1 and the right side depends only on x_2, and yet the equality holds, implies that both sides should be equal to a constant, say c:

$$-\frac{dg}{dx_1} + \frac{P}{2EI} x_1^2 - \frac{(1+\nu) P h^2}{EI} = c, \qquad \frac{df}{dx_2} - \frac{(2+\nu)P}{2EI} x_2^2 = c$$

Integrating the expressions for g and f, we obtain

$$g(x_1) = \frac{P}{6EI}x_1^3 - \frac{(1+\nu)Ph^2}{EI}x_1 - cx_1 + c_1$$

$$f(x_2) = \frac{(2+\nu)P}{6EI}x_2^3 + cx_2 + c_2$$

where c_1 and c_2 are constants of integration, which are to be determined. The displacements are now given by

$$u_1 = -\frac{Px_1^2 x_2}{2EI} + \frac{Px_2^3}{6EI}(2+\nu) + cx_2 + c_2$$

$$u_2 = -\frac{(1+\nu)Ph^2}{EI}x_1 + \frac{\nu Px_1 x_2^2}{2EI} + \frac{Px_1^3}{6EI} - cx_1 + c_1 \qquad (1.5.25)$$

The constants c, c_1, and c_2 can be evaluated using the boundary conditions on the displacements. We impose the following boundary conditions:

$$u_1(L,0) = 0, \qquad u_2(L,0) = 0, \qquad \frac{\partial u_1}{\partial x_2}(L,0) = 0 \qquad (1.5.26)$$

Using Eqs. (1.5.26) in Eqs. (1.5.25), we obtain

$$u_1(L,0) = 0 \rightarrow c_2 = 0$$

$$u_2(L,0) = 0 \rightarrow \frac{PL^3}{6EI} - cL + c_1 = \frac{(1+\nu)Ph^2 L}{EI}$$

$$\frac{\partial u_1(L,0)}{\partial x_2} = 0 \rightarrow -\frac{PL^2}{2EI} + c = 0 \quad \text{or} \quad c = \frac{PL^2}{2EI}$$

Finally, we have

$$u_1 = -\frac{Px_1^2 x_2}{2EI} + \frac{PL^2 x_2}{2EI} + \frac{(2+\nu)Px_2^3}{6EI}$$

$$u_2 = \frac{Ph^2(1+\nu)}{EI}(L - x_1) + \frac{\nu Px_1 x_2^2}{2EI} + \frac{Px_1^3}{6EI} - \frac{PL^2 x_1}{2EI} + \frac{PL^3}{3EI}$$

$$(1.5.27a)$$

The values of the constants c, c_1, and c_2 depend on the boundary conditions used. For example, if one uses $\partial u_2/\partial x_1 = 0$ in place of $\partial u_1/\partial x_2 = 0$ at

$x_1 = L$, $x_2 = 0$, the resulting displacements would be

$$u_1 = -\frac{Px_2}{2EI}\left(x_1^2 - L^2\right) - \frac{\nu Px_2^3}{6EI} + \frac{Px_2^3}{6GI} - \frac{Ph^2 x_2}{2GI}$$

$$u_2 = \frac{\nu Px_1 x_2^2}{2EI} + \frac{Px_1^3}{6EI} - \frac{PL^2 x_1}{2EI} + \frac{PL^3}{3EI} \qquad (1.5.27b)$$

Since we are solving a three-dimensional problem using the two-dimensional assumption, the boundary conditions associated with the third direction are not used.

1.5.4 Transformation of Elastic Moduli

It is of interest, for example in the study of laminated plates of orthotropic layers, to have an explicit form of the transformation equations relating the elastic moduli in one coordinate system to those in another coordinate system. The tensor transformation equation (1.2.16) is valid for c_{ijkl} because they are components of a fourth-order tensor:

$$c'_{ijkl} = a_{im}a_{jn}a_{kp}a_{lq}c_{mnpq} \qquad (1.5.28)$$

where a_{im} denotes the direction cosine associated with the x_i'-axis and the x_m-axis. However, the c_{ij} are not the components of a tensor and, therefore, transformation equations of the form (1.5.28) do not exist for c_{ij}. One can use the stress–strain relation (1.5.1) and the transformation laws (1.2.16) and (1.3.17) for stresses and strains to relate c_{ij} in the x_i-coordinate system to c'_{ij} in the x_i'-coordinate system.

The elastic moduli c_{ij} of an anisotropic material (with the material symmetry axes coinciding with the x_i-axis) are related to the elastic moduli c'_{ij} in the

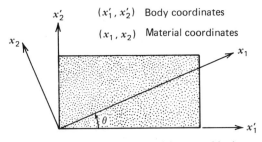

Figure 1.9 Transformation between material axes and body coordinates.

x_i'-coordinate system by the relations (see Fig. 1.9; $m = \cos\theta, n = \sin\theta$)

$$c_{11}' = m^4 c_{11} + 2m^2 n^2 (c_{12} + 2c_{66}) - 4mn(m^2 c_{16} + n^2 c_{26}) + n^4 c_{22}$$

$$c_{12}' = m^2 n^2 (c_{11} + c_{22} - 4c_{66}) + 2mn(m^2 - n^2)(c_{16} - c_{26}) + (m^4 + n^4)c_{12}$$

$$c_{13}' = m^2 c_{13} + n^2 c_{23} - 2mn c_{36}$$

$$c_{14}' = c_{14} m^3 - mn[(2c_{46} - c_{15})m - (c_{24} - 2c_{56})n] + c_{25} n^3$$

$$c_{15}' = c_{15} m^3 - mn[(c_{14} + 2c_{56})m - (c_{25} + 2c_{46})n] - c_{24} n^3$$

$$c_{16}' = m^2(m^2 - 3n^2)c_{16} + mn[m^2 c_{11} - n^2 c_{22} - (m^2 - n^2)(c_{12} + 2c_{66})]$$
$$+ n^2(3m^2 - n^2)c_{26}$$

$$c_{22}' = n^4 c_{11} + 2m^2 n^2 (c_{12} + 2c_{66}) + 4mn(m^2 c_{26} + n^2 c_{16}) + m^4 c_{22}$$

$$c_{23}' = n^2 c_{13} + m^2 c_{23} + 2mn c_{36}$$

$$c_{24}' = c_{24} m^3 + mn[(c_{25} + 2c_{46})m + (c_{14} + 2c_{56})n] + c_{15} n^3$$

$$c_{25}' = c_{25} m^3 - mn[(c_{24} - 2c_{56})m - (c_{15} - 2c_{46})n] - c_{14} n^3$$

$$c_{26}' = m^2(m^2 - 3n^2)c_{26} + mn[n^2 c_{11} - m^2 c_{22} + (m^2 - n^2)(c_{12} + 2c_{66})]$$
$$+ n^2(3m^2 - n^2)c_{16}$$

$$c_{33}' = c_{33}, \qquad c_{34}' = c_{34} m + c_{35} n, \qquad c_{35}' = c_{35} m - c_{34} n$$

$$c_{36}' = (m^2 - n^2)c_{36} - mn(c_{23} - c_{13})$$

$$c_{44}' = m^2 c_{44} + 2mn c_{45} + n^2 c_{55}$$

$$c_{45}' = (m^2 - n^2)c_{45} - mn(c_{44} - c_{55})$$

$$c_{46}' = (c_{46} m + c_{56} n)(m^2 - n^2) + (c_{14} - c_{24})m^2 n + (c_{15} - c_{25})mn^2$$

$$c_{55}' = m^2 c_{55} - 2mn c_{45} + n^2 c_{44}$$

$$c_{56}' = (c_{56} m - c_{46} n)(m^2 - n^2) + (c_{15} - c_{25})m^2 n + (c_{24} - c_{14})mn^2$$

$$c_{66}' = m^2 n^2 (c_{11} + c_{22} - 2c_{12}) - 2mn(m^2 - n^2)(c_{22} - c_{16}) + (m^2 - n^2)^2 c_{66}$$

$$(1.5.29)$$

Note that Eq. (1.5.29) allows one to transform the elastic constants *from* material symmetry-axes *to* a global (or problem) coordinate axes.

1.5.5 Thermoelastic Constitutive Equations

In writing the constitutive relations, we implicitly assumed that the variations in the elastic constants with temperature are negligible. We neglect the variations in the elastic constants, but we are required to consider the thermal expansion of the material. When the thermal expansions (i.e., geometric changes owing to heating) are prevented by boundary conditions or other constraints, the body develops thermal stresses in addition to stresses caused by other mechanical loads. In such cases, the temperature T, measured above an arbitrary reference temperature, enters as a parameter in the linear terms of the strain energy density:

$$U_0 = \tfrac{1}{2} c_{ijkl} e_{ij} e_{kl} - \beta_{ij}(T - T_0) e_{ij} \qquad (1.5.30)$$

and the linear thermoelastic constitutive equations are given by

$$\sigma_{ij} = -\beta_{ij}(T - T_0) + c_{ijkl} e_{kl} \qquad (1.5.31)$$

where T_0 is the reference temperature (hence, $T - T_0$ is the change in temperature) and β_{ij} are the Cartesian components of the tensor $\boldsymbol{\beta}$. For an isotropic body, Eq. (1.5.31) takes the simple form

$$\sigma_{ij} = -\beta \delta_{ij}(T - T_0) + \lambda \delta_{ij} e_{kk} + 2\mu e_{ij} \qquad (1.5.32)$$

where $\beta = (2\mu + 3\lambda)\alpha$, and α is the linear coefficient of thermal expansion.

A linear constitutive relation between heat flux $\mathbf{q}$ and temperature gradient ∇T is given by the Fourier Heat Conduction Law,

$$\mathbf{q} = -\mathbf{k} \cdot \nabla T \quad \text{or} \quad q_i = -k_{ij} \frac{\partial T}{\partial x_j} \qquad (1.5.33)$$

where $\mathbf{k}$ is the thermal conductivity tensor. For an isotropic solid or fluid, the heat flux is necessarily in the same direction as the temperature gradient, giving $k_{ij} = k\delta_{ij}$ and

$$\mathbf{q} = -k \nabla T \quad \text{or} \quad q_i = -k \frac{\partial T}{\partial x_i} \qquad (1.5.34)$$

With $k > 0$, Eq. (1.5.34) implies that heat flows from regions of high temperature toward regions of low temperature. Using the results of Section 1.4.3 and Exercise 1.4.2, we can show that the energy equation (1.4.9) reduces to

$$\rho T \beta_{ij} \dot{\varepsilon}_{ij} + \rho c \frac{\partial T}{\partial t} - \frac{\partial}{\partial x_i}\left(k_{ij} \frac{\partial T}{\partial x_j}\right) = Q \qquad (1.5.35)$$

Exercises 1.5

1. Compute the strains associated with the stresses of the problem in Exercise 1.2.7. Check the compatibility conditions (assume isotropy).

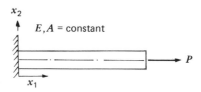

Figure E3

2. Compute the strains corresponding to the stress field of the problem in Exercise 1.2.8. Check the compatibility conditions (assume isotropy).

3. Determine the (small) displacements in a uniaxially loaded isotropic member (see Fig. E3).

4. Determine the displacement field in an isotropic beam bent by pure bending moment, M_0 (disregard the fact that one of the compatibility conditions is not satisfied).

5. Derive the Beltrami–Michell equations by combining the stress equilibrium equations and compatibility equations. Assume isotropy and nonisothermal conditions.

6. Given the following strains for the problem shown in Fig. E6, determine the (small) displacements completely (i.e., free of any undetermined constants):

$$e_{11} = -\frac{e_0}{a}, \qquad e_{22} = 0, \qquad e_{33} = \frac{\nu}{1-\nu}\frac{e_0}{a}, \qquad e_0 = \text{constant}$$

$$e_{12} = e_{13} = e_{23} = 0$$

7. Disregarding the compatibility of strains, determine the form of the displacement field for the problem in Exercise 2.

8. Derive the Navier equations of elasticity for an isotropic medium in the presence of temperature changes.

9. Consider a thin rectangular beam of thickness t, height $2h$, and length L(Fig. E9). Find the distribution of stresses, strains, and displacements associated with an arbitrary variation in temperature throughout the depth, $T = T(x_2)$. Assume that the beam is isotropic and free of surface tractions, and that the body forces are negligible.

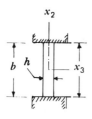
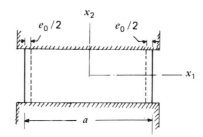

Figure E6

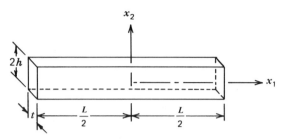

Figure E9

10. An aluminum bar of uniform cross-sectional area of 4 in.^{2}, length of 29.98 in., and temperature of $-20°$F is placed between two rigid immovable supports that are 30 in. apart. The bar is then heated uniformly. (a) Determine the temperature at which the gap first closed. (b) Determine the stress in the bar when the temperature reaches $95°$F. Take the coefficient of thermal expansion to be 12×10^{-6} in./in./°F and modulus of elasticity $E = 10 \times 10^6$ psi.

11. Consider a thin, homogeneous, isotropic plate of dimensions a by b. (a) If the entire plate undergoes a uniform temperature increase T degrees, what is the final size of the plate? (b) If the plate undergoes a temperature gradient $T = T_0(x_1)$, determine the strains and displacements in the plate. Specialize your answer when $T_0 = cx_1$ (see Fig. E11).

12. A uniform bar of cross-sectional area A, modulus E, conductivity k, thermal coefficient of expansion α, and length L is subjected to temperature $T = T_0$ at $x_1 = 0$ and insulated as $x_1 = L$. Determine the strains and displacements in the member.

13. Show that when one plane of elastic symmetry exists, the material coefficient matrix takes the form given in Eq. (1.5.15).

14. Show that for an isotropic material, the constants E, G, and ν are related according to

$$G = \frac{E}{2(1 + \nu)}$$

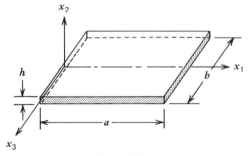

Figure E11

Hint: Use the biaxial case of pure shear stress and stress and strain transformation equations to arrive at the required result.

15. Derive Eq. (1.5.35) using the results of Section 1.4.3 and Exercise 1.4.2.

16. Express the strain energy density U_0 for an isotropic material in terms of the strains e_i of Eq. (1.5.4), temperature T, and the elastic constants E, ν, and β.

17. The complementary (or conjugate) strain energy density U_0^* is defined by

$$U_0^* = \sigma_{ij}e_{ij} - U_0 = \sigma_i e_i - U_0$$

where σ_i and e_i are the stresses and strains in the notation of Eq. (1.5.4). Write U_0^* in terms of σ_i, temperature T, and elastic constants E, ν, and β.

18. Use Eqs. (1.5.16), (1.5.18a), and (1.5.18b) to determine the engineering constants of the aragonite crystal.

1.6   BOUNDARY-VALUE PROBLEMS OF MECHANICS

1.6.1   Summary of Equations

The kinetic and thermodynamic principles, kinematic relations, and constitutive equations presented in the previous sections are valid for any continuous body whose constitutive behavior is linear. We summarize the equations here for future reference.

Equations of Motion $(\sigma_{ij} = \sigma_{ji})$

$$\rho\ddot{\mathbf{u}} = \operatorname{div}\boldsymbol{\sigma} + \mathbf{f} \quad \text{or} \quad \rho\ddot{u}_i = \sigma_{ji,j} + f_i \qquad (1.6.1)$$

Strain–Displacement Equations

$$\varepsilon_{ij} = \tfrac{1}{2}\left(u_{i,j} + u_{j,i} + u_{m,i}u_{m,j}\right) \qquad (1.6.2)$$

Fourier Heat-Conduction Law

$$\rho T\boldsymbol{\beta}:\dot{\mathbf{e}} + \rho c T + \operatorname{div}\mathbf{q} = \rho Q \quad \text{or} \quad \rho T\beta_{ij}\dot{e}_{ij} + \rho c T + q_{i,i} = \rho Q \quad (1.6.3)$$

Linear Constitutive Relations

$$\sigma_{ij} = c_{ijkl}e_{kl} - \beta_{ij}(T - T_0) \qquad (1.6.4)$$

$$q_i = -k_{ij}T_{,j} \qquad (1.6.5)$$

Here, the superposed dot on a variable indicates differentiation with respect to time, $u_{i,j} = \partial u_i/\partial x_j$, and summation on repeated indices is implied. Thus, there are 19 unknowns and 19 equations for a three-dimensional elastic body which should be solved in conjunction with appropriate initial and boundary conditions:

Variables	Number
u_i	3
ε_{ij}	6
σ_{ij}	6
q_i	3
T	1

Under the conditions of infinitesimal strains, the equations (1.6.1), (1.6.2), and (1.6.4), governing a linear elastic body can be expressed in the matrix form as

$$\rho\{\ddot{u}\} = \{F\} + [D]^T\{\sigma\} \tag{1.6.6}$$

$$\{\varepsilon\} = [D]\{u\} \tag{1.6.7}$$

$$\{\sigma\} = [C]\{\varepsilon\} - \{\beta\}(T - T_0) \tag{1.6.8}$$

where $[C]$ is the elasticity matrix, and

$$\{u\} = \begin{Bmatrix} u_1 \\ u_2 \\ u_3 \end{Bmatrix}, \qquad \{F\} = \begin{Bmatrix} f_1 \\ f_2 \\ f_3 \end{Bmatrix},$$

$$\{\varepsilon\} = \begin{Bmatrix} \varepsilon_1 \\ \varepsilon_2 \\ \varepsilon_3 \\ \varepsilon_4 \\ \varepsilon_5 \\ \varepsilon_6 \end{Bmatrix}, \qquad \{\sigma\} = \begin{Bmatrix} \sigma_1 \\ \sigma_2 \\ \sigma_3 \\ \sigma_4 \\ \sigma_5 \\ \sigma_6 \end{Bmatrix}, \qquad \{\beta\} = \begin{Bmatrix} \beta_{11} \\ \beta_{22} \\ \beta_{33} \\ \beta_{23} \\ \beta_{13} \\ \beta_{12} \end{Bmatrix}$$

$$[D]^T = \begin{bmatrix} \dfrac{\partial}{\partial x_1} & 0 & 0 & 0 & \dfrac{\partial}{\partial x_3} & \dfrac{\partial}{\partial x_2} \\[2mm] 0 & \dfrac{\partial}{\partial x_2} & 0 & \dfrac{\partial}{\partial x_3} & 0 & \dfrac{\partial}{\partial x_1} \\[2mm] 0 & 0 & \dfrac{\partial}{\partial x_3} & \dfrac{\partial}{\partial x_2} & \dfrac{\partial}{\partial x_1} & 0 \end{bmatrix} \tag{1.6.9}$$

1.6.2 Initial and Boundary Conditions

For a three-dimensional elastic body, the initial and boundary conditions are of the following form:

Initial Conditions ($t = 0$, x_i inside the body)

$$u_i = u_i^0, \qquad \frac{\partial u_i}{\partial t} = v_i^0, \qquad T = T_0 \qquad\qquad (1.6.10)$$

Boundary Conditions ($t \geq 0$, x_i on the boundary)
    Geometric or essential type:

$$u_i = \hat{u}_i, \qquad T = \hat{T} \qquad\qquad (1.6.11)$$

Forced or natural type:

$$\sigma_{ji} n_j = \hat{t}_i, \qquad q_i n_i = \hat{q} \qquad\qquad (1.6.12)$$

Here, variables with a superscript or subscript zero denote the specified initial values, and those with a caret over them denote the specified boundary values of the variables.

The specification of only one of the two values of each pair, $(u_i, \sigma_{ji} n_j)$ and $(T, q_i n_i)$, is necessary at a boundary point of an elastic body. In the chapter on variational methods, we classify boundary conditions into essential and natural type.

1.6.3 Classification of Boundary-Value Problems

Solving the 19 equations (1.6.1)–(1.6.5) for the 19 unknowns (for isothermal elasticity we have 15 equations in 15 unknowns) in conjunction with the initial and boundary conditions in (1.6.10)–(1.6.12) is of primary interest in solid mechanics. Depending on the type of specified boundary conditions, Eqs. (1.6.1)–(1.6.5) can be expressed in alternative forms. We can classify the formulations into the following three categories (for isothermal case):

1. *Displacement Formulation (Boundary-Value Problem of the First Kind).* This formulation involves the determination of displacements and stresses in the interior of the body under a given body force distribution and specified displacement field over the entire boundary of the body.

2. *Stress Formulation (Boundary-Value Problem of the Second Kind).* This problem entails the determination of displacements and stresses in the interior of the body under the action of a given body force distribution and specified surface forces on the entire boundary.

3. *Mixed Formulation (Boundary-Value Problem of the Third Kind).* This requires the determination of displacements and stresses in the interior of the body under the action of a given body force distribution and

specified displacement field on a portion of the boundary and specified boundary forces on the remaining portion of the boundary of the body.

For boundary-value problems of the first kind, it is convenient to express all of the governing equations in terms of the displacement field. This is done by eliminating stresses in the equations of motion by means of the stress–strain equations and then eliminating the strains by means of the strain–displacement equations. Under isothermal conditions, these equations are given by

$$\rho\{\ddot{u}\} = \{F\} + [D]^T[C][D]\{u\} \tag{1.6.13}$$

The initial and boundary conditions are given by Eqs. (1.6.10) and (1.6.11), respectively. For an isotropic body, Eq. (1.6.13) has the simple form

$$(\lambda + \mu)\text{grad}(\text{div }\mathbf{u}) + \mu\nabla^2\mathbf{u} + \mathbf{f} = \rho\ddot{\mathbf{u}} \tag{1.6.14}$$

These equations are called the *Navier equations* of motion.

For boundary-value problems of the second kind, we find it convenient to write the governing equations in terms of stresses. This is done by substituting for strains, using the strain–stress relations in the compatibility equations, which must be simplified by means of the equations of motion. The strain–stress equations, Eq. (1.5.17), in the notation of Eq. (1.5.4), are given by

$$\{e\} = [S]\{\sigma\} \tag{1.6.15}$$

where $[S]$ is the compliance matrix [see Eq. (1.5.18b)], the inverse of stiffness matrix $[c]$. For an isotropic body, the strain–stress relations have the form

$$e_{ij} = \frac{1}{2\mu}\sigma_{ij} - \frac{\mu + \lambda}{2\mu(3\lambda + \mu)}\sigma_{kk}\delta_{ij} = \frac{1 + \nu}{E}\sigma_{ij} - \frac{\nu}{E}\sigma_{kk}\delta_{ij} \tag{1.6.16}$$

The governing equations for the boundary-value problem of the second kind for an anisotropic body are algebraically very complex and, therefore, are not given here. For an isotropic body, these equations are given by

$$\rho\left(\frac{1}{G}\dot{\sigma}_{ij} - \frac{2\nu}{E}\dot{\sigma}_{kk}\delta_{ij}\right) + f_{i,j} + f_{j,i} = \nabla^2\left(\sigma_{ij} - \frac{\nu}{1 + \nu}\delta_{ij}\sigma_{kk}\right) + \frac{1}{1 + \nu}\sigma_{kk,ij}$$

$$\tag{1.6.17}$$

For the boundary-value problem of the third kind, the governing equations are given by (for $t > 0$)

$$\rho\ddot{\mathbf{u}} = \text{div}\,\boldsymbol{\sigma} + \mathbf{f}, \qquad \sigma_{ij} = \tfrac{1}{2}c_{ijkl}(u_{k,l} + u_{l,k}) \tag{1.6.18}$$

with boundary conditions (1.6.11) on one portion and (1.6.12) on the remainder of the boundary.

Most problems of solid mechanics fall under the third (i.e., mixed) category. Solutions to mixed boundary-value problems are discussed throughout this study.

1.6.4   Existence and Uniqueness of Solutions

We now discuss the existence and uniqueness of solutions to the mixed boundary-value problem of linear elasticity, as described by Eqs. (1.6.10)–(1.6.12) and (1.6.18). A rigorous mathematical proof of the existence of solutions requires concepts from functional analysis that are beyond the scope of the present study. Therefore, this proof is based on the Lax–Milgram theorem, which states that a boundary-value problem has a solution if the strain energy is continuous and positive-definite:

$$U > 0 \quad \text{for all } \mathbf{u}, \boldsymbol{\sigma}$$

$$= 0 \quad \text{if and only if } \mathbf{u} = \boldsymbol{\sigma} = 0. \tag{1.6.19}$$

Here, U is the strain energy of the body. The continuity of U can be easily established for most linear problems of elasticity. For linear elastic bodies, the strain energy density U_0 is a quadratic function of the strains. Therefore, we have

$$U = \int_V U_0 \, dV \tag{1.6.20}$$

Clearly, $U > 0$ because $U_0 = \frac{1}{2}(C_{ijkl}\varepsilon_{ij}\varepsilon_{kl})$ is positive; $U = 0$ if $u_i = 0$ and $\varepsilon_{ij} = 0$ (hence, $\sigma_{ij} = 0$), and if $U = 0$, we have $\varepsilon_{ij} = 0$, which in turn imply that $u_i = \text{constant}$. For homogeneous essential boundary conditions (i.e., $\hat{u}_i = 0$), the constant must be zero, establishing the positive-definiteness of U. All linear problems with nonhomogeneous essential boundary conditions can be recast as boundary-value problems with homogeneous essential boundary conditions. Consequently, the assumption concerning the homogeneous essential boundary conditions in the proof of positive-definiteness is not restrictive. However, for boundary-value problems of the second kind, such an assumption cannot be made, because no displacement boundary conditions are specified. As a result, $U = 0$ even if u_i are not zero. This amounts to saying that the displacements in the boundary-value problems of the second kind can be determined only within a constant (i.e., including rigid body mode).

Next, we consider the question of uniqueness for the dynamic case. Suppose that u_i^1 and u_i^2 are two solutions of the same mixed boundary-value problem. If the solution is unique, we must have $u_i^1 = u_i^2$. To prove this, we assume the following initial conditions and mixed boundary conditions.

$$u_i = u_i^0, \qquad \dot{u}_i = 0 \quad \text{at } t = t_0 \tag{1.6.21a}$$

$$u_i = \hat{u}_i \quad \text{on } S_1, \qquad \sigma_{ji}n_j = \hat{t}_i \quad \text{on } S_2 \tag{1.6.21b}$$

We have

$$\sigma_{ji,j}^1 - \sigma_{ji,j}^2 - \rho\left(\ddot{u}_i^1 - \ddot{u}_i^2\right) = 0$$

or

$$\bar{\sigma}_{ji,j} - \rho\ddot{\bar{u}}_i = 0 \qquad (1.6.22)$$

where

$$\bar{\sigma}_{ij} = \sigma_{ij}^1 - \sigma_{ij}^2, \qquad \bar{u}_i = u_i^1 - u_i^2$$

Multiply Eq. (1.6.22) by $\dot{\bar{u}}_i$ and integrate first over the volume of the body and then with respect to time from initial time t_0 to current time t:

$$0 = \int_{t_0}^t \int_V \left[\frac{\partial}{\partial x_j}\left(\frac{\partial \bar{U}_0}{\partial \bar{e}_{ij}}\right) - \rho\ddot{\bar{u}}_i \right]\dot{\bar{u}}_i \, dV \, dt$$

$$= \int_{t_0}^t \int_V \left[-\frac{\partial \bar{U}_0}{\partial \bar{e}_{ij}} \frac{\partial \dot{\bar{u}}_i}{\partial x_j} - \rho\ddot{\bar{u}}_i\dot{\bar{u}}_i \right] dV \, dt$$

$$+ \int_{t_0}^t \oint_S n_j \frac{\partial \bar{U}_0}{\partial \bar{e}_{ij}}\dot{\bar{u}}_i \, dS \, dt \qquad (1.6.23)$$

The boundary conditions in (1.6.21b) imply that $\dot{\bar{u}}_i = 0$ on S_1 and $(\partial \bar{U}_0/\partial \bar{e}_{ij})n_j = 0$ on S_2. Therefore, the surface integral in Eq. (1.6.23) vanishes. Also, note that

$$\frac{\partial \bar{U}_0}{\partial \bar{e}_{ij}} \frac{\partial \dot{\bar{u}}_i}{\partial x_j} = \frac{\partial \bar{U}_0}{\partial \bar{e}_{ij}} \frac{\partial \bar{e}_{ij}}{\partial t} = \frac{\partial \bar{U}_0}{\partial t} \qquad (1.6.24)$$

We have

$$0 = -\int_{t_0}^t \int_V \frac{d}{dt}\left(\bar{U}_0 + \frac{\rho}{2}\dot{\bar{u}}_i\dot{\bar{u}}_i\right) dV \, dt$$

$$= -\int_V \left(\bar{U}_0 + \frac{\rho}{2}\dot{\bar{u}}_i\dot{\bar{u}}_i\right) dV \Bigg|_t + \int_V \left(\bar{U}_0 + \frac{\rho}{2}\dot{\bar{u}}_i\dot{\bar{u}}_i\right) dV \Bigg|_{t_0} \qquad (1.6.25)$$

The second expression in Eq. (1.6.25) is zero because both U_0 and K are measured from zero initial values, $U_0(t_0) = 0$ and $\dot{u}_i(t_0) = 0$. Thus, we have

$$0 = \int_V \left(\bar{U}_0 + \frac{\rho}{2}\dot{\bar{u}}_i\dot{\bar{u}}_i\right) dV \qquad (1.6.26)$$

Both $\overline{U}_0$ and $(\rho/2)\dot{\bar{u}}_i\dot{\bar{u}}_i$ are quadratic functions of $\bar{\varepsilon}_{ij}$ and $\dot{\bar{u}}_i$, respectively, which implies that

$$\overline{U}_0 = 0 \quad \text{and} \quad \dot{\bar{u}}_i = 0 \text{ everywhere in } V \tag{1.6.27}$$

These in turn require that

$$u_i^1 = u_i^2 + \text{constant} \tag{1.6.28}$$

In other words, the displacements of the mixed boundary value problem are unique within an arbitrary constant (i.e., rigid-body motion). However, this constant must be equal to zero, because both u_i^1 and u_i^2 are equal to $\hat{u}_i$ on S_1. Therefore, we have the uniqueness of solutions.

In the proof of the uniqueness, we made use of the linearity of the differential equation of motion and stress–strain relations. When u_i is not specified at any point of the boundary (like in the boundary-value problems of the second kind), the stresses are unique but the displacements are not; the displacements differ by a rigid-body motion.

The positive-definiteness of the strain energy of a problem is equivalent to the positive-definiteness of the operator in the governing differential equation. In fact, existence theorems, such as the Lax–Milgram theorem, require continuity and positive-definiteness of the operator in the differential equation to prove existence and uniqueness of the solution. A discussion of such ideas is beyond the scope of the present study, and the interested reader can consult the text by Reddy and Rasmussen (1982).

Exercises 1.6

Classify the problems in Exercises 1–5 into one of the three boundary-value problems:

1. Problem in Fig. E6 of Exercises 1.5.
2. The problem in Fig. E2.
3. The problem in Fig. E3.
4. The problem in Fig. E4.
5. Assume biaxial symmetry, and consider a quadrant of the plate with circular hole shown in Fig. E3.

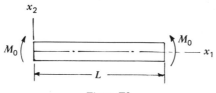

Figure E2

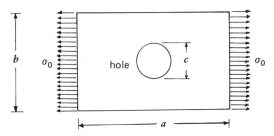

Figure E3

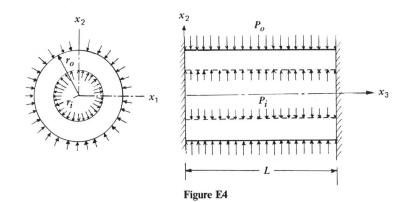

Figure E4

6. Given the stress field (see Fig. E6) $\sigma_{11} = c_1 x_1 x_2 + c_2 x_2$, $\sigma_{12} = c_3(h^2 - x_2^2)$, all other $\sigma_{ij} = 0$, $f_i = 0$:
    (a) Determine the conditions on c_1, c_2, and c_3 for the stress field to be in equilibrium.
    (b) Determine the boundary stresses on the surface of the beam.

7. The stress distribution in a thin elastic plate (Fig. E7) of dimensions $2a$ by b is given by (in the absence of body forces)

$$\sigma_{11} = c_1 x_1^3 x_2 - 2c_2 x_1 x_2 + c_3 x_2$$
$$\sigma_{22} = c_1 x_1 x_2 (x_2^2 - 2x_1^2)$$
$$\sigma_{12} = 3c_4 x_1^2 x_2^2 + c_5 x_2^2 - c_4 x_1^4$$

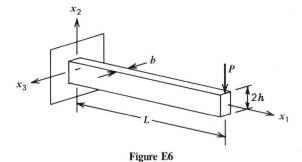

Figure E6

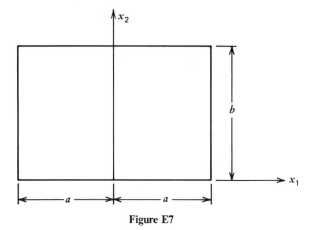

Figure E7

Determine (a) the relationship between c_i so that the stress field corresponds to an equilibrium state, and (b) the boundary stresses.

8. Show that the solution to a clamped beam under uniformly distributed load has a unique solution. (Prove the existence of the solution by showing that the strain energy of the beam is positive-definite).

9. Determine the strain energy density for a two-dimensional orthotropic linear elastic medium.

10. Determine the strain energy density for a linear elastic medium subjected to mechanical loads and temperature.

1.7 EQUATIONS OF BARS, BEAMS, TORSION, AND PLANE ELASTICITY

1.7.1 Introduction

This section is devoted to a review of the governing equations of bars, beams, torsion of prismatic members, and plane elasticity. It is assumed that the reader has completed a basic course that deals with various fundamental modes of loading of simple structural members. The brief review should provide the necessary equations for use in the discussion of variational principles and variational methods of approximation.

All bodies in nature are three-dimensional. However, the solution of these equations for simple structural members is not only a difficult task, but it is not warranted, because the solution of simplified equations is accurate for most practical applications. One can make valid simplifying assumptions concerning stress distributions, loadings, and geometry for certain structural members, such as bars, beams, and plates. In the sections that follow, we make assumptions about loads and their manner of application, displacement fields, stress distributions, and constitutive behavior. These assumptions lead to simplified theories that can be described in terms of fewer unknowns and thus facilitate their solutions analytically and/or numerically.

To keep the scope of the present study within reasonable limits, we do not attempt to develop the theory of curved beams or composite structures. The main emphasis of the present review is on the underlying assumptions used in the derivations and the consequent restrictions placed on the resulting equations. It is important to know when an equation is appropriate and to what extent the restrictions are not met.

1.7.2 Axially Loaded Members: Bars

Consider an axially loaded member of arbitrary but constant cross section and let P be the resultant axial (tensile) force parallel to the x-axis, as shown in Fig. 1.10. At any section along the length of the member, the internal reaction is a resultant of the accumulative effect of the normal tensile stress σ_x distributed over the entire cross-sectional area. Therefore, the equilibrium of axial forces is given by

$$P - \int_A \sigma_x \, dA = 0 \qquad (1.7.1)$$

where the integral can be viewed as the summation of forces $\sigma_x \, dA$ on infinitesimal area dA.

The equations of axially loaded members are based on the assumption of a uniform stress distribution, that is, σ_x is a function of x alone and is independent of the position at any given section. The main restrictions of the theory of axially loaded members, called *bars*, are:

1.   The cross section of the member is arbitrary, but is either uniform or gradually varying in the axial direction; if sudden changes in the cross

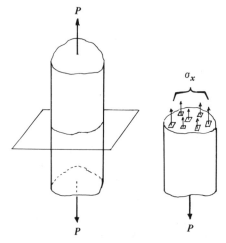

Figure 1.10   Axially loaded members (bars).

section are involved, the centroids of all cross sections can be joined by a straight line parallel to the axis of the member.

2. The material of the member is homogeneous, or the modulus of elasticity E can be a function of the axial coordinate.

3. All applied loads and support points are geometrically positioned in line with the centroidal axis of the member.

4. The load magnitude in compression is less than the critical buckling load of the member.

5. The points of load application and support connections are at reasonable distances from the point of interest.

For uniform cross-section members, in the absence of any distributed forces, the restrictions imply that Eq. (1.7.1) relates the applied external load to the uniform state of axial stress in the member (σ_x = constant)

$$P = \sigma_x A \quad \text{or} \quad \sigma_x = \frac{P}{A} \tag{1.7.2}$$

where A is the cross-sectional area of the member. All other stress components are zero and, therefore, the stress equilibrium equations (1.2.40) are trivially satisfied. The axial displacement $u = u(x)$ in the member can be computed from the strain–displacement and stress–strain equations,

$$\varepsilon_x = \frac{du}{dx}, \quad \sigma_x = E\varepsilon_x \tag{1.7.3}$$

Combining Eqs. (1.7.2) and (1.7.3), we obtain

$$AE\frac{d^2u}{dx^2} = 0 \tag{1.7.4}$$

The differential equations can be integrated and the constants of integration can be evaluated using the boundary conditions.

For gradually varying cross sections, $A = A(x)$, and uniformly distributed axial loads, Eq. (1.7.1) is not valid. To derive associated equilibrium equations, consider an element of length Δx (see Fig. 1.11) of the member. From the free-body diagram of the element for equilibrium we have

$$-\sigma_x A + (\sigma_x + \Delta\sigma_x)(A + \Delta A) + f\Delta x = 0$$

where f is the axially distributed force/unit length. Dividing the equation throughout by Δx, neglecting products of increments (i.e., $\Delta\sigma_x \Delta A \approx 0$), and taking the limit $\Delta x \to 0$, we obtain

$$\sigma_x \frac{dA}{dx} + A\frac{d\sigma_x}{dx} + f = 0$$

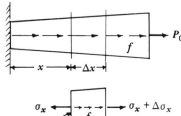

Figure 1.11 Equilibrium of an element of axially loaded members.

or

$$-\frac{d}{dx}(A\sigma_x) = f \tag{1.7.5}$$

Note that Eq. (1.7.5) is valid for any point x along the bar and it is independent of other coordinates. Equation (1.7.5) can be interpreted as a modification of Eq. (1.2.40) for a three dimensional body that meets the restrictions outlined earlier; Eq. (1.7.5) is an integrated form over the cross-sectional area of Eq. (1.2.40). The equilibrium equation can be expressed in terms of the displacement u by

$$-\frac{d}{dx}\left(EA\frac{du}{dx}\right) = f, \qquad 0 < x < L \tag{1.7.6}$$

wherein both E and A, in general, are functions of x.

Equations (1.7.4) and (1.7.6) should be appended with appropriate boundary conditions. There are two types of boundary conditions for bars, and only one of them can be specified at a boundary point: the axial displacement or axial force. The boundary conditions can be expressed in terms of the displacement as

$$u = \hat{u} \quad \text{or} \quad AE\frac{du}{dx} = \hat{P} \tag{1.7.7}$$

For example, if a member is fixed at the left end and subjected to load P_0 at the free end, we have ($\hat{u} = 0$ and $\hat{P} = P_0$)

$$u(0) = 0, \qquad \left(AE\frac{du}{dx}\right)\bigg|_{x=L} = P_0 \tag{1.7.8}$$

1.7.3 Theory of Straight Beams

Transverse loading on bars produces bending, and bars with transverse loads are called *beams*. Beams are probably the most common type of structural members. Generally, when a beam having an arbitrary cross section is subjected to transverse loads, bending, twisting, and/or buckling can occur. In linear analyses, the individual effects of bending, twisting, buckling, and axial deformation can be superimposed. To study the bending effects alone, we place certain initial restrictions on the geometry and loading of beams. The geometric restrictions discussed for bars in Section 1.7.2 are also valid for beams. In addition, we place the following initial restrictions on beams.

1. The cross section of the beam has a longitudinal plane of symmetry.
2. The resultant of the transversely applied loads lies in the longitudinal plane of symmetry.
3. Plane sections originally perpendicular to the longitudinal axis of the beam *remain plane and perpendicular* to the longitudinal axis after bending.
4. In the deformed beam, the planes of cross sections have a common intersection, that is, any line originally parallel to the longitudinal axis of the beam becomes an arc of a circle.

Restrictions 1 and 2 eliminate the possibility that the beam will twist under the action of transverse loads. Restrictions 3 and 4 simplify the kinematic description of the beam. All of the initial restrictions can be removed. For example, the removal of Restrictions 1 and 2 require us to consider the twisting of beams—a topic that is discussed in Section 1.7.4. Restrictions 3 and 4 can be relaxed by removing the normality condition (i.e., plane sections originally perpendicular to the longitudinal axis remain plane, but not necessarily perpendicular, to the longitudinal axis after bending). Beam theory based on the relaxed Restriction 3 is called the *Timoshenko beam theory*; otherwise, the theory is termed *classical beam theory*.

Classical Beam Theory

Consider the beam shown in Fig. 1.12*a*. We shall use the sign convention shown in Fig. 1.12*c*. Let us examine the deformation of an element of length Δx. Under the action of the transverse loads, the upper surface of the beam is shortened and the lower surface is elongated. Therefore, a surface exists between the top and bottom that undergoes no elongation or contraction. This longitudinal surface is called the *neutral surface*. The intersection of the neutral surface with the longitudinal plane of bending is called the *neutral axis* of the beam. We shall demonstrate that the neutral axis is the same as the centroidal axis for symmetric cross sections.

From the free-body diagram of the element shown in Fig. 1.12*c*, we have for the equilibrium of forces in the vertical direction,

$$V - (V + \Delta V) - f(x_0)\Delta x = 0$$

where x_0 is a point between x and $x + \Delta x$. Dividing by Δx and taking the

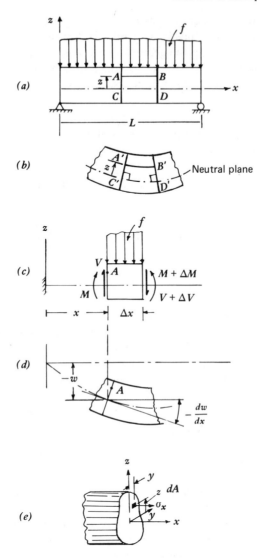

Figure 1.12   Geometry and deformation of a beam element.

limit $\Delta x \to 0 \; (x_0 \to x)$, we have

$$\frac{dV}{dx} = -f(x) \tag{1.7.9}$$

Equation (1.7.9) states that, where the loading is continuous, the rate of change of the shear force V is equal to the negative of the loading.

Taking sum of the moments about the y-axis, we obtain

$$M + \Delta M + f(x')\,\Delta x\, c\Delta x - M - V\Delta x = 0$$

where x' is some value of x between x and $x + \Delta x$, and c is a number between 0 and 1. Dividing throughout by Δx and taking the limit $\Delta x \to 0$, we obtain

$$\frac{dM}{dx} = V \tag{1.7.10}$$

Equation (1.7.10) states that, where loading is continuous, the rate of change of the bending moment is equal to the shear force. Combining Eqs. (1.7.9) and (1.7.10) we obtain,

$$\frac{d^2 M}{dx^2} = -f(x) \tag{1.7.11}$$

This completes the discussion of equilibrium conditions. In deriving the equilibrium equations, we assumed that the geometric changes are small enough that one can use the undeformed element for equilibrium. Next, we determine the kinematic relations for the beam.

Consider a line element AB of length Δx located at a distance z from the neutral axis (see Fig. 1.12a). We *assume* that the distance between the neutral axis and the line element $A'B'$ in the deformed body did not change significantly (or, Poisson's ratio is zero). Let the radius of curvature of the line element $CD = C'D'$ be κ. Then, the radius of curvature of $A'B'$ is $\kappa - z$. From the geometry of the deformed element, the axial strain in the line element $A'B'$, which is located between sections x and $x + \Delta x$ and a vertical distance z, is

$$e_x = \frac{A'B' - AB}{AB} = \frac{(\kappa - z)\, d\theta}{\kappa\, d\theta} - 1$$

$$= -\frac{z}{\kappa} \tag{1.7.12}$$

By Restriction 4, $1/\kappa$ is a constant and, therefore, Eq. (1.7.12) states that the longitudinal strain is proportional to the distance from the neutral axis. The negative sign indicates that the strain is compressive for positive z. The curvature $1/\kappa$ can be related to the derivatives of the transverse deflection by

$$\frac{1}{\kappa} = \frac{(d^2 w / dx^2)}{\left[1 + (dw/dx)^2\right]^{3/2}} \approx \frac{d^2 w}{dx^2} \tag{1.7.13}$$

The approximation holds for small gradients of w. Thus, the axial strain at point (x, z) is given by

$$e_x = -z \frac{d^2 w}{dx^2} \tag{1.7.14}$$

An alternative procedure to that described above can be used to obtain the result in Eq. (1.7.14). For small deformations, the line element $AB = \Delta x$ in the undeformed beam is assumed to be displaced by an amount $-w(x)$ in the negative z-direction and to rotate in the xz-plane by an amount equal to the slope $(-dw/dx)$ of the deflection $(-w)$ at x (see Fig. 1.12d). By Restrictions 3 and 4, the normals AC and BD also rotate as the amount $-dw/dx$. Thus, the axial displacement u at x is given by

$$u = -z\frac{dw}{dx} \qquad (1.7.15)$$

Therefore, the axial strain e_x at a point (x, z) in the vertical plane is given by

$$e_x \equiv \frac{\partial u}{\partial x} = -z\frac{d^2 w}{dx^2}$$

which coincides with Eq. (1.7.14).

The transverse normal strains e_y and e_z are small (equal to $-\nu e_x$, where ν is Poisson's ratio) and are assumed to be negligible:

$$e_y = 0, \qquad e_z = 0 \qquad (1.7.16)$$

The displacement v along the y direction is purely due to the Poisson's effect and is, therefore, negligible. The displacement along the z-direction is simply $w(x)$ (so that $e_z = \partial w/\partial z = 0$). Thus, the complete displacement at point (x, y, z) in the beam is given by

$$u = -z\frac{dw}{dx}, \qquad v = 0, \qquad w = w(x) \qquad (1.7.17)$$

The stresses (which can be obtained from the stress–strain relations by neglecting Poisson's effects) are assumed to be

$$\sigma_x = Ee_x = -Ez\frac{d^2 w}{dx^2}, \qquad \sigma_y = \sigma_z = 0 \qquad (1.7.18)$$

The normal stress caused by bending is known as the *flexure stress*.

Next, we derive the internal equilibrium relations—the relationship between internal stresses, moments, and forces. The equilibrium of moments about the x-axis is satisfied because there are no twisting moments, and the equilibrium of moments about the z-axis is satisfied because the z-axis is the vertical axis of symmetry. The equilibrium of moments about the y-axis gives (see Fig. 1.12e)

$$M + \int_A \sigma_x z \, dA = 0 \qquad (1.7.19)$$

The equilibrium of axial forces gives

$$\int_A \sigma_x \, dA = 0 \tag{1.7.20a}$$

which, in view of Eq. (1.7.18), implies that

$$-E\frac{d^2w}{dx^2}\int_A z \, dA = 0 \tag{1.7.20b}$$

Equation (1.7.20b) requires that the neutral axis (from which z is measured) be the centroidal axis. Thus, for any given cross section, the neutral axis location is given by its centroidal axis. Substitution of Eq. (1.7.18) into Eq. (1.7.19) gives the relationship between the internal moment about the y-axis and the transverse deflection and axial stress:

$$M = -\int_A \sigma_x z \, dA$$

$$= E\frac{d^2w}{dx^2}\int_A z^2 \, dA$$

$$= EI\frac{d^2w}{dx^2} \tag{1.7.21a}$$

$$= -\frac{\sigma_x I}{z} \tag{1.7.21b}$$

where I is the moment of inertia of the cross section about the y-axis.

The equilibrium of transverse internal forces shows that (see Fig. 1.13) a transverse stress σ_{zx} exists such that the element $ABEF$ is in equilibrium:

$$\sigma_{xz} h \, \Delta x - \int_\blacksquare \sigma_x \, dA + \int_\blacksquare (\sigma_x + \Delta\sigma_x) \, dA = 0 \tag{1.7.22}$$

where $\blacksquare$ denotes the area shaded, which is located at a distance z from the neutral surface. Using Eq. (1.7.21b) in Eq. (1.7.22), we get

$$\sigma_{zx} h \, \Delta x - \frac{\Delta M}{I}\int_\blacksquare z \, dA = 0$$

Dividing throughout by Δx and taking the limit $\Delta x \to 0$, we have

$$\sigma_{zx} = \frac{dM}{dx}\frac{Q}{Ih}$$

$$= \frac{VQ}{Ih} \tag{1.7.23}$$

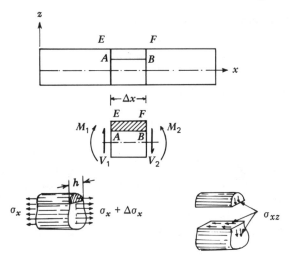

Figure 1.13   Shearing stresses in a beam.

where V is the resultant shear force and Q is the moment of the area bounded by line z and the top surface of the beam about the neutral axis:

$$Q = \int_{\blacksquare} z \, dA \qquad (1.7.24)$$

In the absence of the twisting moment, all other shear stresses are zero.

Finally, the equilibrium equation (1.7.11) can be expressed in terms of the transverse deflection by using Eq. (1.7.21a),

$$\frac{d^2}{dx^2}\left(EI\frac{d^2w}{dx^2}\right) + f = 0, \qquad 0 < x < L \qquad (1.7.25)$$

where L denotes the length of the beam (see Fig. 1.12). There are two pairs of boundary conditions for the problem, and only one element of each pair can be specified at a point:

$$\text{Specify:} \quad w \text{ or } V \quad \text{and} \quad \frac{dw}{dx} \text{ or } M \qquad (1.7.26)$$

For a beam clamped at the left end and subjected to moment M_0 and shear force V_0 at the free end, the boundary conditions are

$$w(0) = 0, \qquad \frac{dw}{dx}(0) = 0, \qquad V(L) = V_0, \qquad M(L) = M_0 \quad (1.7.27)$$

The Timoshenko Beam Theory

When we assume that plane sections originally perpendicular to the longitudinal axis of the beam remain plane, but not necessarily perpendicular, to the longitudinal axis of the beam, the displacement field in Eq. (1.7.17) is modified.

To derive the new displacement field, we begin with the line element CAE normal to the centerline of the beam in the undeformed state (see Fig. 1.14). After deformation, the line moves vertically down by an amount $w(x)$ and rotates by an angle $\psi(x)$. In general, $\psi(x)$ is not equal to the slope dw/dx of the deflection curve and, therefore, the line element $C'A'E'$ is not perpendicular to the deformed centerline. The difference $\psi - (-dw/dx)$, which denotes the rotation of elements along the centerline, is a measure of the shear angle at all points at position x. This amounts to assuming that the shear stress is uniform through the thickness. The displacement field now becomes

$$u = -z\psi(x), \qquad v = 0, \qquad w = w(x) \tag{1.7.28}$$

The nonzero stresses of the Timoshenko theory are given by

$$\sigma_x = -Ez\frac{d\psi}{dx}, \qquad \sigma_{zx} = Gk\left(\psi + \frac{dw}{dx}\right) \tag{1.7.29}$$

where G is the shear modulus and k is a constant that accounts for nonuniform shear stress distribution through the thickness. The value of k depends on the cross section and on Poisson's ratio.

Thus, in the Timoshenko beam theory, we must determine two displacement functions, w and ψ. The bending moment M and shear force V are related to

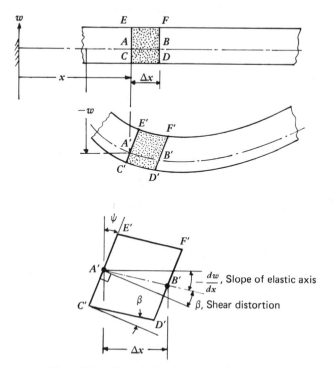

Figure 1.14 Shear deformation in a beam.

w and ψ by

$$M = \int_A \sigma_x z\, dA = -EI\frac{d\psi}{dx}$$

$$V = \int_A \sigma_{zx}\, dA = kGA\left(\psi + \frac{dw}{dx}\right) \qquad (1.7.30)$$

and the governing equations for displacements can be obtained from Eqs. (1.7.30), (1.7.9), and (1.7.10):

$$\frac{d}{dx}\left[kGA\left(\psi + \frac{dw}{dx}\right)\right] + f = 0 \qquad (1.7.31a)$$

$$\frac{d}{dx}\left(EI\frac{d\psi}{dx}\right) + kGA\left(\psi + \frac{dw}{dx}\right) = 0 \qquad (1.7.31b)$$

This completes the review of the beam equations.

1.7.4   Torsion of Prismatic Members

In this section, we derive the equations governing stresses and deformations in prismatic members subjected to equal and opposite end torques. The main restrictions of the torsion theory are:

1.   The material is homogeneous and obeys Hooke's law.
2.   The cross section of the member is arbitrary but constant along the length.
3.   The warping of cross sections is permitted, but is taken to be independent of axial location.
4.   The projection of any warped cross section rotates as a rigid body, and the angle of twist/unit length is constant.

Consider a prismatic bar of constant arbitrary cross section subjected to equal and opposite twisting moment, M_z, as shown in Fig. 1.15. The origin of the coordinate system is located at the *center of twist*, about which the cross section rotates during twisting. The center of twist is the point that experiences no rotation. The warping is described by the deflection w,

$$w = w(x, y) \qquad (1.7.32)$$

Under the action of the applied twisting moments, an arbitrary point $P : (x, y)$ on the cross section moves to the point $P' : (x + u, y + v)$. Assuming that no rotation occurs at end $z = 0$, and that the angle of twist θ is small, the displacements of the point (x, y) can be expressed as

$$u = -(r\theta z)\sin\alpha = -y\theta z$$
$$v = r\theta z\cos\alpha = x\theta z$$
$$w = w(x, y) \qquad (1.7.33)$$

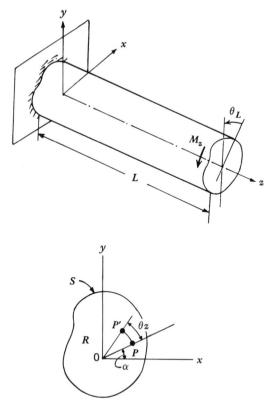

Figure 1.15 Torsion of a prismatic member.

where θz is the angular displacement of a point located at radial distance r and axial distance z, α is the angle between the line element OP and the x-axis, and w is the warping function.

The strains can be computed from Eq. (1.3.11)

$$e_x = e_y = e_z = e_{xy} = 0$$

$$2e_{xz} = \frac{\partial w}{\partial x} - \theta y, \qquad 2e_{yz} = \frac{\partial w}{\partial y} + \theta x \qquad (1.7.34)$$

and the stresses from Eq. (1.5.22)

$$\sigma_x = \sigma_y = \sigma_z = \sigma_{xy} = 0$$

$$\sigma_{xz} = G\left(\frac{\partial w}{\partial x} - \theta y\right), \qquad \sigma_{yz} = G\left(\frac{\partial w}{\partial y} + \theta x\right) \qquad (1.7.35)$$

The equations of equilibrium (1.2.40), in the absence of body forces, take the form

$$\frac{\partial \sigma_{xz}}{\partial z} = 0, \qquad \frac{\partial \sigma_{yz}}{\partial z} = 0, \qquad \frac{\partial \sigma_{xz}}{\partial x} + \frac{\partial \sigma_{yz}}{\partial y} = 0 \qquad (1.7.36)$$

The fifth compatibility equation (1.3.28) takes the form (all others are trivially satisfied)

$$\frac{\partial \sigma_{xz}}{\partial y} - \frac{\partial \sigma_{yz}}{\partial x} = -2G\theta \qquad (1.7.37)$$

Equations (1.7.36) and (1.7.37) define the torsion of a prismatic member.

The solution of Eqs. (1.7.36) and (1.7.37) can be achieved by introducing the so-called *Prandtl stress function*, Φ, which is defined by

$$\sigma_{xz} = \frac{\partial \Phi}{\partial y}, \qquad \sigma_{yz} = -\frac{\partial \Phi}{\partial x} \qquad (1.7.38)$$

This particular choice of Φ identically satisfies the equilibrium equation (1.7.36). Equation (1.7.37) takes the form

$$-\left(\frac{\partial^2 \Phi}{\partial x^2} + \frac{\partial^2 \Phi}{\partial y^2} \right) = 2G\theta \quad \text{in } R \qquad (1.7.39a)$$

The Poisson equation (1.7.39a) for Φ should be adjoined by appropriate boundary conditions. The boundary for the domain of Eq. (1.7.39a) is the lateral surface of the member. At any point on the boundary, the member should satisfy the stress-free condition,

$$\sigma_{xz} n_x + \sigma_{yz} n_y = 0$$

where $n_x = dy/ds$ and $n_y = -dx/ds$ are the direction cosines of the unit normal. In terms of Φ, the boundary condition becomes

$$\frac{\partial \Phi}{\partial y} \frac{dy}{ds} + \frac{\partial \Phi}{\partial x} \frac{dx}{ds} = 0 \quad \text{or} \quad \frac{d\Phi}{ds} = 0 \text{ on } S$$

In other words, Φ must be an arbitrary constant on the boundary. We take this constant to be zero because the stresses in Eq. (1.7.38) do not depend on the additive constant. Thus, we require that

$$\Phi = 0 \text{ on } S \qquad (1.7.39b)$$

The applied twisting moment M_z can be related to the stress function Φ by summing the moments about the z-axis.

$$M_z = \int_A (\sigma_{yz} x - \sigma_{xz} y) \, dA = 2 \int_A \Phi \, dx \, dy \qquad (1.7.40)$$

This completes the discussion of the torsion problem.

1.7.5 Plane Elasticity

The theory of bars and beams presented in Sections 1.7.2 and 1.7.3 is a simplification of three-dimensional elasticity theory to one-dimensional theory. Here, we discuss two-dimensional simplifications of three-dimensional elasticity problems.

Consider a linear elastic solid of uniform thickness h (say, in the z-direction) bounded by two parallel planes $z = -h/2$ and $z = h/2$ and by any closed boundary S. When the thickness h is very large, the problem is considered to be a *plane-strain problem*, and when the thickness is small compared with the lateral dimensions (x, y in this case), the problem is considered to be a *plane-stress problem*. Both of these problems are simplifications of three-dimensional elasticity problems under the following assumptions on loading:

1. The body forces and applied boundary forces, if any, do not have components in the thickness direction.
2. The body forces do not depend on the thickness coordinate, and the applied boundary forces are uniformly distributed across the thickness.
3. No loads are applied on the parallel planes bounding the top and bottom surfaces.

The assumption that the forces are zero on the parallel planes implies that the stresses associated with the z-direction are negligibly small for plane stress problems

$$\sigma_z = \sigma_{yz} = \sigma_{xz} = 0 \qquad (1.7.41)$$

For plane-strain problems, the assumption is that the strains associated with the z-direction are zero (in generalized plane-strain problems, they are assumed to be constant)

$$e_z = e_{xz} = e_{yz} = 0 \qquad (1.7.42)$$

An example that illustrates the difference between plane-stress and plane-strain problems is provided by the bending of a rectangular cross-section beam. If the beam is narrow, it is considered a plane-stress problem; if the beam is very wide, it is considered a plane-strain problem. Other examples of plane stress and plane strain are provided, respectively, by a thin plate subjected to inplane forces and a long cylinder held between two rigid and smooth blocks and subjected to uniform internal and/or external pressure (see Fig. 1.16). The equations governing the two types of plane elasticity problems discussed above can be obtained from the three-dimensional equations by using the assumptions in Eqs. (1.7.41) and (1.7.42). These are summarized below.

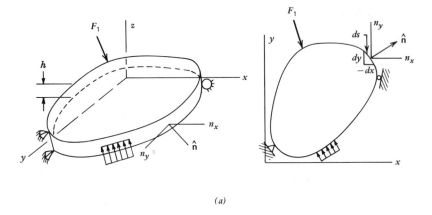

(a)

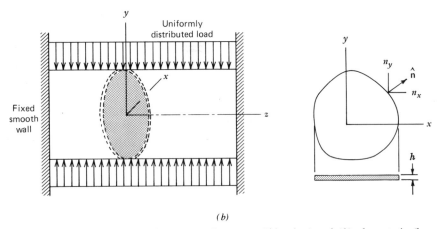

(b)

Figure 1.16 Plane elasticity problems: (*a*) plane-stress (thin plate) and (*b*) plane-strain (long cylinder).

Equilibrium Equations in Terms of Stresses

$$\frac{\partial \sigma_x}{\partial x} + \frac{\partial \sigma_{xy}}{\partial y} + f_x = 0$$

$$\text{in } R \qquad (1.7.43)$$

$$\frac{\partial \sigma_{xy}}{\partial x} + \frac{\partial \sigma_y}{\partial y} + f_y = 0$$

where f_x and f_y denote the body forces along the x and y directions, respectively.

Strain–Displacement Relations

$$e_x = \frac{\partial u}{\partial x}, \qquad e_y = \frac{\partial v}{\partial y}, \qquad 2e_{xy} = \frac{\partial u}{\partial y} + \frac{\partial v}{\partial x} \qquad (1.7.44)$$

Stress–Strain (or Constitutive) Relations

$$\sigma_x = c_{11}\varepsilon_x + c_{12}\varepsilon_y$$

$$\sigma_y = c_{12}\varepsilon_x + c_{22}\varepsilon_y$$

$$\sigma_{xy} = c_{66}\gamma_{xy} = 2c_{66}e_{xy} \qquad (1.7.45)$$

where c_{11}, c_{12}, and so on are the elasticity (material) constants. For an orthotropic elastic body, these constants are given as follows:

Plane Stress

$$c_{11} = \frac{E_1}{1 - \nu_{12}\nu_{21}}, \qquad c_{22} = \frac{E_2}{1 - \nu_{12}\nu_{21}}, \qquad c_{12} = \nu_{12}c_{22} = \nu_{21}c_{11}. \qquad c_{66} = G_{12}$$

$$(1.7.46a)$$

Plane Strain

$$c_{11} = \frac{E_1(1 - \nu_{12})}{(1 + \nu_{12})(1 - \nu_{12} - \nu_{21})}, \qquad c_{22} = \frac{E_2(1 - \nu_{21})}{(1 + \nu_{21})(1 - \nu_{12} - \nu_{21})},$$

$$c_{12} = \nu_{12}c_{22} = \nu_{21}c_{11}, \qquad c_{66} = G_{12} \qquad (1.7.46b)$$

where E_1 and E_2 are Young's moduli in one and two material directions, ν_{ij} is Poisson's ratio for transverse strain in the jth direction when stressed in the ith direction, and G_{12} is the shear moduli in the 1-2 plane. For an isotropic material, we have $E_1 = E_2 = E$, $\nu_{12} = \nu_{21} = \nu$, and $G_{12} = G$, and c_{ij} become

Plane Stress

$$c_{11} = c_{22} = \frac{E}{1 - \nu^2}, \qquad c_{12} = \frac{\nu E}{1 - \nu^2}, \qquad c_{66} = \frac{E}{2(1 + \nu)}$$

Plane Strain

$$c_{11} = c_{22} = \frac{E(1 - \nu)}{(1 + \nu)(1 - 2\nu)}, \qquad c_{12} = \frac{E\nu(1 - \nu)}{(1 + \nu)(1 - 2\nu)}, \qquad c_{66} = \frac{E}{2(1 + \nu)}$$

$$(1.7.47)$$

Boundary Conditions

$$natural: \quad \left. \begin{matrix} \sigma_x n_x + \sigma_{xy} n_y = \hat{t}_x \\ \sigma_{xy} n_x + \sigma_y n_y = \hat{t}_y \end{matrix} \right\} \quad \text{on } S_2 \qquad (1.7.48a)$$

$$essential: \quad \left. \begin{matrix} u = \hat{u} \\ v = \hat{v} \end{matrix} \right\} \quad \text{on } S_1 \qquad (1.7.48b)$$

where $\hat{\mathbf{n}} = (n_x, n_y)$ denotes unit normal to the boundary S, S_1 and S_2 are portions of the boundary S, $\hat{t}_x$ and $\hat{t}_y$ denote specified boundary forces, and $\hat{u}$ and $\hat{v}$ are specified displacements.

Equations (1.7.43) and (1.7.48a) can be expressed in terms of only the displacements u and v by substituting Eq. (1.7.44) into Eq. (1.7.45), and the result into Eqs. (1.7.43) and (1.7.48a):

$$\left. \begin{matrix} \dfrac{\partial}{\partial x}\left(c_{11}\dfrac{\partial u}{\partial x} + c_{12}\dfrac{\partial v}{\partial y} \right) + c_{66}\dfrac{\partial}{\partial y}\left(\dfrac{\partial u}{\partial y} + \dfrac{\partial v}{\partial x} \right) = f_x \\[2mm] c_{66}\dfrac{\partial}{\partial x}\left(\dfrac{\partial u}{\partial y} + \dfrac{\partial v}{\partial x} \right) + \dfrac{\partial}{\partial y}\left(c_{12}\dfrac{\partial u}{\partial x} + c_{22}\dfrac{\partial v}{\partial y} \right) = f_y \end{matrix} \right\} \quad \text{in } R \qquad (1.7.49)$$

$$\left. \begin{matrix} t_x \equiv \left(c_{11}\dfrac{\partial u}{\partial x} + c_{12}\dfrac{\partial v}{\partial y} \right)n_x + c_{66}\left(\dfrac{\partial u}{\partial y} + \dfrac{\partial v}{\partial x} \right)n_y = \hat{t}_x \\[2mm] t_y \equiv c_{66}\left(\dfrac{\partial u}{\partial y} + \dfrac{\partial v}{\partial x} \right)n_x + \left(c_{12}\dfrac{\partial u}{\partial x} + c_{22}\dfrac{\partial v}{\partial y} \right)n_y = \hat{t}_y \end{matrix} \right\} \quad \text{on } S_2 \quad (1.7.50)$$

We return to these equations in Section 3.4 on the finite element method. This completes the review of the equations of bars, beams, torsion of prismatic members, and plane elasticity.

Exercises 1.7 (see also the problems on beams in Exercises 1.5).

1. For a bar of uniform cross-sectional area A, length L, modulus of elasticity E, and mass density ρ, determine the deformation caused by its own weight (see Fig. E1). Do the stresses satisfy the boundary conditions?

2. The composite member made of aluminum ($E_a = 10^7$ psi) and steel ($E_s = 30 \times 10^6$ psi) is subjected to the axial forces shown in Fig. E2. Determine the total elongation of the composite rod ($A = 4$ in.^{2}).

3. If the right end of the composite rod in Exercise 2 is also constrained, determine the stresses developed in the aluminum and steel members.

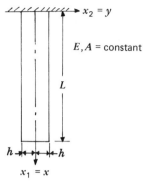

Figure E1

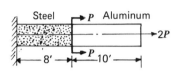

Figure E2

4. A homogeneous right circular cone shown in Fig. E4 has a specific weight of γ dynes/cm³. Determine the longitudinal deformation of the axis of the cone under its own weight.

5. Determine the expressions for flexure and shear stress as functions of x for the beam shown in Fig. E5.

6. Determine the transverse deflection w for the problem in Exercise 5.

7. Determine the transverse deflection of a clamped, simply supported beam under a uniformly distributed load.

8. Consider a narrow cantilever beam of rectangular cross section, loaded at its free end by a concentrated load P (see Example 1.6). Compare the maximum deflections obtained by the elasticity theory (in Example 1.6) and the classical theory of beams. Determine the ratio of the shear deflection to the bending deflection of the free end, and comment on the effect of shear deformation on the deflection.

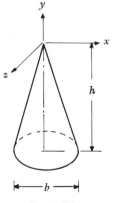

Figure E4

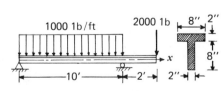

Figure E5

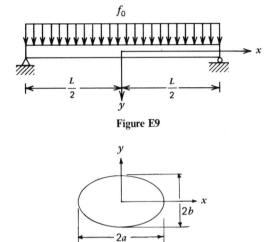

Figure E9

Figure E10

9. Repeat Exercise 8 for a simply supported beam subjected to a uniformly distributed load (see Fig. E9).

10. Consider the solid bar of elliptical cross section shown in Fig. E10. Determine the Prandtl stress function, shearing stresses, and the warping function.

11. The stress function for the torsion of a member of equilateral section (see Fig. E11) is given by

$$\Phi = c\left(x - \sqrt{3}\,y - \tfrac{2}{3}a\right)\left(x + \sqrt{3}\,y - \tfrac{2}{3}a\right)\left(x + \tfrac{1}{3}a\right)$$

where c is a constant. Determine the shearing stresses and the twisting moment for a given angle of twist, θ.

12. Consider a propellar blade that has a cross section as shown in Fig. E12. The thickness (symmetric about the y-axis) varies according to the equation, $h = b[1 - (2x/a)^2]$. Determine the stress function Φ, shearing stresses, and the twisting moment for a given angle of twist, θ.

13. For a rectangular cross-section beam, $-a \leq x \leq a$, and $-b \leq y \leq b$, determine the stress function in terms of the constants a, b, G, and θ.

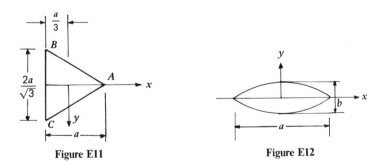

Figure E11   Figure E12

14. Show that the strain–stress relations for isotropic plane elastic bodies are given by

Plane Stress

$$e_x = \frac{1}{E}(\sigma_x - \nu\sigma_y)$$

$$e_y = \frac{1}{E}(\sigma_y - \nu\sigma_x)$$

$$e_{xy} = \frac{1+\nu}{E}\sigma_{xy}$$

Plane Strain

$$e_x = \frac{1-\nu^2}{E}\left(\sigma_x - \frac{\nu}{1-\nu}\sigma_y\right)$$

$$e_y = \frac{1-\nu^2}{E}\left(\sigma_y - \frac{\nu}{1-\nu}\sigma_x\right)$$

$$e_{xy} = \frac{1+\nu}{E}\sigma_{xy}$$

15. Show that for an isotropic plane-stress problem, the compatibility equations can be combined with the equilibrium equations to obtain

$$\nabla^2(\sigma_x + \sigma_y) = -(1+\nu)\left(\frac{\partial f_x}{\partial x} + \frac{\partial f_y}{\partial y}\right)$$

16. For an isotropic plane-stress problem, in the absence of body forces, equilibrium equations are identically satisfied if the stress field can be derivable from the potential function ϕ, called the *Airy stress function*,

$$\sigma_x = \frac{\partial^2 \phi}{\partial y^2}, \qquad \sigma_y = \frac{\partial^2 \phi}{\partial x^2}, \qquad \sigma_{xy} = -\frac{\partial^2 \phi}{\partial x\,\partial y}$$

Derive the differential equation for ϕ such that the compatibility conditions are satisfied.

17. Consider a long hollow cylinder with axially restrained ends (see Fig. E17). The inside and the outside radii are, respectively, r_i and r_o. The cylinder is subjected to internal and external pressures, p_i and p_o, respectively. The stress equilibrium (in the radial direction) and kinematic equations are given by

$$\frac{d\sigma_r}{dr} + \frac{\sigma_r - \sigma_\theta}{r} = 0, \qquad e_r = \frac{du}{dr}, \qquad e_\theta = \frac{u}{r}$$

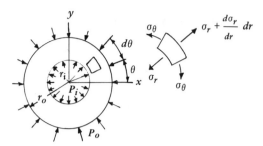

where r is the radial coordinate, and (σ_r, e_r) and $(\sigma_\theta, e_\theta)$ are the radial and tangential stresses and strains. Assuming that the problem is one of plane strain, express the equilibrium equation in terms of the radial displacement u, and solve the resulting equation for u.

18. *A higher-order theory of beams.*   In Eq. (1.7.28), assume that the axial displacement is given by

$$u(x, z) = z\left[\psi - \frac{4}{3}\left(\frac{z}{h}\right)^2\left(\psi + \frac{dw}{dx}\right)\right]$$

where h is the height of the beam. Derive the linear strains and stresses corresponding to the displacement field. As shown in Chapter 2, the equilibrium equations of the new theory are different from those in Eq. (1.7.31). Note that the particular form of the displacement given above results in the satisfaction of the stress-free condition at the top and bottom of the beam [i.e., $\sigma_{zx}(x, \pm h/2) = 0$].

2

ENERGY AND VARIATIONAL
PRINCIPLES

2.1 PRELIMINARY CONCEPTS

2.1.1 Introduction

In the formulation as well as the solution of many engineering problems, we encounter situations in which we are required to seek extreme (i.e., minimum or maximum) values of functions. These may be functions of points in Euclidean space, functions of functions in Euclidean space, or other geometrical and physical entities. For example, the determination of a curve $y = y(x)$ that joins two given points in two-dimensional Euclidean space, and that has a given length such that it encompasses the maximum possible area, constitutes a variational problem, called the *isoperimetric problem*. It involves maximizing the expression

$$\int_a^b y(x)\, dx$$

subject to the condition

$$\int_a^b \sqrt{1 + \left(\frac{dy}{dx}\right)^2}\, dx = \text{given length, } L$$

A circular arc turns out to be the desired solution for the isoperimetric problem. In the present case, the expression to be extremized is a function of $y(x)$, which is a function of position x in the Euclidean space. Such expressions are called *functionals*, when they take on a real value for any given y.

We recall the uses of variational formulations. They can be utilized to determine the governing equations of problems that are posed in terms of determining maxima or minima of a functional. Problems that are formulated by other methods, such as vector mechanics, can also be formulated by means of variational statements in the interest of solving them by variational methods. Also, variational statements of physical problems form a powerful basis for obtaining solutions to many physical problems, many of which are intractable otherwise. The remainder of this book deals with these uses of variational formulations. We begin with the concepts of work and energy.

2.1.2 Work and Energy

Consider a material particle, moving from point A to point B along some path in space and subjected to a force $\mathbf{F}$. The work dW performed by the force in moving the particle by infinitesimal displacement $d\mathbf{u}$ along the path is defined as (see Fig. 2.1a)

$$dW = \mathbf{F} \cdot d\mathbf{u}$$
$$= |\mathbf{F}| \, |d\mathbf{u}| \cos \alpha \quad \text{or } dW = F_1 \, du_1 + F_2 \, du_2 + F_3 \, du_3 \qquad (2.1.1)$$

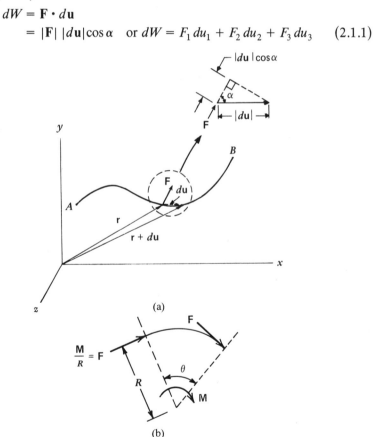

Figure 2.1 (a) Motion of a particle under the action of a force; (b) motion of a particle under the action of a moment.

where $\cdot$ denotes the dot product of two vectors. In other words, work is the product of force and displacement in the direction of the force. The work done W by the force $\mathbf{F}$ in moving the particle from A to B is given by

$$W = \int_A^B \mathbf{F} \cdot d\mathbf{u} = \int_A^B (F_1 \, du_1 + F_2 \, du_2 + F_3 \, du_3) \qquad (2.1.2)$$

The work done dW by a couple or moment $\mathbf{M}$ in rotating a particle about a point through an angle $d\theta$ is (see Fig. 2.1b)

$$dW = \mathbf{M} \cdot d\theta = M_1 \, d\theta_1 + M_2 \, d\theta_2 + M_3 \, d\theta_3 \qquad (2.1.3)$$

and the total work done by $\mathbf{M}$ as the particle moves from A to B is given by

$$W = \int_A^B \mathbf{M} \cdot d\theta = \int_A^B (M_1 \, d\theta_1 + M_2 \, d\theta_2 + M_3 \, d\theta_3) \qquad (2.1.4)$$

The definition of work can be extended to material bodies. If $\mathbf{F} = \mathbf{F}(\mathbf{x})$ is the distributed force acting on a particle occupying position $\mathbf{x}$ in the body, and $\mathbf{u} = \mathbf{u}(\mathbf{x})$ is the displacement of the particle, then the work done on the particle is $\mathbf{F} \cdot \mathbf{u}$, and the total work done on the body is the sum of work done on all particles of the body:

$$W = \int_V \mathbf{F} \cdot \mathbf{u} \, dV = \int_V (F_1 u_1 + F_2 u_2 + F_3 u_3) \, dV \qquad (2.1.5)$$

where V denotes the volume of the body. If a body is subjected to point forces $\mathbf{F}^1$, $\mathbf{F}^2$, etc. that displace the points of action by displacements $\mathbf{u}^1, \mathbf{u}^2$, etc., respectively, then the work done by the forces on the body is the sum of the work done by individual forces

$$W = \mathbf{F}^1 \cdot \mathbf{u}^1 + \mathbf{F}^2 \cdot \mathbf{u}^2 + \cdots = \sum_{i=1} \mathbf{F}^i \cdot \mathbf{u}^i \qquad (2.1.6)$$

Note that in this case, $\mathbf{F}^i$ and $\mathbf{u}^i$ are not vector functions but constant vectors. Equations (2.1.5) and (2.1.6) apply for distributed and point moments when $\mathbf{F}$ is replaced by $\mathbf{M}$ and $\mathbf{u}$ by θ.

Energy is the capacity to do work. It is a measure of the capacity of all forces that can be associated with matter to perform work. Work is performed on a body through a change in energy. The energy E of a body acted upon by time-dependent forces $\mathbf{F} = \mathbf{F}(\mathbf{x}, t)$ is given by the expression

$$E = \int_{t_0}^{t_1} \int_V \mathbf{F} \cdot \mathbf{u} \, dV \, dt \qquad (2.1.7)$$

Both work and energy are scalars that are independent of the coordinate system used to express them. The choice of the coordinate system only dictates the components of force and displacement, but their product is the same in any coordinate system.

2.1.3 Strain Energy and Complementary Strain Energy

The first law of thermodynamics gives rise to the energy equation [see Eq. (1.4.9)],

$$\frac{\partial U_0}{\partial t} = \sigma_{ij}\dot{\varepsilon}_{ij} - q_{i,i} + \rho Q$$

where U_0 is the internal energy, q_i are the components of heat flux vector, and Q is the internal heat generation. For elastic bodies under isothermal conditions, the internal energy consists of only stored elastic strain energy, as discussed in Section 1.5.2.

The existence of a scalar function U_0 of strains such that the stresses are derivable from U_0 is of special importance. Such stresses satisfy the energy equation and, consequently, they are said to be *conservative*. The value of the strain energy potential function U_0 is called the *strain energy density* of the internal forces. Under isothermal conditions, we have

$$\frac{\partial U_0}{\partial \varepsilon_{ij}} = \sigma_{ij} \quad \text{or} \quad dU_0 = \sigma_{ij}\, d\varepsilon_{ij} \tag{2.1.8}$$

The strain energy density is given by integrating Eq. (2.1.8)

$$U_0 = \int_0^{\varepsilon_{ij}} \sigma_{ij}\, d\varepsilon_{ij} \tag{2.1.9}$$

The expression in Eq. (2.1.9) is valid for all elastic bodies with linear or nonlinear strain–displacement relations.

Equation (2.1.9) can also be derived by considering the work done by internal forces in causing the deformation. Consider the rectangular parallele-piped element (taken from inside an elastic body) of sides dx_1, dx_2, and dx_3 shown in Fig. 2.2a. Suppose that the element is subjected to a system of external forces that vary slowly until they reach their final values, so that equilibrium is maintained at all times. The normal forces owing to the applied normal stresses are (negative of the same forces act on the opposite faces)

$$\sigma_{11}\, dx_2\, dx_3, \sigma_{22}\, dx_3\, dx_1, \sigma_{33}\, dx_1\, dx_2 \tag{2.1.10a}$$

and the shear forces resulting from the shear stresses on the six faces are

$$\sigma_{12}\, dx_2\, dx_3, \sigma_{13}\, dx_3\, dx_2, \sigma_{21}\, dx_1\, dx_3, \sigma_{23}\, dx_3\, dx_1, \sigma_{31}\, dx_1\, dx_2, \sigma_{32}\, dx_2\, dx_1$$

$$\tag{2.1.10b}$$

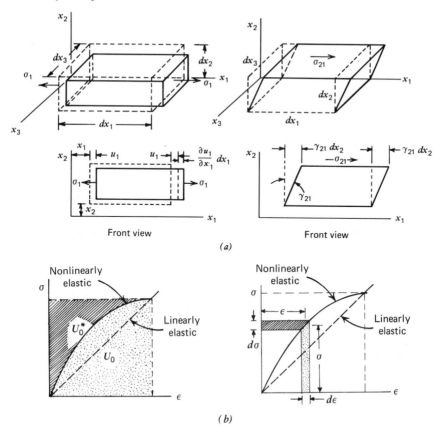

Figure 2.2 (a) Internal work done and (b) strain energy and complementary strain energy densities for the uniaxial case.

At any stage during the action of these forces, the faces of the parallelepiped will undergo displacements in the normal directions by the amounts $d\varepsilon_{11}\,dx_1$, $d\varepsilon_{22}\,dx_2$, and $d\varepsilon_{33}\,dx_3$, and distort by the amounts $2\,d\varepsilon_{12}$, $2\,d\varepsilon_{23}$, and $2\,d\varepsilon_{13}$. As examples, the deformations caused by the normal force $\sigma_{11}\,dx_2\,dx_3$ and the shear force $\sigma_{12}\,dx_2\,dx_3$, each acting alone, are shown in Fig. 2.2a. The work done by individual forces can be summed to obtain the total work done by the simultaneous application of all the forces, because, for example, an x_1-directed force does no work in the x_2 or x_3 directions. The work done, for instance, during the application of the force $\sigma_{11}\,dx_2\,dx_3$ is given by

$$\sigma_{11}\,dx_2\,dx_3 \cdot d\varepsilon_{11}\,dx_1 = \sigma_{11}\,d\varepsilon_{11}\,dV$$

Similarly, the work done by the shear force $\sigma_{21}\,dx_1\,dx_3$ is given by

$$\sigma_{21}\,dx_1\,dx_3 \cdot d\varepsilon_{21}\,dx_2 = \sigma_{21}\,d\varepsilon_{21}\,dV$$

The *net* work done by all forces in varying slowly from zero to their final value is given by

$$dW_I = \left(\int_0^{\varepsilon_{11}} \sigma_{11}\, d\varepsilon_{11} + \int_0^{\varepsilon_{12}} \sigma_{12}\, d\varepsilon_{12} + \cdots + \int_0^{\varepsilon_{32}} \sigma_{32}\, d\varepsilon_{32} \right) dV$$

$$= \left(\int_0^{\varepsilon_{ij}} \sigma_{ij}\, d\varepsilon_{ij} \right) dV \tag{2.1.11}$$

where a sum on repeated indices is implied. The *total* work done is given by

$$W_I = \int_V dW_I \tag{2.1.12}$$

This total work represents the mechanical energy stored in the body and is called the *strain energy* of the body, $U = W_I$. The strain energy/unit volume (i.e., strain energy density) of the body is given by the expression in the parenthesis of Eq. (2.1.11),

$$U_0 = \int_0^{\varepsilon_{ij}} \sigma_{ij}\, d\varepsilon_{ij}$$

which is the same as that in Eq. (2.1.9). This quantity represents the dotted area under the stress–strain curve in Fig. 2.2b. The area above the stress–strain curve represents the so-called *complementary strain energy density*, U_0^*, which can be computed from

$$U_0^* = \int_0^{\sigma_{ij}} \varepsilon_{ij}\, d\sigma_{ij} \tag{2.1.13}$$

or

$$U_0^* = \tfrac{1}{2} S_{ijkl} \sigma_{ij} \sigma_{kl} + \cdots \tag{2.1.14}$$

where S_{ijkl} are the compliance coefficients of the material. For linear elastic materials, we have

$$U_0 = \tfrac{1}{2} c_{ijkl} \varepsilon_{ij} \varepsilon_{kl}, \qquad U_0^* = \tfrac{1}{2} S_{ijkl} \sigma_{ij} \sigma_{kl} \tag{2.1.15}$$

and $U_0 = U_0^*$.

The strain energy U and complementary strain energy U^* of an elastic body are given by

$$U = \int_V U_0\, dV \tag{2.1.16a}$$

$$U^* = \int_V U_0^*\, dV \tag{2.1.16b}$$

Analogous to Eq. (2.1.8), the strains are computed from

$$\varepsilon_{ij} = \frac{\partial U_0^*}{\partial \sigma_{ij}} \tag{2.1.17}$$

It should be noted that both strain energy density and complementary strain energy density can be expressed either in terms of displacements or in terms of forces.

Example 2.1

Consider the pin-connected structure shown in Fig. 2.3. The members of the truss are made of an elastic material whose uniaxial stress–strain behavior in tension and compression are given by

$$\sigma = \begin{cases} E\sqrt{\varepsilon}, & \varepsilon \geq 0 \\ -E\sqrt{-\varepsilon}, & \varepsilon \leq 0 \end{cases} \qquad (2.1.18)$$

We wish to compute the strain energy and complementary strain energy densities of the structure. To compute the stresses and strains in the two members, we first compute the forces. From the free-body diagram of joint A, we obtain

$$F_1 = -\sqrt{3}\,P, \qquad F_2 = 2P$$

$$\sigma_1 = -\frac{\sqrt{3}\,P}{a_1}, \qquad \sigma_2 = \frac{2P}{a_2}$$

$$\varepsilon_1 = -\frac{3P^2}{E^2 a_1^2}, \qquad \varepsilon_2 = \frac{4P^2}{E^2 a_2^2} \qquad (2.1.19)$$

We have

$$U_{01} = \int_0^{\varepsilon_1} \sigma\, d\varepsilon = \frac{2E}{3}(-\varepsilon_1)^{3/2} = \frac{2}{3}\left(\frac{3\sqrt{3}\,P^3}{E^2 a_1^3}\right)$$

$$U_{02} = \int_0^{\varepsilon_2} \sigma\, d\varepsilon = \frac{2E}{3}(\varepsilon_2)^{3/2} = \frac{2}{3}\left(\frac{8P^3}{E^2 a_2^3}\right)$$

$$U_{01}^* = \int_0^{\sigma_1} \varepsilon\, d\sigma = -\frac{1}{3}\left(\frac{\sigma_1^3}{E^2}\right) = \frac{1}{3}\left(\frac{3\sqrt{3}\,P^3}{E^2 a_1^3}\right)$$

$$U_{02}^* = \int_0^{\sigma_2} \varepsilon\, d\sigma = \frac{1}{3}\left(\frac{\sigma_2^3}{E^2}\right) = \frac{1}{3}\left(\frac{8P^3}{E^2 a_2^3}\right)$$

$$U = a_1 L_1 U_{01} + a_2 L_2 U_{02} = \frac{2}{3}\left[\frac{3\sqrt{3}\,P^3 L_1}{E^2 a_1^2} + \frac{8P^3 L_2}{E^2 a_2^2}\right]$$

$$U^* = a_1 L_1 U_{01}^* + a_2 L_2 U_{02}^* = \frac{1}{3}\left[\frac{3\sqrt{3}\,P^3 L_1}{E^2 a_1^2} + \frac{8P^3 L_2}{E^2 a_2^2}\right] \qquad (2.1.20)$$

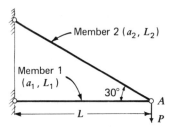

Member 2 (a_2, L_2)

Member 1
(a_1, L_1)

$30°$

A

L

P **Figure 2.3**   Pin-connected structure.

The next example is concerned with the determination of the strain energy for beams in bending alone.

Example 2.2

Under the usual assumptions of the classical beam theory (see Section 1.7), the strain energy of a beam in bending alone is only due to the axial stress, σ_x. The strain energy density is given by

$$U_0 = \int_0^{e_x} \sigma_x \, de_x$$

For a linear elastic material, we get $U_0 = \frac{1}{2} E e_x^2$. Using Eq. (1.7.14), we obtain

$$U_0 = \frac{Ez^2}{2} \left(\frac{d^2 w}{dx^2} \right)^2$$

The total strain energy becomes

$$U = \int_V U_0 \, dV = \int_0^L \int_A \frac{Ez^2}{2} \left(\frac{d^2 w}{dx^2} \right)^2 dA \, dx$$

$$= \frac{1}{2} \int_0^L EI \left(\frac{d^2 w}{dx^2} \right)^2 dx \qquad\qquad (2.1.21a)$$

where I is the moment of inertia.

The complementary strain energy density of a linear elastic beam in bending alone is given by

$$U_0^* = \int_0^{\sigma_x} e_x \, d\sigma_x = \frac{1}{2} \frac{\sigma_x^2}{E}$$

and the total complementary strain energy is obtained by using Eq. (1.7.21b),

$$U^* = \frac{1}{2} \int_0^L \int_A \frac{M^2 z^2}{EI^2} dA \, dx$$

$$= \frac{1}{2} \int_0^L \frac{M^2}{EI} dx \qquad\qquad (2.1.21b)$$

Note that U in Eq. (2.1.21a) is equal to U^* in Eq. (2.1.21b).

2.1.4 Virtual Work

The use of vector methods to determine the governing equations of a mechanical system requires the isolation of a particle or an element of the system with all its applied and reactive forces (i.e., the free-body diagram). For simple mechanical systems, the vector methods provide an easy and direct way of deriving the equations. However, for complicated systems, the procedure becomes more cumbersome and intractable. In such cases, variational methods provide alternative means for the determination of the governing equations. The principles of virtual work and stationary total potential energy can be used to derive the equations of solid mechanics. To this end, we first study the concept of virtual work.

From purely geometrical considerations, a given mechanical system can take many possible configurations consistent with the geometric constraints (i.e., essential boundary conditions). Of all the possible configurations, only one corresponds to the actual configuration; it is this configuration that also satisfies Newton's second law (i.e., equilibrium of forces and moments). The set of configurations that satisfies the geometric constraints is called the *set of admissible configurations*. These configurations are restricted to a neighborhood of the true configuration and are obtained from *variations* of the true configuration. During such variations, the geometric constraints of the system are not violated and all the forces are fixed at their actual equilibrium values. When a mechanical system experiences such variations in its displacement configuration, it is said to undergo *virtual displacements* from its equilibrium configuration. These displacements need not have any relationship with the actual displacements that might occur because of a change in the loads applied. The displacements are called virtual because they are *imagined* to take place (i.e., hypothetical) with the actual loads acting at their fixed values. For example, an admissible virtual displacement of a cantilevered beam is of the form

$$\delta w = c x^n$$

where c is a constant and $n \geq 2$. The displacement satisfies the geometric constraints that w and dw/dx be zero at $x = 0$. Thus, the virtual displacements at the boundary points at which the geometric conditions are specified (to be zero or not) are necessarily zero.

The work done by the actual forces through a virtual displacement of the actual configuration is called *virtual work*. If we denote the virtual displacement by $\delta \mathbf{u}$, the virtual work done by forces $\mathbf{F}$ is given by

$$\delta W = \int_V \mathbf{F} \cdot \delta \mathbf{u} \, dV$$

The external virtual work done on an elastic body subjected to body forces $\mathbf{f}$/unit volume and surface stresses $\hat{\mathbf{t}}$/unit area on boundary S_2 is given by

$$\delta W_E = -\int_V \mathbf{f} \cdot \delta \mathbf{u} \, dV - \int_{S_2} \hat{\mathbf{t}} \cdot \delta \mathbf{u} \, dS \qquad (2.1.22)$$

The negative sign implies that work is performed on the body. It is assumed that the displacements on boundary S_1 are specified. Therefore, $\delta \mathbf{u} = \mathbf{0}$ on S_1 (irrespective of whether $\mathbf{u}$ is specified to be zero or not). The internal virtual work stored owing to the elastic deformation is given by

$$\delta W_I = \int_V \delta \left(\int_0^{\varepsilon_{ij}} \sigma_{ij} \, d\varepsilon_{ij} \right) dV \qquad (2.1.23)$$

For a linear elastic body, this becomes

$$\delta W_I = \int_V \delta \left(\frac{1}{2} \sigma_{ij} \varepsilon_{ij} \right) dV$$

$$= \int_V \sigma_{ij} \delta \varepsilon_{ij} \, dV \qquad (2.1.24)$$

For a linear elastic body with small strains and subjected to a specified temperature distribution ($\delta T = 0$), the virtual work of internal forces is given by

$$\delta W_I = \int_V \sigma_{ij} \, \delta e_{ij} \, dV = \int_V \rho \, \delta \Psi \left(e_{ij}, T \right) dV \qquad (2.1.25)$$

where Ψ is the free-energy function defined in Eq. (1.5.8). We can also express δW_I in terms of the strain energy density $U_0(e_{ij}, T)$ in Eq. (1.5.30). We have

$$\frac{\partial U_0}{\partial e_{ij}} = \sigma_{ij} - \beta_{ij}(T - T_0) \qquad (2.1.26)$$

and

$$\delta W_I = \int_V \frac{\partial U_0}{\partial e_{ij}} \delta e_{ij} \, dV + \int_V \beta_{ij}(T - T_0)\delta e_{ij} \, dV$$

$$= \delta U + \delta \int_V \beta_{ij}(T - T_0) e_{ij} \, dV \qquad (2.1.27)$$

where, in arriving at the last line, we assumed that $\delta T = 0$, because T is a *given* function of position in thermally uncoupled problems. The temperature

T is obtained independently by solving the uncoupled Fourier heat conduction equation (1.5.35):

$$-\frac{\partial}{\partial x_i}\left(k_{ij}\frac{\partial T}{\partial x_j}\right) = Q \tag{2.1.28}$$

In this book, we are concerned mainly with *ideal systems*. An ideal system is one in which no work is dissipated by friction.

Example 2.3

In this example, we first consider a structure made of rigid members and then an elastic structure.

Consider the equilibrium of a rigid lever under the action of vertical loads F_1, F_2, and P (see Fig. 2.4a). The virtual work done by the actual forces in rotating the member through a virtual displacement (here, a rotation) $\delta\theta$, as shown in Fig. 2.4a, is given by

$$\delta W = F_1\delta(a_1\theta) + F_2\delta(a_2\theta) - P\delta(b\theta)$$

$$= (F_1 a_1 + F_2 a_2 - Pb)\,\delta\theta$$

Here δW represents the external virtual work done on the body; the internal virtual work done is zero, because the member experiences no internal forces

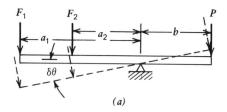

(a)

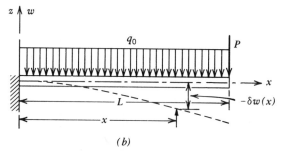

(b)

Figure 2.4 Equilibrium of a rigid lever and a cantilever beam discussed in Example 2.3.

(since it is assumed to be rigid). As we shall see, the principle of virtual displacements states that $\delta W = 0$, from which we obtain the relation between the three forces:

$$F_1 a_1 + F_2 a_2 - Pb = 0$$

Consider the cantilever beam shown in Fig. 2.4b. The virtual work done by the actual forces in moving through virtual displacement $\delta w(x)$ is given by

$$\delta W_E = \int_0^L q_0 \, \delta w \, dx + P \delta w(L)$$

$$\delta W_I = \int_V \delta U_0 \, dV = \int_0^L \int_A \left(\delta \int_0^{\varepsilon_{ij}} \sigma_{ij} \, d\varepsilon_{ij} \right) dA \, dx$$

$$= \int_0^L \int_A E \varepsilon_{11} \, \delta \varepsilon_{11} \, dA \, dx$$

$$= \int_0^L E \frac{d^2 w}{dx^2} \frac{d^2 \delta w}{dx^2} \, dx \left(\int_A z^2 \, dA \right)$$

$$= \int_0^L EI \frac{d^2 w}{dx^2} \frac{d^2 \delta w}{dx^2} \, dx$$

The total virtual work done

$$\delta W = \int_0^L \left(EI \frac{d^2 w}{dx^2} \frac{d^2 \delta w}{dx^2} + q_0 \delta w \right) dx + P \delta w(L)$$

The admissible virtual displacements are subjected to the geometric constraints

$$\delta w(0) = 0, \qquad \frac{d \delta w}{dx}(0) = 0$$

The principle of virtual displacements, $\delta W = 0$, gives rise to the equilibrium equation as well as to the boundary conditions on the forces (i.e., bending moments and shear force):

$$\frac{d^2}{dx^2} \left(EI \frac{d^2 w}{dx^2} \right) + q_0 = 0$$

$$\left(EI \frac{d^2 w}{dx^2} \right) \Bigg|_{x=L} = 0, \qquad \left[\frac{d}{dx} \left(EI \frac{d^2 w}{dx^2} \right) \right]_{x=L} = P$$

Exercises 2.1

Express the problems in Exercises 1–3 in mathematical terms as one of minimizing appropriate quantities, subjected to certain end conditions, and possibly some constraints.

1. *The Brachistochrone Problem.* Determine the curve $y = y(x)$ connecting two points, $(0,0)$ and (a, b), in a vertical plane such that a particle, sliding without friction under its own weight, travels from point $(0,0)$ to point (a, b) along the curve in the *shortest time*.

2. *Geodesic Problem.* Find the curve $y = y(x)$ of *minimum length* joining two given points (a, y_a) and (b, y_b) in the xy-plane.

3. *Isoperimetric Problem.* Among all curves with a continuous derivative that join two given points, (a, y_a) and (b, y_b), and have the length L, find the one that encompasses the largest possible area.

4. Determine the complementary strain energy of a cylindrical member subjected to axial force $N(x)$, transverse shear force $V(x)$, and twisting moment T. Assume that the member is of length L, area of cross section A, moment of inertia about the axis of bending I, and polar moment of inertia J.

5. Find the strain energy and complementary strain energy of the structural members shown in Fig. E5.

6. Two rigid links are pin-connected at A and to a rigid support at 0, as shown in Fig. E6. Compute the virtual work done by the forces caused by the virtual changes $\delta\theta_1$ and $\delta\theta_2$ in the generalized coordinates θ_1 and θ_2.

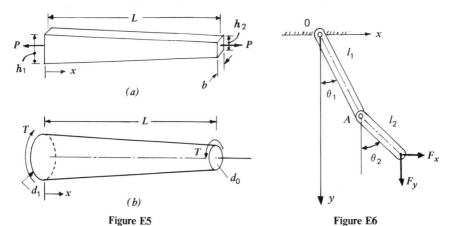

Figure E5   Figure E6

7. Determine the virtual work done by internal and external forces of a
   cantilevered beam subjected to uniform loading throughout the span,
   and a point load and bending moment at the free end (see the figure for
   Exercise 1.2.7).

2.2 CALCULUS OF VARIATIONS

2.2.1 The Variational Operator

The delta operator used in conjunction with virtual displacements has special
importance in variational methods. The operator is called the *variational
operator* because it is used to denote a variation (or change) in a given
quantity. Here we discuss certain operational properties of the operator.

Let $u = u(x)$ be the true configuration (i.e., displacement vector) of a given
mechanical system, and suppose that $u = \hat{u}$ on boundary S_1 of the total
boundary S. An admissible configuration is of the form

$$\bar{u} = u + \alpha v, \qquad \alpha \text{ small} \tag{2.2.1}$$

everywhere in the body, where v is an arbitrary function that satisfies the
homogeneous geometric boundary conditions of the system,

$$v = 0 \quad \text{on } S_1 \tag{2.2.2}$$

Clearly, αv is a variation of the given configuration u; the variation is
consistent with the geometric constraint of the system. Equation (2.2.1) defines
a set of varied configurations (see Fig. 2.5); an infinite number of paths $\bar{u}$ can
be generated for a fixed v by assigning values to α. All of these paths satisfy
the specified geometric boundary conditions on boundary S_1. For any v, all the
paths reduce to the actual path when α is set to zero. Therefore, for any *fixed*
x, αv can be viewed as a change or *variation* in the actual path u. This
variation is often denoted by δu

$$\delta u = \alpha v, \qquad \delta\left(\frac{du}{dx}\right) = \alpha\frac{dv}{dx} \tag{2.2.3}$$

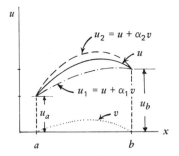

Figure 2.5 Set of admissible configurations and variations.

The operator δ is called the *variational operator* and δu is called the *first variation* of u.

Next, consider a function of the dependent variable u and its derivative $u' = du/dx$

$$F = F(x, u, u') \tag{2.2.4}$$

For fixed x, the change in F associated with a variation in u (and hence u') is given by

$$\Delta F = F(x, u + \delta u, u' + \delta u') - F(x, u, u')$$

$$= F(x, u + \alpha v, u' + \alpha v') - F(x, u, u') \tag{2.2.5}$$

Expanding the first term on the right-hand side in Taylor's series in terms of the parameter α, we get

$$\Delta F = \left[F + \frac{\partial F}{\partial u} \alpha v + \frac{\partial F}{\partial u'} \alpha v' + \frac{1}{2!} \frac{\partial^2 F}{\partial u^2} (\alpha v)^2 \right.$$

$$\left. + \frac{1}{2!} \frac{\partial^2 F}{\partial u \, \partial u'} (\alpha v')(\alpha v) + \cdots \right] - F$$

$$= \frac{\partial F}{\partial u} \alpha v + \frac{\partial F}{\partial u'} \alpha v' + O(\alpha^2) \tag{2.2.6}$$

where $O(\alpha^2)$ denote terms involving α^2 and higher powers of α. The first total variation of F is defined by

$$\delta F = \alpha \lim_{\alpha \to 0} \left[\frac{F(x, u + \alpha v, u' + \alpha v') - F}{\alpha} \right] = \alpha \left[\frac{dF(u + \alpha v, u' + \alpha v')}{d\alpha} \right]\Bigg|_{\alpha = 0}$$

$$= \frac{\partial F}{\partial u} \alpha v + \frac{\partial F}{\partial u'} \alpha v' \tag{2.2.7a}$$

$$= \frac{\partial F}{\partial u} \delta u + \frac{\partial F}{\partial u'} \delta u' \tag{2.2.7b}$$

There is an analogy between the first variation of F and the total differential of F. The total differential of F is

$$dF = \frac{\partial F}{\partial x} dx + \frac{\partial F}{\partial u} du + \frac{\partial F}{\partial u'} du' \tag{2.2.8}$$

Since x is fixed during the variation of u to $u + \delta u$, $dx = 0$, and the analogy between δF in Eq. (2.2.7) and dF in Eq. (2.2.8) becomes apparent. That is, the variational operator, δ, acts like a differential operator with respect to the dependent variable. Indeed, the laws of variation of sums, products, ratios, powers, and so forth are completely analogous to the corresponding laws of differentiation; that is, the variational calculus resembles the differential calculus. For example, if $F_1 = F_1(u)$ and $F_2 = F_2(u)$, we have

(i)
$$\delta(F_1 \pm F_2) = \delta F_1 \pm \delta F_2$$

(ii)
$$\delta(F_1 F_2) = \delta F_1 \, F_2 + F_1 \, \delta F_2$$

(iii)
$$\delta\left(\frac{F_1}{F_2}\right) = \frac{\delta F_1 \, F_2 - F_1 \, \delta F_2}{F_2^2}$$

(iv)
$$\delta(F_1)^n = n(F_1)^{n-1} \delta F_1 \qquad (2.2.9a)$$

If $G = G(u, v, w)$ is a function of several dependent variables (and possibly their derivatives), the total variation is the sum of partial variations:

$$\delta G = \delta_u G + \delta_v G + \delta_w G \qquad (2.2.9b)$$

where, for example, δ_u denotes the partial variation with respect to u. The variational operator can be interchanged with differential and integral operators:

(i)
$$\frac{d}{dx}(\delta u) = \frac{d}{dx}(\alpha v) = \alpha \frac{dv}{dx} = \delta\left(\frac{du}{dx}\right)$$

(ii)
$$\delta\left(\int_0^a u \, dx\right) = \alpha \int_0^a v \, dx = \int_0^a \alpha v \, dx = \int_0^a \delta u \, dx \qquad (2.2.10)$$

Example 2.4

Given the functions

(i)
$$F(u, u', u'') = c_1 u^2 + c_2 \left(\frac{du}{dx}\right)^2 + c_3 \left(\frac{d^2 u}{dx^2}\right)^2 + c_4 u \frac{du}{dx}$$

(ii)
$$G(u, v, u', v') = c_1 u^2 + c_2 v^2 + c_3 uv + c_4 \left(\frac{du}{dx}\right)^2$$

$$+ c_5 \left(\frac{dv}{dx}\right)^2 + c_6 \frac{du}{dx}\frac{dv}{dx}$$

find their first variations. Here c_i denote functions of x only.

We have

(i)
$$\delta F = \frac{\partial F}{\partial u}\delta u + \frac{\partial F}{\partial u'}\delta u' + \frac{\partial F}{\partial u''}\delta u''$$

$$= \left(2c_1 u + c_4 \frac{du}{dx}\right)\delta u + \left(2c_2 \frac{du}{dx} + c_4 u\right)\delta\left(\frac{du}{dx}\right)$$

$$+ 2c_3 \frac{d^2 u}{dx^2}\delta\left(\frac{d^2 u}{dx^2}\right)$$

(ii)
$$\delta_u G = \frac{\partial G}{\partial u}\delta u + \frac{\partial G}{\partial u'}\delta u'$$

$$= (2c_1 u + c_3 v)\delta u + \left(2c_4 \frac{du}{dx} + c_6 \frac{dv}{dx}\right)\delta\left(\frac{du}{dx}\right)$$

$$\delta_v G = \frac{\partial G}{\partial v}\delta v + \frac{\partial G}{\partial v'}\delta v'$$

$$= (2c_2 v + c_3 u)\delta v + \left(2c_5 \frac{dv}{dx} + c_6 \frac{du}{dx}\right)\delta\left(\frac{dv}{dx}\right)$$

This completes the example.

2.2.2 The First Variation of a Functional

The first variation of an integral function of u and its derivatives can be calculated using the definition in Eq. (2.2.7). Consider the integral function

$$I(u) = \int_a^b F(x, u, u')\, dx. \tag{2.2.11}$$

The integral $I(u)$, which assumes a real value for every u, is called a *functional*. The first variation of the functional I is defined to be

$$\delta I = \alpha\left[\frac{dI\,(u + \alpha v, u' + \alpha v')}{d\alpha}\right]\Bigg|_{\alpha=0} \tag{2.2.12a}$$

or, equivalently,

$$\delta I = \int_a^b \delta F\, dx$$

$$= \int_a^b \left(\frac{\partial F}{\partial u}\delta u + \frac{\partial F}{\partial u'}\delta u'\right) dx \tag{2.2.12b}$$

Thus, the variational operator can be used to compute the variation of any functional. Note that one can compute the first variation of a functional either using

$$\delta F = \alpha \left[\frac{dF\left(u + \alpha v, u' + \alpha v'\right)}{d\alpha} \right] \Bigg|_{\alpha = 0}$$

or using

$$\delta F = \frac{\partial F}{\partial u} \delta u + \frac{\partial F}{\partial u'} \delta u'$$

The former case (i.e., differential form) gives the result in terms of the variation αv and the latter (variational form) in terms of δu. The variational operator form is less cumbersome to use than the differential operator form.

A functional I is said to have a *minimum* at u_0 if and only if

$$I(u) \geq I(u_0) \qquad (2.2.13)$$

for all u. The equality holds only if $u = u_0$. The minimum property plays a crucial role in the proof of existence and uniqueness of solutions to problems posed in terms of variational statements.

2.2.3 Extremum of a Functional

From elementary calculus, we know that a differentiable function $f(x)$, $a < x < b$, has an extremum (i.e., minimum or maximum) at a point x_0 in the interval (a, b) only if (necessary condition)

$$\frac{df}{dx} \bigg|_{x = x_0} = 0 \qquad (2.2.14a)$$

The sufficient condition for maximum (or minimum) is that

$$\frac{d^2 f}{dx^2} < 0 \qquad \left(\text{or} \ \frac{d^2 f}{dx^2} > 0 \right) \qquad (2.2.14b)$$

Similarly, for a differential function $f(x, y)$ in two dimensions, the necessary condition for an extremum at the point (x_0, y_0) is that its total differential be zero at (x_0, y_0):

$$df \equiv \frac{\partial f}{\partial x} dx + \frac{\partial f}{\partial y} dy = 0 \quad \text{at } x = x_0 \ \text{and} \ y = y_0 \qquad (2.2.15)$$

If x and y are linearly independent (i.e., x and y are not related), Eq. (2.2.15) implies

$$\frac{\partial f}{\partial x} = 0 \quad \text{and} \quad \frac{\partial f}{\partial y} = 0 \qquad (2.2.16)$$

at $x = x_0$ and $y = y_0$.

Analogous to the case of ordinary functions, the necessary condition for a functional $I(u)$ to attain its extremum at a configuration u_0 from the set of admissible configurations is that its first variation be zero

$$\delta I \equiv \alpha \left[\frac{dI\left(u_0 + \alpha v, u_0' + \alpha v'\right)}{d\alpha} \right]_{\alpha=0} = 0 \qquad (2.2.17\text{a})$$

or

$$\left. \frac{dI\left(u_0 + \alpha v, u_0' + \alpha v'\right)}{d\alpha} \right|_{\alpha=0} = 0 \qquad (2.2.17\text{b})$$

Equation (2.2.17b) is equivalent to (2.2.13) in that, for a given u and v, I is an ordinary function of α and, therefore, $dI/d\alpha = 0$ is a necessary condition for I to attain an extremum. However, the extremum of I at u_0 is attained only for $\alpha = 0$. Thus, Eq. (2.2.17b) contains the condition $\alpha = 0$ and condition (2.2.13).

2.2.4 The Euler Equations

Consider the problem of determining the minimum of the functional

$$I(u) = \int_a^b F(x, u, u') \, dx \qquad (2.2.18)$$

subject to the end conditions

$$u(a) = u_a, \qquad u(b) = u_b \qquad (2.2.19)$$

It is clear that any candidate for the minimizing function should satisfy the end conditions in Eq. (2.2.19), and be sufficiently differentiable. The set of all such functions is called the *set of admissible functions*. Clearly, any element u from the set of admissible functions has the form

$$u = u_0 + \alpha v \qquad (2.2.20)$$

where α is a real number, v is a sufficiently differentiable function that satisfies the homogeneous form of the end conditions in Eq. (2.2.19),

$$v(a) = v(b) = 0 \qquad (2.2.21)$$

and u_0 is the actual function that minimizes the functional in Eq. (2.2.18). The set of all functions v such that Eqs. (2.2.20) and (2.2.21) hold is called the *set of admissible variations*.

Assuming that, for each admissible function u, $F(x, u, u')$ exists and is continuously differentiable with respect to its arguments, and that $I(u)$ takes one and only one real value, we seek the particular function $u_0(x)$ that makes the integral a minimum. The necessary condition for I to attain a minimum gives

$$0 = \frac{dI(u_0 + \alpha v)}{d\alpha}\bigg|_{\alpha=0}$$

$$= \frac{d}{d\alpha}\int_a^b F(x, u, u')\,dx\bigg|_{\alpha=0}$$

$$= \int_a^b \left(\frac{\partial F}{\partial u}\frac{\partial u}{\partial \alpha} + \frac{\partial F}{\partial u'}\frac{\partial u'}{\partial \alpha}\right)\bigg|_{\alpha=0}\,dx$$

$$= \int_a^b \left(\frac{\partial F}{\partial u_0}v + \frac{\partial F}{\partial u_0'}v'\right)dx \qquad (2.2.22)$$

where $u = u_0 + \alpha v$. Integrating the second term in the last equation by parts to transfer differentiation from v to u_0, we get

$$0 = \int_a^b v\left[\frac{\partial F}{\partial u_0} - \frac{d}{dx}\left(\frac{\partial F}{\partial u_0'}\right)\right]dx + \left(\frac{\partial F}{\partial u_0'}v\right)\bigg|_a^b \qquad (2.2.23)$$

The boundary term vanishes because v is zero at $x = a$ and $x = b$ [see Eq. (2.2.21)]. Because v is arbitrary inside the interval (a, b), and yet the equation should hold, implies that the expression in the square brackets is zero identically:

$$\frac{\partial F}{\partial u_0} - \frac{d}{dx}\left(\frac{\partial F}{\partial u_0'}\right) = 0 \quad \text{in } a < x < b \qquad (2.2.24)$$

The argument given here to arrive at Eq. (2.2.24) from Eq. (2.2.23) is known as the *fundamental lemma of calculus of variations*. The lemma states that if

$$\int_a^b v\Phi\left(x, u_0, u_0'\right)dx = 0 \qquad (2.2.25)$$

holds for *any* v, then it follows that $\Phi = 0$. This can be readily seen by setting $v = \Phi$ in Eq. (2.2.25).

Equation (2.2.24) is called the *Euler equation* of the functional in Eq. (2.2.18). Of all the admissible functions, the one that satisfies (i.e., the solution of) Eq. (2.2.24) is the true minimizer of the functional I.

2.2.5 Natural and Essential Boundary Conditions

Consider the problem of minimizing the functional in Eq. (2.2.18), subject to no end conditions (hence $\delta v \neq 0$ at $x = a, b$). The necessary condition for I to attain a minimum yields, as before, equation (2.2.23). Now suppose that

$\partial F/\partial u_0'$ and v are selected such that

$$\frac{\partial F}{\partial u_0'}v = 0 \quad \text{for} \quad x = a \quad \text{and} \quad x = b \qquad (2.2.26)$$

Using the fundamental lemma of the calculus of variations, we obtain the Euler equation in Eq. (2.2.24).

Let us consider Eq. (2.2.26). It is satisfied identically for any of the following combinations.

(i) $\qquad\qquad\qquad v(a) = 0, \qquad v(b) = 0$

(ii) $\qquad\qquad\qquad v(a) = 0, \qquad \dfrac{\partial F}{\partial u_0'}(b) = 0$

(iii) $\qquad\qquad \dfrac{\partial F}{\partial u_0'}(a) = 0, \qquad v(b) = 0$

(iv) $\qquad\qquad \dfrac{\partial F}{\partial u_0'}(a) = 0, \qquad \dfrac{\partial F}{\partial u_0'}(b) = 0 \qquad (2.2.27)$

The requirement that $v = 0$ at an end point is equivalent to the requirement that u_0 is specified (to be some value) at that point. The end conditions in (2.2.27) are classified into two types: *essential boundary conditions*, which require that the variation v (or δu) and possibly its derivatives vanish at the boundary; and *natural boundary conditions*, which require that the coefficients of the variation v and its derivatives be specified. Thus, we have

Essential Boundary Conditions

$$\text{Specify } v = 0 \text{ or } u = \hat{u} \text{ on the boundary} \qquad (2.2.28)$$

Natural Boundary Conditions

$$\text{Specify } \frac{\partial F}{\partial u_0'} = 0 \text{ on the boundary} \qquad (2.2.29)$$

The specified essential boundary conditions of a problem are not included in the functional, $I(u)$ and, therefore, they cannot be derived from the condition $\delta I = 0$. It is possible to include the essential boundary conditions by treating them as constraints on $I(u)$, as seen in the sequel.

In a given problem, only one of the four combinations given in Eq. (2.2.27) can be specified. Problems in which all of the boundary conditions are the essential type are called *Dirichlet boundary-value problems*, and those in which all of the boundary conditions are the natural type are called *Neumann boundary-value problems*. *Mixed boundary-value problems* are those in which both essential and natural boundary conditions are specified on complementary portions of the boundary. Essential boundary conditions are also known as *Dirichlet* or *geometric* boundary conditions, and natural boundary conditions are known as *Neumann* or *dynamic* boundary conditions.

As a general rule, the vanishing of the variation v (or δu)—equivalently, the specification of u—on the boundary constitutes the essential boundary condition, and the vanishing of the coefficient of the variation constitutes the natural boundary condition. This rule applies to any functional in one, two, or three dimensions, and to integrands that are functions of one or more dependent variables and their derivatives of any order. In the following, we consider a more general functional than that considered above.

2.2.6   A More General Functional

Consider the functional

$$I = \int_R F(x_i, u_i, u_{i,j}) \, dx_1 \, dx_2, \qquad i, j = 1, 2 \qquad (2.2.30)$$

where R is a two-dimensional domain, x_i are the coordinates, u_i are the dependent variables, and $u_{i,j} = \partial u_i / \partial x_j$. The first variation of I is given by

$$\delta I = \alpha \int_R \left. \frac{dF(x_i, u_i + \alpha v_i, u_{i,j} + \alpha v_{i,j})}{d\alpha} \right|_{\alpha=0} dx_1 \, dx_2$$

$$= \alpha \int_R \left(\frac{\partial F}{\partial u_i} v_i + \frac{\partial F}{\partial u_{i,j}} v_{i,j} \right) dx_1 \, dx_2$$

$$= \int_R \left(\frac{\partial F}{\partial u_i} \delta u_i + \frac{\partial F}{\partial u_{i,j}} \delta u_{i,j} \right) dx_1 \, dx_2 \qquad (2.2.31)$$

wherein summation on repeated subscripts is implied. Consider the second term in Eq. (2.2.31):

$$\int_R \frac{\partial F}{\partial u_{i,j}} \delta u_{i,j} \, dx_1 \, dx_2 = \int_R \frac{\partial F}{\partial u_{i,j}} \delta \left(\frac{\partial u_i}{\partial x_j} \right) dx_1 \, dx_2$$

$$= \int_R \frac{\partial F}{\partial u_{i,j}} \frac{\partial \delta u_i}{\partial x_j} \, dx_1 \, dx_2$$

$$= \int_R \left[\frac{\partial}{\partial x_j} \left(\frac{\partial F}{\partial u_{i,j}} \delta u_i \right) - \delta u_i \frac{\partial}{\partial x_j} \left(\frac{\partial F}{\partial u_{i,j}} \right) \right] dx_1 \, dx_2$$

$$= \oint_S n_j \frac{\partial F}{\partial u_{i,j}} \delta u_i \, dS - \int_R \delta u_i \frac{\partial}{\partial x_j} \left(\frac{\partial F}{\partial u_{i,j}} \right) dx_1 \, dx_2$$

$$(2.2.32)$$

where S denotes the boundary of R, and n_j are the components of the unit normal to the boundary. The last line is obtained from the third line by using the component form of the gradient theorem, Eq. (1.1.15). Substituting Eq. (2.2.32) into Eq. (2.2.31), we get

$$0 = \delta I = \int_R \delta u_i \left[\frac{\partial F}{\partial u_i} - \frac{\partial}{\partial x_j} \left(\frac{\partial F}{\partial u_{i,j}} \right) \right] dx_1\, dx_2 + \oint_S n_j \frac{\partial F}{\partial u_{i,j}} \delta u_i\, dS$$

$$(2.2.33)$$

The Euler equations are given by setting the coefficient of δu_i in R to zero:

$$\frac{\partial F}{\partial u_i} - \frac{\partial}{\partial x_j} \left(\frac{\partial F}{\partial u_{i,j}} \right) = 0, \qquad i = 1, 2 \qquad (2.2.34)$$

The natural boundary conditions are of the form given by setting the coefficient of δu_i on the portion of the boundary where δu_i is *not* specified. If u_i is not specified on any portion of S, then from Eq. (2.2.33), we obtain

$$\text{Specify: } n_j \frac{\partial F}{\partial u_{i,j}} = 0, \qquad i = 1, 2 \qquad (2.2.35)$$

All of the foregoing discussion can be applied to the general case of a functional involving p dependent variables with mth order partial derivatives with respect to n independent variables. In this case, there will be p Euler equations involving $2m$th order derivatives in n independent variables. Next, we consider some illustrative examples.

Example 2.5

Consider the total potential energy functional of an elastic bar of length L, modulus of elasticity E, area of cross section A, which is fixed at the left end and subjected to load P at the right end:

$$I(u) = \int_0^L \frac{EA}{2} \left(\frac{du}{dx} \right)^2 dx - Pu(L) \qquad (2.2.36)$$

where u denotes the longitudinal deformation of the bar. The first term in $I(u)$ represents the strain energy stored in the bar, and the second term denotes the work done on the bar by the load P in displacing the end $x = L$ through displacement $u(L)$. We wish to determine the Euler equation of the bar by requiring that $I(u)$ be a minimum.

The first variation of I is given by

$$\delta I = \int_0^L \frac{EA}{2} 2 \left(\frac{du}{dx} \right) \delta \left(\frac{du}{dx} \right) dx - P \delta u(L) \qquad (2.2.37)$$

Note that E, A, and P are fixed quantities that do not depend on u and, therefore, their variations with respect to u are zero. Integrating the first term by parts, we get

$$\delta I = \int_0^L -\frac{d}{dx}\left(EA\frac{du}{dx}\right)\delta u\,dx + EA\frac{du}{dx}\delta u\bigg|_0^L - P\delta u(L)$$

$$= -\int_0^L \delta u\frac{d}{dx}\left(EA\frac{du}{dx}\right)dx + \left[\left(EA\frac{du}{dx}\right)\bigg|_{x=L} - P\right]\delta u\,(L)$$

$$- \left(EA\frac{du}{dx}\right)\bigg|_{x=0}\delta u\,(0)$$

The last term is zero because of the specified essential boundary condition, which implies $\delta u(0) = 0$. Setting the coefficients of δu in $(0, L)$ and δu at $x = L$ to zero separately, we obtain

Euler Equation

$$-\frac{d}{dx}\left(EA\frac{du}{dx}\right) = 0, \qquad 0 < x < L$$

Natural Boundary Condition

$$EA\frac{du}{dx} - P = 0 \quad \text{at } x = L \tag{2.2.38}$$

Thus, the solution of Eq. (2.2.38) that satisfies $u(0) = 0$ is the minimizer of the energy functional in Eq. (2.2.36).

Example 2.6

The total potential energy of a linear elastic body in three dimensions is given by (sum on repeated subscripts is implied)

$$\Pi(\mathbf{u}) = \int_V \left(\tfrac{1}{2}\sigma_{ij}e_{ij} - f_i u_i\right)dV - \int_{S_2}\hat{t}_i u_i\,dS \tag{2.2.39}$$

where $\hat{t}_i$ are the specified boundary tractions on portion S_2 of the total boundary S, and f_i are the components of the body force vector. On the remaining portion of the boundary, $S_1 \equiv S - S_2$, the displacements are assumed to be zero. The first term under the volume integral represents the strain energy density of the elastic body, the second term represents the work done by the body force at point $\mathbf{x}$ of the body, and the surface integral denotes the work done by the specified external forces. For an isotropic body, the stress–strain relations are given by Eq. (1.5.22),

$$\sigma_{ij} = 2\mu e_{ij} + \lambda \delta_{ij} e_{kk}$$

The strain–displacement relations of the linear theory are given by Eq. (1.3.11),

$$e_{ij} = \tfrac{1}{2}(u_{i,j} + u_{j,i})$$

Substituting the stress–strain relations and strain–displacement equations into Eq. (2.2.39), we can write Π completely in terms of the displacements

$$\Pi(\mathbf{u}) = \int_V \left[\frac{\mu}{4}(u_{i,j} + u_{j,i})(u_{i,j} + u_{j,i}) + \frac{\lambda}{2} u_{j,j} u_{k,k} - f_i u_i \right] dV$$

$$- \int_{S_2} \hat{t}_i u_i \, dS \tag{2.2.40}$$

Now we wish to derive the Euler equations associated with the total potential energy functional in Eq. (2.2.40). Setting the first variation of Π to zero we get

$$0 = \int_V \left[\frac{\mu}{2}(\delta u_{i,j} + \delta u_{j,i})(u_{i,j} + u_{j,i}) + \lambda \delta u_{j,j} u_{k,k} - f_i \delta u_i \right] dV$$

$$- \int_{S_2} \hat{t}_i \delta u_i \, dS \tag{2.2.41}$$

wherein the product rule of variation is used and similar terms are combined. Using integration-by-parts in the manner described in Eq. (2.2.32), we obtain

$$0 = \int_V \left[-\frac{\mu}{2}(u_{i,j} + u_{j,i})_{,j} \delta u_i - \frac{\mu}{2}(u_{i,j} + u_{j,i})_{,i} \delta u_j - \lambda u_{k,kj} \delta u_j \right.$$

$$\left. - f_i \delta u_i \right] dV + \oint_S \left[\frac{\mu}{2}(u_{i,j} + u_{j,i})(n_j \delta u_i + n_i \delta u_j) + \right.$$

$$\left. + \lambda u_{k,k} n_j \delta u_j \right] dS - \int_{S_2} \delta u_i \hat{t}_i \, dS \tag{2.2.42}$$

After some algebraic manipulations (that involve renaming the subscripts and combining like terms), we obtain

$$0 = \int_V \left[-\mu(u_{i,jj} + u_{j,ij}) - \lambda u_{j,ji} - f_i \right] \delta u_i \, dV$$

$$+ \int_{S_1} n_j \left[\mu(u_{i,j} + u_{j,i}) + \lambda u_{k,k} \delta_{ij} \right] \delta u_i \, dS + \int_{S_2} (t_i - \hat{t}_i) \delta u_i \, dS$$

$$\tag{2.2.43}$$

where

$$t_i = n_j \left[\mu (u_{i,j} + u_{j,i}) + \lambda \delta_{ij} u_{k,k} \right] \qquad (2.2.44)$$

and the boundary integral over S is split into two parts, one on S_1 and the other on S_2. The boundary integral on S_1 vanishes because u_i is specified (hence, $\delta u_i = 0$) there. Setting the coefficients of δu_i in V and on S_2 to zero separately, we obtain

$$\mu (u_{i,jj} + u_{j,ij}) + \lambda u_{j,ji} + f_i = 0 \quad \text{in } V$$

$$t_i = \hat{t}_i \quad \text{on } S_2 \qquad (2.2.45)$$

for $i = 1, 2, 3$. Of course, the essential boundary conditions are $u_i = 0$ on S_1. The Euler equations are the same as the Navier equations.

2.2.7 Minimization with Linear Equality Constraints

Recall from Section 2.1.1 that minimization (or, in general, extremization) of functionals can involve subsidary conditions, called constraint conditions. The isoperimetric problem involves the determination of a function u such that

(i) $\qquad\qquad I(u) \equiv \displaystyle\int_a^b u(x)\, dx$ is a minimum, and

(ii) $\qquad\qquad \displaystyle\int_a^b \sqrt{1 + \left(\frac{du}{dx}\right)^2}\; dx - L = 0$

In the preceding sections, we devoted our attention to the determination of a function that minimizes a given functional. The necessary condition leads to a differential equation (the Euler equation) and boundary conditions governing the function. It turns out that it is also a sufficient condition for linear problems. In the present section, we consider ways to determine the minimum of quadratic functionals (i.e., functionals that involve quadratic terms of the dependent variables) with linear equality constraints. The Lagrange multiplier method and the penalty function method are discussed.

Consider the problem of finding the minimum of a functional (assumed to be quadratic in u and v),

$$I(u,v) = \int_a^b F(x, u, u', v, v')\, dx \qquad (2.2.46)$$

subject to the constraint

$$G(u, u', v, v') = 0 \qquad (2.2.47)$$

In addition, u and v should satisfy, as before, certain end conditions (or essential boundary conditions) of the problem. For the sake of the discussion, end conditions of the form in Eq. (2.2.19) are assumed for u and v.

In the constrained minimization problems, the admissible functions should not only satisfy the specified end conditions and be sufficiently continuous, but they should satisfy the constraint conditions. Furthermore, the admissible variations should be such that the constraint conditions are not violated.

Lagrange Multiplier Method

The necessary condition for the minimum of $I(u,v)$ in Eq. (2.2.46) is given by Eq. (2.2.17). We have

$$0 = \delta I = \int_a^b \left(\frac{\partial F}{\partial u} \delta u + \frac{\partial F}{\partial u'} \delta u' + \frac{\partial F}{\partial v} \delta v + \frac{\partial F}{\partial v'} \delta v' \right) dx \quad (2.2.48)$$

Since u and v must satisfy the constraint condition (2.2.47), the variations δu and δv are related by

$$0 = \delta G = \frac{\partial G}{\partial u} \delta u + \frac{\partial G}{\partial u'} \delta u' + \frac{\partial G}{\partial v} \delta v + \frac{\partial G}{\partial v'} \delta v' \quad (2.2.49)$$

The Lagrange multiplier method consists of multiplying Eq. (2.2.49) with an arbitrary parameter (λ), integrating over the interval (a, b), and adding the result to Eq. (2.2.48). The multiplier λ is called the Lagrange multiplier. We have

$$0 = \int_a^b \left[\frac{\partial F}{\partial u} \delta u + \frac{\partial F}{\partial u'} \delta u' + \frac{\partial F}{\partial v} \delta v + \frac{\partial F}{\partial v'} \delta v' \right.$$

$$\left. + \lambda \left(\frac{\partial G}{\partial u} \delta u + \frac{\partial G}{\partial u'} \delta u' + \frac{\partial G}{\partial v} \delta v + \frac{\partial G}{\partial v'} \delta v' \right) \right] dx$$

$$= \int_a^b \left\{ \left[\frac{\partial F}{\partial u} + \lambda \frac{\partial G}{\partial u} - \frac{d}{dx} \left(\frac{\partial F}{\partial u'} + \lambda \frac{\partial G}{\partial u'} \right) \right] \delta u \right.$$

$$\left. + \left[\frac{\partial F}{\partial v} + \lambda \frac{\partial G}{\partial v} - \frac{d}{dx} \left(\frac{\partial F}{\partial v'} + \lambda \frac{\partial G}{\partial v'} \right) \right] \delta v \right\} dx \quad (2.2.50)$$

The boundary terms vanish because $\delta v(a) = \delta v(b) = \delta u(a) = \delta u(b) = 0$. Since the variations δu and δv are not both independent, we cannot conclude that their coefficients are zero. Suppose that δu is independent and δv is related to δu by Eq. (2.2.49). We choose λ such that the coefficient of δv is zero. Then, by the fundamental lemma of variational calculus, it follows that (because δu is arbitrary) the coefficient of δu is also zero. Thus, we have

$$\frac{\partial}{\partial u}(F + \lambda G) - \frac{d}{dx}\left[\frac{\partial}{\partial u'}(F + \lambda G)\right] = 0$$

$$\frac{\partial}{\partial v}(F + \lambda G) - \frac{d}{dx}\left[\frac{\partial}{\partial v'}(F + \lambda G)\right] = 0 \qquad (2.2.51)$$

Equations (2.2.51) and (2.2.47) furnish three equations for the determination of u, v, and λ.

It is clear from Eq. (2.2.50) that Lagrange's method can be viewed as one of determining u, v, and λ by setting the first variation of the *modified* functional

$$L(u, v, \lambda) \equiv I(u, v) + \int_a^b \lambda G(u, u', v, v') \, dx$$

$$= \int_a^b (F + \lambda G) \, dx \qquad (2.2.52)$$

to zero. The Euler equations of the functional are precisely Eqs. (2.2.47) and (2.2.51). Indeed, we have

$$0 = \delta L = \int_a^b \delta(F + \lambda G) \, dx$$

$$= \int_a^b (\delta F + \delta \lambda G + \lambda \delta G) \, dx$$

$$= \int_a^b \left[\left(\frac{\partial F}{\partial u} + \lambda \frac{\partial G}{\partial u}\right)\delta u + \left(\frac{\partial F}{\partial u'} + \lambda \frac{\partial G}{\partial u'}\right)\delta u' + \left(\frac{\partial F}{\partial v} + \lambda \frac{\partial G}{\partial v}\right)\delta v \right.$$

$$\left. + \left(\frac{\partial F}{\partial v'} + \lambda \frac{\partial G}{\partial v'}\right)\delta v' + \delta \lambda G\right] dx$$

$$= \int_a^b \left\{\left[\frac{\partial F}{\partial u} + \lambda \frac{\partial G}{\partial u} - \frac{d}{dx}\left(\frac{\partial F}{\partial u'} + \lambda \frac{\partial G}{\partial u'}\right)\right]\delta u \right.$$

$$\left. + \left[\frac{\partial F}{\partial v} + \lambda \frac{\partial G}{\partial v} - \frac{d}{dx}\left(\frac{\partial F}{\partial v'} + \lambda \frac{\partial G}{\partial v'}\right)\right]\delta v + G\,\delta \lambda\right\} dx \qquad (2.2.53)$$

from which we obtain Eqs. (2.2.51) and (2.2.47) by setting, respectively, the coefficients of δu, δv, and $\delta \lambda$ to zero.

Penalty Function Method

The penalty function method involves the reduction of conditional extremum problems to extremum problems without constraints by the introduction of a penalty function associated with the constraints. As applied to the problem in

(2.2.46) and (2.2.47), the technique involves seeking the minimum of a modified functional obtained by adding a quadratic term associated with the constraint in (2.2.47) (i.e., the constraint is approximately satisfied in the least-squares sense):

$$P(u,v) = I(u,v) + \frac{\gamma}{2} \int_a^b [G(u, u', v, v')]^2 \, dx \qquad (2.2.54)$$

where γ is the penalty parameter (a preassigned real number). Setting the first variation of P to zero, we get the Euler equations for the functional:

$$\frac{\partial F}{\partial u} - \frac{d}{dx}\left(\frac{\partial F}{\partial u'}\right) + \gamma\left[G\frac{\partial G}{\partial u} - \frac{d}{dx}\left(G\frac{\partial G}{\partial u'}\right)\right] = 0$$

$$\frac{\partial F}{\partial v} - \frac{d}{dx}\left(\frac{\partial F}{\partial v'}\right) + \gamma\left[G\frac{\partial G}{\partial v} - \frac{d}{dx}\left(G\frac{\partial G}{\partial v'}\right)\right] = 0 \qquad (2.2.55)$$

For successively large values of γ, the solution of Eq. (2.2.55) gets closer to the true solution. In the penalty function method, the constraint is satisfied only approximately, and no additional unknowns are introduced into the variational formulation. Further, in this method an approximation to the Lagrange multiplier can be computed from the solution u and v of Eq. (2.2.55) by the formula [compare Eq. (2.2.55) with Eq. (2.2.51)]

$$\lambda = \gamma G(u, u', v, v') \qquad (2.2.56)$$

Both the Lagrange multiplier method and the penalty function method can be generalized for functionals with many variables and constraints. When the constraint involves an integral, such as the one in the isoperimetric problem, the Lagrange multiplier is a constant.

Example 2.7

Consider the problem of minimizing the functional

$$\Pi(w, \psi) = \int_0^L \left[\frac{EI}{2}\left(\frac{d\psi}{dx}\right)^2 + fw\right] dx - Pw(L) \qquad (2.2.57)$$

subject to the constraint

$$G(w', \psi) \equiv \frac{dw}{dx} + \psi = 0 \qquad (2.2.58)$$

The functional represents the total potential energy of a cantilever beam subjected to uniformly distributed load f and point load P at its free end. The constraint in Eq. (2.2.58) represents the relation between the transverse deflection w and bending slope ψ. The essential boundary conditions for the problem are

$$w(0) = \psi(0) = 0 \tag{2.2.59}$$

In the Lagrange multiplier method, the modified functional for the problem is given by

$$\Pi_L(w, \psi, \lambda) = \Pi(w, \psi) + \int_0^L \lambda\left(\frac{dw}{dx} + \psi\right) dx \tag{2.2.60}$$

where λ denotes the Lagrange multiplier. The Euler equations of the problem are obtained by setting the partial variations of Π_L with respect to w, ψ, and λ equal to zero:

$$\delta_w \Pi_L = 0: \qquad \begin{aligned} -\frac{d\lambda}{dx} + f &= 0, \qquad\qquad 0 < x < L \\ \lambda(L) - P &= 0 \end{aligned} \tag{2.2.61a}$$

$$\delta_\psi \Pi_L = 0: \qquad \begin{aligned} -\frac{d}{dx}\left(EI\frac{d\psi}{dx}\right) + \lambda &= 0, \qquad 0 < x < L \\ EI\frac{d\psi}{dx}(L) &= 0 \end{aligned} \tag{2.2.61b}$$

$$\delta_\lambda \Pi_L = 0: \qquad \frac{dw}{dx} + \psi = 0, \qquad\qquad 0 < x < L \tag{2.2.61c}$$

The Lagrange multiplier in the present case turns out to be the shear force.

In the penalty function method, the modified functional is given by

$$\Pi_p(w, \psi) = \Pi(w, \psi) + \frac{\gamma}{2}\int_0^L \left(\frac{dw}{dx} + \psi\right)^2 dx \tag{2.2.62}$$

where γ is the penalty parameter. The Euler equations are given by

$$\delta_w \Pi_p = 0: \qquad \begin{aligned} -\gamma\frac{d}{dx}\left(\frac{dw}{dx} + \psi\right) + f &= 0, \qquad 0 < x < L \\ \gamma\left(\frac{dw}{dx} + \psi\right)\bigg|_{x=L} - P &= 0 \end{aligned} \tag{2.2.63a}$$

$$\delta_\psi \Pi_p = 0: \qquad \begin{aligned} -\frac{d}{dx}\left(EI\frac{d\psi}{dx}\right) + \gamma\left(\frac{dw}{dx} + \psi\right) &= 0, \qquad 0 < x < L \\ EI\frac{d\psi}{dx}(L) &= 0 \end{aligned} \tag{2.2.63b}$$

A comparison of Eq. (2.2.63) with Eq. (2.2.61) shows that the Lagrange multiplier is related to ψ and w by

$$\lambda = \gamma\left(\frac{dw}{dx} + \psi\right) \qquad (2.2.64)$$

For $\gamma = kGA$, Eq. (2.2.64) represents the shear force and shear strain relation where k is the shear correction coefficient and A is the area of cross–section.

Exercises 2.2

Find the first variation of the functionals in Exercises 1–5. The arguments of the functionals denote the dependent variables, x and y denote the independent variables, and all other functions/constants are known.

1. $I(u) = \int_0^L \left[\frac{EA}{2}\left(\frac{du}{dx}\right)^2 + f\cos u\right] dx - Pu(L), \qquad u(0) = 0$

2. $I(u, v) = \int_0^L \left[\left(\frac{du}{dx}\right)^2 + 2\frac{du}{dx}\frac{dv}{dx} + \left(\frac{dv}{dx}\right)^2 + 2fu + 2gv\right] dx,$
   $u(0) = v(0) = 0$

3. $I(u) = \int_V \left[\frac{1}{2}\text{grad } u \cdot \text{grad } u - fu\right] dV - \oint_S u\hat{u}\, dS$

4. $I(w) = \frac{D}{2}\int_0^a \int_0^b \left[\left(\frac{\partial^2 w}{\partial x^2}\right)^2 + \left(\frac{\partial^2 w}{\partial y^2}\right)^2 + 2\nu\frac{\partial^2 w}{\partial x^2}\frac{\partial^2 w}{\partial y^2}\right.$
   $\left. + 2(1 - \nu)\left(\frac{\partial^2 w}{\partial x \partial y}\right)^2 + kw^2\right] dx\, dy$
   $- \int_0^a \int_0^b q_0 w\, dx\, dy, \qquad w = 0 \text{ on the boundary}$

   This functional arises in the classical theory of isotropic plates in bending.

5. $I(u_1, u_2, P) = \int_R \left[\frac{\mu}{2}(u_{i,j} + u_{j,i})u_{i,j} - Pu_{i,i}\right] dx_1\, dx_2$
   $- \int_R f_i u_i\, dx_1\, dx_2 + \oint_S \hat{t}_i u_i\, dS, \qquad (i, j = 1, 2)$

   This functional arises in connection with the slow, laminar, steady flows of a viscous fluid in two dimensions, where u_1 and u_2 denote the velocity components and P denotes the pressure.

6–10. Find the Euler equations, and identify the natural and essential boundary conditions of the problems in Exercises 1–5.

11. Derive the Euler equations and the natural and essential boundary conditions associated with the Lagrange multiplier functional of the

following problem. Minimize the functional

$$I(w, \psi_x, \psi_y) = \frac{D}{2} \int_0^a \int_0^b \left[\left(\frac{\partial \psi_x}{\partial x} \right)^2 + \left(\frac{\partial \psi_y}{\partial y} \right)^2 + 2\nu \frac{\partial \psi_x}{\partial x} \frac{\partial \psi_y}{\partial y} \right.$$

$$+ (1 - \nu) \left(\frac{\partial \psi_x}{\partial y} + \frac{\partial \psi_y}{\partial x} \right)^2 \right] dx\, dy$$

$$- \int_0^a \int_0^b qw\, dx\, dy$$

subject to the constraints

$$\frac{\partial w}{\partial x} + \psi_x = 0, \qquad \frac{\partial w}{\partial y} + \psi_y = 0$$

The resulting functional represents the total potential energy for isotropic Mindlin plates.

12. Repeat Exercise 11 using the penalty function method.

13. Consider the problem of maximizing the functional

$$I(u) = \int_a^b u(x)\, dx, \qquad u(a) = u_a, \qquad u(b) = u_b$$

subject to the constraint

$$G(u') \equiv \int_a^b \sqrt{1 + \left(\frac{du}{dx} \right)^2}\, dx - L = 0$$

Use (a) the Lagrange multiplier method and (b) the penalty method to introduce the constraint into $I(u)$ and derive associated Euler equations. Note that the constraint here involves an integral.

14. Consider the problem of minimizing the functional

$$I(u) = \int_a^b u \sqrt{1 + \left(\frac{du}{dx} \right)^2}\, dx, \qquad u(a) = u_a, \qquad u(b) = u_b$$

subject to the constraint

$$\int_a^b \sqrt{1 + \left(\frac{du}{dx} \right)^2}\, dx = L$$

Use (a), the Lagrange multiplier method, and (b), the penalty method to introduce the constraint into the functional and derive the Euler equations.

15. Show that the solution to the Euler equation of Exercise 14 is

$$u = c_1 + c_2 \cosh\left(\frac{x + c_3}{c_1}\right)$$

where c_1, c_2, and c_3 are constants.

2.3 VIRTUAL WORK AND ENERGY PRINCIPLES

2.3.1 Introduction

This section is devoted to the virtual work and energy principles of solid mechanics. We study the principles of virtual work and complementary virtual work and derive the principles of total potential energy and total complementary energy.

The use of variational principles in the derivation and solution of equations of elastic bodies forms the major part of the remaining study. More specifically, we use the variational principles developed in this section to study beams, plane elasticity problems, and plates and shells. Both direct variational methods and the finite element method are used to find approximate solutions to the problems.

2.3.2 Principle of Virtual Displacements

Recall from Section 2.1.4 that virtual work is the work done on a particle or a deformable body by actual forces in displacing the particle or the body through a hypothetical displacement that is consistent with the geometric constraints. The applied forces are kept constant during the virtual displacement. The principle of virtual displacements states that the virtual work done by actual forces in moving through virtual displacements is zero if the body is in equilibrium.

It is informative to analyze the principle of virtual displacements for a particle in static equilibrium. Consider a particle under the action of n concurrent forces $\mathbf{F}_1, \mathbf{F}_2, \ldots, \mathbf{F}_n$. Suppose that the particle is given an arbitrary virtual displacement $\delta\mathbf{u}$, during which all forces and their directions are fixed. The total virtual work done by all forces is given by

$$\delta W = \mathbf{F}_1 \cdot \delta\mathbf{u} + \mathbf{F}_2 \cdot \delta\mathbf{u} + \cdots + \mathbf{F}_n \cdot \delta\mathbf{u}$$

$$= \left(\sum_{i=1}^{n} \mathbf{F}_i\right) \cdot \delta\mathbf{u} \tag{2.3.1}$$

Note that the expression in the parenthesis is the vector sum of all forces acting on the particle. From vector mechanics, we know that the sum is zero if the particle is in equilibrium, giving $\delta W = 0$. Conversely, if $\delta W = 0$ and $\delta \mathbf{u}$ is arbitrary, it follows that $\sum_i^n \mathbf{F}_i = 0$; that is, the particle is in equilibrium. In other words, the particle is in equilibrium if and only if

$$\delta W = 0 \qquad (2.3.2)$$

for *any* choice of $\delta \mathbf{u}$. Equation (2.3.2) is the mathematical statement of the principle of virtual displacements for a particle.

This discussion can easily be extended to rigid bodies, because a rigid body is merely a collection of particles. We consider a generalization of the principle of virtual displacements to deformable bodies. In deformable bodies, material points can move relative to one another and can do work in addition to that done by external forces. Thus, we should consider the virtual work done by *internal forces* as well as that done by *external forces*.

Consider a continuous body V in equilibrium under the action of body force $\mathbf{f}$ and surface tractions $\mathbf{t}$. Suppose that over portion S_1 of the boundary S the displacements are $\hat{u}_i$, and on portion S_2 the tractions are $\hat{t}_i$. The boundary portions S_1 and S_2 are disjoint (i.e., do not overlap), and their sum is the total boundary, S. Let $\mathbf{u} = (u_1, u_2, u_3)$ be the displacement vector corresponding to the equilibrium configuration of the body, and let σ_{ij} and ε_{ij} be the associated stress and strain components, respectively.

The set of admissible configurations are defined by sufficiently differentiable functions that satisfy the geometric boundary conditions: $\mathbf{u} = \hat{\mathbf{u}}$ on S_1. Of all such admissible configurations, the actual one corresponding to the equilibrium configuration with the prescribed loads is given by the principle of virtual work. To determine the true displacement field $\mathbf{u}$, we let the body experience a virtual displacement $\delta \mathbf{u}$ from the true configuration. The virtual displacements are arbitrary continuous functions, except that they satisfy the homogeneous form of geometric constraints of the body; that is, they must belong to the set of admissible variations. The principle of virtual work states that *a continuous body is in equilibrium if the virtual work of all forces acting on the body is zero in a virtual displacement*:

$$\delta W_I + \delta W_E = 0 \qquad (2.3.3a)$$

where

$$\delta W_I = \text{virtual work resulting from internal forces}$$

$$\delta W_E = \text{virtual work resulting from external forces} \qquad (2.3.3b)$$

The principle of virtual work is independent of any constitutive law and applies to elastic (linear and nonlinear) and inelastic continuum problems. Here we use the principle to derive the equilibrium equations of a linear elastic body undergoing large displacements but small strains.

From Eqs. (2.1.22), (2.1.24), and (2.3.3), we have

$$-\int_V f_i\,\delta u_i\,dV - \int_{S_2} \hat{t}_i\,\delta u_i\,dS + \int_V \sigma_{ij}\,\delta\varepsilon_{ij}\,dV = 0 \qquad (2.3.4)$$

For a continuous medium, the kinematic relations are given by $\varepsilon_{ij} = \frac{1}{2}(u_{i,j} + u_{j,i} + u_{m,i}u_{m,j})$. For $\sigma_{ij} = \sigma_{ji}$ we have

$$\sigma_{ij}\,\delta\varepsilon_{ij} = \sigma_{ij}\,\delta u_{i,j} + \sigma_{ij}u_{m,i}\,\delta u_{m,j}$$

$$= (\sigma_{ij} + \sigma_{mj}u_{i,m})\,\delta u_{i,j} = \sigma_{mj}(\delta_{im} + u_{i,m})\,\delta u_{i,j}$$

where δ_{ij} denotes the Kronecker delta. Using integration by parts, we write

$$\int_V \sigma_{mj}(\delta_{im} + u_{i,m})\,\delta u_{i,j}\,dV = \int_V -\left[\sigma_{mj}(\delta_{im} + u_{i,m})\right]_{,j}\delta u_i\,dV$$

$$+\oint_S n_j\sigma_{mj}(\delta_{im} + u_{i,m})\,\delta u_i\,dS \quad (2.3.5a)$$

The boundary integral can be expressed as the sum of two integrals, one on S_1 and the other on S_2. The integral on S_1 is zero, because u_i is specified there (and, therefore, $\delta u_i = 0$ on S_1):

$$\oint_S n_j\sigma_{mj}(\delta_{im} + u_{i,m})\,\delta u_i\,dS = \int_{S_2} n_j\sigma_{mj}(\delta_{im} + u_{i,m})\,\delta u_i\,dS \quad (2.3.5b)$$

Substituting Eq. (2.3.5) for $\sigma_{ij}\,\delta\varepsilon_{ij}$ into Eq. (2.3.4), we obtain

$$-\int_V \left\{\left[\sigma_{mj}(\delta_{im} + u_{i,m})\right]_{,j} + f_i\right\}\delta u_i\,dV$$

$$+\int_{S_2}\left[n_j\sigma_{mj}(\delta_{im} + u_{i,m}) - \hat{t}_i\right]\delta u_i\,dS = 0 \qquad (2.3.6)$$

Because the virtual displacements are arbitrary and independent in V and on S_2, Eq. (2.3.6) implies

$$\left[\sigma_{mj}(\delta_{im} + u_{i,m})\right]_{,j} + f_i = 0 \text{ in } V$$

$$n_j\sigma_{mj}(\delta_{im} + u_{i,m}) - \hat{t}_i = 0 \text{ on } S_2 \qquad (2.3.7)$$

Equations (2.3.7) are the equations of equilibrium for solids experiencing large displacements. Note that the principle of virtual work gives the equilibrium equations as well as the natural boundary conditions. Equations (1.2.40), which were derived using vector methods, are the special case of Eqs. (2.3.7) for small displacements. It should be noted again that no constitutive equation is used in arriving at Eqs. (2.3.7).

Example 2.8

Consider an isotropic, elastic, cantilever beam of constant flexural rigidity EI and length L, which is subjected to the loads shown in Fig. 2.6. We wish to use the virtual work principle to determine the governing equation and natural boundary conditions. We have

$$W_E = -(-P_0)w(b) - M_0\frac{dw}{dx}(L) - \int_0^a (-f)w\,dx$$

$$W_I = \int_V \tfrac{1}{2}\sigma_{ij}e_{ij}\,dV \qquad\qquad (2.3.8)$$

Using the virtual work principle, we get

$$\int_V \sigma_{ij}\,\delta e_{ij}\,dV + \int_0^a f\delta w\,dx + P_0\delta w(b) - M_0\delta\frac{dw}{dx}(L) = 0 \quad (2.3.9)$$

Because only $\sigma_{11} = \sigma_x$ and $e_{11} = e_x$ are nonzero, and e_x is given by the kinematic relation (for small displacements)

$$e_x = -z\frac{d^2w}{dx^2} \qquad\qquad (2.3.10)$$

we get (because $dV = dA\,dx$, $dA = dy\,dz$),

$$0 = -\int_0^L \left(\int_{\text{area}} \sigma_x z\frac{d^2\,\delta w}{dx^2}\,dA\right)dx + \int_0^a f\delta w\,dx + P_0\delta w(b) - M_0\delta\frac{dw}{dx}(L)$$

$$= -\int_0^L M\frac{d^2\,\delta w}{dx^2}\,dx$$

$$+ \int_0^L [fH(a-x) + P_0 D(b-x)]\,\delta w\,dx - M_0\delta\left[\frac{dw}{dx}(L)\right] \qquad (2.3.11)$$

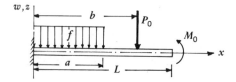

Figure 2.6 Cantilever beam subjected to various loads.

where $H(a - x)$ is the Heaviside step function

$$H(a - x) = \begin{cases} 1 & \text{if } x \le a \\ 0 & \text{of } x > a \end{cases} \qquad (2.3.12)$$

$D(b - x)$ is the Dirac delta function,

$$\int_{-\infty}^{\infty} D(b - x)f(x)\, dx = f(b) \qquad (2.3.13)$$

and M is the bending moment,

$$M = \int_{\text{area}} \sigma_x z\, dA$$

$$= \int_{\text{area}} E e_x z\, dA$$

$$= -E \frac{d^2 w}{dx^2} \int_{\text{area}} z^2\, dA$$

$$= -EI \frac{d^2 w}{dx^2} \qquad (2.3.14)$$

and I is the moment of inertia. Integrating this first term in Eq. (2.3.11) twice by parts, we get

$$0 = \int_0^L \left[\frac{d^2}{dx^2}\left(EI \frac{d^2 w}{dx^2} \right) + fH(a - x) + P_0 D(b - x) \right] \delta w\, dx$$

$$- \left(M\delta \frac{dw}{dx} - \frac{dM}{dx} \delta w \right)\Big|_0^L - M_0 \delta \frac{dw}{dx}(L) \qquad (2.3.15)$$

It is clear from the boundary terms that the specification of w and dw/dx at $x = 0$ and L constitute the essential boundary conditions and the specification of M and dM/dx at $x = 0$ and L constitute the natural boundary conditions for beam bending problems. Thus, the virtual displacement δw must satisfy the constraints

$$\delta w(0) = \delta \left(\frac{dw}{dx} \right)\Big|_{x=0} = 0 \qquad (2.3.16)$$

Substituting Eq. (2.3.16) into Eq. (2.3.15) and setting the coefficients of δw,

$\delta(dw/dx)(L)$, and $\delta w(L)$ to zero, we obtain

$$\frac{d^2}{dx^2}\left(EI\frac{d^2w}{dx^2}\right) + fH(a - x) + P_0D(b - x) = 0, \qquad 0 < x < L$$

$$EI\frac{d^2w}{dx^2}(L) - M_0 = 0$$

$$\left.\frac{d}{dx}\left(EI\frac{d^2w}{dx^2}\right)\right|_{x=L} = 0 \qquad\qquad (2.3.17)$$

This completes the example.

2.3.3 Unit-Dummy-Displacement Method

From the discussions of the previous section we know that the principle of virtual displacements gives the equilibrium equations. The principle of virtual work can also be used to directly determine reaction forces (and displacements) in structural problems. If R_0 is the reaction force (or moment) at point 0 in an elastic structure, we can prescribe a virtual displacement (or rotation) δu_0 at the point, but keep all other external forces on the structure stationary. The virtual strains owing to the virtual displacements are determined from kinematic considerations. Then the method of virtual work, Eq. (2.3.4), reads as

$$R_0\delta u_0 = \int_V \delta U_0 \, dV$$

$$= \int_V \sigma_{ij}\delta\varepsilon_{ij}^0 \, dV \qquad\qquad (2.3.18)$$

where σ_{ij} are the actual stresses and $\delta\varepsilon_{ij}^0$ are the virtual strains of the entire structure, consistent with the geometric constraints. Since δu_0 is arbitrary, one can take $\delta u_0 = 1$ (and $\delta\varepsilon_{ij}^0$ denote the virtual strains corresponding to the unit displacement at point 0). This procedure is called the *unit–dummy–displacement method*, which can be used to determine point loads or deflections in structures. We now consider two applications of this method.

Example 2.9

We wish to determine the vertical displacement of pin D under load P in the pin-connected structure shown in Fig. 2.7. We assume that members 1 and 3 are of the same material and dimensions and that member 2 is of different material and dimensions. We further assume that the material of the members

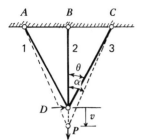

Figure 2.7 A pin-connected truss: A_i = area of member i; L_i = length of member i; $L_1 = L_3$, $A_1 = A_3$, $E_1 = E_3$.

is linearly elastic and that the stress is related to the strain by the one-dimensional form of Hooke's law:

$$\sigma = E\varepsilon \tag{2.3.19}$$

Under the applied vertical load P, pin D displaces only vertically and one can use either v or α as the degree of freedom (i.e., generalized coordinate) to describe the deformation. For a virtual displacement δv of pin D, we have from Eqs. (2.3.18)

$$P\delta v = 2A_1 L_1 \sigma_1 \delta\varepsilon_1 + A_2 L_2 \sigma_2 \delta\varepsilon_2 \tag{2.3.20}$$

where A_i and L_i are the area of cross section and length of the ith member, and σ_i are the axial stresses and $\delta\varepsilon_i$ are the virtual strains due to virtual displacements δv in the ith member.

The actual strains ε_1 and ε_2 are given by

$$\varepsilon_1 = \frac{\sqrt{(L_2 + v)^2 + (L_2\tan\theta)^2}}{L_1} - 1 = \frac{\sqrt{v^2 + 2L_2 v + L_1^2}}{L_1} - 1$$

$$\varepsilon_2 = \frac{(L_2 + v)}{L_2} - 1 = \frac{v}{L_2} \tag{2.3.21}$$

The actual stresses in members 1 and 2 are given by

$$\sigma_1 = E_1\varepsilon_1, \qquad \sigma_2 = E_2\varepsilon_2 \tag{2.3.22}$$

The strains due to the virtual displacement δv are given by

$$\delta\varepsilon_1 = \sqrt{\frac{(\delta v)^2 + 2L_2\delta v + L_1^2}{L_1^2}} - 1$$

$$\delta\varepsilon_2 = \frac{\delta v}{L_2} \tag{2.3.23}$$

It should be noted that the strains and displacements are not assumed to be small.

Next, suppose that the displacements are small. Then the strains in Eq. (2.3.21) can be simplified by neglecting squares of v:

$$\varepsilon_1 \simeq \sqrt{1 + 2\frac{L_2 v}{L_1^2}} - 1 \approx \frac{L_2 v}{L_1^2}$$

$$\varepsilon_2 = \frac{v}{L_2} \tag{2.3.24}$$

Similarly, the linearized virtual strains are given by

$$\delta\varepsilon_1 = \frac{\delta v\, L_2}{L_1^2}$$

$$\delta\varepsilon_2 = \frac{\delta v}{L_2} \tag{2.3.25}$$

Note that the virtual strains can also be obtained directly from Eq. (2.3.24) by taking variations of ε_1 and ε_2. Substituting Eqs. (2.3.24) and (2.3.25) into Eq. (2.3.20), we obtain

$$P\delta v = 2A_1 L_1 E_1\left(\frac{L_2 v}{L_1^2}\right)\left(\frac{L_2\,\delta v}{L_1^2}\right) + A_2 L_2 E_2\left(\frac{v}{L_2}\right)\left(\frac{\delta v}{L_2}\right) \tag{2.3.26}$$

Using the relation $L_2 = L_1\cos\theta$, we obtain

$$v = \frac{P L_2}{\left(2A_1 E_1\cos^3\theta + A_2 E_2\right)} \tag{2.3.27}$$

This completes the example.

In the next example we illustrate the application of the unit-dummy-displacement method to an axially loaded member.

Example 2.10

Consider the axially loaded member shown in Fig. 2.8. We wish to determine a relationship between the load P and the displacement e. The actual strains and stresses are given by

$$\varepsilon_1 = \frac{e}{a}, \qquad \varepsilon_2 = -\frac{e}{b}$$

$$\sigma_1 = E_1\varepsilon_1, \qquad \sigma_2 = E_2\varepsilon_2 \tag{a}$$

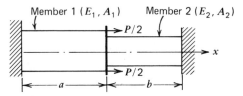

Figure 2.8   Axially loaded composite member.

The virtual strains are

$$\delta\varepsilon_1 = \frac{\delta e}{a}, \qquad \delta\varepsilon_2 = -\frac{\delta e}{b} \qquad \text{(b)}$$

The unit-dummy-displacement method gives

$$P\delta e = \delta U = \delta U_1 + \delta U_2$$

$$= \int_{V_1} \sigma_1 \,\delta\varepsilon_1 \, dV + \int_{V_2} \sigma_2 \,\delta\varepsilon_2 \, dV$$

$$= \left(A_1 E_1 \frac{e}{a} + A_2 E_2 \frac{e}{b} \right) \delta e$$

or,

$$P = \left(\frac{A_1 E_1}{a} + \frac{A_2 E_2}{b} \right) e \qquad \text{(c)}$$

2.3.4   Principle of Total Potential Energy

The principle of virtual work discussed in the previous section is applicable to any continuous body with arbitrary constitutive behavior (i.e., elastic or inelastic). A special case of the principle of virtual work that deals with elastic (linear as well as nonlinear) bodies is known as the principle of total potential energy, which we now describe.

For elastic bodies (in the absence of temperature variations), there exists a strain energy density function U_0 such that

$$\sigma_{ij} = \frac{\partial U_0}{\partial\varepsilon_{ij}} \qquad (2.3.28)$$

The strain energy density U_0 is a function of strains at a point and is assumed to be positive definite. Now the principle of virtual work statement, Eq. (2.3.4), can be expressed in terms of the strain energy density U_0:

$$\int_V f_i\,\delta u_i\,dV + \int_{S_2} \hat{t}_i\,\delta u_i\,dS = \int_V \frac{\partial U_0}{\partial \varepsilon_{ij}}\,\delta\varepsilon_{ij}\,dV$$

$$= \int_V \delta U_0\,dV$$

$$= \delta U \tag{2.3.29}$$

where U is the strain energy

$$U = \int_V U_0\,dV \tag{2.3.30}$$

Note that we have used a constitutive equation (2.3.28) to arrive at Eq. (2.3.29). If the temperature changes are considered, the strain energy density U_0 should be replaced by $\rho\Psi$, where Ψ is the free-energy function defined in Section 1.5.

If we define the potential energy of the applied loads, V, as a function of the displacements u_i

$$V = -\int_V f_i u_i\,dV - \int_{S_2} \hat{t}_i u_i\,dS \tag{2.3.31}$$

the principle of virtual work in Eq. (2.3.29) takes the form

$$\delta(V + U) \equiv \delta\Pi = 0 \tag{2.3.32}$$

The sum $V + U \equiv \Pi$ is called the *total potential energy* of the elastic body and is given by Eq. (2.2.39). The statement in Eq. (2.3.32) is known as the *principle of minimum total potential energy*.

An example of application of the principle of minimum potential energy is considered next.

Example 2.11

Here we present an alternative formulation of torsion of prismatic members (see Section 1.7.4). Consider the displacement field in Eq. (1.7.33) and the associated stresses in Eq. (1.7.35). The strain energy function U, for the linear elastic case, is given by

$$U = \int_V \tfrac{1}{2}\sigma_{ij}\varepsilon_{ij}\,dV$$

$$= L\int_R (\sigma_{xz}e_{xz} + \sigma_{yz}e_{yz})\,dx\,dy$$

$$= \frac{L}{2}\int_R \left[G\left(\frac{\partial w}{\partial x} - \theta y\right)^2 + G\left(\frac{\partial w}{\partial y} + \theta x\right)^2 \right] dx\,dy \tag{a}$$

If M_z denotes the twisting couple on the member (see Fig. 1.15), the potential energy of applied loads is given by

$$V = -M_z \theta L \tag{b}$$

Then the principle of minimum potential energy gives

$$\delta \Pi = \delta(U + V) = 0$$

or

$$
0 = \delta \left\{ \frac{L}{2} \int_R \left[G\left(\frac{\partial w}{\partial x} - \theta y \right)^2 + G\left(\frac{\partial w}{\partial y} + \theta x \right)^2 \right] dx\, dy - M_z \theta L \right\}
$$

$$
= \int_R GL \left[\frac{\partial \delta w}{\partial x} \left(\frac{\partial w}{\partial x} - \theta y \right) + \frac{\partial \delta w}{\partial y} \left(\frac{\partial w}{\partial y} + \theta x \right) + (-y\delta\theta) \left(\frac{\partial w}{\partial x} - \theta y \right) \right.
$$

$$
\left. + (x\,\delta\theta) \left(\frac{\partial w}{\partial y} + \theta x \right) \right] dx\, dy - M_z L\, \delta\theta
$$

$$
= GL \int_R \left[-\frac{\partial}{\partial x} \left(\frac{\partial w}{\partial x} - \theta y \right) - \frac{\partial}{\partial y} \left(\frac{\partial w}{\partial y} + \theta x \right) \right] \delta w\, dx\, dy
$$

$$
+ L \int_R \left\{ G \left[-y \left(\frac{\partial w}{\partial x} - \theta y \right) + x \left(\frac{\partial w}{\partial y} + \theta x \right) \right] \delta\theta\, dx\, dy \right.
$$

$$
+ GL \oint_S \left[\left(\frac{\partial w}{\partial x} - \theta y \right) n_x + \left(\frac{\partial w}{\partial y} + \theta x \right) n_y \right] \delta w\, dS - M_z L\, \delta\theta \tag{c}
$$

Setting the coefficients of δw in R and on S, and the coefficient of $\delta\theta$ in R to zero separately, we obtain

$$
\delta w: \quad GL \left(\frac{\partial^2 w}{\partial x^2} + \frac{\partial^2 w}{\partial y^2} \right) = 0 \quad \text{in } R
$$

$$
\delta w: \quad GL \left[\left(\frac{\partial w}{\partial x} - \theta y \right) n_x + \left(\frac{\partial w}{\partial y} + \theta x \right) n_y \right] = 0 \quad \text{on } S \tag{d}
$$

$$
\delta\theta: \quad M_z = G \int_R \left[x \frac{\partial w}{\partial y} - y \frac{\partial w}{\partial x} + \theta(x^2 + y^2) \right] dx\, dy \quad \text{in } R
$$

The function w and twist angle θ can be determined from Eq. (d) for any given twisting moment M_z, shear modulus G, and a given cross section of the member. We shall return to these equations in Chapter 3.

2.3.5   Principles of Virtual Forces and Complementary Potential Energy

In Section 2.3.3, we discussed the principle of virtual displacements. Naturally, the virtual work done by virtual forces in moving through actual displacements should have similar use. Here we formulate the complementary principle of virtual forces. The basic idea is that each particle in the body undergoes actual displacements until the stresses developed satisfy the equations of equilibrium.

Consider single-valued, differentiable variations of a stress field $\delta\sigma_{ij}$ and body forces δf_i that satisfy the linear equilibrium equations both within the body and on its boundaries:

$$\frac{\partial \delta\sigma_{ij}}{\partial x_j} + \delta f_i = 0 \quad \text{in } V \tag{2.3.33}$$

$$n_j \delta\sigma_{ij} = \delta t_i \quad \text{on } S_2 \tag{2.3.34}$$

We call such a stress field a *statically admissible field of variation*. These virtual stresses and forces, except for *self-equilibrating* ones, are completely arbitrary and independent of the true stresses and forces. Examples of possible virtual forces for a beam are given in Fig. 2.9.

The external *complementary virtual work* is defined by

$$\delta W_E^* = -\int_V u_i \,\delta f_i \,dV - \int_{S_1} \hat{u}_i \,\delta t_i \,dS \tag{2.3.35}$$

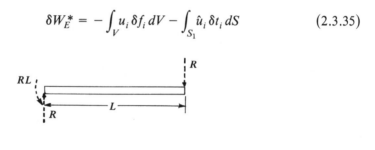

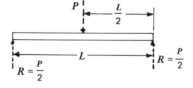

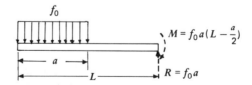

Figure 2.9   A possible set of virtual forces.

where u_i are displacement components, and δf_i and δt_i satisfy Eqs. (2.3.33) and (2.3.34). The internal complementary virtual work is given by

$$\delta W_I^* = \int_V e_{ij}\, \delta\sigma_{ij}\, dV$$

$$= \int_V \delta U_0^*\, dV, \qquad \left(U_0^* = \sigma_{ij} e_{ij} - U_0 \right) \qquad (2.3.36)$$

where e_{ij} are actual linear strains and $\delta\sigma_{ij}$ is the virtual stress field that satisfies Eqs. (2.3.33) and (2.3.34). If the temperature effects are considered, U_0^* should be replaced by $\rho\Psi^* = \rho\Psi - \sigma_{ij} e_{ij}$, where Ψ^* is the *Gibbs function*.

The principle of complementary virtual work (or virtual forces) states that *the strains and displacements in a deformable body are compatible and consistent with the constraints if and only if the total complementary virtual work is zero*

$$\delta W_I^* + \delta W_E^* = 0 \qquad (2.3.37)$$

Using Eqs. (2.3.35) and (2.3.36) in Eq. (2.3.37), we obtain

$$-\int_V u_i\, \delta f_i\, dV - \int_{S_1} \hat{u}_i\, \delta t_i\, dS + \int_V e_{ij}\, \delta\sigma_{ij}\, dV = 0$$

$$-\int_V u_i \left(-\delta\sigma_{ij,j}\right) dV - \int_{S_1} \hat{u}_i n_j\, \delta\sigma_{ij}\, dS + \int_V e_{ij}\, \delta\sigma_{ij}\, dV = 0$$

$$-\int_V \left[\tfrac{1}{2}\left(u_{i,j} + u_{j,i}\right) - e_{ij}\right] \delta\sigma_{ij}\, dV + \int_{S_1} \left(u_i - \hat{u}_i\right) n_j\, \delta\sigma_{ij}\, dS = 0$$

$$(2.3.38)$$

Because $\delta\sigma_{ij}$ is arbitrary, we obtain the strain–displacement equations and displacement boundary conditions

$$e_{ij} - \tfrac{1}{2}\left(u_{i,j} + u_{j,i}\right) = 0 \quad \text{in } V$$

$$u_i - \hat{u}_i = 0 \quad \text{on } S_1 \qquad (2.3.39)$$

The principle of virtual displacements gives the equilibrium equations and stress boundary conditions, and the principle of virtual forces gives the strain–displacement equations and displacement boundary conditions.

Noting that the complementary potential energy owing to virtual loads and complementary strain energy are given by

$$\delta V^* = \delta W_E^*, \qquad \delta U^* = \delta W_I^* \qquad (2.3.40)$$

we can arrive at the *principle of complementary potential energy* from Eq. (2.3.37):

$$\delta \Pi^* \equiv \delta(U^* + V^*) = 0 \qquad (2.3.41)$$

where Π^* denotes the total complementary potential energy.

Example 2.12

Consider the beam problem of Example 2.8. The complementary virtual works are

$$\delta W_E^* = w(b)\,\delta P - \frac{dw}{dx}(L)\delta M_0 + \int_0^a w\,\delta f\,dx$$

$$\delta W_I^* = \int_V e_x\,\delta\sigma_x\,dV \qquad (2.3.42)$$

where δP, δM_0, δf, and $\delta\sigma_x$ satisfy the equilibrium equations

$$\frac{d^2\,\delta M}{dx^2} = \delta f H(a - x) + \delta P\, D(b - x)$$

$$\delta M(L) = \delta M_0 = 0$$

$$\delta M = \int_A \delta\sigma_x\,z\,dA \qquad (2.3.43)$$

Substituting Eq. (2.3.42) into Eq. (2.3.37) and simplifying the expression, we get

$$
\begin{aligned}
0 &= \int_V e_x\,\delta\sigma_x\,dV + \int_0^L w\frac{d^2\,\delta M}{dx^2}\,dx \\
&= \int_V e_x\,\delta\sigma_x\,dV + \int_0^L \frac{d^2 w}{dx^2}\,\delta M\,dx + \left(w\frac{d\,\delta M}{dx} - \frac{dw}{dx}\,\delta M\right)_0^L \\
&= \int_V \left(e_x + z\frac{d^2 w}{dx^2}\right)\delta\sigma_x\,dV - w(0)\frac{d\,\delta M(0)}{dx} + \frac{dw}{dx}(0)\,\delta M(0) \qquad (2.3.44)
\end{aligned}
$$

In arriving at the last equation, we used the identity

$$
\begin{aligned}
\int_0^L \frac{d^2 w}{dx^2}\,\delta M\,dx &= \int_0^L \frac{d^2 w}{dx^2}\left(\int_A \delta\sigma_x\,z\,dA\right)dx \\
&= \int_0^L \int_A z\frac{d^2 w}{dx^2}\,\delta\sigma_x\,dA\,dx \\
&= \int_V z\frac{d^2 w}{dx^2}\,\delta\sigma_x\,dV \qquad (2.3.45)
\end{aligned}
$$

which is made possible because w is only a function of x. From Eq. (2.3.44) it follows, because $\delta\sigma_x$ and δM are arbitrary, that

$$\left(e_x + z\frac{d^2 w}{dx^2}\right) = 0, \qquad 0 < x < L$$

$$w(0) = 0, \qquad \frac{dw}{dx}(0) = 0 \qquad\qquad (2.3.46)$$

Thus, the principle of complementary virtual work gives the kinematic equation and geometric boundary conditions of the problem. This completes the example.

2.3.6 Unit-Dummy-Load Method

The virtual work principles are useful for deriving governing equations of structures, but they are not helpful for calculating actual displacements and forces. Both the unit-dummy-displacement and unit-dummy-load methods provide the means to determine point forces and displacements. The unit-dummy-load method is a special case of the complementary virtual work. The method can be used to determine displacements (and forces) in structures. The basic idea can be described in analogy with the unit-dummy-displacement method. If u_0 is the true displacement at point 0 in an elastic structure, we can prescribe a virtual force δR_0 at that point. The application of virtual force induces a system of virtual stresses $\delta\sigma_{ij}$ that satisfy the equilibrium equations. Then, from the principle of virtual forces (2.3.37), we have ($\delta W_E^* = -u_0\,\delta R_0$),

$$u_0\,\delta R_0 = \int_V \varepsilon_{ij}\,\delta\sigma_{ij}^0\,dV \qquad\qquad (2.3.47)$$

Once again, one can take $\delta R_0 = 1$ and calculate corresponding virtual internal stresses $\delta\sigma_{ij}^0$.

Example 2.13

Consider the cantilevered beam shown in Fig. 2.10. We wish to determine the tip displacement w_0 under the action of the point load P at the center of the

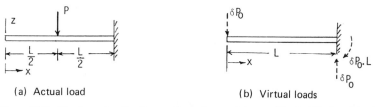

(a) Actual load   (b) Virtual loads

Figure 2.10 Configuration for the cantilever beam: (a) actual load; (b) virtual load.

beam. Let the virtual force at the tip be δP_0. Then

$$\delta W_E^* = -w_0 \delta P_0$$

$$\delta W_I^* = \int_V \varepsilon_{ij}\, \delta\sigma_{ij}^0 \, dV = \int_0^L -\frac{d^2w}{dx^2}\, \delta M_0 \, dx = \int_0^L \frac{M}{EI}\, \delta M_0 \, dx \quad (2.3.48)$$

where δM_0 is the virtual bending moment resulting from the virtual load δP_0. For $\delta P_0 = 1$, we have, from the unit-dummy-load method,

$$w_0 = \int_0^L \frac{M}{EI}\, \delta M_0 \, dx \qquad\qquad (2.3.49)$$

where M and δM_0 for the problem are given by

$$M = \begin{cases} 0, & 0 \le x \le \dfrac{L}{2} \\[2mm] -P\left(x - \dfrac{L}{2}\right), & \dfrac{L}{2} \le x \le L \end{cases}$$

$$\delta M_0 = (-1)x, \qquad 0 \le x \le L \qquad\qquad (2.3.50)$$

Substituting into Eq. (2.3.49), we obtain the tip deflection.

$$w_0 = \int_0^{L/2} 0(-x)\, dx + \int_{L/2}^L \frac{P}{EI}\left(x - \frac{L}{2}\right) x \, dx = \frac{5}{48}\frac{PL^3}{EI}$$

Exercises 2.3

1. Consider the beam problem shown in Fig. E1, which represents the free-body diagram of a typical beam. Give the expression for total virtual work caused by the virtual displacement $\delta w(x)$.

2. The total potential energy of the large-deflection bending and stretching of a cantilevered ($x = 0$) beam of length L, area of cross section A, modulus of elasticity E, moment of inertia I, which is subjected to distributed transverse load f, axial load P, and

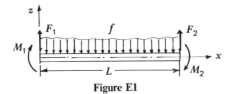

Figure E1

bending moment M_0 is given by (neglecting the shear deformation)

$$\Pi(u,w) = \int_0^L \left\{ \frac{EA}{2} \left[\frac{du}{dx} + \frac{1}{2} \left(\frac{dw}{dx} \right)^2 \right]^2 + \frac{EI}{2} \left(\frac{d^2w}{dx^2} \right)^2 + fw \right\} dx$$

$$- Pu(L) - M_0 \frac{dw}{dx}(L)$$

where u is the axial displacement and w is the transverse deflection. Determine the Euler equations by using the principle of total minimum potential energy.

3–5. Use the principle of virtual displacements to obtain equilibrium conditions and natural boundary conditions for the beam structures shown in the figures below (see Figs. E3–E5).

6. The total potential energy of the linear bending of a clamped orthotropic annular plate, neglecting the shear deformation, is given by

$$\Pi(w) = \frac{1}{2} \int_a^b \left[D_{11} \left(\frac{d^2w}{dr^2} \right)^2 + \frac{2D_{12}}{r} \frac{dw}{dr} \frac{d^2w}{dr^2} + D_{22} \left(\frac{1}{r} \frac{dw}{dr} \right)^2 \right] r\, dr$$

$$- \int_a^b wq(r) r\, dr$$

where r is the radial coordinate, a is the internal radius, b is the outside radius, q is the distributed load, and D_{ij} are the material constants. Using the principle of total potential energy, determine the governing differential equation of the plate.

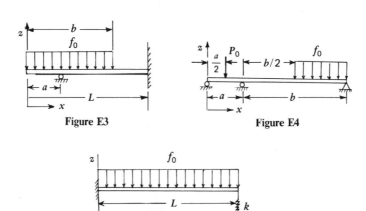

Figure E3   Figure E4

Figure E5

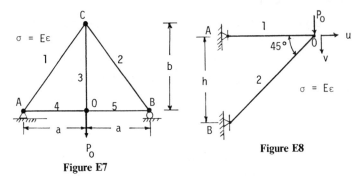

Figure E7

Figure E8

7–8. Use the unit-dummy-displacement method to determine the displacements at point 0 in each of the structures shown in Figs. E7 and E8. Assume that all members are of same material and cross sections.

9. Use the principle of virtual forces to determine the nonlinear strain–displacement equations of an elastic body.

10. Use the principle of virtual forces to determine a relationship between the tip deflection w_0 and deflection w_c at the center of the beam shown in Fig. E10. Use the unit-dummy-load method.

11–12. Use the unit-dummy-load method to determine the displacements at point 0 in each of the structures shown in Figs. E11 and E12.

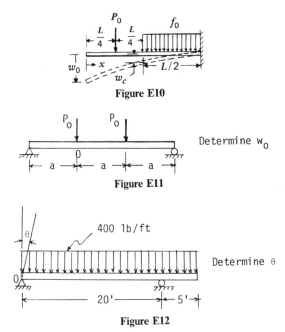

Figure E10

Figure E11

Figure E12

13. Use the total potential energy principle to derive the equilibrium equations of the higher-order beam theory presented in Exercise 1.7.18.

2.4 STATIONARY VARIATIONAL PRINCIPLES

2.4.1 Introduction

The principles of total potential energy and total complementary energy are both extremum principles. Recall from Eq. (2.2.13) that a functional $\Pi(\mathbf{u})$ is a minimum at $\mathbf{u}$ if and only if

$$\Pi(\bar{\mathbf{u}}) \geq \Pi(\mathbf{u}) \quad \text{for all } \bar{\mathbf{u}} \tag{2.4.1}$$

and the equality holds only if $\bar{\mathbf{u}} = \mathbf{u}$. To show that the total potential energy principle of linear elasticity is a minimum principle, let $\mathbf{u}$ be the true displacement field and $\bar{\mathbf{u}}$ be any arbitrary but admissible displacement field. Then $\bar{\mathbf{u}}$ is of the form

$$\bar{\mathbf{u}} = \mathbf{u} + \alpha\mathbf{v} \tag{2.4.2}$$

where $\mathbf{v}$ is a sufficiently differentiable function that satisfies the homogeneous form of the essential boundary conditions. From Eq. (2.2.39), we have

$$\Pi(\mathbf{u} + \alpha\mathbf{v}) = \int_V \left[\tfrac{1}{2}C_{ijkl}(e_{kl} + \alpha g_{kl})(e_{ij} + \alpha g_{ij}) - f_i(u_i + \alpha v_i)\right] dV$$

$$- \int_{S_2} \hat{t}_i(u_i + \alpha v_i)\, dS \tag{2.4.3}$$

where $g_{ij} = \tfrac{1}{2}(v_{i,j} + v_{j,i})$. Collecting the terms, we get (because $C_{ijkl} = C_{klij}$)

$$\Pi(\mathbf{u} + \alpha\mathbf{v}) = \Pi(\mathbf{u}) + \alpha\left[\int_V \left(-f_i v_i + C_{ijkl}e_{kl}g_{ij} + \frac{\alpha}{2}C_{ijkl}g_{ij}g_{kl}\right) dV\right.$$

$$\left. - \int_{S_2} \hat{t}_i v_i\, dS\right] \tag{2.4.4}$$

Using the equilibrium equations of an homogeneous elastic body, we write

$$-\int_V f_i v_i\, dV = \int_V \sigma_{ij,j} v_i\, dV$$

$$= \int_V C_{ijkl}e_{kl,j}v_i\, dV$$

$$= -\int_V C_{ijkl}e_{kl}v_{i,j}\, dV + \int_{S_2} C_{ijkl}e_{kl}v_i n_j\, dS$$

$$= -\int_V C_{ijkl}e_{kl}g_{ij}\, dV + \int_{S_2} \hat{t}_i v_i\, dS \tag{2.4.5}$$

Substituting Eq.(2.4.5) into Eq. (2.4.4), we arrive at

$$\Pi(\bar{\mathbf{u}}) = \Pi(\mathbf{u}) + \frac{\alpha^2}{2} \int_V C_{ijkl} g_{ij} g_{kl} \, dV \qquad (2.4.6)$$

In view of the nonnegative nature of the second term on the right-hand side of Eq. (2.4.6), it follows that

$$\Pi(\bar{\mathbf{u}}) \geq \Pi(\mathbf{u})$$

and $\Pi(\bar{\mathbf{u}}) = \Pi(\mathbf{u})$ only if the quadratic expression $\frac{1}{2}C_{ijkl} g_{ij} g_{kl}$ is zero. Because of the positive-definitness of the strain energy density, the quadratic expression is zero only if $v_i = 0$, which in turn implies that $\bar{u}_i = u_i$. Thus, Eqs. (2.3.32) and (2.3.47) imply that, of all admissible displacement fields, the true one makes the total potential energy a minimum. Therefore, the total potential energy principle is a minimum principle. Similar arguments can be made for the total complementary energy.

It should be noted that the consideration of whether a functional is minimum or maximum does not enter the calculation of the Euler equations. Such consideration is needed in analyzing the stability of structures.

2.4.2 The Hellinger–Reissner Variational Principle

A stationary principle is one in which the functional attains neither a minimum nor a maximum in its arguments. In fact, the functional can attain a maximum with respect to one set of variables (while others are fixed) and a minimum with respect to another set of variables involved in the functional. An example of such functionals is provided by the Lagrange multiplier functional in Eq. (2.2.60). The functional attains a minimum with respect to (w, ψ) and a maximum with respect to λ. Stationary principles, also called *mixed* variational principles, are of special importance in the analysis of structures. Here we consider the *Hellinger–Reissner variational principle* for an elastic body. The stationary functional is constructed, by an inverse procedure, from the equations of an elastic continuum by treating all the dependent variables as independent of each other. The stationary conditions of the principle are the strain–displacement equations, stress–strain equations, stress–equilibrium equations, and both natural and essential boundary conditions—in short, all of the governing equations of elasticity (see Oden and Reddy, 1982). Here we consider the principle for an elastic body undergoing large displacements (static case).

Consider the total potential energy functional Π defined in Eq. (2.3.32). In deriving the total potential energy principle, it was assumed that the displacements and strains are related by a kinematic relationship [i.e., Eq. (1.3.8)] and that the displacement field satisfies the specified boundary conditions on S_1. The principle gives the equilibrium equations and traction boundary conditions as the Euler equations. Now suppose that we wish to include the strain–displacement equations (1.3.8) and displacement boundary conditions (1.6.11) in the variational statement by treating the equations as constraints.

Let λ_{ij} $(i, j = 1, 2, 3)$ and μ_i $(i = 1, 2, 3)$ be the Lagrange multipliers associated with the strain–displacement equations and the displacement boundary conditions, respectively. We have

$$H(u_i, \varepsilon_{ij}, \lambda_{ij}, \mu_i) = \Pi(\varepsilon_{ij}) + \int_V \left[\tfrac{1}{2}(u_{i,j} + u_{j,i} + u_{m,i}u_{m,j}) - \varepsilon_{ij} \right] \lambda_{ij} \, dV$$

$$- \int_{S_1} \mu_i(u_i - \hat{u}_i) \, dS$$

$$= \int_V \left\{ \left[\tfrac{1}{2}(u_{i,j} + u_{j,i} + u_{m,i}u_{m,j}) - \varepsilon_{ij} \right) \right] \lambda_{ij} \right.$$

$$\left. + U_0(\varepsilon_{ij}) - f_i u_i \right\} \, dV - \int_{S_1} \mu_i(u_i - \hat{u}_i) \, dS$$

$$- \int_{S_2} \hat{t}_i u_i \, dS \tag{2.4.7}$$

where $S_1 + S_2 = S$ is the total boundary, $\hat{u}_i$ and $\hat{t}_i$ are prescribed displacements and stresses, respectively, on S_1 and S_2, and U_0 is the strain energy density in terms of the strains.

Setting the first variation of H to zero, we get

$$0 = \delta H \equiv \delta_u H + \delta_\varepsilon H + \delta_\lambda H + \delta_\mu H \tag{2.4.8a}$$

Since the variations in **u**, **ε**, **λ**, and **μ** are arbitrary, it follows that

$$\delta_u H = 0, \; \delta_\varepsilon H = 0, \; \delta_\lambda H = 0, \; \delta_\mu H = 0 \tag{2.4.8b}$$

We have

$$\delta H = \int_V \left[\tfrac{1}{2}(\delta u_{i,j} + \delta u_{j,i} + \delta u_{m,i}u_{m,j} + u_{m,i}\delta u_{m,j})\lambda_{ij} - f_i \delta u_i \right] dV$$

$$- \int_{S_1} \mu_i \delta u_i \, dS - \int_{S_2} \hat{t}_i \delta u_i \, dS + \int_V \left(-\delta \varepsilon_{ij}\lambda_{ij} + \frac{\partial U_0}{\partial \varepsilon_{ij}} \delta \varepsilon_{ij} \right) dV \tag{2.4.9a}$$

$$+ \int_V \left[\tfrac{1}{2}(u_{i,j} + u_{j,i} + u_{m,i}u_{m,j}) - \varepsilon_{ij} \right] \delta \lambda_{ij} \, dV - \int_{S_1} \delta \mu_i (u_i - \hat{u}_i) \, dS$$

$$= \int_V \left[\delta u_{i,j}(\delta_{im} + u_{i,m})\lambda_{mj} - f_i \delta u_i \right] dV - \int_{S_1} \mu_i \delta u_i \, dS - \int_{S_2} \hat{t}_i \delta u_i \, dS$$

$$+ \int_V \left(\frac{\partial U_0}{\partial \varepsilon_{ij}} - \lambda_{ij} \right) \delta \varepsilon_{ij} \, dV$$

$$+ \int_V \left[\tfrac{1}{2}(u_{i,j} + u_{j,i} + u_{m,i}u_{m,j}) - \varepsilon_{ij} \right] \delta \lambda_{ij} \, dV - \int_{S_1} \delta \mu_i (u_i - \hat{u}_i) \, dS$$

$$\tag{2.4.9b}$$

Using the integration by parts, the first integral in Eq. (2.4.9b) can be
expressed as

$$\int_V \delta u_{i,j}(\delta_{im} + u_{i,m})\lambda_{mj}\,dV = -\int_V [\lambda_{mj}(\delta_{im} + u_{i,m})]_{,j}\,\delta u_i\,dV$$

$$+ \oint_S n_j\lambda_{mj}(\delta_{im} + u_{i,m})\delta u_i\,dS \qquad (2.4.10)$$

Substituting Eq. (2.4.10) into Eq. (2.4.9b), and setting the coefficients of δu_i,
$\delta\varepsilon_{ij}$, and $\delta\lambda_{ij}$ in V, δu_i on S_1 and S_2, and $\delta\mu_i$ on S_1 to zero, we obtain

$$\delta u_i: \quad [\lambda_{mj}(\delta_{im} + u_{i,m})]_{,j} + f_i = 0 \quad \text{in } V \qquad (2.4.11a)$$

$$\delta\varepsilon_{ij}: \quad \frac{\partial U_0}{\partial\varepsilon_{ij}} - \lambda_{ij} = 0 \quad \text{in } V \qquad (2.4.11b)$$

$$\delta\lambda_{ij}: \quad \tfrac{1}{2}(u_{i,j} + u_{j,i} + u_{m,i}u_{m,j}) - \varepsilon_{ij} = 0 \quad \text{in } V \qquad (2.4.11c)$$

$$\delta u_i: \quad n_j\lambda_{mj}(\delta_{im} + u_{i,m}) - \mu_i = 0 \quad \text{on } S_1 \qquad (2.4.12)$$

$$\delta u_i: \quad n_j\lambda_{mj}(\delta_{im} + u_{i,m}) - \hat{t}_i = 0 \quad \text{on } S_2 \qquad (2.4.13a)$$

$$\delta\mu_i: \quad u_i - \hat{u}_i = 0 \quad \text{on } S_1 \qquad (2.4.13b)$$

Equations (2.4.11b) and (2.4.12) indicate that the Lagrange multipliers λ_{ij} and
μ_i are equal to stress components and boundary tractions, respectively.

2.4.3   The Reissner Variational Principle

A special case of the Hellinger–Reissner principle is the Reissner principle, in
which strains are eliminated by assuming that the strains are related to the
stresses by

$$\varepsilon_{ij} = \frac{\partial U_0^*}{\partial\sigma_{ij}} \qquad (2.4.14)$$

We get

$$R(u_i, \sigma_{ij}) = \int_B \left[\tfrac{1}{2}(u_{i,j} + u_{j,i} + u_{m,i}u_{m,j})\sigma_{ij} - U_0^*(\sigma_{ij}) - f_i u_i\right]dV$$

$$- \int_{S_1} t_i(u_i - \hat{u}_i)\,dS - \int_{S_2} \hat{t}_i u_i\,dS \qquad (2.4.15)$$

The Euler equations are given by Eqs. (2.4.11a), (2.4.12), (2.4.13), and

$$\delta\sigma_{ij}: \quad \tfrac{1}{2}(u_{i,j} + u_{j,i} + u_{m,i}u_{m,j}) - \frac{\partial U_0^*}{\partial\sigma_{ij}} = 0 \quad \text{in } V \qquad (2.4.16)$$

The advantage of stationary principles lies in incorporating all of the governing equations into a single functional. Such functionals form the basis of a variety of new finite-element models, known as the mixed and hybrid models, that are believed to yield better accuracies for stresses than the displacement finite-element models (which are based on the total potential energy functional).

Example 2.14

Consider the equations of the classical theory of beams.

$$\textit{Kinematics:} \quad \kappa = \frac{d^2w}{dx^2}, \qquad w(0) = w_0 \qquad \frac{dw}{dx}(0) = \theta_0$$

$$\textit{Constitutive equation:} \quad M = EI\kappa$$

$$\textit{Kinetics (equilibrium):} \quad \frac{d^2M}{dx^2} + f = 0, \qquad M(L) = M_L, \qquad \frac{dM}{dx}(L) = V_L$$

$$(2.4.17)$$

The Hellinger–Reissner type variational principle for beams is given by

$$H(w, \kappa, \lambda, \mu_1, \mu_2) = \Pi(\kappa) + \int_0^L \left(\frac{d^2w}{dx^2} - \kappa\right)\lambda \, dx + \mu_1[w(0) - w_0]$$

$$+\mu_2\left[\frac{dw}{dx}(0) - \theta_0\right] \qquad (2.4.18)$$

where $\Pi(\kappa)$ is the total potential energy functional expressed in terms of the curvature κ. The functional Π accounts for the moment and shear force boundary conditions because they are the natural boundary conditions for Π. We have

$$H(w, \kappa, \lambda, \mu_1, \mu_2) = \int_0^L \left[\frac{EI}{2}\kappa^2 + \left(\frac{d^2w}{dx^2} - \kappa\right)\lambda + fw\right] dx$$

$$- M_L\frac{dw}{dx}(L) - V_L w(L) + \mu_1[w(0) - w_0]$$

$$+\mu_2\left[\frac{dw}{dx}(0) - \theta_0\right] \qquad (2.4.19)$$

The Euler equations of this functional are

$$\delta\kappa: \quad EI\kappa - \lambda = 0$$

$$\delta w: \quad \frac{d^2\lambda}{dx^2} + f = 0, \quad \text{in } 0 < x < L \qquad (2.4.20a)$$

$$\delta\lambda: \quad \frac{d^2w}{dx^2} - \kappa = 0$$

$$\frac{d\delta w}{dx}: \quad \lambda(0) = \mu_2, \qquad \lambda(L) = M_L$$

$$\delta w: \quad \frac{d\lambda}{dx}(0) = -\mu_1, \qquad \frac{d\lambda}{dx}(L) = V_L$$

$$\delta\mu_1: \quad w(0) = w_0 \qquad (2.4.20b)$$

$$\delta\mu_2: \quad \frac{dw}{dx}(0) = \theta_0$$

From a comparison of Eqs. (2.4.20) with (2.4.17), we note that the Lagrange multipliers λ, μ_1, and μ_2 are given by

$$\lambda(x) = M(x), \qquad \mu_1 = -\frac{dM}{dx}(0), \qquad \mu_2 = M(0) \qquad (2.4.21)$$

Exercises 2.4

Derive the Euler equations of the stationary functionals in Exercises 1–4.

1. $I(u, \mathbf{v}) = \int_V \left(-\frac{1}{2k}\mathbf{v} \cdot \mathbf{v} + \mathbf{v} \cdot \operatorname{grad} u - fu \right) dV$

    $\qquad - \int_{S_2} gu \, dS - \int_{S_1} \mathbf{n} \cdot \mathbf{v}(u - \hat{u}) \, dS$

2. $I(w, \psi, \lambda) = \int_0^L \left[\frac{EI}{2}\left(\frac{d\psi}{dx}\right)^2 + fw + \lambda\left(\frac{dw}{dx} + \psi\right) \right] dx - V_0 w(L)$

3. $I(u_1, u_2, P) = \int_R \left[\frac{\mu}{2}(u_{i,j} + u_{j,i})u_{i,j} - Pu_{i,i} \right] dx_1 \, dx_2 - \int_{S_2} \hat{t}_i u_i \, dS$

4. $I(w, \psi_1, \psi_2, \lambda_1, \lambda_2)$

$$= \frac{D}{2} \int_R \left[\left(\frac{\partial \psi_1}{\partial x_1} \right)^2 + \left(\frac{\partial \psi_2}{\partial x_2} \right)^2 + 2\nu \frac{\partial \psi_1}{\partial x_1} \frac{\partial \psi_2}{\partial x_2} \right.$$

$$+ (1-\nu) \left(\frac{\partial \psi_1}{\partial x_2} + \frac{\partial \psi_2}{\partial x_1} \right)^2 \Bigg] dx_1 \, dx_2$$

$$+ \int_R \left[\lambda_1 \left(\frac{\partial w}{\partial x_1} + \psi_1 \right) + \lambda_2 \left(\frac{\partial w}{\partial x_2} + \psi_2 \right) - fw \right] dx_1 \, dx_2$$

5. Derive the Hellinger–Reissner functional for the cantilevered beam problem of Example 2.5.

6. Derive the stationary functional corresponding to the problem of minimizing the functional

$$I(u_i) = \int_V \left[\tfrac{1}{2}(u_{i,j} + u_{j,i})u_{i,j} - f_i u_i \right] dV - \int_{S_2} \hat{t}_i u_i \, dS$$

subject to the constraint $u_{i,i} = 0$.

7. *Washizu's Principle.* Derive the following generalized Washizu functional,

$$W(u_i, e_{ij}, \sigma_{ij}) = \int_V \left\{ \left[\tfrac{1}{2}(u_{i,j} + u_{j,i} + u_{m,i}u_{m,j}) - e_{ij} \right] \sigma_{ij} \right.$$

$$+ \rho \Psi - f_i u_i \right\} dV - \int_{S_1} (u_i - \hat{u}_i) t_i \, dS$$

$$- \int_{S_2} \hat{t}_i u_i \, dS$$

where e_{ij}, σ_{ij}, u_i, and t_i are subject to independent variation, and $\Psi = \Psi(e_{ij}, T)$ is a free-energy function.

8. Derive the Euler equations of the Washizu functional by requiring that the actual state of equilibrium of a body subjected to mechanical and thermal loads is associated with a stationary value of the functional W in Exercise 7.

9. Modify the Hellinger–Reissner variational principle for elastic bodies acted upon by mechanical and thermal loads.

2.5 HAMILTON'S PRINCIPLE

2.5.1 Introduction

The principles of virtual work and their special cases are limited to static equilibrium of solids. Hamilton's principle is a generalization of the principle of virtual displacements to dynamics of systems of particles, rigid bodies, or deformable bodies. The principle assumes that the system under consideration is characterized by two energy functions: a *kinetic energy K* and a *potential energy V*. For *discrete* systems (i.e., systems with a finite number of degrees of freedom), these energies can be described in terms of a finite number of generalized coordinates and their derivatives, with respect to time *t*. For *continuous* systems (i.e., systems that cannot be described by a finite number of generalized coordinates), the energies can be expressed in terms of the dependent variables (which are functions of position) of the problem. Hamilton's principle becomes the principle of virtual displacements for systems that are in static equilibrium.

The present section is devoted to the application of Hamilton's principle to dynamics of discrete systems and continuous systems. We begin the discussion with Hamilton's principle for particles.

2.5.2 Hamilton's Principle for Particles and Rigid Bodies

Consider a single particle of mass *m* moving under the influence of a force $\mathbf{F} = \mathbf{F}(\mathbf{r})$. The path $\mathbf{r}(t)$ followed by the particle is related to the force $\mathbf{F}$ and mass *m* by Newton's second law of motion,

$$m\frac{d^2\mathbf{r}}{dt^2} - \mathbf{F}(\mathbf{r}) = \mathbf{0} \tag{2.5.1}$$

A path that differs from the actual path is expressed as $\mathbf{r} + \delta\mathbf{r}$, where $\delta\mathbf{r}$ is the variation of the path for any *fixed* time *t*. We suppose that the actual path $\mathbf{r}$ and the *varied* path differ, except at two distinct times t_1 and t_2; that is, $\delta\mathbf{r}(t_1) = \delta\mathbf{r}(t_2) = \mathbf{0}$. Taking the scalar product of Eq. (2.5.1) with the variation $\delta\mathbf{r}$, and integrating with respect to time between t_1 and t_2, we obtain

$$\int_{t_1}^{t_2}\left[m\frac{d^2\mathbf{r}}{dt^2} - \mathbf{F}(\mathbf{r})\right]\cdot\delta\mathbf{r}\,dt = 0 \tag{2.5.2}$$

Integration by parts of the first term in Eq. (2.5.2) yields

$$-\int_{t_1}^{t_2}\left(m\frac{d\mathbf{r}}{dt}\cdot\frac{d\delta\mathbf{r}}{dt} + \mathbf{F}(\mathbf{r})\cdot\delta\mathbf{r}\right)dt + \left(m\frac{d\mathbf{r}}{dt}\cdot\delta\mathbf{r}\right)\Bigg|_{t_1}^{t_2} = 0 \tag{2.5.3}$$

The last term in Eq. (2.5.3) vanishes because $\delta\mathbf{r}(t_1) = \delta\mathbf{r}(t_2) = \mathbf{0}$. Also, note that

$$m\frac{d\mathbf{r}}{dt} \cdot \frac{d\delta\mathbf{r}}{dt} = \delta\left[\frac{m}{2}\frac{d\mathbf{r}}{dt} \cdot \frac{d\mathbf{r}}{dt}\right] \equiv \delta K \qquad (2.5.4)$$

where K is the kinetic energy of the particle

$$K = \frac{m}{2}\frac{d\mathbf{r}}{dt} \cdot \frac{d\mathbf{r}}{dt} \qquad (2.5.5)$$

Equation (2.5.3) now takes the form

$$\int_{t_1}^{t_2}(\delta K + \mathbf{F} \cdot \delta\mathbf{r})\, dt = 0 \qquad (2.5.6)$$

which is known as the *general form of Hamilton's principle* for a single particle.

Suppose that the force $\mathbf{F}$ is conservative (that is, the sum of the potential and kinetic energies is conserved) such that it can be replaced by the gradient of a potential,

$$\mathbf{F} = -\operatorname{grad}V \qquad (2.5.7)$$

where $V = V(\mathbf{r})$ is the potential energy of the particle. Then Eq. (2.5.6) can be expressed in the form

$$\delta\int_{t_1}^{t_2}(K - V)\, dt = 0 \qquad (2.5.8)$$

because

$$\operatorname{grad}V \cdot \delta\mathbf{r} = \frac{\partial V}{\partial x_1}\delta x_1 + \frac{\partial V}{\partial x_2}\delta x_2 + \frac{\partial V}{\partial x_3}\delta x_3 = \delta V(x_1, x_2, x_3)$$

The difference between the kinetic and potential energies is called the *Lagrangian* function:

$$L \equiv K - V \qquad (2.5.9)$$

Equation (2.5.8) represents Hamilton's principle for the conservative motion of a particle. It states that *the motion of a particle acted on by conservative forces between two arbitrary instants of time t_1 and t_2 is such that the line integral over the Lagrangian function is an extremum for the path motion.* Stated in other words, of all possible paths that the particle could travel from its position at time t_1 to its position at time t_2, its actual path will be one for which the integral

$$I \equiv \int_{t_1}^{t_2}L\, dt \qquad (2.5.10)$$

is an extremum (i.e., a minimum, a maximum, or an inflection).

If the path $\mathbf{r}$ can be expressed in terms of the generalized coordinates q_i ($i = 1, 2, 3$), the Lagrangian can be written in terms of q_i and their time derivatives.

$$L = L(q_1, q_2, q_3, \dot{q}_1, \dot{q}_2, \dot{q}_3) \qquad (2.5.11)$$

Then recall from Section 2.2 that the condition for the extremum of I results in the equation

$$\delta I = \delta \int_{t_1}^{t_2} L(q_1, q_2, q_3, \dot{q}_1, \dot{q}_2, \dot{q}_3)\, dt = 0$$

$$= \int_{t_1}^{t_2} \sum_{i=1}^{3} \left[\frac{\partial L}{\partial q_i} - \frac{d}{dt}\left(\frac{\partial L}{\partial \dot{q}_i} \right) \right] \delta q_i\, dt \qquad (2.5.12)$$

When all q_i are linearly independent (i.e., no constraints among q_i), the variations δq_i are independent for all t, except $\delta q_i = 0$ at t_1 and t_2. Therefore, the coefficients of δq_1, δq_2, and δq_3 vanish separately:

$$\frac{\partial L}{\partial q_i} - \frac{d}{dt}\left(\frac{\partial L}{\partial \dot{q}_i} \right) = 0, \qquad i = 1, 2, 3 \qquad (2.5.13)$$

These equations are called the *Lagrange equations of motion*. In Section 2.2, these equations were also called the Euler equations. In the discussions to follow, we refer to them as the Euler–Lagrange equations.

When the forces are not conservative, we must deal with the general form of Hamilton's principle in Eq. (2.5.6). Recognizing that $\mathbf{F} \cdot \delta \mathbf{r}$ represents the virtual work δW owing to forces $\mathbf{F}$, we can write Eq. (2.5.6) in the form

$$\delta \int_{t_1}^{t_2} K\, dt + \int_{t_1}^{t_2} \delta W\, dt = 0 \qquad (2.5.14)$$

In this case, there exists no functional I that must be an extremum. If the virtual work can be expressed in terms of the generalized coordinates q_i by

$$\delta W = Q_1\, \delta q_1 + Q_2\, \delta q_2 + Q_3\, \delta q_3 \qquad (2.5.15)$$

where Q_i are the *generalized forces*, then we can write Eq. (2.5.14) as

$$\int_{t_1}^{t_2} \sum_{i=1}^{3} \left[\frac{\partial K}{\partial q_i} - \frac{d}{dt}\left(\frac{\partial K}{\partial \dot{q}_i} \right) + Q_i \right] \delta q_i\, dt = 0 \qquad (2.5.16)$$

and the Euler–Lagrange equations for the nonconservative forces are given by

$$\frac{\partial K}{\partial q_i} - \frac{d}{dt}\left(\frac{\partial K}{\partial \dot{q}_i} \right) + Q_i = 0, \qquad i = 1, 2, 3 \qquad (2.5.17)$$

Example 2.15

Consider the motion of a pendulum in a plane. Ideally, the pendulum is imagined to be a mass m attached at the end of a rigid massless rod of length l that pivots about a fixed point 0, as shown in Fig. 2.11. We choose $q_1 = l$ and $q_2 = \theta$ as the generalized coordinates (polar coordinates). But $q_1 = l$ is fixed (constraint condition), we have

$$\dot{q}_1 = 0$$

and $q_2 = \theta$ is the only independent generalized coordinate. The coordinate θ is measured from the vertical position. The force $\mathbf{F}$ acting on the bob is the component of the gravitational force,

$$\mathbf{F} = -mg(\sin\theta\,\hat{\mathbf{e}}_\theta + \cos\theta\,\hat{\mathbf{e}}_r) \qquad (2.5.18a)$$

The component along $\hat{\mathbf{e}}_r$ does no work (because $l = $ constant). The first component is derivable from the potential

$$V = mgl(1 - \cos\theta)$$

The gradient operator in the present case is given by ($r = l$)

$$\mathrm{grad} = \hat{\mathbf{e}}_r\frac{\partial}{\partial r} + \frac{\hat{\mathbf{e}}_\theta}{r}\frac{\partial}{\partial\theta}$$

Thus, the kinetic and potential energies are given by

$$K = \frac{m}{2}(l\dot{\theta})^2, \qquad V = mgl(1 - \cos\theta) \qquad (2.5.18b)$$

Therefore, the Lagrangian function L is a function of θ and $\dot{\theta}$. The

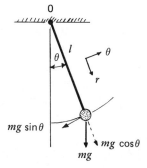

Figure 2.11   Configuration for the pendulum.

Euler–Lagrange equation is given by

$$\frac{\partial L}{\partial \theta} - \frac{d}{dt}\left(\frac{\partial L}{\partial \dot\theta}\right) = 0$$

$$-mgl\sin\theta - \frac{d}{dt}(ml^2\dot\theta) = 0$$

$$\ddot\theta + \frac{g}{l}\sin\theta = 0 \qquad (2.5.19)$$

Now suppose that the mass experiences a resistance force **F*** proportional to its speed, according to Stoke's law,

$$\mathbf{F^*} = -6\pi\mu al\dot\theta\hat{\mathbf{e}}_\theta \equiv Q\dot\theta\hat{\mathbf{e}}_\theta \qquad (2.5.20)$$

where μ is the viscosity of the surrounding medium, a is the radius of the pendulum bob, and $\hat{\mathbf{e}}_\theta$ is the unit vector tangential to the circular path. The resistance of the massless rod supporting the bob is neglected. The force **F*** is not derivable from a potential function. Thus, one part of the force is conservative and the other is nonconservative. The conservative part is included in V and the nonconservative part is included in Q. We have

$$\frac{\partial L}{\partial \theta} - \frac{d}{dt}\left(\frac{\partial L}{\partial \dot\theta}\right) + Q = 0$$

$$\ddot\theta + \frac{g}{l}\sin\theta + \frac{6\pi a\mu}{m}\dot\theta = 0 \qquad (2.5.21)$$

The coefficient $c = 6\pi a\mu/m$ is called the *damping coefficient*. This completes the example.

Hamilton's principle in Eqs. (2.5.8) and (2.5.14) can be easily extended to a system of N particles, hence to rigid bodies. We have

$$\delta\int_{t_1}^{t_2}L(q_i,\dot q_i)\,dt = 0 \quad \text{for conservative forces} \qquad (2.5.22a)$$

$$\delta\int_{t_1}^{t_2}K\,dt + \int_{t_1}^{t_2}\delta W\,dt = 0 \quad \text{for nonconservative forces} \qquad (2.5.22b)$$

where

$$K = \sum_{i=1}^{N}\frac{m_i}{2}\frac{d\mathbf{r}_i}{dt}\cdot\frac{d\mathbf{r}_i}{dt}, \qquad \delta W = \sum_{i=1}^{N}\mathbf{F}_i\cdot\delta\mathbf{r}_i \qquad (2.5.23)$$

and m_i and $\mathbf{r}_i$ denote the mass and path of the ith particle. Of course, Eqs. (2.5.22) and (2.5.23) apply to a rigid body, because a rigid body can be viewed as a set of particles with fixed positions $\mathbf{r}_i$.

2.5.3   Hamilton's Principle for a Continuum

Following essentially the same procedure as that used for discrete systems, we can develop Hamilton's principle for the dynamics of deformable bodies. Newton's second law of motion also applies to deformable bodies; it expresses the global statement of the principle of conservation of linear momentum. However, it should be noted that Newton's second law of motion for continuous media is not sufficient to determine its motion $\mathbf{u} = \mathbf{u}(\mathbf{x}, t)$; the kinematic conditions and constitutive equations derived in Chapter 1 are needed to completely determine the motion.

Newton's law of motion for a continuous body is given by

$$\mathbf{F} - m\mathbf{a} = \mathbf{0} \qquad (2.5.24)$$

where m is the mass, $\mathbf{a}$ is the acceleration vector, and $\mathbf{F}$ is the resultant of *all* forces acting on the body. The actual path $\mathbf{u} = \mathbf{u}(\mathbf{x}, t)$ followed by a material particle in position $\mathbf{x}$ in the body is varied, consistent with kinematic (essential) boundary conditions, to $\mathbf{u} + \delta\mathbf{u}$, where $\delta\mathbf{u}$ is the admissible variation (or virtual displacement) of the path. We suppose that the varied path differs from the actual path, except at initial and final times, t_1 and t_2, respectively. Thus, an admissible variation $\delta\mathbf{u}$ satisfies the conditions,

$$\delta\mathbf{u} = 0 \quad \text{on } S_1 \text{ for all } t$$

$$\delta\mathbf{u}(\mathbf{x}, t_1) = \delta\mathbf{u}(\mathbf{x}, t_2) = 0 \quad \text{for all } \mathbf{x} \qquad (2.5.25)$$

where S_1 denotes the portion of the surface of the body where the displacement vector $\mathbf{u}$ is specified. Note that the scalar multiplication of Eq. (2.5.24) with $\delta\mathbf{u}$ gives work done at point $\mathbf{x}$, because $\mathbf{F}$, $\mathbf{a}$, and $\mathbf{u}$ are functions of position (whereas the work done should be a number). Integration of the product over the volume (and surface) of the body gives the work done on all points (i.e., total work done).

The *work done on the body* at time t by the resultant force in moving through the virtual displacement $\delta\mathbf{u}$ is given by

$$\int_V \mathbf{f} \cdot \delta\mathbf{u} \, dV + \int_{S_2} \mathbf{t} \cdot \delta\mathbf{u} \, dS - \int_V \boldsymbol{\sigma} : \delta\boldsymbol{\varepsilon} \, dV \qquad (2.5.26)$$

where $\mathbf{f}$ is the body force vector, $\mathbf{t}$ is the specified surface stress vector, and $\boldsymbol{\sigma}$ and $\boldsymbol{\varepsilon}$ are the stress and strain tensors. The last term in Eq. (2.5.26) represents the *virtual work* of internal forces *stored in the body* (hence, it is negative). The strains $\delta\boldsymbol{\varepsilon}$ are assumed to be compatible in the sense that the strain–displacement relations (1.3.8) are satisfied. The work done by the inertia force $m\mathbf{a}$ in moving through the virtual displacement $\delta\mathbf{u}$ is given by

$$\int_V \rho \frac{\partial^2 \mathbf{u}}{\partial t^2} \cdot \delta\mathbf{u} \, dV \tag{2.5.27}$$

where ρ is the mass density (can be a function of position) of the medium. We have from Eq. (2.5.24), analogous to Eq. (2.5.1) for discrete systems,

$$\int_{t_1}^{t_2}\left[\int_V \rho \frac{\partial^2 \mathbf{u}}{\partial t^2} \cdot \delta\mathbf{u}\,dV - \left(\int_V \mathbf{f}\cdot\delta\mathbf{u}\,dV + \int_{S_2}\mathbf{t}\cdot\delta\mathbf{u}\,dS - \int_V \boldsymbol{\sigma}:\delta\boldsymbol{\varepsilon}\,dV\right)\right]dt = 0$$

$$-\int_{t_1}^{t_2}\left[\int_V \rho \frac{\partial \mathbf{u}}{\partial t} \cdot \frac{\partial\,\delta\mathbf{u}}{\partial t}\,dV + \int_V \mathbf{f}\cdot\delta\mathbf{u}\,dV + \int_{S_2}\mathbf{t}\cdot\delta\mathbf{u}\,dS - \int_V \boldsymbol{\sigma}:\delta\boldsymbol{\varepsilon}\,dV\right]dt = 0$$

$$\tag{2.5.28}$$

To arrive at the last expression, integration by parts is used on the first term; the integrated terms vanish because of the initial and final conditions in Eq. (2.5.25). Equation (2.5.28) is known as the general form of Hamilton's principle for a continuous medium (conservative or not and elastic or not).

For an ideal elastic body, we recall from the previous discussions that the forces $\mathbf{f}$ and $\mathbf{t}$ are conservative, and that a strain energy density function U_0 exists such that

$$\delta V = -\int_V \mathbf{f}\cdot\delta\mathbf{u}\,dV - \int_{S_2}\mathbf{t}\cdot\delta\mathbf{u}\,dS$$

$$\sigma_{ij} = \frac{\partial U_0}{\partial \varepsilon_{ij}}, \qquad U_0 = U_0(\varepsilon_{ij}) \tag{2.5.29}$$

Substituting Eq. (2.5.29) into Eq. (2.5.28), we obtain

$$\delta\int_{t_1}^{t_2}[K - (V + U)]\,dt = 0 \tag{2.5.30}$$

where K and U are the kinetic and strain energies:

$$K = \int_V \frac{\rho}{2}\frac{\partial \mathbf{u}}{\partial t}\cdot\frac{\partial \mathbf{u}}{\partial t}\,dV, \qquad U = \int_V U_0\,dV \tag{2.5.31}$$

Equation (2.5.30) represents Hamilton's principle for an elastic body (linear or nonlinear). Recall that the sum of the strain energy and potential energy of external forces, $U + V$, is called the total potential energy, Π, of the body. For bodies involving no motion (i.e., forces are applied sufficiently slowly that the motion is independent of time and the inertia forces are negligible), Hamilton's principle (2.5.30) reduces to the principle of virtual displacements (2.3.32).

The Euler–Lagrange equations associated with the Lagrangian, $L = K - \Pi$, can be obtained from Eq. (2.5.30):

$$0 = \delta \int_{t_1}^{t_2} L(u_i, \dot{u}_i)\, dt$$

$$= \int_{t_1}^{t_2} \left[\int_V \left(\rho \frac{\partial^2 \mathbf{u}}{\partial t^2} - \operatorname{div} \boldsymbol{\sigma} - \mathbf{f} \right) \cdot \delta \mathbf{u}\, dV \right.$$

$$\left. + \int_{S_2} (\mathbf{t} - \hat{\mathbf{t}}) \cdot \delta \mathbf{u}\, dS \right] dt \qquad (2.5.32)$$

where integration by parts, gradient theorems, and Eq. (2.5.25) were used in arriving at Eq. (2.5.32) from Eq. (2.5.30). Because $\delta \mathbf{u}$ is arbitrary for t $(t_1 < t < t_2)$, for $\mathbf{x}$ in V, and also on S_2, it follows that

$$\rho \frac{\partial^2 \mathbf{u}}{\partial t^2} - \operatorname{div} \boldsymbol{\sigma} - \mathbf{f} = 0 \quad \text{in } V$$

$$\mathbf{t} - \hat{\mathbf{t}} = 0 \quad \text{on } S_2 \qquad (2.5.33)$$

Equations (2.5.33) are the Euler–Lagrange equations of an elastic body.

Using the results of this section and the previous sections, one can derive statements of Hamilton's principle corresponding to the principle of virtual forces (or complementary potential energy) and stationary principles. These are included as exercises for the reader.

Example 2.16

Consider the axial motion of an elastic bar of length L, area of cross section A, modulus of elasticity E, and mass density ρ, which subjected to distributed force f/unit length and an end load P. We wish to determine the equations of motion for the bar. The kinetic and total potential energies of the system are

$$K = \int_V \frac{\rho}{2} \left(\frac{\partial u}{\partial t} \right)^2 dV = \int_0^L \frac{\rho A}{2} \left(\frac{\partial u}{\partial t} \right)^2 dx$$

$$\Pi = \int_V \frac{1}{2} \sigma_{ij} \varepsilon_{ij}\, dV - \int_0^L fu\, dx - Pu(L)$$

$$= \int_0^L \frac{A}{2} \sigma \varepsilon\, dx - \int_0^L fu\, dx - Pu(L) \qquad (2.5.34)$$

wherein u, σ, and ε are assumed to be functions of x only, and

$$u(0, t) = 0 \quad \text{(bar is fixed at } x = 0\text{)}$$

$$\varepsilon = \frac{\partial u}{\partial x} \quad \text{(strain–displacement relation)} \qquad (2.5.35)$$

Substituting for K and Π from Eq. (2.5.34) into Eq. (2.5.30), we obtain

$$0 = \int_{t_1}^{t_2} \left\{ \int_0^L \left[A\rho \frac{\partial u}{\partial t} \frac{\partial \delta u}{\partial t} - A\sigma\, \delta\!\left(\frac{\partial u}{\partial x}\right) + f\delta u \right] dx + P\delta u(L) \right\} dt$$

$$= \int_0^L \left[\int_{t_1}^{t_2} - \frac{\partial}{\partial t}\!\left(\rho A \frac{\partial u}{\partial t}\right) \delta u\, dt + \rho A \frac{\partial u}{\partial t}\delta u \Big|_{t_1}^{t_2} \right] dx$$

$$+ \int_{t_1}^{t_2}\left\{ \int_0^L \left[\frac{\partial}{\partial x}(A\sigma) + f\right]\delta u\, dx - (A\sigma\delta u)|_0^L + P\delta u(L) \right\} dt$$

$$= -\int_{t_1}^{t_2}\left\{ \int_0^L \left[\frac{\partial}{\partial t}\!\left(\rho A \frac{\partial u}{\partial t}\right) - \frac{\partial}{\partial x}(A\sigma) - f\right]\delta u\, dx$$

$$- (A\sigma - P)|_{x=L}\delta u(L) \right\} dt \qquad (2.5.36)$$

where $\delta u(0, t) = 0$ and $\delta u(x, t_1) = \delta u(x, t_2) = 0$ are used to simplify the expression. The Euler–Lagrange equations are obtained by setting the coefficients of δu in $(0, L)$ and at $x = L$ to zero separately:

$$\frac{\partial}{\partial t}\!\left(\rho A \frac{\partial u}{\partial t}\right) - \frac{\partial}{\partial x}(A\sigma) - f = 0, \quad 0 < x < L$$

$$(A\sigma)|_{x=L} - P = 0 \qquad (2.5.37)$$

for all t, $t_1 < t < t_2$. For linear elastic materials, we have

$$\sigma = E\varepsilon = E\frac{\partial u}{\partial x} \qquad (2.5.38)$$

and Eq. (2.5.37) becomes

$$\frac{\partial}{\partial t}\!\left(\rho A \frac{\partial u}{\partial t}\right) - \frac{\partial}{\partial x}\!\left(EA \frac{\partial u}{\partial x}\right) - f = 0, \quad 0 < x < L$$

$$\left(AE\frac{\partial u}{\partial x}\right)\Big|_{x=L} - P = 0 \qquad (2.5.39)$$

Now suppose that the bar also experiences a nonconservative (viscous damping) force proportional to the velocity,

$$F^* = -\mu \frac{\partial u}{\partial t} \qquad (2.5.40)$$

where μ is the damping coefficient (a constant). Then the Euler–Lagrange equations from Eq. (2.5.28) are given by

$$\frac{\partial}{\partial t}\left(\rho A \frac{\partial u}{\partial t}\right) - \frac{\partial}{\partial x}\left(EA \frac{\partial u}{\partial x}\right) - f + \mu \frac{\partial u}{\partial t} = 0, \qquad 0 < x < L$$

$$\left(AE \frac{\partial u}{\partial x}\right)\bigg|_{x=L} - P = 0 \qquad (2.5.41)$$

This completes the example.

2.5.4 Hamilton's Principle for Constrained Systems

The kinematic relations that restrict a motion are called *constraints*. When the relations are between generalized coordinates, and possibly between position and time, in the form

$$G(t, q) = 0, \qquad \text{for a discrete system}$$
$$G(\mathbf{x}, t, \mathbf{u}, \text{grad } u) = 0, \quad \text{for a continuous system} \qquad (2.5.42)$$

they are called *holonomic constraints*. When the constraints represent an inequality or relate q_i to $\dot{q}_i$, they are called *nonholonomic constraints*.

As discussed in Section 2.2.7, constraints can be included in the Lagrangian function either by means of the Lagrange multipliers or by the penalty function method. To fix the ideas, we consider the motion of a single particle whose motion is constrained by holonomic constraints:

$$G_1(t, q_1, q_2, q_3) = 0$$

$$G_2(t, q_1, q_2, q_3) = 0 \qquad (2.5.43)$$

The variation of these equations yields

$$\frac{\partial G_1}{\partial q_1}\delta q_1 + \frac{\partial G_1}{\partial q_2}\delta q_2 + \frac{\partial G_1}{\partial q_3}\delta q_3 = 0$$

$$\frac{\partial G_2}{\partial q_1}\delta q_1 + \frac{\partial G_2}{\partial q_2}\delta q_2 + \frac{\partial G_2}{\partial q_3}\delta q_3 = 0 \qquad (2.5.44)$$

Since there are two constraints and three degrees of freedom, only one of the variations δq_1, δq_2, and δq_3 is independent and the other two are related to the independent one by Eq. (2.5.44). We multiply the first part of Eq. (2.5.44) by the Lagrange multiplier λ_1 and the second part by λ_2, integrate each equation over t_1 to t_2, and add the results to Eq. (2.5.16):

$$\int_{t_1}^{t_2} \sum_{i=1}^{3} \left[\frac{\partial K}{\partial q_i} - \frac{d}{dt}\left(\frac{\partial K}{\partial \dot{q}_i} \right) + Q_i + \lambda_1 \frac{\partial G_1}{\partial q_i} + \lambda_2 \frac{\partial G_2}{\partial q_i} \right] \delta q_i \, dt = 0$$

$$(2.5.45)$$

Since the variations δq_1, δq_2, and δq_3 are not all independent, we cannot use the usual argument to set the coefficients of δq_i ($i = 1, 2, 3$) to zero. However, the Lagrange multipliers λ_1 and λ_2 are arbitrary. Therefore, we choose λ_1 and λ_2 such that the coefficients of two of the variations out of δq_1, δq_2, and δq_3 vanish. Since the remaining variation is linearly independent, its coefficient should be zero. Thus, we obtain

$$\frac{\partial K}{\partial q_i} - \frac{d}{dt}\left(\frac{\partial K}{\partial \dot{q}_i} \right) + Q_i + \lambda_1 \frac{\partial G_1}{\partial q_i} + \lambda_2 \frac{\partial G_2}{\partial q_i} = 0 \qquad (2.5.46)$$

for $i = 1, 2, 3$. Equations (2.5.46) together with (2.5.43) provide five equations for the five unknowns ($q_1, q_2, q_3, \lambda_1, \lambda_2$).

In the penalty function method, Hamilton's principle in Eq. (2.5.14) is modified to read

$$\delta \int_{t_1}^{t_2} K_p \, dt + \int_{t_1}^{t_2} \delta W \, dt + \delta \int_{t_1}^{t_2} \left(\frac{\gamma_1}{2} G_1^2 + \frac{\gamma_2}{2} G_2^2 \right) dt = 0 \qquad (2.5.47)$$

where γ_1 and γ_2 are the penalty parameters and K_p is the kinetic energy of the constrained system. Since the constraint conditions (2.5.43) are now included, in the least-squares sense, we suppose that the variations $\delta q_1, \delta q_2, \delta q_3$ are independent of each other. Performing the variation indicated in Eq. (2.5.47), we obtain (after the usual argument) the Euler–Lagrange equations

$$\frac{\partial K_p}{\partial q_i} - \frac{d}{dt}\left(\frac{\partial K_p}{\partial \dot{q}_i} \right) + Q_i + \gamma_1 G_1 \frac{\partial G_1}{\partial q_i} + \gamma_2 G_2 \frac{\partial G_2}{\partial q_i} = 0 \qquad (2.5.48)$$

A comparison of Eq. (2.5.48) with Eq. (2.5.46) shows that the Lagrange multipliers λ_1 and λ_2 can be computed from

$$\lambda_1 = \gamma_1 G_1, \qquad \lambda_2 = \gamma_2 G_2 \qquad (2.5.49)$$

where G_1 and G_2 are given by Eq. (2.5.43) for the generalized coordinates q_i computed from Eq. (2.5.48); G_1 and G_2 thus computed are not identically zero, because q_i computed from Eq. (2.5.48) are different from the true values, and the error is inversely proportional to the penalty parameters γ_1 and γ_2.

Example 2.17

To illustrate the ideas presented in the preceding paragraphs, we reconsider the damped motion of a pendulum of Example 2.15. The generalized coordinates are

$$q_1 = r, \qquad q_2 = \theta \tag{2.5.50a}$$

and the constraint is

$$G_1(q_1) \equiv q_1 - l = 0 \tag{2.5.50b}$$

The work done by external forces in moving through virtual displacement $\delta\mathbf{q}$ is given by

$$\delta W = \mathbf{F} \cdot \delta\mathbf{q} = \left[mg(-\sin\theta\hat{\mathbf{e}}_\theta + \cos\theta\hat{\mathbf{e}}_r) - 6\mu ar\dot{\theta}\hat{\mathbf{e}}_\theta \right] \cdot (\delta q_1\hat{\mathbf{e}}_r + r\delta q_2\hat{\mathbf{e}}_\theta)$$

$$= mg\cos\theta\,\delta q_1 - r(mg\sin\theta + 6\pi\mu ar\dot{\theta})\delta q_2$$

$$\equiv Q_1\,\delta q_1 + Q_2\,\delta q_2 \tag{2.5.51}$$

The kinetic energy is given by

$$K = \frac{m}{2}\left[(\dot{r})^2 + (l\dot{\theta})^2 \right]$$

$$= \frac{m}{2}\left(\dot{q}_1^2 + l^2\dot{q}_2^2 \right) \tag{2.5.52}$$

Substituting Eq. (2.5.52) and Q_i from Eq. (2.5.51) into Eq. (2.5.46), we obtain

$$-m\ddot{r} + mg\cos\theta + \lambda_1 = 0$$

$$-ml^2\ddot{\theta} - r(mg\sin\theta + 6\pi\mu ar\dot{\theta}) = 0 \tag{2.5.53a}$$

In view of the constraint condition (2.5.50b), we have $\ddot{r} = 0$ and

$$\lambda_1 = -mg\cos\theta$$

$$\ddot{\theta} + \frac{g}{l}\sin\theta + \frac{6\pi\mu a}{m}\dot{\theta} = 0 \tag{2.5.53b}$$

The Lagrange multiplier λ_1 can be interpreted as the force exerted on the pendulum bob by the massless rod. It is the force necessary to oppose gravity and maintain the motion of the bob in a circular arc.

In the penalty function method, the Euler–Lagrange equations are obtained by substituting Eqs. (2.5.51) and (2.5.18b) into Eq. (2.5.48):

$$mg\cos\theta + \gamma_1(r - l) = 0$$

$$-ml^2\ddot{\theta} - r(mg\sin\theta + 6\pi\mu ar\dot{\theta}) = 0 \tag{2.5.54}$$

A comparison of the first equation in (2.5.54) with the first equation in (2.5.53a) shows that the approximate Lagrange multiplier is given by

$$\lambda_1(\gamma_1) = \gamma_1(r - l) = -mg\cos\theta$$

The error in the constraint is given by

$$r - l = -\frac{mg\cos\theta}{\gamma_1}$$

which goes to zero as γ_1 goes to infinity.

We close this section with two remarks on Hamilton's principle. First, the principle assumes that the state of the system at two distinct times t_1 and t_2 is known; second, the principle does not include the initial conditions of the problem. Variational principles that include the initial conditions for various initial-value problems in mechanics can be found in the monograph by Oden and Reddy (1982).

Exercises 2.5

1–3. Derive the Lagrangian function for the linear spring, mass, and linear dash-pot systems shown in Figs. E1–E3. Assume that motion starts from rest and the contact surfaces are free of friction. The constitutive equations for the spring and dash pot are given by equations of the form

Force = $R \times$ displacement,   for springs ($R =$ spring constant)

Force = $\eta \times$ velocity,   for dash pots ($\eta =$ dash-pot constant)

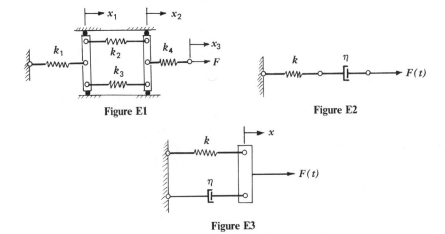

Figure E1   Figure E2

Figure E3

4. Consider the double pendulum shown in the figure of Exercise 1.1.4. Determine the Euler–Lagrange equations of motion of the pendulum when it oscillates in a plane under the action of gravity.

5. Consider a pendulum of mass m_1 with a flexible suspension, as shown in Fig. E5. The hinge of the pendulum is in a block of mass m_2, which can move up and down between the frictionless guides. The block is connected by a linear spring (of spring constant k) to an immovable support. The coordinate x is measured from the position of the block in which the system remains stationary. Derive the Euler–Lagrange equations of motion for the system.

6. Figure E6 represents a double pendulum that is suspended from a block which moves horizontally with a prescribed motion, $x = f(t)$. Derive the Euler–Lagrange equations of motion of the pendulum when it oscillates in the (x, y)-plane under the action of gravity and prescribed motion.

7. Two masses, m_1 and m_2, are attached to the ends of inextensible cord that is suspended over a frictionless stationary pulley, as shown in Exercise 1.1.4. Find the equations of motion of the system.

8. Repeat Exercise 7 for the case in which a monkey of mass m_3 is climbing up the cord above mass m_1 with a speed v_0 relative to mass m_1 (see the figure in Exercise 1.1.4).

9. Determine the motion of all masses in the suspended double-pulley problem represented in the figure of Exercise 1.1.4.

10. Consider a clamped-clamped beam of length L, area of cross section A, moment of inertia I, modulus of elasticity E, and mass density ρ. The beam is subjected to distributed load

$$f(x, t) = f_0(t) \sin \frac{\pi x}{L}$$

Determine the equation of motion.

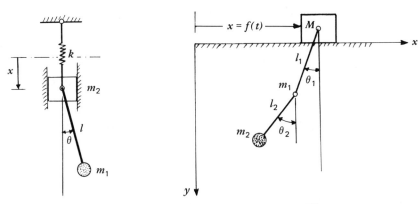

Figure E5   Figure E6

11. Consider a rectangular membrane of dimensions a by b, and mass density ρ, which is subjected to distributed transverse load $f(x,t)$. The membrane is fixed at all points of its boundary. If the strain energy stored in the membrane is given by

$$U = \int_0^a \int_0^b \frac{T_0}{2}\left[\left(\frac{\partial u}{\partial x}\right)^2 + \left(\frac{\partial u}{\partial y}\right)^2\right] dx\, dy$$

where T_0 is the tension in the membrane and u is the transverse displacement, determine the equation of motion.

12. The Lagrange function of a cantilevered beam with a point load P at the free end is given by

$$L = \int_0^L \int_A \left[\frac{\rho}{2}\left(\frac{\partial u}{\partial t} + z\frac{\partial \psi}{\partial t}\right)^2 + \frac{\rho}{2}\left(\frac{\partial w}{\partial t}\right)^2 - \frac{E}{2}\left(\frac{\partial u}{\partial x} + z\frac{\partial \psi}{\partial x}\right)^2\right] dA\, dx$$
$$+ Pw(L)$$

where u is the axial displacement, w is the transverse displacement, and ψ is the bending slope, all of which are functions of x and t. If w and ψ are constrained by the relation

$$w + \frac{\partial \psi}{\partial x} = 0$$

determine the equations of motion of the beam.

13. Show that the application of Hamilton's principle to linear coupled thermoelastic solids leads to

$$\frac{\partial L}{\partial q_i} + \frac{\partial D}{\partial \dot{q}_i} - \frac{d}{dt}\left(\frac{\partial L}{\partial \dot{q}_i}\right) + Q_i = 0$$

where q_i are the generalized coordinates, L is the Lagrangian function, D is the dissipation function, and Q_i is the generalized thermoelastic force.

14. Derive the governing equations of a linear coupled thermoelastic medium using the results of Exercise 13.

15. Derive the dynamic version of the Hellinger–Reissner variational principle for isothermal elasticity.

16. Derive the dynamic version of the Washizu variational principle.

2.6 ENERGY THEOREMS OF STRUCTURAL MECHANICS

2.6.1 Introduction

Except for the unit-dummy-displacement and force methods, the variational principles presented in this chapter do not yield solutions for structural problems; they yield differential equations and boundary conditions governing the solutions. As we illustrate in Chapter 3, these variational principles are very useful in obtaining approximate solutions of the differential equations. In this section, we derive additional theorems, analogous to the unit-dummy-displacement and force methods, that can be used to determine the unknown point loads and displacements of structural systems (i.e., systems composed of discrete structural members). The study is confined to stable structural systems at rest. Castigliano (1847–1884), an Italian mathematician and railroad engineer, was mainly concerned with linear elastic materials, for which the strain and complementary strain energies are equal. Thus, the strain energy U of a linear elastic body is expressed in terms of either forces F_i acting on the body or displacements u_i produced by the forces. There seems to be some disagreement between various authors on the order of Castigliano's theorems. Some authors refer to the statement

$$\frac{\partial U(F_i)}{\partial F_i} = u_i$$

as the first theorem of Castigliano. In the present study, we consider Castigliano's theorems in a generalized form that are applicable to both linear and nonlinear elastic materials. Therefore, we must distinguish between the strain energy and complementary strain energy of an elastic body. We call the theorem based on strain energy the first theorem and that based on complementary energy the second theorem. The generalization of Castigliano's original theorems to the case in which displacements are nonlinear functions of external forces is attributed to Engesser. Engesser substituted complementary energy for strain energy in the second theorem of Castigliano.

In the remainder of this chapter we neglect, unless otherwise explicitly stated, the energy caused by shearing forces. Of course, this is not a restriction imposed by any of the theorems discussed in this section.

2.6.2 Castigliano's First Theorem

Consider the uniaxially loaded elastic member shown in Fig. 1.10. Suppose that the member is loaded slowly to some load. The area *under* the force–deformation curve gives the work done by the load. The shaded area *above* the curve is the complementary work done. Let the load F be increased to $F + \Delta F$, where ΔF is a small increment in the load. The small increase in load results in a small increase δu in displacement u. The corresponding change in the work done is given by

$$-\delta V = F\delta u + \tfrac{1}{2}\Delta F\delta u + \cdots \tag{2.6.1}$$

The corresponding change in the strain energy is denoted by δU. We assume that U can be expressed in terms of u. By the theorem of total potential energy, we have

$$\delta \Pi \equiv \delta U + \delta V = 0$$

$$\left(\frac{\partial U}{\partial u} - F\right)\delta u = 0$$

or

$$\frac{\partial U}{\partial u} = F \qquad (2.6.2)$$

Equation (2.6.2) expresses the first theorem of Castigliano, which states that for an elastic body, the rate of change of strain energy with respect to the displacement is equal to the load causing the displacement. Now we consider a more general derivation of the theorem.

Consider a general three-dimensional structure that is in equilibrium under the action of N forces $F_i, i = 1, 2, \ldots, N$ (see Fig. 2.12). Let u_i be the displacement corresponding to the force F_i. The potential energy of the external forces is equal to

$$V = -\sum_{i-1}^{N} F_i u_i \qquad (2.6.3)$$

Here the word force is used to imply both forces and moments.

The displacements u_i are point values, not functions of position. We assume that the displacement functions of the body can be expressed in terms of u_i. Therefore, u_i serve as the generalized coordinates and the strain energy U of the body can be expressed in terms of $u_i, i = 1, 2, \ldots, N$. The total potential energy of the body is given by

$$\Pi = U + V$$

$$= U(u_1, u_2, \ldots, u_N) - \sum_{i=1}^{N} F_i u_i \qquad (2.6.4)$$

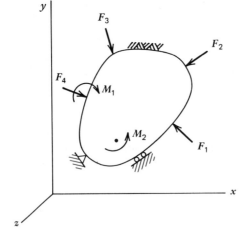

Figure 2.12 Configuration of a three-dimensional body acted upon by point forces and moments.

For any virtual variation δu_i in the displacement u_i, the variation $\delta\Pi$ in the total potential energy Π must vanish (by the principle of total potential energy):

$$\delta\Pi = \frac{\partial U}{\partial u_i}\delta u_i - F_i\,\delta u_i$$

$$= \left(\frac{\partial U}{\partial u_i} - F_i\right)\delta u_i = 0 \qquad (2.6.5)$$

Since the variations $\delta u_1, \delta u_2, \ldots, \delta u_N$ are independent of each other, it follows that

$$\frac{\partial U}{\partial u_i} - F_i = 0, \qquad i = 1, 2, \ldots, N \qquad (2.6.6)$$

Equation (2.6.6) is the general statement of Castigliano's first theorem: *If the strain energy of a structural system can be expressed in terms of N independent displacements $u_1, u_2, \ldots, u_N$ corresponding to N specified forces $F_1, F_2, \ldots, F_N$, the first partial derivative of the strain energy with respect to any displacement u_i (under the load F_i) is equal to the force F_i in the direction of u_i.*

It is clear from the derivation that Castigliano's first theorem is a special case of the principle of total potential energy and, hence, the principle of virtual displacements. Indeed, the theorem is equivalent to the unit-dummy-displacement method. To see this, consider Eq. (2.3.3) with $\delta W_E = \delta V$ and $\delta W_I = \delta U$:

$$-\delta V = \delta U$$

$$F_i\,\delta u_i = \delta\int_V U_0(u_1, u_2, \ldots, u_N)\,dV$$

$$= \int_V \frac{\partial U_0}{\partial u_i}\delta u_i\,dV$$

For $\delta u_i = 1$, and all other δu_j ($j = 1, 2, \ldots, i-1, i+1, \ldots, N$) zero, we get

$$F_i = \int_V \frac{\partial U_0}{\partial u_i}\,dV$$

$$= \frac{\partial U}{\partial u_i}, \qquad i = 1, 2, \ldots, N$$

which is the same as Eq. (2.6.6).

For future reference, we now give the expressions for the strain energy and complementary energy of a beam experiencing bending moment $M(x)$, axial force $N(x)$, transverse displacement w, and axial displacement u (but neglecting the shear force):

$$U = \frac{1}{2} \int_0^L \left[EI \left(\frac{d^2 w}{dx^2} \right)^2 + EA \left(\frac{du}{dx} \right)^2 \right] dx \qquad (2.6.7)$$

$$U^* = \frac{1}{2} \int_0^L \left(\frac{M^2}{EI} + \frac{N^2}{EA} \right) dx \qquad (2.6.8)$$

where E, I, and A are the modulus, the moment of inertia, and the area of cross section and L is the length of the beam.

Applications of Castigliano's first theorem are given in the following examples.

Example 2.18

Consider the pin-connected truss of Fig. 2.13. We wish to determine the horizontal and vertical displacements of joint B under the applied load P using Castigliano's theorem. The bars are made of nonlinear elastic material whose stress–strain behavior is given by relation (2.1.18), and the cross-sectional area of each member is A.

From the kinematics (i.e., geometry of deformation) of the truss, the strains in each member are given by

$$\varepsilon_1 = \frac{AB' - AB}{AB} = \sqrt{\frac{(h+u)^2 + v^2}{h^2}} - 1$$

$$\varepsilon_2 = \frac{CB' - CB}{CB} = \sqrt{\frac{(h+u)^2 + (h-v)^2}{2h^2}} - 1$$

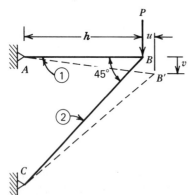

Figure 2.13 Displacement of a pin-connected structure; $\sigma = K\sqrt{\varepsilon}$ ($\varepsilon \geq 0$) and $-K\sqrt{-\varepsilon}$ ($\varepsilon \leq 0$).

Under the assumption of small displacements (i.e., $u^2 \approx 0$ and $v^2 \approx 0$), we obtain

$$\varepsilon_1 = \sqrt{1 + 2\frac{u}{h}} - 1 \approx \frac{u}{h}$$

$$\varepsilon_2 = \sqrt{1 + \frac{u - v}{h}} - 1 \approx \frac{1}{2}\left(\frac{u - v}{h}\right)$$

The strain energy density of the structure is given by ($U_0 = U_{01} + U_{02}$)

$$U_0 = \int_0^{\varepsilon_1} \sigma\, d\varepsilon + \int_0^{\varepsilon_2} \sigma\, d\varepsilon = \int_0^{\varepsilon_1} K\sqrt{\varepsilon}\, d\varepsilon + \int_0^{\varepsilon_2} - K\sqrt{-\varepsilon}\, d\varepsilon$$

$$= \frac{2K}{3}(\varepsilon_1)^{3/2} + \frac{2K}{3}(-\varepsilon_2)^{3/2}$$

$$= \frac{2}{3}K\left[\left(\frac{u}{h}\right)^{3/2} + \frac{1}{2\sqrt{2}}\left(\frac{v - u}{h}\right)^{3/2}\right]$$

where the fact that member 2 is in compression is used in writing the strain energy density of member 2. The total strain energy is given by ($U = U_1 + U_2$)

$$U = Ah\frac{2}{3}K\left(\frac{u}{h}\right)^{3/2} + A\sqrt{2}\,h\frac{2}{3}K\frac{1}{2\sqrt{2}}\left(\frac{v - u}{h}\right)^{3/2}$$

$$= \frac{AK}{3\sqrt{h}}\left[2u\sqrt{u} + (v - u)\sqrt{v - u}\right]$$

Using Castigliano's first theorem, we obtain

$$\frac{\partial U}{\partial u} = 0 = \frac{AK}{2\sqrt{h}}\left(2\sqrt{u} - \sqrt{v - u}\right)$$

$$\frac{\partial U}{\partial v} = P = \frac{AK}{2\sqrt{h}}\sqrt{v - u}$$

Solving for u and v, we get

$$u = \frac{P^2 h}{A^2 K^2}, \qquad v = \frac{5P^2 h}{A^2 K^2}$$

This completes the example.

Example 2.19

We wish to solve the problem in Example 2.10 using Castigliano's first theorem. We have

$$U = U_1 + U_2$$

$$= \frac{1}{2} \int_{V_1} \sigma_1 \varepsilon_1 \, dV + \frac{1}{2} \int_{V_2} \sigma_2 \varepsilon_2 \, dV$$

$$= \frac{1}{2} \left(E_1 A_1 \frac{e^2}{a^2} \cdot a + E_2 A_2 \frac{e^2}{b^2} \cdot b \right)$$

$$P = \frac{\partial U}{\partial e} = \left(\frac{E_1 A_1}{a} + \frac{E_2 A_2}{b} \right) e$$

which is the same relation as Eq. (c) of Example 2.10.

2.6.3 Castigliano's Second Theorem

Castigliano's first theorem is based on the total potential energy principle. Castigliano's second theorem is based on the total complementary energy principle. For the uniaxial member in Fig. 1.11, we have

$$\delta \Pi^* \equiv \delta U^* + \delta V^*$$

$$0 = \frac{\partial U^*}{\partial F} \delta F - u \, \delta F$$

so that

$$\frac{\partial U^*}{\partial F} = u \tag{2.6.9}$$

where $U^* = U^*(F)$ is the complementary strain energy of the member.

The generalization of Eq. (2.6.9) to a structural system that is in equilibrium under the action of N forces F_i follows the same procedure as presented for the first theorem. We have for $U^* = U^*(F_1, F_2, \ldots, F_N)$, the equation

$$\frac{\partial U^*}{\partial F_i} = u_i \tag{2.6.10}$$

which represents Castigliano's second theorem.

Equations (2.6.6) and (2.6.10) are valid for structures that are linearly elastic as well as nonlinearly elastic. When they are linearly elastic, we have $U = U^*$, and one can express the strain energy in terms of displacements or forces. Hence, the unit–dummy–load method is equivalent to Castigliano's second theorem.

Example 2.20

Consider the pin-connected structure consisting of two bars (see Fig. 2.3) and supporting a load P. We assume that the bars have the same cross-sectional area A and are made of nonlinear elastic material whose stress–strain behavior is given by Eq. (2.1.18):

$$\sigma = \begin{cases} E\sqrt{\varepsilon}, & \varepsilon \geq 0 \\ -E\sqrt{-\varepsilon}, & \varepsilon \leq 0 \end{cases}$$

We wish to determine the vertical deflection using Castigliano's second theorem.

From Example 2.1, the forces in the horizontal and the inclined members are given, respectively, by

$$F_1 = -\sqrt{3}\,P, \qquad F_2 = 2P$$

and the complementary strain energy is given by Eq. (2.1.20)

$$U^* = \frac{P^3 L}{3E^2 A^2}\left(3\sqrt{3} + \frac{16}{3}\sqrt{3}\right) \tag{2.6.11}$$

Then the displacement v under the load P is given by

$$v = \frac{\partial U^*}{\partial P} = \frac{25}{\sqrt{3}}\frac{P^2 L}{E^2 A^2} \tag{2.6.12}$$

In the next example, we illustrate the use of Castigliano's second theorem in the determination of reactions of an indeterminate beam.

Example 2.21

Consider the indeterminate beam structure shown in Fig. 2.14. We wish to use Castigliano's second theorem to determine the reaction of the central support of the beam. The vertical member is assumed to be elastic with modulus E_1, which can be different from the modulus E of the beam. We assume linear elastic behavior.

Suppose that R is the reaction at the center support. Since the center member is replaced with its reaction, we must consider the energy of the beam only. We have

$$U^* = \int_0^L \frac{M^2}{2EI}\,dx = \int_0^{L/2}\frac{M_1^2}{2EI}\,dx + \int_{L/2}^L \frac{M_2^2}{2EI}\,dx \tag{2.6.13a}$$

where

$$M(x) = \begin{cases} M_1 \equiv R_1 x - \dfrac{f_0 x^2}{2}, & 0 \leq x \leq \dfrac{L}{2} \\[2mm] M_2 \equiv R_1 x - f_0 \dfrac{x^2}{2} + R\left(x - \dfrac{L}{2}\right), & \dfrac{L}{2} \leq x \leq L \end{cases}$$

$$\tag{2.6.13b}$$

Now we must express the unknown reaction in terms of R (because we are

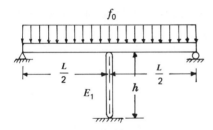

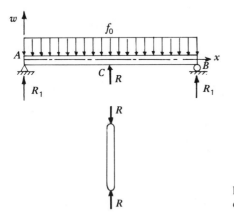

Figure 2.14 Configuration of the beam with a central support.

interested in finding R). In view of the symmetry, we have

$$R_1 = \frac{f_0 L - R}{2}$$

Using Castigliano's second theorem, we get (because of symmetry one can compute the first integral and multiply it by 2)

$$
\begin{aligned}
w_c = \frac{\partial U^*}{\partial R} &= \int_0^{L/2} \frac{M_1}{EI} \frac{\partial M_1}{\partial R}\, dx + \int_{L/2}^{L} \frac{M_2}{EI} \frac{\partial M_2}{\partial R}\, dx \\
&= \int_0^{L/2} \frac{1}{4EI} \left(f_0 L x - R x - f_0 x^2 \right)(-x)\, dx \\
&\quad + \int_{L/2}^{L} \frac{1}{4EI} \left[f_0 L x - R x - f_0 x^2 + R(2x - L) \right](x - L)\, dx \\
&= \frac{1}{4EI} \left(\frac{-10}{384} f_0 L^4 + \frac{RL^3}{24} - \frac{10}{384} f_0 L^4 + \frac{RL^3}{24} \right) \\
&= \frac{1}{EI} \left(-\frac{5}{384} f_0 L^4 + \frac{RL^3}{48} \right) \qquad (2.6.14)
\end{aligned}
$$

where the center deflection w_c is the compression in the center post, and is given by

$$w_c = -\frac{Rh}{E_1 A_1} \qquad (2.6.15)$$

From Eqs. (2.6.14) and (2.6.15), we have the reaction

$$R = \frac{5 f_0 L^4}{384 EI \left(h/A_1 E_1 + L^3/48 EI \right)} \tag{2.6.16}$$

Note that if the central support is rigid (i.e., $E_1 = \infty$), Eq. (2.6.16) gives the reaction ($R = 5 f_0 L/8$) at the midspan of a three-point supported beam.

It is simple, as illustrated in Example 2.21, to determine the deflections and slopes under applied point forces and moments in a beam. However, when a beam is subjected to a distributed load, the determination of the displacement and slope at a point requires the use of a fictitious load. For example, consider a beam that is subjected to uniformly distributed load f_0 and point loads $F_1, F_2, \ldots$, and so on. Suppose that we wish to determine the vertical deflection w_0 at a point where no point load is applied. We introduce a fictitious vertical load R at the point and then write the complementary strain energy in terms of f_0, R, and $F_1, F_2, \ldots$. We then use Castigliano's second theorem to determine the desired displacement,

$$w_0 = \left. \frac{\partial U^*}{\partial R} \right|_{R=0} \tag{2.6.17}$$

This procedure is illustrated in the following example.

Example 2.22

Consider a simply supported beam under uniformly distributed load. We wish to determine the center deflection. To this end, we introduce a fictitious concentrated load R at the center of the beam. We have

$$U^* = \int_0^{L/2} \frac{M_1^2}{2EI}\, dx + \int_{L/2}^{L} \frac{M_2^2}{2EI}\, dx$$

In view of the symmetry, these two integrals yield the same value. We have

$$w_0 = \left. \frac{\partial U^*}{\partial R} \right|_{R=0} = \left. \left(\frac{2}{EI} \int_0^{L/2} M_1 \frac{\partial M_1}{\partial R}\, dx \right) \right|_{R=0}$$

$$= \left. \left\{ \frac{2}{EI} \int_0^{L/2} \left[\left(\frac{f_0 L}{2} + \frac{R}{2} \right) x - f_0 \frac{x^2}{2} \right] \frac{x}{2}\, dx \right\} \right|_{R=0}$$

$$= \frac{5 f_0 L^4}{384 EI}$$

2.6.4   Betti's and Maxwell's Reciprocity Theorems

The *principle of superposition* is said to hold for a solid body if the displacements obtained under a given set of forces is equal to the sum of the individual displacements that would be obtained by applying the single forces separately. Consider, for example, the indeterminate beam shown in Fig. 2.15. The beam can be replaced by the beams shown there. At point A, the beam experiences the deflections w_f and w_s, owing to the distributed load f_0 and spring force S, respectively. Within the restrictions of small displacement theory and linear

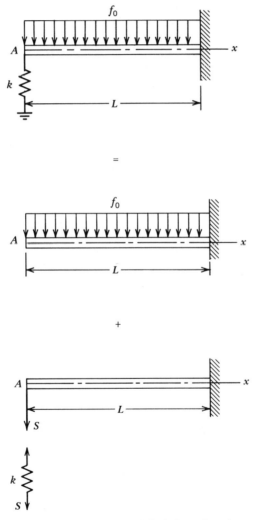

Figure 2.15   Configuration of the indeterminate beam.

elasticity, the deflections (and stresses) are linear functions of the loads, and therefore, superposition is valid. The deflection at point A is

$$w_A = w_f + w_s$$

$$= -\frac{f_0 L^4}{8EI} - \frac{SL^3}{3EI} \tag{2.6.18}$$

Because the spring force S is equal to kw_A, we can calculate w_A from Eq. (2.6.18).

$$w_A = \frac{-f_0 L^4}{8EI(1 + kL^3/3EI)} \tag{2.6.19}$$

The procedure can be used, in principle, to represent complicated linear problems as a linear combination of simple linear problems.

The principle of superposition is not valid for strain and potential energies, because they are quadratic functions of displacements and/or forces. In other words, when a linear elastic body is subjected to more than one external force, the total work caused by external forces is not equal to the sum of the works that are obtained by applying the single forces seperately. The present section is devoted to the discussion of two reciprocity theorems for work done on structural systems.

Consider a linear elastic solid that is in equilibrium under the action of two external forces, F_1 and F_2 (see Fig. 2.16). Since the order of application of the forces is arbitrary, we suppose that force F_1 is applied first. Let W_1 be the work produced by F_1. Then, we apply force F_2, which produces work W_2. This work is the same as that produced by force F_2, if it alone were acting on the body. When force F_2 is applied, force F_1 (which is already acting on the body) does additional work, because its point of application is displaced, owing to the deformation caused by force F_2. Let us denote this work by W_{12}. Thus, the total work done by the application of forces F_1 and F_2 (F_1 first and F_2 next) is

$$W = W_1 + W_2 + W_{12} \tag{2.6.20}$$

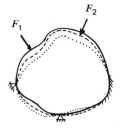

Figure 2.16 Configurations of an elastic body resulting from the application of loads F_1 and F_2: undeformed configuration (—); deformed configuration after the application of load F_1 (---); deformed configuration after the application of load F_2 (····).

Work W_{12}, which can be positive or negative, is zero if and only if the displacement of the point of application of force F_1 produced by force F_2 is zero or is perpendicular to the direction of F_1.

Now suppose that we change the order of application. Then the total work done is equal to

$$\overline{W} = W_1 + W_2 + W_{21} \tag{2.6.21}$$

where W_{21} is the work done by force F_2, because of the application of force F_1. The work done in both cases should be the same because, at the end, the elastic body is loaded by the same pair of external forces. Thus, we have

$$W = \overline{W} \quad \text{or} \quad W_{12} = W_{21} \tag{2.6.22}$$

Equation (2.6.22) is a mathematical statement of Betti's (1823–1892) reciprocity theorem. Applied to a three-dimensional elastic body, Eq. (2.6.22) takes the form

$$\int_V f_i \bar{u}_i \, dV + \int_{S_2} t_i \bar{u}_i \, dS = \int_V \bar{f}_i u_i \, dV + \int_{S_2} \bar{t}_i u_i \, dS \tag{2.6.23}$$

where $\bar{u}_i$ are the displacements produced by body forces $\bar{f}_i$ and surface forces $\bar{t}_i$, and u_i are the displacements produced by body forces f_i and surface forces t_i. The left-hand side of Eq. (2.6.23), for example, denotes the work done by forces f_i and t_i in moving through the displacements $\bar{u}_i$ produced by forces $\bar{f}_i$ and $\bar{t}_i$. We now state Betti's reciprocity theorem: if a linear elastic body is subjected to two different sets of forces, the work done by the first system of forces in moving through the displacements produced by the second system of forces is equal to the work done by the second system of forces in moving through the displacements produced by the first system of forces.

The following example verifies Betti's theorem.

Example 2.23

Consider a cantilevered beam of length L subjected to a concentrated load P at the free end and to a uniformly distributed load of intensity f_0 (see Fig. 2.17). The deflection equation resulting from the concentrated load alone is

$$w_p(x) = -\frac{P}{6EI}(x^3 - 3L^2 x + 2L^3)$$

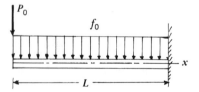

Figure 2.17 Configuration of a cantilever beam.

and the deflection equation from the distributed load is

$$w_f(x) = -\frac{f_0}{24EI}(x^4 - 4L^3x + 3L^4)$$

The work done by P because of the application of f_0 is

$$W_{12} = Pw_f(0) = -\frac{Pf_0L^4}{8EI}$$

The work done by f_0 because of the application of P is

$$W_{21} = \int_0^L -\frac{P}{6EI}(x^3 - 3L^2x + 2L^3)f_0\,dx = -\frac{Pf_0L^4}{8EI}$$

which is in agreement with W_{12}.

An important special case of Betti's reciprocity theorem is given by Maxwell's (1831–1879) reciprocity theorem. It is informative to note that Maxwell's theorem was given in 1864 and Betti's theorem was given in 1872, yet we derive Maxwell's theorem from Betti's theorem.

Consider a linear elastic solid subjected to force $\mathbf{F}^1$ of unit magnitude acting at point A, and force $\mathbf{F}^2$ of unit magnitude acting at a different point B of the body. Let $\mathbf{u}_{AB}$ be the displacement of point A in the direction of force $\mathbf{F}^1$ produced by unit force $\mathbf{F}^2$, and $\mathbf{u}_{BA}$ be the displacement of point B in the direction of force $\mathbf{F}^2$ produced by unit force $\mathbf{F}^1$ (see Fig. 2.18). From Betti's theorem it follows that

$$\mathbf{F}^1 \cdot \mathbf{u}_{AB} = \mathbf{F}^2 \cdot \mathbf{u}_{BA}$$

or

$$u_{AB} = u_{BA} \qquad\qquad (2.6.24)$$

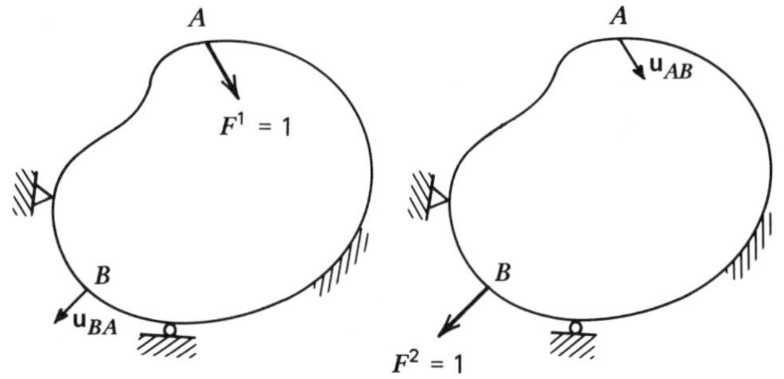

Figure 2.18 Configurations of the body discussed in the Maxwell theorem.

Equation (2.6.24) is a statement of Maxwell's theorem. If $\hat{e}_1$ and $\hat{e}_2$ denote the unit vectors along forces $\mathbf{F}^1$ and $\mathbf{F}^2$, respectively, Maxwell's theorem states that the displacement of point A in the $\hat{e}_1$-direction produced by a unit force acting at point B in the $\hat{e}_2$-direction is equal to the displacement of point B in the $\hat{e}_2$-direction produced by a unit force acting at point A in the $\hat{e}_1$-direction.

We close this section with the following two examples that illustrate the usefulness of Maxwell's theorem.

Example 2.24

Consider a cantilevered beam ($E = 24 \times 10^6$ psi, $I = 120$ in.^{4}) of length 12 ft, subjected to a point load 4000 lb at the free end. We wish to find the deflection at a point 3 ft from the free end (see Fig. 2.19).

By Maxwell's theorem, the displacement w_{CB} at $x = 3$ ft produced by the 4000-lb load at $x = 0$ is equal to the deflection w_{BC} at $x = 0$ produced by applying the 4000-lb load at $x = 3$ ft. Let w_C and θ_C denote the deflection and slope, respectively, at point C ($x = 3$ ft) owing to load $P = 4000$ lb applied at point C. Then, the deflection at point C ($x = 3$ ft) caused by load $P_0 = 4000$ lb at point B ($x = 0$) is ($w_C = PL^3/3EI$ and $\theta_C = PL^2/2EI$)

$$w_{CB} = w_{BC} = w_C + (3 \times 12)\theta_C$$

$$= \frac{4000(9 \times 12)^3}{3EI} + \frac{(3 \times 12)4000(9 \times 12)^2}{2EI}$$

$$= \frac{243 \times 6000 \times (12)^3}{24 \times 10^6 \times 120}\,\text{in.}$$

$$= 0.8748 \text{ in.}$$

This completes the example.

Example 2.25

Consider a simply supported beam subjected to a point load P_B at point B. The load produces deflections w_A, w_C, and w_D at points A, C, and D, respectively. We wish to find the deflection w_B at point B produced by loads P_A, P_C, and P_D (see Fig. 2.20).

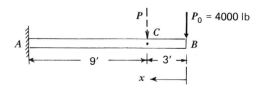

Figure 2.19 Configuration of the cantilever beam.

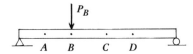

Figure 2.20 Configuration of the simply supported beam

A unit load acting at point B produces displacements w_A/P_B, w_C/P_B, and w_D/P_B at points A, C, and D. But by Maxwell's theorem, w_A/P_B, w_C/P_B, and w_D/P_B are the displacements of point B caused by a unit load acting at points A, C, and D, respectively. Therefore, the displacement at point B owing to forces P_A, P_C, and P_D is given by

$$w_B = P_A\left(\frac{w_A}{P_B}\right) + P_C\left(\frac{w_C}{P_B}\right) + P_D\left(\frac{w_D}{P_B}\right)$$

$$= \frac{P_A w_A + P_C w_C + P_D w_D}{P_B} \qquad (2.6.25)$$

Exercises 2.6

1. Determine the load P and its direction in the pin-connected structure of Fig. 2.7 for the given displacement v of point D. Assume that all members are of equal cross-sectional area and are made of the same linear elastic material, and $\theta = 30°$.

2. Repeat Exercise 1 for the case in which the structural members are made of a nonlinear elastic material whose stress–strain relation is given by Eq. (2.1.18).

3–8. Using Castigliano's second theorem determine the displacements under the applied point load(s); in Exercise 8, find the compression in the spring.

9. Prove Betti's theorem for a three-dimensional elastic solid, that is, establish Eq. (2.6.23).

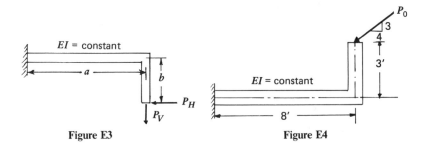

Figure E3   Figure E4

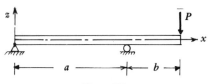

Figure E5

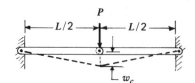

Figure E6

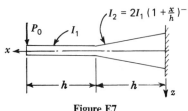

Figure E7

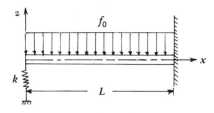

Figure E8

10. Consider a simply supported beam of length L subjected to a concentrated load P at the midspan and a bending moment M_0 at the left end. Verify that Betti's theorem holds.

11. Determine the slope at the left end of the beam shown in Fig. E5.

12. Determine the reactions in the indeterminate beam of Exercise 8; see Fig. E8.

13. A load $P = 4000$ lb acting at a point A of a beam produces 0.25 in. at point B and 0.75 in. at point C of the beam. Find the deflection of point A produced by loads 4500 and 2000 lb acting at points B and C, respectively.

14. Determine the deflection at the midspan of a cantilevered beam subjected to uniformly distributed load and point load at the free end. Use Maxwell's theorem and superposition.

15. Determine the vertical and horizontal displacements of the structure shown in Fig. 2.13 using Castigliano's second theorem.

3

VARIATIONAL METHODS OF APPROXIMATION

3.1 SOME PRELIMINARIES

3.1.1 Introduction

In Chapter 2, we saw how variational statements can be used to obtain governing equations, associated boundary conditions, and, in certain simple cases, solutions for displacements and forces at selective points of a structure. However, the variational methods considered in Sections 2.3 and 2.5 cannot be used to determine continuous solutions to problems.

The present chapter deals with approximate methods that employ the variational statements (i.e., either variational principles or weak formulations) to determine continuous solutions of problems in mechanics. In Chapter 2, we obtained the governing differential equations and boundary conditions of various problems by applying the virtual work principles or their derivatives. These principles involved setting the first variation of an appropriate functional with respect to the dependent variables to zero. The procedures of the calculus of variations were then used to obtain the governing (Euler–Lagrange) equations of the problem. In contrast, the methods described in this chapter seek a solution in terms of adjustable parameters that are determined by substituting the assumed solutions into the functional and finding its stationary value with respect to the parameters. Such solution methods are called *direct methods*, because the approximate solutions are obtained directly by applying the same variational principle that was used to derive the governing equations.

The assumed solutions in the variational methods are in the form of a finite linear combination of *undetermined parameters* with appropriately chosen functions. This amounts to representing a continuous function by a finite linear combination of functions. Since the solution of a continuum problem in general cannot be represented by a finite set of functions, error is introduced into the solution. Therefore, the solution obtained is an *approximation* of the true solution for the equations describing a physical problem. As the number of linearly independent terms in the assumed solution is increased, the error in the approximation will be reduced, and the assumed solution converges to the desired solution of Euler's equations.

The equations governing a physical problem themselves are approximate. The approximations are introduced via several sources, including the geometry, the representation of specified loads and displacements, and the material constitution. In the present study, our primary concern is to determine accurate approximate solutions to appropriate analytical descriptions of physical problems.

The variational methods of approximation described here include those of Rayleigh and Ritz, Galerkin, Petrov–Galerkin (weighted-residuals), Kantorovitch, Trefftz, and the finite element method, which is a "piecewise" application of the Ritz–Galerkin method. Examples of applications of these methods are drawn from the problems of bars, beams, torsion, and plane elasticity. Applications to plates and shells are considered in Chapter 4.

3.1.2   Some Mathematical Concepts

Before we discuss the variational methods of approximation, it is essential to equip ourselves with certain simple mathematical concepts. These include the concepts of linear independence, orthogonality, and linear and bilinear forms of functions.

In the following discussions, we assume that all dependent functions satisfy the rules (i.e., addition of two functions and scalar multiple of a function are defined) of a linear space (see Reddy and Rasmussen, 1982). A set (or space) of square-integrable functions u defined over a domain Ω is called the L_2-*space*. A real-valued function $u(x)$ is said to be *square-integrable* in the domain Ω if the integrals

$$\int_\Omega u(x)\, dx, \qquad \int_\Omega |u(x)|^2\, dx$$

exist and are finite.

Linear Independence

Recall the concepts of coplanar and collinear vectors in Euclidean space. These can be generalized to functions of independent variables. An expression of the form

$$\alpha_1 u_1 + \alpha_2 u_2 + \cdots = \sum_{i=1}^{n} \alpha_i u_i \qquad (3.1.1)$$

for all functions $u_i(\mathbf{x})$ and scalars α_i in R (real number field), is called a *linear combination* of u_i. The linear combination $\sum_i^n \alpha_i u_i = 0$ is called a *linear relation* among the functions u_i. A set of n functions, $u_1, u_2, \ldots, u_n$, is said to be *linearly dependent* if a set of n numbers, $\alpha_1, \alpha_2, \ldots, \alpha_n$, not all of which are zero, can be found such that the following linear combination holds:

$$\sum_{i=1}^{n} \alpha_i u_i = 0 \qquad (3.1.2)$$

If there does not exist at least one nonzero number among α_i such that the above relation is satisfied, the vectors are said to be *linearly independent*.

Example 3.1

Consider the following set of polynomials,

$$p_1(x) = 1 + x, \qquad p_2(x) = 1 + x^2, \qquad p_3(x) = 1 + x + x^3$$

The linear combination

$$\alpha_1 p_1 + \alpha_2 p_2 + \alpha_3 p_3 = 0$$

holds for *any* α_i, if (collecting the coefficients of various powers of x)

$$\alpha_1 + \alpha_2 + \alpha_3 = 0, \qquad \alpha_1 + \alpha_3 = 0, \qquad \alpha_2 = 0, \qquad \alpha_3 = 0$$

The solution to these equations is trivial (i.e., all $\alpha_i = 0$) hence, the set $\{ p_1, p_2, p_3 \}$ is linearly independent.

If p_3 is replaced by $p_4 = 2 + x + x^2$, we see that the linear combination

$$\alpha_1 p_1 + \alpha_2 p_2 + \alpha_4 p_4 = 0$$

requires that

$$\alpha_1 + \alpha_2 + 2\alpha_4 = 0, \qquad \alpha_1 + \alpha_4 = 0, \qquad \alpha_2 + \alpha_4 = 0$$

An infinite number of solutions to the above set of equations exists. For example,

$$\alpha_4 = 1, \qquad \alpha_1 = \alpha_2 = -1$$

is a solution. Hence, the set $\{ p_1, p_2, p_4 \}$ is linearly dependent. Indeed, p_4 can be expressed as a linear combination of p_1 and p_2:

$$p_4 = \alpha_1 p_1 + \alpha_2 p_2, \qquad \alpha_1 = \alpha_2 = 1$$

Inner Product and Norm

The concepts of distance between two points and length of a vector can be generalized to functions. An *inner product* of a pair of functions u and v is defined to be a real number, denoted (u, v), which satisfies the following axioms:

(i) Symmetry: $(u, v) = (v, u)$

(iia) Homogeneous: $(\alpha u, v) = \alpha(u, v)$

(iib) Additive: $(u_1 + u_2, v) = (u_1, v) + (u_2, v)$

(iii) Positive–definite: $(u, u) > 0$ for all $u \neq 0$ (3.1.3)

for every u_1, u_2, u and v in V and α in R.

Analogous to the scalar product of an ordinary vector with itself, the *norm* of a function defines a measure of its "magnitude." One can measure the magnitude of a function in a variety of ways, but all must satisfy certain axioms, called *axioms of a norm*. The norm of a function u, denoted by $\|u\|$, is a real number that satisfies the following rules:

(i) Nonnegative: (a) $\|u\| \geq 0$ for all u
                         (b) $\|u\| = 0$ only if $u = 0$

(ii) Homogeneous: $\|\alpha u\| = |\alpha|\,\|u\|$

(iii) Triangle inequality: $\|u + v\| \leq \|u\| + \|v\|$.   (3.1.4)

If $\|u\|$ satisfies (ia), (ii), and (iii), it is called the *seminorm*, and is denoted by $|u|$. The distance between two functions u and v can be measured as $\|u - v\|$.

One can define a number of inner products and associated *natural norms* for pairs of functions that, along with their derivatives, are square-integrable:

$$(u, v)_0 = \int_\Omega uv\, dx\, dy, \qquad \|u\|_0 = \sqrt{(u, u)_0} \tag{3.1.5}$$

$$(u, v)_1 = \int_\Omega \left(uv + \frac{\partial u}{\partial x}\frac{\partial v}{\partial x} + \frac{\partial u}{\partial y}\frac{\partial v}{\partial y} \right) dx\, dy, \qquad \|u\|_1 = \sqrt{(u, u)_1} \tag{3.1.6}$$

$$(u, v)_2 = \int_\Omega \left(uv + \frac{\partial u}{\partial x}\frac{\partial v}{\partial x} \right.$$

$$\left. + \frac{\partial u}{\partial y}\frac{\partial v}{\partial y} + \frac{\partial^2 u}{\partial x\,\partial y}\frac{\partial^2 v}{\partial x\,\partial y} + \frac{\partial^2 u}{\partial x^2}\frac{\partial^2 v}{\partial x^2} + \frac{\partial^2 u}{\partial y^2}\frac{\partial^2 v}{\partial y^2} \right) dx\, dy$$

$$\|u\|_2 = \sqrt{(u, u)_2} \tag{3.1.7}$$

Orthogonality of Functions

Two functions u and v are said to be *orthogonal* if

$$(u,v) = 0 \qquad (3.1.8)$$

Note that the concept of orthogonality is a generalization of the familiar notion of perpendicularity of one vector to another in Euclidean space. A set of mutually orthogonal functions is called an orthogonal set.

A sequence of functions $\{\phi_i\}$ in $L_2(\Omega)$ is called *orthonormal* if

$$(\phi_i,\phi_j)_0 = \delta_{ij} = \begin{cases} 1, & \text{if } i = j \\ 0, & \text{if } i \neq j \end{cases} \qquad (3.1.9)$$

Here δ_{ij} denotes the Kronecker delta. It can be shown that every orthonormal system is linearly independent.

A set of functions $\{\phi_i\}$ is said to be *complete* in $L_2(\Omega)$ if every piecewise continuous function f can be approximated in Ω by the sum $\sum_{j=1}^n c_j\phi_j$ in such a way that

$$\mathscr{E}_n \equiv \int_\Omega \left(f - \sum_{j=1}^n c_j\phi_j\right)^2 dx \qquad (3.1.10)$$

can be made as small as we wish (by increasing n). This property of the coordinate functions is the key to the proof of the convergence of the Ritz and Galerkin approximations (see Sections 3.2 and 3.3).

Linear and Bilinear Forms

Recall that an operator A mapping one set (or space) of functions to another is said to be *linear* if it satisfies the following relation:

$$A(\alpha u_1 + \beta u_2) = \alpha A(u_1) + \beta A(u_2) \qquad (3.1.11)$$

for functions u_1 and u_2 and scalars α and β. When A does not satisfy the above relation we call it a *nonlinear operator*. Operators that map functions into real numbers are of special interest in the present study. Such operators are called *functionals*. A linear operator that maps functions into a real number is called *linear functional*. Similarly, a linear operator that maps pairs of functions into real numbers is called a *bilinear form*. Examples of linear and bilinear forms are provided by

$$l(u) = \int_a^b fu\,dx, \qquad B(u,v) = \int_a^b \left(\frac{du}{dx}\frac{dv}{dx} + uv\right) dx \qquad (3.1.12)$$

A bilinear form is said to be *symmetric* if it is symmetric in its arguments:

$$B(u,v) = B(v,u) \qquad (3.1.13)$$

Quadratic Functionals

A *quadratic functional* is one that is quadratic in its arguments and yields a real number. Every bilinear form B can be converted to a quadratic form Q by setting

$$Q(u) = B(u, u) \qquad (3.1.14)$$

The variational principles discussed in Chapter 2 provide a number of examples of quadratic forms.

Exercises 3.1

In Exercises 1–5, determine whether the sets of functions are linearly independent.

1. $\left\{\sin\dfrac{n\pi x}{L}\right\}_{n=1}^{3}$, $0 \le x \le L$.

2. $\left\{\cos\dfrac{n\pi x}{L}\right\}_{n=1}^{3}$, $0 \le x \le L$.

3. $\{x^n(1 - x)\}_{n=0}^{3}$, $0 \le x \le 1$.

4. $\{x, 5x, x^2\}$, $0 \le x \le 1$.

5. $\{x(3 - 2x), x(2 - x^2), x^2(4 - 3x)\}$, $0 \le x \le 1$.

6. Let $\mathbf{u} = \mathbf{u}(x)$ and $\mathbf{v} = \mathbf{v}(x)$ be two vector functions of x. Define the products

   (i) $(\mathbf{u}, \mathbf{v}) \equiv \displaystyle\int_0^L \mathbf{u} \cdot \mathbf{v}\, dx$

   (ii) $(\mathbf{u}, \mathbf{v}) \equiv \displaystyle\int_0^L \mathbf{u}(L - x) \cdot \mathbf{v}(x)\, dx$

   Which one of these products qualifies as an inner product?

   Compute the norm of the functions in Exercises 7–9 using the inner products indicated.

7. $u = x^3 - 3L^2 x + 2L^3$, $0 \le x \le L$, use $\|u\|_0$ in Eq. (3.1.5).

8. $u = x^3 - 3L^2 x + 2L^3$, $0 \le x \le L$, use the one-dimensional version of Eq. (3.1.6).

9. $u = (x^2 - a^2)(y^2 - b^2)$, $0 \le x \le a$ and $0 \le y \le b$, Eq. (3.1.6).

10. Find the distance between the following pair of functions using the inner product in Eq. (3.1.5): $u = x^3 - 3x + 2$ and $v = (x - 1)^2$, $0 \le x \le 1$.

11. Check whether the following pair of functions are orthogonal in the $L_2(0, 1)$-space: $u = 2 + 3x^2 - x$ and $v = \frac{1}{3} + 3x - 5x^2$.

12. Determine the constants a and b such that $w = a + bx + 3x^2$ is orthogonal to both u and v of Exercise 11.

In Exercises 13–17, which operators qualify as linear operators. Identify the linear and bilinear functionals.

13. $A(u) = -\dfrac{d}{dx}\left(a\dfrac{du}{dx}\right).$

14. $T(u) = \nabla^2 u + 1.$

15. $I(u) = \displaystyle\int_0^a K(x)\dfrac{du}{dx}\,dx - u(0).$

16. $I(u,v) = \displaystyle\int_\Omega \nabla u \cdot \nabla v\,dx\,dy.$

17. $I(u,v) = \displaystyle\int_0^a \left(b\dfrac{d^2 u}{dx^2}\dfrac{d^2 v}{dx^2} + fv\right)dx$

18. Show that if $\Pi(u)$ is a quadratic functional, its first variation is a bilinear functional of u and δu.

A set of n linear algebraic equations

$$a_{11}x_1 + a_{12}x_2 + \cdots + a_{1n}x_n = f_1$$

$$a_{21}x_1 + a_{22}x_2 + \cdots + a_{2n}x_n = f_2$$

$$\vdots$$

$$a_{n1}x_1 + a_{n2}x_2 + \cdots + a_{nn}x_n = f_n$$

is said to be linearly independent if the determinant of the coefficient matrix is nonzero: $\det|a_{ij}| \neq 0$. Determine whether the equations in Exercises 19–21 are linearly independent.

19. $\begin{aligned} 2x_1 - x_2 &= 0. \\ -x_1 + 2x_2 - x_3 &= 0. \\ x_2 - x_3 &= 1. \end{aligned}$

20. $\begin{aligned} x_1 + x_2 + x_3 &= 2. \\ x_1 - x_2 - 3x_3 &= 3. \\ 3x_1 + x_2 - x_3 &= 1. \end{aligned}$

21. $\begin{aligned} 2x_1 + 3x_2 - x_3 &= 4. \\ x_1 - x_2 + 2x_3 &= 2. \\ x_1 + 2x_2 - x_3 &= 1. \end{aligned}$

3.2 THE RITZ METHOD

3.2.1 Introduction

In the principle of virtual displacements, the Euler equations are the equilibrium equations, whereas in the principle of virtual forces, they are the compatibility equations. These Euler equations are in the form of differential equations that are not always tractable by exact methods of solution. A number of approximate methods exist for solving differential equations (e.g., finite-difference methods, perturbation methods, etc.). The most direct methods bypass the derivation of the Euler equations and go directly from a variational statement of the problem to the solution of the Euler equations. One such direct method was proposed by Lord Rayleigh (whose actual name was John William Strutt, 1842–1919). A generalization of the method was proposed independently by Ritz (1878–1909).

3.2.2 Description of the Method

The basic idea of the Ritz method is described here using the principle of virtual displacements. In this method, we approximate the displacements u_1, u_2, and u_3 by a finite linear combination of the form

$$u_1 \approx U_1 = \sum_{j=1}^{n} c_j^1 \phi_j^1 + \phi_0^1 \qquad (3.2.1a)$$

$$u_2 \approx U_2 = \sum_{j=1}^{n} c_j^2 \phi_j^2 + \phi_0^2 \qquad (3.2.1b)$$

$$u_3 \approx U_3 = \sum_{j=1}^{n} c_j^3 \phi_j^3 + \phi_0^3 \qquad (3.2.1c)$$

and then determine the parameters c_j^1, c_j^2, and c_j^3 by requiring that the principle of virtual displacements hold for arbitrary variations of the parameters. In Eqs. (3.2.1) c_j^1, c_j^2, and c_j^3 denote undetermined parameters, and ϕ_0^α and ϕ_j^α ($\alpha = 1, 2, 3$) denote appropriate functions of position (x_1, x_2, x_3). We will discuss the properties of these functions shortly.

Equation (3.2.1a), for example, can be viewed as a representation of u in a component form; the parameters c_j are the *components* (or coordinates) and ϕ_j are the *coordinate functions*. Another interpretation of Eq. (3.2.1a) is provided by the finite Fourier series, and c_j are known as the *Fourier coefficients*.

The coordinate functions ϕ_0^α and ϕ_j^α ($\alpha = 1, 2, 3$) should be such that, when Eqs. (3.2.1) are substituted into the principle of virtual displacements, Eq. (2.3.3) [or, equivalently, Eq. (2.3.32)], one obtains $3n$ linearly independent equations for the parameters c_j^α ($\alpha = 1, 2, 3$ and $j = 1, 2, \ldots, n$) that have a solution. To insure that the algebraic equations resulting from the Ritz approximation have a solution, and the approximate solution converges to the true solution of the problem as the value of n is increased, we should choose ϕ_j^α ($j = 1, 2, \ldots, n$) and ϕ_0^α ($\alpha = 1, 2, 3$) that satisfy the following requirements:

1. ϕ_0^α ($\alpha = 1, 2, 3$) should satisfy the specified (in their actual form) essential boundary conditions associated with u_α ($\alpha = 1, 2, 3$), respectively.

2. ϕ_j^α ($j = 1, 2, \ldots, n$) should satisfy the following three conditions:
   (a) be continuous, as required by the variational principle being used;
   (b) satisfy the homogeneous form of the specified essential boundary conditions;
   (c) be linearly independent and complete.   (3.2.2)

Since the natural boundary conditions of the problem are included in the variational statement, we require the assumed displacements U_α ($\alpha = 1, 2, 3$) to satisfy only the essential boundary conditions. For convenience, we select ϕ_i^α to satisfy the homogeneous form and ϕ_0^α to satisfy the actual form of the essential boundary conditions. For instance, if u_1 is specified to be $\hat{u}_1$ on the boundary S_1, we require $\phi_0^1 = \hat{u}_1$ on S_1 and $\phi_i^1 = 0$ on S_1 for all $i = 1, 2, \ldots, n$, so that $U_1 = \hat{u}_1$ on S_1.

The conditions in Eq. (3.2.2) provide guidelines for selecting the coordinate functions; they do not, however, give any formulae for generating the functions. Thus, apart from the guidelines, the selection of the coordinate functions is largely arbitrary. As a general rule, the coordinate functions ϕ_i should be selected from an admissible set, from the lowest order to a desirable order, without missing any intermediate terms [i.e., the completeness property in the sense of Eq. (3.1.10) should be imposed]. Also, ϕ_0 should be any lowest order (including zero) that satisfies the specified essential boundary conditions of the problem.

Equation (3.2.1) is used to compute approximate strains which are then used to compute the virtual work expressions in Eq. (2.3.32). We obtain

$$\delta\Pi(u_1, u_2, u_3) = \delta\Pi\left(c_1^1, c_2^1, \ldots, c_n^1; c_1^2, c_2^2, \ldots, c_n^2; c_1^3, c_2^3, \ldots, c_n^3\right)$$

$$= \sum_{i=1}^{n}\left(\frac{\partial\Pi}{\partial c_i^1}\delta c_i^1 + \frac{\partial\Pi}{\partial c_i^2}\delta c_i^2 + \frac{\partial\Pi}{\partial c_i^3}\delta c_i^3\right)$$

where the variations δu_α ($\alpha = 1, 2, 3$) were replaced by

$$\delta u_1 = \sum_{i=1}^{n}\delta c_i^1\,\phi_i^1, \qquad \delta u_2 = \sum_{i=1}^{n}\delta c_i^2\,\phi_i^2, \qquad \delta u_3 = \sum_{i=1}^{n}\delta c_i^3\,\phi_i^3 \quad (3.2.3)$$

If the system is in equilibrium, we have

$$\delta\Pi = \sum_{i=1}^{n}\left(\frac{\partial\Pi}{\partial c_i^1}\delta c_i^1 + \frac{\partial\Pi}{\partial c_i^2}\delta c_i^2 + \frac{\partial\Pi}{\partial c_i^3}\delta c_i^3\right) = 0 \qquad (3.2.4)$$

for arbitrary variations δc_i^1, δc_i^2, and δc_i^3 ($i = 1, 2, \ldots, n$). Since δc_i^1, δc_i^2, and

δc_i^3 $(i = 1, 2, \ldots, n)$ are arbitrary and independent, it follows that

$$\frac{\partial \Pi}{\partial c_i^1} = 0, \qquad i = 1, 2, \ldots, n$$

$$\frac{\partial \Pi}{\partial c_i^2} = 0, \qquad i = 1, 2, \ldots, n$$

$$\frac{\partial \Pi}{\partial c_i^3} = 0, \qquad i = 1, 2, \ldots, n \qquad (3.2.5)$$

Thus, we obtain $3n$ linearly independent simultaneous equations for the $3n$ unknowns. Once c_i^1, c_i^2, and c_i^3 $(i = 1, 2, \ldots, n)$ are determined from Eq. (3.2.5), the approximate displacements of the problem are given by Eq. (3.2.1). These displacements can be used to evaluate strains and stresses.

Some general features of the Ritz approximations based on the principle of virtual displacements are listed below:

1. If the coordinate functions ϕ_i^α are selected to satisfy Eq. (3.2.2), the assumed approximation for displacements normally converges with an increase in the number of parameters (i.e., $n \to \infty$). A mathematical proof of such an assertion is not given here, but interested readers can consult Reddy and Rasmussen (1982).

2. For increasing values of n, the previously computed coefficients of the algebraic equations (3.2.5) remain unchanged (provided the previously selected coordinate functions are not changed), and one must add newly computed coefficients to the system of equations.

3. If the variational form used in the Ritz approximation is such that its bilinear form is symmetric (in u and δu), the resulting algebraic equations are also symmetric and, therefore, one need to compute only upper or lower diagonal elements in the coefficient matrix.

4. If $\delta \Pi$ is nonlinear in u_i, the resulting algebraic equations will also be nonlinear in the parameters c_i. To solve such nonlinear equations, a variety of numerical methods are available (e.g., the Newton–Raphson method). Generally, there is more than one solution to the equations.

5. Since the strains are computed from the approximate displacements, the strains and stresses are generally less accurate than the displacements.

6. The equilibrium equations of the problem are satisfied only in the energy sense ($\delta \Pi = 0$), not in the differential equation sense. Therefore the displacements obtained from the approximation generally do not satisfy the equations of equilibrium point-wise.

7. Since a continuous system is approximated by a finite number of coordinates (or degrees of freedom), the approximate system is less

flexible than the actual system. Consequently, the displacements obtained in the Ritz method converge to the exact displacements from below:

$$U_1^1 < U_1^2 < \cdots < U_1^n < U_1^m \cdots < u_1, \quad \text{for } m > n$$

where U_1^n denotes the n-parameter Ritz approximation of u_1 obtained from the total potential energy principle. The displacements obtained from the Ritz approximations based on the total complementary energy principle provide an upper bound for the exact solution.

8. Although the discussion of the Ritz method in this section is confined to solid mechanics, the method can be employed for any equation that admits a variational formulation (see Section 3.2.5).

3.2.3 Matrix Form of the Ritz Equations

Before we consider some illustrative examples, we return to the Ritz procedure and derive the matrix form of the Ritz equations (3.2.5), which permits easy computation of the coefficients of the algebraic equations. The scheme developed here for the Ritz approximation is based on the total potential energy of a linear elastic body.

Recall that the total potential energy of a linear elastic body undergoing small displacements is given by Eq. (2.2.39). The functional is of the form

$$\Pi(u_1, u_2, u_3) = Q(u_1, u_2, u_3) - L(u_1, u_2, u_3) \tag{3.2.6}$$

where Q and L are the quadratic and linear forms in $\mathbf{u} = (u_1, u_2, u_3)$:

$$Q(u_1, u_2, u_3) = \frac{1}{2} \sum_{i,j=1}^{3} \int_V \varepsilon_{ij} \sigma_{ij} \, dV$$

$$L(u_1, u_2, u_3) = \sum_{i=1}^{3} \left(\int_V f_i u_i \, dV + \int_{S_2} \hat{t}_i u_i \, dS \right) \tag{3.2.7}$$

Here, the strains and stresses ε_{ij} and σ_{ij} are known in terms of the displacements via the kinematic and constitutive relations. Substituting Eq. (3.2.1) into Eq. (3.2.6) and invoking the equilibrium conditions in Eq. (3.2.5), we obtain

$$\delta\Pi = \delta Q - \delta L$$

$$= \sum_{i=1}^{n} \left[\left(\frac{\partial Q}{\partial c_i^1} - \frac{\partial L}{\partial c_i^1} \right) \delta c_i^1 + \left(\frac{\partial Q}{\partial c_i^2} - \frac{\partial L}{\partial c_i^2} \right) \delta c_i^2 + \left(\frac{\partial Q}{\partial c_i^3} - \frac{\partial L}{\partial c_i^3} \right) \delta c_i^3 \right]$$

or

$$\frac{\partial Q}{\partial c_i^1} - \frac{\partial L}{\partial c_i^1} = 0, \qquad \frac{\partial Q}{\partial c_i^2} - \frac{\partial L}{\partial c_i^2} = 0, \qquad \frac{\partial Q}{\partial c_i^3} - \frac{\partial L}{\partial c_i^3} = 0 \tag{3.2.8}$$

Since Q is quadratic and L is linear in u_1, u_2, and u_3 (hence, in c_i^1, c_i^2, and c_i^3), their partial derivatives with respect to c_i^1, c_i^2, and c_i^3 are bilinear and linear functionals of ϕ_i^α ($\alpha = 1, 2, 3$), respectively. We can write

$$\frac{\partial Q}{\partial c_i^\alpha} = \sum_{\beta=1}^{3} \sum_{j=1}^{n} B^{\alpha\beta}\left(\phi_i^\alpha, \phi_j^\beta\right) c_j^\beta$$

$$\frac{\partial L}{\partial c_i^\alpha} = F^\alpha(\phi_i^\alpha) \tag{3.2.9}$$

so that we have

$$\sum_{\beta=1}^{3} \sum_{j=1}^{n} B^{\alpha\beta}\left(\phi_i^\alpha, \phi_j^\beta\right) c_j^\beta - F^\alpha(\phi_i^\alpha) = 0 \tag{3.2.10a}$$

We can write Eq. (3.2.10a) in the matrix form

$$\begin{bmatrix} [B^{11}] & [B^{12}] & [B^{13}] \\ [B^{21}] & [B^{22}] & [B^{23}] \\ [B^{31}] & [B^{32}] & [B^{33}] \end{bmatrix} \begin{Bmatrix} \{c^1\} \\ \{c^2\} \\ \{c^3\} \end{Bmatrix} = \begin{Bmatrix} \{F^1\} \\ \{F^2\} \\ \{F^3\} \end{Bmatrix} \tag{3.2.10b}$$

where the coefficients $b_{ij}^{\alpha\beta}$ of the matrix $[B^{\alpha\beta}]$ can be identified in terms of the volume integral of the material coefficients and coordinate functions. The coefficients f_i^α are given by

$$f_i^\alpha = \int_V f_\alpha \phi_i^\alpha \, dV + \int_{S_2} \hat{t}_\alpha \phi_i^\alpha \, dS \tag{3.2.11}$$

The matrix form in Eq. (3.2.10) is a general feature of the Ritz approximation based on quadratic functionals. Equations (3.2.10) and (3.2.1) give the n-parameter Ritz approximation for the displacements.

3.2.4 Illustrative Examples

Here we illustrate the Ritz method by considering some simple but typical problems. Additional applications of the Ritz method are considered in the subsequent chapters.

Example 3.2

Consider axial deformation of an isotropic elastic member of length L. The total potential energy associated with the unconstrained member is given by

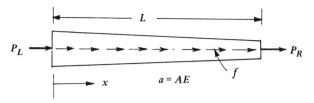

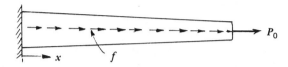

Figure 3.1 A free-free bar and a bar fixed at the left end and subjected to point load P_0 at the right end: A = area of cross section; E = modulus of elasticity.

(see Fig. 3.1)

$$\Pi(u) = \int_0^L \left[\frac{a}{2} \left(\frac{du}{dx} \right)^2 - fu \right] dx - u(0) P_L - u(L) P_R \qquad (3.2.12)$$

where u is the axial displacement, $a = EA$ is the product of the modulus of elasticity and area of cross section, f is the distributed axial force, and P_L and P_R are the axial forces at the left and right end of the member. Once the boundary conditions are specified, the boundary terms in Eq. (3.2.12) can be simplified. For example, if the member is fixed at the left end and subjected to a force P_0 at the right end, then the functional becomes

$$\Pi(u) = \int_0^L \left[\frac{a}{2} \left(\frac{du}{dx} \right)^2 - fu \right] dx - u(L) P_0 \qquad (3.2.13)$$

The force P_L becomes the reaction force at the left end. Although we do not need the Euler equations to find a Ritz approximation of u, it is informative to know that they are of the form

$$-\frac{d}{dx} \left(a \frac{du}{dx} \right) = f, \qquad 0 < x < L$$

$$\left. \left(a \frac{du}{dx} \right) \right|_{x=L} = P_0 \qquad (3.2.14)$$

with the essential boundary condition, $u(0) = 0$.

First, we select the number of parameters in the approximation. Suppose that we wish to determine a two-parameter Ritz approximation of the axial displacement u:

$$u(x) \approx U(x) = c_1\phi_1 + c_2\phi_2 + \phi_0 \tag{3.2.15}$$

Property 1 of Eq. (3.2.2) tells us that ϕ_0 should be a function that satisfies the specified essential boundary condition:

$$\phi_0(0) = 0$$

Obviously, $\phi_0 = 0$ satisfies the condition. Although the choice $\phi_0 = ax^m$, for any given a and $m > 0$, satisfies the requirement, we should select the minimum-order admissible function.

Property 2 of Eq. (3.2.2) tells us that ϕ_1 and ϕ_2 should be continuous, as required by the functional $\Pi(u)$, satisfy the homogeneous form of the specified essential boundary condition, and be linearly independent. From Eq. (3.2.13) it is clear that u should have a nonzero first derivative (otherwise, the strain energy is not represented), implying that ϕ_1 and ϕ_2 should have nonzero first derivatives. Let us begin with the linear polynomial for ϕ_1

$$\phi_1 = a + bx$$

In view of Property 2(b), ϕ_1 should satisfy the condition $\phi_1(0) = 0$. It should be pointed out that even in the case $u(0) \neq 0$, ϕ_i should satisfy the homogeneous form, $\phi_i(0) = 0$. The condition $\phi_1(0) = 0$ gives us $a = 0$. Since b is arbitrary, we take it to be unity (any nonzero constant will be absorbed into c_1). Property 2(c) requires, only if $n \geq 2$, that ϕ_2 should be selected such that it is linearly independent and makes the set complete. This is done by choosing ϕ_2 to be x^2. Clearly, $\phi_2 = x^2$ meets the conditions $\phi_2(0) = 0$, linearly independent of $\phi_1 = x$ (i.e., ϕ_2 cannot be related to ϕ_1 by a constant multiple), and the set (x, x^2) is complete (i.e., no other admissible term up to quadratic is omitted). The choice $\phi_1 = x^2$, $\phi_2 = x^3$ would satisfy all the properties except the completeness. In other words, in selecting coordinate functions of a given degree, one should not omit any lower-order terms that are admissible. Otherwise, the approximate solution will never converge, no matter how many terms are used, to the desired solution. This should be obvious, because if the exact solution contains the first-order term, it is not accounted for in the approximation.

From the above discussions we have

$$\phi_0 = 0, \qquad \phi_1 = x, \qquad \phi_2 = x^2 \tag{3.2.16}$$

If we wish to find higher-order ($n > 2$) approximations, it is clear that we can take $\phi_k = x^k$, $k = 1, 2, \ldots$.

Without regard to the specific form of ϕ_0 and ϕ_i, we can obtain from Eqs. (3.2.13) and (3.2.15) the matrix equations of the form found in Eq. (3.2.10).

Substituting Eq. (3.2.15) into Eq. (3.2.13), we obtain

$$\Pi(U) = \int_0^L \left[\frac{a}{2} \left(c_1 \frac{d\phi_1}{dx} + c_2 \frac{d\phi_2}{dx} + \frac{d\phi_0}{dx} \right)^2 - f(c_1\phi_1 + c_2\phi_2 + \phi_0) \right] dx$$

$$- \left[c_1\phi_1(L) + c_2\phi_2(L) + \phi_0(L) \right] P_0 = \Pi(c_1, c_2) \qquad (3.2.17)$$

From the principle of minimum total potential energy, we have ($\delta\Pi = 0$),

$$\frac{\partial\Pi}{\partial c_1} = 0 = \int_0^L \left[a \left(c_1 \frac{d\phi_1}{dx} + c_2 \frac{d\phi_2}{dx} + \frac{d\phi_0}{dx} \right) \frac{d\phi_1}{dx} - f\phi_1 \right] dx - \phi_1(L)P_0$$

$$\frac{\partial\Pi}{\partial c_2} = 0 = \int_0^L \left[a \left(c_1 \frac{d\phi_1}{dx} + c_2 \frac{d\phi_2}{dx} + \frac{d\phi_0}{dx} \right) \frac{d\phi_2}{dx} - f\phi_2 \right] dx - \phi_2(L)P_0$$

$$(3.2.18)$$

It should be noted that the same equations can be obtained directly from Eq. (2.2.37) by substituting Eq. (3.2.15) for u and $(\delta c_1\phi_1 + \delta c_2\phi_2)$ for δu, and then setting the coefficients of δc_1 and δc_2 to zero separately. Let

$$b_{ij} = \int_0^L a \frac{d\phi_i}{dx} \frac{d\phi_j}{dx} dx, \qquad f_i = \int_0^L \left(-a \frac{d\phi_0}{dx} \frac{d\phi_i}{dx} + f\phi_i \right) dx + \phi_i(L)P_0$$

$$(3.2.19)$$

Then we have, from Eq. (3.2.18),

$$\begin{aligned} b_{11}c_1 + b_{12}c_2 &= f_1 \\ b_{21}c_1 + b_{22}c_2 &= f_2 \end{aligned} \quad \text{or} \quad [B]\{c\} = \{F\} \qquad (3.2.20)$$

For the choice in Eq. (3.2.16), and taking

$$a = a_0 \left(2 - \frac{x}{L} \right), \qquad f = f_0$$

we obtain

$$b_{11} = \frac{3a_0 L}{2}, \qquad b_{12} = b_{21} = \frac{4}{3} a_0 L^2, \qquad b_{22} = \frac{5a_0 L^3}{3}$$

$$f_1 = f_0 \frac{L^2}{2} + P_0 L, \qquad f_2 = \frac{f_0 L^3}{3} + P_0 L^2 \qquad (3.2.21a)$$

Substituting the coefficients into Eq. (3.2.20) and solving (by Cramer's rule) the resulting equations for c_1 and c_2, we obtain

$$\frac{a_0 L}{6} \begin{bmatrix} 9 & 8L \\ 8L & 10L^2 \end{bmatrix} \begin{Bmatrix} c_1 \\ c_2 \end{Bmatrix} = \frac{L}{6} \begin{Bmatrix} 3f_0 L + 6P_0 \\ 2f_0 L^2 + 6P_0 L \end{Bmatrix} \qquad (3.2.21b)$$

$$c_1 = \frac{7f_0 L + 6P_0}{13a_0}, \qquad c_2 = \frac{3(P_0 - f_0 L)}{13a_0 L}$$

$$U(x) = \frac{7f_0 L + 6P_0}{13a_0} x + \frac{3(P_0 - f_0 L)}{13a_0 L} x^2 \qquad (3.2.22)$$

The exact solution of Eq. (3.2.14) with $u(0) = 0$, $a = a_0[2 - (x/L)]$, and $f = f_0$ is

$$u(x) = \frac{f_0 L}{a_0} x + \frac{(f_0 L - P_0)L}{a_0} \log\left(1 - \frac{x}{2L}\right) \qquad (3.2.23a)$$

$$\approx \frac{f_0 L + P_0}{2a_0} x + \frac{P_0 - f_0 L}{8a_0 L} x^2 + \frac{P_0 - f_0 L}{24a_0 L^2} x^3 + \cdots \qquad (3.2.23b)$$

The one-term approximation can be obtained from Eq. (3.2.20) by setting $b_{12} = b_{22} = f_2 = 0$:

$$c_1 = \frac{f_1}{b_{11}} = \frac{f_0 L + 2P_0}{3a_0}$$

For the n-parameter approximation of the form

$$U_n = \sum_{i=1}^{n} c_i x^i \qquad (3.2.24)$$

The coefficients in Eq. (3.2.19) are given by

$$b_{ij} = \frac{ij(1 + i + j)}{(i + j - 1)(i + j)} (L)^{i+j-1}$$

$$f_i = \frac{f_0}{i + 1}(L)^{i+1} + P_0(L)^i \qquad (3.2.25)$$

Table 3.1 contains a comparison of the Ritz coefficients c_i for $n = 1, 2, \ldots, 8$ with the exact coefficients in Eq. (3.2.23b) for $L = 10$ ft, $a_0 = 180 \times 10^6$ lb, $f_0 = 0$, and $P_0 = 10$ kip. Clearly, the Ritz coefficients converge to the exact ones as n goes from 1 to 8. This completes the example.

The next example is concerned with the natural vibration of a bar with an end spring.

Table 3.1 The Ritz Coefficients for the Axial Deformation of an Isotropic Bar Subjected to Axial Force

n	$c_1 \times 10^6$	$c_2 \times 10^7$	$c_3 \times 10^8$	$c_4 \times 10^9$	$c_5 \times 10^{10}$	$c_6 \times 10^{11}$	$c_7 \times 10^{12}$	$c_8 \times 10^{13}$
1	37.037							
2	25.641	12.821						
3	28.219	4.409	5.879					
4	27.691	7.788	0.000	3.029				
5	27.794	6.701	3.389	-1.040	1.664			
6	27.775	7.009	1.904	2.012	-1.142	0.952		
7	27.778	6.929	2.453	0.320	1.447	-0.980	0.560	
8	27.778	6.948	2.272	1.094	-0.287	1.136	-0.769	0.336
Exact	27.778	6.944	2.315	0.868	0.347	0.145	0.062	0.027

Example 3.3

Consider a uniform cross-section bar of length L, with the left end fixed and the right end connected to a rigid support via a linear elastic spring (with spring constant k). The kinetic energy and potential energy associated with the axial motion of the member are given by [see Eq. (2.5.34)]

$$K = \int_0^L \frac{\rho A}{2} \left(\frac{\partial u}{\partial t} \right)^2 dx$$

$$U = \int_0^L \frac{EA}{2} \left(\frac{\partial u}{\partial x} \right)^2 dx + \frac{k}{2} [u(L)]^2 \qquad (3.2.26)$$

Substituting for K and U from Eq. (3.2.26), and $V = 0$ in Eq. (2.5.30), we obtain

$$0 = \int_{t_1}^{t_2} \delta(K - U) \, dt$$

$$= \frac{1}{2} \delta \int_{t_1}^{t_2} \left\{ \int_0^L \left[\rho A \left(\frac{\partial u}{\partial t} \right)^2 - EA \left(\frac{\partial u}{\partial x} \right)^2 \right] dx - k [u(L)]^2 \right\} dt \qquad (3.2.27)$$

Here we are interested in finding the periodic motion of the form

$$u(x, t) = u_0(x) e^{i\omega t} \qquad (3.2.28)$$

where ω is called the *frequency* of natural vibration, and $u_0(x)$ is the amplitude. Substituting Eq. (3.2.28) into the Euler equation of (3.2.27) and constructing the variational form of the equation, we obtain

$$0 = \delta \left\{ \int_0^L \frac{1}{2} \left[\rho A \omega^2 (u_0)^2 - EA \left(\frac{du_0}{dx} \right)^2 \right] dx - \frac{k}{2} [u_0(L)]^2 \right\} \qquad (3.2.29)$$

where $\int_{t_1}^{t_2} e^{2i\omega t}\, dt$, being nonzero, is factored out. We use Eq. (3.2.29) to determine the values of ω. The Euler equation of Eq. (3.2.29) is given by

$$\frac{d}{dx}\left(EA\frac{du_0}{dx}\right) + \rho A\omega^2 u_0 = 0, \qquad 0 < x < L$$

$$\left(EA\frac{du_0}{dx} + ku_0\right)\bigg|_{x=L} = 0 \qquad\qquad (3.2.30)$$

A nondimensionalization of the variables ($\bar{x} = x/L$, $\bar{u} = u_0/L$) gives

$$0 = \delta\left\{\frac{1}{2}\int_0^1\left[\lambda u^2 - \left(\frac{du}{dx}\right)^2\right]dx - \frac{\alpha}{2}[u(1)]^2\right\}, \qquad \alpha = \frac{kL}{EA}, \qquad \lambda = \frac{\omega^2\rho L^2}{E}$$

$$(3.2.31)$$

where the bar over the nondimensional variables is omitted in the interest of brevity. In the following discussions, we assume that $\alpha = 1$.

Suppose that we wish to find a two-parameter Ritz approximation of u. Following the procedure outlined in the previous example, we select

$$\phi_0 = 0, \qquad \phi_1 = x, \qquad \phi_2 = x^2 \qquad\qquad (3.2.32)$$

Substituting $u = c_1 x + c_2 x^2$ into Eq. (3.2.31), we get

$$0 = \int_0^1\left[\lambda\left(c_1 x + c_2 x^2\right)\left(\delta c_1 x + \delta c_2 x^2\right) - \left(c_1 + 2c_2 x\right)\left(\delta c_1 + 2\delta c_2 x\right)\right]dx$$

$$- \left(c_1 + c_2\right)\left(\delta c_1 + \delta c_2\right)$$

Collecting the coefficients of δc_1 and δc_2 and setting them to zero separately, we have

$$\delta c_1: 0 = \int_0^1\left[\lambda\left(c_1 x + c_2 x^2\right)x - \left(c_1 + 2c_2 x\right)\right]dx - \left(c_1 + c_2\right)$$

$$\delta c_2: 0 = \int_0^1\left[\lambda\left(c_1 x + c_2 x^2\right)x^2 - \left(c_1 + 2c_2 x\right)2x\right]dx - \left(c_1 + c_2\right)$$

$$(3.2.33a)$$

or, in general,

$$(\lambda[M] - [B])\{c\} = \{0\} \qquad\qquad (3.2.33b)$$

where

$$m_{ij} = \int_0^1 \phi_i\phi_j\, dx, \qquad b_{ij} = \int_0^1\frac{d\phi_i}{dx}\frac{d\phi_j}{dx}\, dx + \phi_i(1)\phi_j(1) \qquad (3.2.33c)$$

For the choice of ϕ_i in Eq. (3.2.32), we have

$$m_{11} = \tfrac{1}{3}, \qquad m_{12} = \tfrac{1}{4}, \qquad b_{11} = 2, \qquad b_{12} = 2, \qquad m_{22} = \tfrac{1}{5}, \qquad b_{22} = \tfrac{7}{3}$$

$$\left(\begin{bmatrix} 2 & 2 \\ 2 & \tfrac{7}{3} \end{bmatrix} - \lambda \begin{bmatrix} \tfrac{1}{3} & \tfrac{1}{4} \\ \tfrac{1}{4} & \tfrac{1}{5} \end{bmatrix} \right) \begin{Bmatrix} c_1 \\ c_2 \end{Bmatrix} = \begin{Bmatrix} 0 \\ 0 \end{Bmatrix} \tag{3.2.34}$$

A nontrivial solution ($c_1 \neq 0$, $c_2 \neq 0$) to Eq. (3.2.34) exists, only if the determinant of the coefficient matrix is zero:

$$\begin{vmatrix} 2 - \dfrac{\lambda}{3} & 2 - \dfrac{\lambda}{4} \\[2mm] 2 - \dfrac{\lambda}{4} & \dfrac{7}{3} - \dfrac{\lambda}{5} \end{vmatrix} = 0$$

or

$$15\lambda^2 - 640\lambda + 2400 = 0$$

The quadratic equation has two roots:

$$\lambda_1 = 4.1545, \qquad \lambda_2 = 38.512$$

$$\omega_1 = \frac{2.038}{L} \sqrt{\frac{E}{\rho}}, \qquad \omega_2 = \frac{6.206}{L} \sqrt{\frac{E}{\rho}} \tag{3.2.35}$$

The exact solution is given by the equation

$$\lambda + \tan \lambda = 0$$

whose first two roots are

$$\omega_{01} = \frac{2.02875}{L} \sqrt{\frac{E}{\rho}}, \qquad \omega_{02} = \frac{4.91318}{L} \sqrt{\frac{E}{\rho}} \tag{3.2.36}$$

Note that the higher-order approximate frequencies deviate more from the exact ones.

If one selects ϕ_i to satisfy the natural boundary condition also, the degree of approximation inevitably goes up. For example, the lowest-order function that satisfies the homogeneous form of the natural boundary condition $u'(1) + u(1) = 0$ is

$$\phi_1 = 3x - 2x^2 \tag{3.2.37}$$

This choice gives $\lambda_1 = 50/12 = 4.1667$, which is no better than that computed using the same degree polynomial (but with two parameters).

Next, we consider the Ritz approximation of a simply supported beam.

Example 3.4

Consider a simply-supported beam of length L. We wish to find the transverse deflection of the beam under uniformly distributed transverse loading. The principle of virtual displacements for the problem becomes,

$$0 = \int_0^L \left(EI \frac{d^2 \delta w}{dx^2} \frac{d^2 w}{dx^2} + \delta w f_0 \right) dx \qquad (3.2.38)$$

The kinematic boundary conditions are

$$w(0) = w(L) = 0$$

We choose a two-parameter approximation,

$$w = c_1 \phi_1 + c_2 \phi_2 + \phi_0, \qquad \delta w = \delta c_1 \phi_1 + \delta c_2 \phi_2 \qquad (3.2.39a)$$

where

$$\phi_0 = 0, \qquad \phi_1 = x(L-x), \qquad \phi_2 = x^2(L-x) \qquad (3.2.39b)$$

Substituting Eq. (3.2.39a) into Eq. (3.2.38), we obtain

$$0 = \int_0^L \left[EI \left(\delta c_1 \phi_1'' + \delta c_2 \phi_2'' \right) \left(c_1 \phi_1'' + c_2 \phi_2'' \right) + \left(\delta c_1 \phi_1 + \delta c_2 \phi_2 \right) f_0 \right] dx$$

$$= \sum_{i=1}^{2} \left(\sum_{j=1}^{2} b_{ij} c_j - f_i \right) \delta c_i$$

or

$$0 = \sum_{j=1}^{2} b_{ij} c_j - f_i \qquad (3.2.40a)$$

where

$$b_{ij} = \int_0^L EI \frac{d^2 \phi_i}{dx^2} \frac{d^2 \phi_j}{dx^2} \, dx, \qquad f_i = -\int_0^L f_0 \phi_i \, dx \qquad (3.2.40b)$$

For the particular choice of ϕ_i in Eq. (3.2.39b), we have

$$EIL \begin{bmatrix} 4 & 2L \\ 2L & 4L^2 \end{bmatrix} \begin{Bmatrix} c_1 \\ c_2 \end{Bmatrix} = -\frac{f_0 L^3}{12} \begin{Bmatrix} 2 \\ L \end{Bmatrix}$$

$$c_1 = -\frac{f_0 L^2}{24 EI}, \qquad c_2 = 0$$

$$w(x) = -\frac{f_0 L^3}{24 EI} x + \frac{f_0 L^2}{24 EI} x^2 \qquad (3.2.41)$$

The exact solution is given by

$$w_0(x) = -\frac{f_0 L^3 x}{24EI} + \frac{f_0 L x^3}{12EI} - \frac{f_0 x^4}{24EI} \qquad (3.2.42)$$

The three-parameter approximation ($\phi_3 = x^3(L - x)$) yields

$$EIL\begin{bmatrix} 4 & 2L & 2L^2 \\ 2L & 4L^2 & 4L^3 \\ 2L^2 & 4L^3 & 4.8L^4 \end{bmatrix}\begin{Bmatrix} c_1 \\ c_2 \\ c_3 \end{Bmatrix} = -\frac{f_0 L^3}{12}\begin{Bmatrix} 2 \\ L \\ 0.6L^2 \end{Bmatrix} \qquad (3.2.43a)$$

$$c_1 = c_2 L = -c_3 L^2 = -\frac{f_0 L^2}{24EI} \qquad (3.2.43b)$$

and the three-parameter Ritz approximation of w coincides with the exact solution.

Suppose that instead of the uniformly distributed load, the beam is subjected to a point load P at $x = L/2$. The virtual work statement for this case is given by

$$0 = \int_0^L EI\frac{d^2\delta w}{dx^2}\frac{d^2 w}{dx^2}\,dx - P\delta w\left(\frac{L}{2}\right) \qquad (3.2.44)$$

Using the three-parameter approximation, we obtain the coefficient matrix as in Eq. (3.2.43a), with the right-hand side given by

$$-\frac{PL^2}{4}\begin{Bmatrix} 1 \\ \dfrac{L}{2} \\ \dfrac{L^2}{4} \end{Bmatrix} \qquad (3.2.45)$$

The solution of the equations is given by

$$c_1 = -\frac{PL}{12EI}, \qquad c_2 = \frac{P}{12EI}, \qquad c_3 = \frac{5P}{64EIL} \qquad (3.2.46)$$

The exact solution is given by

$$w_0(x) = \begin{cases} -\dfrac{PL}{12EI}x(L-x) - \dfrac{P}{12EI}x^2(L-x) + \dfrac{PxL^2}{48EI}, & 0 \le x \le \dfrac{L}{2} \\[4mm] -\dfrac{PL}{12EI}x(L-x) - \dfrac{P}{12EI}x^2(L-x) + \dfrac{PxL^2}{48EI} - \dfrac{P\left(x - \dfrac{L}{2}\right)^3}{6EI}, \\[4mm] \hspace{6cm} \dfrac{L}{2} \le x \le L \end{cases}$$

$$(3.2.47)$$

The maximum deflection given by the Ritz approximation is

$$w_R\left(\frac{L}{2}\right) = -\frac{7}{384}\frac{PL^3}{EI} = -\frac{PL^3}{54.86EI}$$

whereas the exact value is

$$w_0\left(\frac{L}{2}\right) = -\frac{PL^3}{48EI}$$

The Ritz solution is defined by a single expression, whereas the exact solution is defined by two expressions in two parts of the beam. The exact bending moment is discontinuous at $x = L/2$, but the approximate one is continuous. This is because the Ritz solution is based on the total potential energy functional, which does not account for the point discontinuity in d^2w/dx^2.

Another remark on the present example is in order. One can also use trigonometric polynomials in place of algebraic polynomials for the coordinate functions. However, such functions can make the evaluation of b_{ij} and/or f_i a difficult task. Further, for nontrigonometric loads, the Ritz solution does not converge to the true solution for a finite number of parameters. The reason is that loads represented by algebraic polynomials [e.g., $f = f_0(a_1 + a_2x + a_3x^2 + \cdots)$] cannot be represented by a finite trigonometric series. For instance, the simply-supported beam under uniform loading can be modeled by the approximation,

$$w = c_1\sin\frac{\pi x}{L} + c_2\sin\frac{3\pi x}{L} + \cdots + c_n\sin(2n + 1)\frac{\pi x}{L} \qquad (3.2.48)$$

The functions $\phi_i = \sin(2i + 1)(\pi x/L)$ are linearly independent, and are complete if all lower functions up to $\sin(2n + 1)(\pi x/L)$ are included. However, the Ritz solution (3.2.48) will not coincide with the exact solution (3.2.42) for any finite value of n, because the sine-series representation of a uniform load is an infinite series:

$$f = \sum_{i=1,3,\ldots}^{\infty} \frac{16f_0}{\pi i}\sin\frac{i\pi x}{L} \qquad (3.2.49)$$

The Ritz solution (3.2.48) converges rapidly, giving almost an exact solution, especially away from the ends, for only finitely large values of n.

Consider, for example, a simply supported beam subjected to uniformly distributed load of intensity f_0 and point load P at the midspan. A one-term approximation, $w = c_1\sin(\pi x/L)$, gives the Ritz solution

$$w_1 = \frac{2L^3(2f_0L + \pi P)}{EI\pi^5}\sin\frac{\pi x}{L} \qquad (3.2.50a)$$

The exact solution is given by the superposition of the expressions in Eqs. (3.2.42) and (3.2.47). Evaluating Eq. (3.2.50a) at $x = L/2$, we get

$$w_1\left(\frac{L}{2}\right) = \frac{f_0 L^4}{76.5EI} + \frac{PL^3}{48.7EI} \qquad (3.2.50b)$$

whereas the exact solution is

$$w_0\left(\frac{L}{2}\right) = \frac{f_0 L^4}{76.8EI} + \frac{PL^3}{48EI}$$

The center deflection is 0.39% in error in the case of uniform load and 1.46% in error for point load at the center. The two-term approximation gives the deflection

$$w_2 = w_1 + \left(\frac{4f_0 L^4}{243EI\pi^5} - \frac{2PL^3}{81EI\pi^4}\right)\sin\frac{3\pi x}{L} \qquad (3.2.51a)$$

Evaluating w_2 at $x = L/2$, we get

$$w_2\left(\frac{L}{2}\right) = \frac{f_0 L^4}{76.8EI} + \frac{PL^3}{48.1EI} \qquad (3.2.51b)$$

Thus, the two-term approximation gives a center deflection that is exact for uniform load and 0.21% in error for point load at the center.

The last example of this section is concerned with the Ritz approximation of the transverse deflection of a beam using the Reissner type variational principle.

Example 3.5

Consider the transverse deflection of a cantilevered beam under uniformly distributed load f_0. The governing equations in terms of the deflection and bending moment are given by

$$\text{Kinematic relation:} \quad \frac{d^2 w}{dx^2} - \frac{M}{EI} = 0 \qquad (3.2.52a)$$

$$\text{Equilibrium equation:} \quad \frac{d^2 M}{dx^2} + f_0 = 0 \qquad (3.2.52b)$$

$$\text{Boundary conditions:} \quad w(0) = \frac{dw}{dx}(0) = 0, \quad M(L) = \frac{dM}{dx}(L) = 0$$

$$(3.2.52c)$$

In the minimum total potential energy principle, Eqs. (3.2.52a) and (3.2.52b) are combined into a fourth-order equation in w alone:

$$\delta\Pi(w) \equiv \int_0^L \left(EI \frac{d^2\delta w}{dx^2} \frac{d^2 w}{dx^2} + f_0 \delta w \right) dx = 0 \qquad (3.2.53)$$

A two-parameter Ritz approximation of the type ($\phi_0 = 0$)

$$w_2 = c_1\phi_1 + c_2\phi_2 = c_1 x^2 + c_2 x^3 \qquad (3.2.54)$$

which satisfies the essential boundary conditions of the functional $w = dw/dx = 0$ at $x = 0$, yields the solution

$$w_2(x) = -\frac{5f_0}{24} \frac{L^2}{EI} x^2 + \frac{f_0 L}{12 EI} x^3, \qquad w_2(L) = -\frac{f_0 L^4}{8EI} \qquad (3.2.55)$$

The bending moment computed from the approximate deflection is

$$M_2(x) = -\frac{5}{12} f_0 L^2 + \frac{1}{2} f_0 Lx, \qquad M_2(0) = -\frac{5}{12} f_0 L^2 \qquad (3.2.56)$$

The exact solution for the deflection and bending moment are given by

$$w_0(x) = -\frac{f_0}{24EI}(x^4 - 4Lx^3 + 6L^2 x^2), \qquad w_0(L) = -\frac{f_0 L^4}{8EI}$$

$$M_0(x) = -\frac{f_0}{2}(L - x)^2, \qquad M_0(0) = -\frac{f_0 L^2}{2} \qquad (3.2.57)$$

Thus, the two-parameter Ritz solution, based on the total potential energy principle, gives exact maximum deflection and a less accurate maximum bending moment. The maximum bending moment is 16.67% in error.

Now consider the stationary functional that yields Eqs. (3.2.52) as the Euler equations:

$$R(w, M) = \int_0^L \left(-\frac{dw}{dx} \frac{dM}{dx} - \frac{M^2}{2EI} + f_0 w \right) dx \qquad (3.2.58)$$

In this case, $w(0) = M(L) = 0$ are the essential boundary conditions and $dw/dx(0) = dM/dx(L) = 0$ are the natural boundary conditions.

Let us consider the two-parameter Ritz approximations of the form ($\phi_0 = \psi_0 = 0$)

$$w_2 = a_1\phi_1 + a_2\phi_2 = a_1 x + a_2 x^2$$

$$M_2 = b_1\psi_1 + b_2\psi_2 = b_1(L - x) + b_2(L - x)^2 \qquad (3.2.59)$$

which satisfy the essential boundary conditions. Now there are four parameters to be determined. Substituting Eq. (3.2.59) into Eq. (3.2.58), and taking the variation of the functional with respect to a_1, a_2, b_1, and b_2, we obtain

$$R(a_1, a_2, b_1, b_2) = \int_0^L \left[-\left(\sum_{j=1}^2 a_j \frac{d\phi_j}{dx} \right)\left(\sum_{j=1}^2 b_j \frac{d\psi_j}{dx} \right) \right.$$

$$\left. -\frac{1}{2EI}\left(\sum_{j=1}^2 b_j\psi_j \right)\left(\sum_{j=1}^2 b_j\psi_j \right) + f_0\left(\sum_{j=1}^2 a_j\phi_j \right) \right] dx$$

$$(3.2.60\text{a})$$

$$\frac{\partial R}{\partial a_i} = \int_0^L \left[-\frac{d\phi_i}{dx}\left(\sum_{j=1}^2 b_j\frac{d\psi_j}{dx} \right) + f_0\phi_i \right] dx = 0, \qquad i = 1, 2$$

$$\frac{\partial R}{\partial b_i} = \int_0^L \left[-\frac{d\psi_i}{dx}\left(\sum_{j=1}^2 a_j\frac{d\phi_j}{dx} \right) - \frac{1}{EI}\psi_i\left(\sum_{j=1}^2 b_j\psi_j \right) \right] dx = 0, \qquad i = 1, 2$$

$$(3.2.60\text{b})$$

In matrix form, we have,

$$\sum_{j=1}^2 K_{ij}b_j = f_i, \qquad \sum_{j=1}^2 K_{ji}a_j + \sum_{j=1}^2 G_{ij}b_j = 0 \qquad (3.2.61\text{a})$$

where

$$K_{ij} = \int_0^L \frac{d\phi_i}{dx}\frac{d\psi_j}{dx}\, dx, \qquad G_{ij} = \frac{1}{EI}\int_0^L \psi_i\psi_j\, dx, \qquad f_i = \int_0^L f_0\phi_i\, dx$$

$$(3.2.61\text{b})$$

For the choice of the approximation functions in Eq. (3.2.59), we have

$$[K] = \begin{bmatrix} -L & -L^2 \\ -L^2 & -\frac{2}{3}L^3 \end{bmatrix}, \qquad [G] = \frac{1}{EI}\begin{bmatrix} \dfrac{L^3}{3} & \dfrac{L^4}{4} \\ \dfrac{L^4}{4} & \dfrac{L^5}{5} \end{bmatrix}, \qquad \{f\} = \begin{Bmatrix} \dfrac{f_0 L^2}{2} \\ \dfrac{f_0 L^3}{3} \end{Bmatrix}$$

$$(3.2.62)$$

Solving the first equation for b's yields

$$b_1 = 0, \qquad b_2 = -\frac{f_0}{2} \qquad (3.2.63)$$

Substituting for b's in the second equation in Eq. (3.2.61a), and solving for a's we obtain

$$a_1 = -\frac{f_0 L^3}{20EI}, \qquad a_2 = -\frac{3}{40}\frac{f_0 L^2}{EI} \qquad (3.2.64)$$

Table 3.2 Comparison of the Scaled Deflections and Forces Obtained by the Two-Parameter Ritz Method Based on Potential Energy Functional (PEF) and Reissner's Variational Functional (RVF)[a]

x	Deflection $[\bar{w}(x)]$			Slope $(d\bar{w}/dx)$			Bending Moment $[\bar{M}(x)]$		Shear Force $(d\bar{M}/dx)$	
	Exact	PEF	RVF	Exact	PEF	RVF	Exact (RVF)	PEF	Exact (RVF)	PEF
0.0	0.0	0.0	0.0	0.0	0.0	−0.050	−0.500	−0.4167	1.0	0.5
0.1	−0.0023	−0.0020	−0.0057	−0.0452	−0.0392	−0.065	−0.405	−0.3667	0.9	0.5
0.2	−0.0087	−0.0077	−0.0130	−0.0813	−0.0733	−0.080	−0.320	−0.3167	0.8	0.5
0.3	−0.0183	−0.0165	−0.0218	−0.1095	−0.1025	−0.095	−0.245	−0.2667	0.7	0.5
0.4	−0.0304	−0.0280	−0.0320	−0.1307	−0.1267	−0.110	−0.180	−0.2167	0.6	0.5
0.5	−0.0443	−0.0417	−0.0438	−0.1458	−0.1458	−0.125	−0.125	−0.1667	0.5	0.5
0.6	−0.0594	−0.0570	−0.0570	−0.1560	−0.1600	−0.140	−0.080	−0.1167	0.4	0.5
0.7	−0.0753	−0.0735	−0.0717	−0.1622	−0.1692	−0.155	−0.045	−0.0667	0.3	0.5
0.8	−0.0917	−0.0907	−0.0880	−0.1653	−0.1733	−0.170	−0.020	−0.0167	0.2	0.5
0.9	−0.1083	−0.1080	−0.1057	−0.1665	−0.1725	−0.185	−0.005	−0.0333	0.1	0.5
1.0	−0.1250	−0.1250	−0.1250	−0.1667	−0.1667	−0.200	0.0	0.0833	0.0	0.5

[a] $\bar{w} = w(EI/f_0 L^4)$; $d\bar{w}/dx = (dw/dx)(EI/f_0 L^3)$; $\bar{M} = M/f_0 L^2$; $d\bar{M}/dx = (dM/dx)/f_0 L$.

The two-parameter Ritz solutions based on the stationary variational principle are

$$w_2(x) = -\frac{f_0 L^3 x}{20EI} - \frac{3}{40}\frac{f_0 L^2 x^2}{EI}, \qquad w_2(L) = -\frac{f_0 L^4}{8EI}$$

$$M_2(x) = -\frac{f_0}{2}(L-x)^2, \qquad M_2(0) = -\frac{f_0 L^2}{2} \qquad (3.2.65)$$

Note that maximum values of both deflection and moment coincide with the exact values, and that the expression for the bending moment coincides with the exact solution. Although the maximum deflection predicted by the approximation coincides with the exact solution, the deflection obtained violates the slope boundary condition at $x = 0$ (which is a natural boundary condition in the present case). Thus, the stationary variational statements give more accurate solutions for the forces because they are approximated independently; however, the accuracy of the displacements is somewhat degraded. Table 3.2 contains a comparison of pointwise deflections, slope, bending moment, and shear force obtained by the Ritz method and the exact method.

3.2.5 Application to General Boundary-Value Problems

All the examples presented in Section 3.2.4 utilized the associated total potential energy principle. For problems outside the field of solid mechanics, the construction of an analog of the total potential energy functional (or its first variation) is needed for the implementation of the Ritz method. Here the inverse procedure of constructing variational statements from differential equations and use of these statements in the Ritz approximation are studied.

Variational Formulation

The steps in the variational formulation of differential equations are described with the aid of a model problem: consider the problem of determining the solution u to the partial differential equation,

$$-\frac{\partial}{\partial x}\left(a_1 \frac{\partial u}{\partial x}\right) - \frac{\partial}{\partial y}\left(a_2 \frac{\partial u}{\partial y}\right) + a_0 u = f \quad \text{in } R \qquad (3.2.66)$$

in a two-dimensional domain R. Here a_0, a_1, a_2 and f are known functions of position (x, y) in R. The function u is required to satisfy, in addition to the differential equation (3.2.66), certain boundary conditions on the boundary S of R. The variational formulation to be presented tells us the precise form of the essential and natural boundary conditions of the equation. Equation (3.2.66) arises in many fields of engineering (see Table 3.3), including torsion.

Table 3.3 Some Examples of the Second-Order Partial Differential Equation

$$-\frac{\partial}{\partial x}\left(a_1 \frac{\partial u}{\partial x}\right) - \frac{\partial}{\partial y}\left(a_2 \frac{\partial u}{\partial y}\right) = f$$

Natural B.C.: $a_1 \frac{\partial u}{\partial x} n_x + a_2 \frac{\partial u}{\partial y} n_y = q$; essential B.C.: $u = \hat{u}$.

Field of application	Primary Variable u	Material Constants a_1, a_2	Source Variable f	Secondary Variables q
1. Heat transfer	temperature T	conductivities of an orthotropic medium: k_1, k_2	heat source Q	heat flux q
2. Irrotational flow of an ideal fluid	stream function ψ velocity potential ϕ	density ρ	mass production σ	$\dfrac{\partial \psi}{\partial x} = -v,\ \dfrac{\partial \psi}{\partial y} = u$ $\dfrac{\partial \phi}{\partial x} = u,\ \dfrac{\partial \phi}{\partial y} = v$ u and v are velocities

			recharge Q (or pumping $-Q$)	seepage q
3.	Groundwater flow	piezometric head, ϕ	permeabilities K_1, K_2	recharge Q (or pumping $-Q$)	seepage q $q = K_1 \dfrac{\partial \phi}{\partial x} n_x + K_2 \dfrac{\partial \phi}{\partial y} n_y$ $\left. \begin{array}{l} u = -K_1 \dfrac{\partial \phi}{\partial x} \\ v = -K_2 \dfrac{\partial \phi}{\partial y} \end{array} \right\}$ (velocities)
4.	Torsion of constant cross-section members	stress function Φ warping function w	$a_1 = a_2 = 1/G$, G = shear modulus $a_1 = a_2 = G\alpha$	$Q = 2\theta$, θ = angle of twist per unit length $f = 0$	$\left. \begin{array}{l} \dfrac{\partial \phi}{\partial x} = -\tau_{zy} \\ \dfrac{\partial \phi}{\partial y} = \tau_{zx} \end{array} \right\}$ shear stresses
5.	Electrostatics	scalar potential ϕ	dielectric constants $\varepsilon_1, \varepsilon_2$	charge density ρ	displacement flux density D_n
6.	Magnetostatics	magnetic potential ϕ	permeabilities μ_1, μ_2	charge density ρ	magnetic flux density B_n
7.	Transverse deflection of elastic membranes	transverse deflection u	$a_1 = a_2 = T$, T = tension in the membrane	transversely distributed load	q = normal force

The first step in the procedure is to multiply the given equation (with all nonzero terms on one side of the equality in the equation) with the variation δu and integrate over the domain:

$$0 = \int_R \delta u \left[-\frac{\partial}{\partial x}\left(a_1 \frac{\partial u}{\partial x} \right) - \frac{\partial}{\partial y}\left(a_2 \frac{\partial u}{\partial y} \right) + a_0 u - f \right] dx\, dy \quad (3.2.67)$$

Next, we trade the differentiation, using integration-by-parts (or use of the Green–Gauss theorem), between u and δu so that both u and δu have the same order derivatives. This is done to relax the continuity of the approximation functions ϕ_i in the Ritz approximation, $u = \sum_{i=1}^{n} c_i \phi_i + \phi_0$. If it is not possible to distribute differentiation equally between u and δu in any term of a given differential equation, we do not integrate those terms by parts. In the model problem at hand, the first two terms contain second-order derivatives, and therefore we can transfer one derivative to δu:

$$0 = \int_R \left(\frac{\partial \delta u}{\partial x} a_1 \frac{\partial u}{\partial x} + \frac{\partial \delta u}{\partial y} a_2 \frac{\partial u}{\partial y} + \delta u\, a_0 u - \delta u f \right) dx\, dy$$

$$- \oint_S \delta u \left(a_1 \frac{\partial u}{\partial x} n_x + a_2 \frac{\partial u}{\partial y} n_y \right) dS \quad (3.2.68)$$

The boundary term shows that the specification of u is the essential boundary condition and the specification of

$$a_1 \frac{\partial u}{\partial x} n_x + a_2 \frac{\partial u}{\partial y} n_y$$

is the natural boundary condition for the problem.

The last step in the procedure is to impose the specified boundary conditions. Suppose that u is specified on portion S_1 and the natural boundary condition is specified on the remaining portion S_2 of the boundary:

$$u = \hat{u} \quad \text{on } S_1$$

$$a_1 \frac{\partial u}{\partial x} n_x + a_2 \frac{\partial u}{\partial y} n_y = g \quad \text{on } S_2 \quad (3.2.69)$$

Then δu is arbitrary on S_2 and equal to zero on S_1. Consequently, Eq. (3.2.68) simplifies to

$$0 = \int_R \left(a_1 \frac{\partial \delta u}{\partial x} \frac{\partial u}{\partial x} + a_2 \frac{\partial \delta u}{\partial y} \frac{\partial u}{\partial y} + a_0 \delta u\, u - \delta u f \right) dx\, dy$$

$$- \int_{S_2} \delta u\, g\, dS \quad (3.2.70)$$

Equation (3.2.70) is called the *weak* or *variational form* of the differential equation (3.2.66). This form always includes the data of the natural boundary conditions of the problem. Therefore the weak form is used as a basis of the Ritz method. One can obtain a functional from the weak form if the expression involving both u and δu is symmetric in u and δu. In the present case, we have the symmetry. In such cases, we can pull the variational operator outside the entire expression (inverse of taking variation—or integration with respect to the variational operator) to obtain

$$
0 = \delta\left\{ \int_R \left\{ \frac{1}{2}\left[a_1\left(\frac{\partial u}{\partial x}\right)^2 + a_2\left(\frac{\partial u}{\partial y}\right)^2 + a_0 u^2 \right] - uf \right\} dx\,dy - \int_{S_2} ug\,dS \right\}
$$

$$
\equiv \delta I \tag{3.2.71}
$$

where $I(u)$ is the functional

$$
I(u) = \frac{1}{2}\int_R \left[a_1\left(\frac{\partial u}{\partial x}\right)^2 + a_2\left(\frac{\partial u}{\partial y}\right)^2 + a_0 u^2 \right] dx\,dy
$$

$$
- \int_R uf\,dx\,dy - \int_{S_2} gu\,dS \tag{3.2.72}
$$

This completes the variational formulation of the model problem.

Ritz Approximation

The Ritz method can be applied directly to the weak form (3.2.70). We substitute

$$
u = \sum_{j=1}^n c_j\phi_j + \phi_0, \qquad \delta u = \sum_{i=1}^n \delta c_i\,\phi_i \tag{3.2.73}
$$

into Eq. (3.2.70) and collect the coefficients of δc_i:

$$
0 = \sum_{i=1}^n \delta c_i \left\{ \int_R \left[a_1 \frac{\partial \phi_i}{\partial x}\left(\sum_{j=1}^n c_j\frac{\partial \phi_j}{\partial x} + \frac{\partial \phi_0}{\partial x} \right) + a_2 \frac{\partial \phi_i}{\partial y}\left(\sum_{j=1}^n c_j\frac{\partial \phi_j}{\partial y} + \frac{\partial \phi_0}{\partial y} \right) \right. \right.
$$

$$
\left. \left. + a_0\phi_i\left(\sum_{j=1}^n c_j\phi_j + \phi_0 \right) - \phi_i f \right] dx\,dy - \int_{S_2} \phi_i g\,dS \right\} \tag{3.2.74}
$$

Since δc_i are arbitrary variations of c_i and c_i are linearly independent, we must have the coefficients of δc_i for $i = 1, 2, \ldots, n$ should be zero. Rearranging the terms of the expressions in the curl brackets of Eq. (3.2.74), we obtain a system of n equations for the n parameters:

$$0 = \sum_{j=1}^{n} \left[\int_R \left(a_1 \frac{\partial \phi_i}{\partial x} \frac{\partial \phi_j}{\partial x} + a_2 \frac{\partial \phi_i}{\partial y} \frac{\partial \phi_j}{\partial y} + a_0 \phi_i \phi_j \right) dx \, dy \right] c_j$$

$$- \left[\int_R \left(\phi_i f - a_1 \frac{\partial \phi_i}{\partial x} \frac{\partial \phi_0}{\partial x} - a_2 \frac{\partial \phi_i}{\partial y} \frac{\partial \phi_0}{\partial y} - a_0 \phi_i \phi_0 \right) dx \, dy + \int_{S_2} \phi_i g \, dS \right]$$

$$= \sum_{j=1}^{n} K_{ij} c_j - F_i \tag{3.2.75}$$

where

$$K_{ij} = \int_R \left(a_1 \frac{\partial \phi_i}{\partial x} \frac{\partial \phi_j}{\partial x} + a_2 \frac{\partial \phi_i}{\partial y} \frac{\partial \phi_j}{\partial y} + a_0 \phi_i \phi_j \right) dx \, dy$$

$$F_i = \int_R \left(\phi_i f - a_1 \frac{\partial \phi_i}{\partial x} \frac{\partial \phi_0}{\partial x} - a_2 \frac{\partial \phi_i}{\partial y} \frac{\partial \phi_0}{\partial y} - a_0 \phi_i \phi_0 \right) dx \, dy + \int_{S_2} \phi_i g \, dS$$

$$\tag{3.2.76}$$

Once ϕ_i and ϕ_0 are selected, subjected to the conditions stated in Section 3.2.2, the coefficients of matrix $[K]$ and column vector $\{F\}$ can be computed, and the linear algebraic equations in Eq. (3.2.75) can be solved for the Ritz coefficients. We consider a specific example.

Example 3.6

We wish to solve the partial differential equation

$$-\left(\frac{\partial^2 u}{\partial x^2} + \frac{\partial^2 u}{\partial y^2} \right) = 0 \quad \text{in a unit square} \tag{3.2.77a}$$

$$u = 0 \quad \text{on } x = 0, 1 \text{ and } y = 0.$$

$$u = \sin \pi x \quad \text{on } y = 1 \tag{3.2.77b}$$

The origin of the coordinate system is taken at the lower left corner of the domain.

The variational form of the equation can be obtained as a special case from Eq. (3.2.75) by setting $f = 0$, $a_1 = a_2 = 1$, $a_0 = 0$ and $S_2 = 0$ ($S_1 = S$):

$$0 = \int_0^1 \int_0^1 \left(\frac{\partial \delta u}{\partial x} \frac{\partial u}{\partial x} + \frac{\partial \delta u}{\partial y} \frac{\partial u}{\partial y} \right) dx \, dy$$

The Ritz equations are given by Eq. (3.2.75) with K_{ij} and F_i given by

$$K_{ij} = \int_0^1 \int_0^1 \left(\frac{\partial \phi_i}{\partial x} \frac{\partial \phi_j}{\partial x} + \frac{\partial \phi_i}{\partial y} \frac{\partial \phi_j}{\partial y} \right) dx\, dy$$

$$F_i = -\int_0^1 \int_0^1 \left(\frac{\partial \phi_i}{\partial x} \frac{\partial \phi_0}{\partial x} + \frac{\partial \phi_i}{\partial y} \frac{\partial \phi_0}{\partial y} \right) dx\, dy \qquad (3.2.78)$$

Next we select ϕ_0 and ϕ_i for the problem. The function ϕ_0 is required to satisfy the boundary conditions in Eq. (3.2.77b) because all of them are of essential type. The following choice for ϕ_0 meets the conditions:

$$\phi_0 = y \sin \pi x \qquad (3.2.79a)$$

The functions ϕ_i ($i = 1, 2, \ldots, n$) are required to satisfy the homogeneous form of Eq. (3.2.77b) and be linearly independent. We choose

$$\phi_1 = \sin \pi x \sin \pi y, \qquad \phi_2 = \sin \pi x \sin 2\pi y, \qquad \phi_3 = \sin 2\pi x \sin \pi y, \quad \text{etc.}$$

$$(3.2.79b)$$

From Eq. (3.2.78) we obtain

$$K_{ij} = \begin{cases} \dfrac{\pi^2}{2} & \text{if } i = j \\ 0 & \text{if } i \neq j \end{cases} \quad \text{(i.e., } [K] \text{ is diagonal)}$$

$$F_i = \begin{cases} -\dfrac{\pi}{2} & \text{if } i = 1 \\ 0 & \text{if } i \neq 1 \end{cases} \qquad (3.2.80)$$

Hence, the solution becomes

$$u(x, y) = \sin \pi x \left(y - \frac{\sin \pi y}{\pi} \right) \qquad (3.2.81)$$

The exact solution of Eq. (3.2.77) is given by

$$u(x, y) = \frac{\sin \pi x \sinh \pi y}{\sinh \pi} \qquad (3.2.82)$$

We close this section with a few remarks on the Ritz method. The method can be applied, in principle, to any physical problem that can be cast in a variational form. The variational form does not have to be a quadratic functional, but it should be a form that includes the natural boundary conditions of the problem. Therefore, the method can be applied to even nonlinear problems that can be cast in a variational form in the sense discussed above. Of course, the resulting simultaneous algebraic equations are nonlinear and there can be more than one solution to the equations (see Exercise 3.3.11). Although most examples in this section were restricted to one-dimensional equations, the method can be applied to two- and three-dimensional problems. The selection of coordinate functions becomes increasingly difficult with the dimension and shape of the domain. We later consider some two-dimensional problems for geometrically simple domains. The finite-element method, a peicewise application of the Ritz type methods, is considered in Sec. 3.4.

Exercises 3.2

Give admissible coordinate functions, either algebraic or trigonometric, for a two-parameter Ritz approximation of Problems 1–7. Assume that the total potential energy principle is used to construct the Ritz approximation:

1. A cable suspended between points $A:(0,0)$ and $B:(L,h)$ and subjected to uniformly distributed transverse load of intensity f_0.

2. A cantilever beam subjected to uniformly distributed load of intensity f_0.

3. The symmetric half of the simply supported beam problem considered in Example 3.4 ($0 \le x \le L/2$).

4. A beam clamped at the left end and simply-supported at the right end, and subjected to point load P at $x = L/2$.

5. The spring-supported cantilever beam of Exercise 2.6.8.

6. A square elastic membrane fixed on all its sides and subjected to a uniformly distributed load of intensity f_0.

7. A quadrant model (because of the biaxial symmetry) of the membrane problem of Exercise 6.

8–14. Find the two-parameter Ritz approximation of the problems in Exercises 1–7 and compare the solutions with the exact solutions.

15. Find a one-parameter approximation of deflection and bending moment of the beam problem of Example 3.4, using Reissner's variational principle. Use uniformly distributed load.

16. Find the first two natural frequencies of a cantilever beam. Take $EI = (a + bx)^{-1}$ where a and b are constants.

17. Find a two-parameter Ritz approximation of the transverse deflection of a simply-supported beam on an elastic foundation that is subjected to uniformly distributed load. Use (a) algebraic polynomials, and (b) trigonometric polynomials.

18. Compute the norm of the error, $E = w - w_0$, where w and w_0 are the functions in Eqs. (3.2.41) and (3.2.42), using (a) the norm defined in Eq. (3.1.5) and (b) the norm $(u, v) = \int_0^L u''v'' \, dx$.

19. Use a two-parameter Ritz approximation with trigonometric functions to determine the critical buckling load P of a simply-supported beam.

20. Derive the matrix equations corresponding to the n-parameter Ritz approximation

$$w = c_1 x^2 + c_2 x^3 + \cdots + c_n x^{n+1}$$

of a cantilever beam with a uniformly distributed load. Compute b_{ij} and f_j in explicit form in terms of i, j, L, EI, and f_0 (see Fig. 2.4 for the geometry; set $P = 0$ for the present case).

For the following differential equations with associated boundary conditions, give the weak form and a one-parameter Ritz solution:

21. $-\dfrac{d}{dx}\left(u\dfrac{du}{dx}\right) + 1 = 0, \qquad 0 < x < 1$

$\dfrac{du}{dx}(0) = 0, \qquad u(1) = \sqrt{2}$

22. $\dfrac{d^2}{dx^2}\left(EI\dfrac{d^2 w}{dx^2}\right) + P\dfrac{d^2 w}{dx^2} = 0, \qquad 0 < x < L$

$w = 0, \quad EI\dfrac{d^2 w}{dx^2} = 0 \quad \text{at } x = 0, L$

Assume that $EI = $ constant and determine P from the resulting eigenvalue problem. The problem arises in connection with the buckling of a simply-supported beam.

23. $-\left(\dfrac{\partial^2 u}{\partial x^2} + \dfrac{\partial^2 u}{\partial y^2}\right) = 1$ in a unit square

$u(1, y) = u(x, 1) = 0$

$\dfrac{\partial u}{\partial x}(0, y) = \dfrac{\partial u}{\partial y}(x, 0) = 0$

The origin of the coordinate system is taken at the lower left corner of the unit square.

3.3 WEIGHTED-RESIDUAL METHODS

3.3.1 Introduction

The Ritz method can be employed in the approximate solution of any differential equation that has either a functional (i.e. variational principle) or a weak form. In both cases the variational statement contains the natural boundary conditions of the problem. Therefore, the Ritz approximations are required to satisfy only the specified essential boundary conditions of the problem. A generalization of the Ritz method to problems that do not have either a variational principle or a weak form (e. g., first-order equations) is due to Galerkin and Petrov. In the Galerkin method the parameters c_j of the approximation are determined using a weighted-integral form of the given equation, the weight functions being the approximation functions. For example, the Galerkin solution of Eq. (3.2.66) is given by $u = \Sigma c_j \phi_j$, where c_j are determined by substituting $u = \Sigma c_j \phi_j$ and $\delta u = \phi_i$ into Eq. (3.2.67). The Petrov–Galerkin method differs from the Galerkin method in the choice of the weight functions; the weight functions are different from the approximation functions. The present section is devoted to the discussion of these methods and other related methods, all of which are grouped as the weighted-residual methods.

Consider an operator equation of the form

$$A(u) = f \text{ in } R$$

$$B_1(u) = \hat{u} \text{ on } S_1, \qquad B_2(u) = \hat{g} \text{ on } S_2 \qquad (3.3.1)$$

A is a linear or nonlinear differential operator, u is the dependent variable, f is a given force term in the domain R, B_1 and B_2 are boundary operators associated with the operator A, and $\hat{u}$ and $\hat{g}$ are specified functions on the portions S_1 and S_2 of the boundary S of the domain. An example of Eq. (3.3.1) is given by

$$A(u) \equiv -\operatorname{div}(k \operatorname{grad} u), \qquad k = k(\mathbf{x})$$

$$B_1(u) \equiv u, \qquad B_2(u) \equiv k\frac{\partial u}{\partial n} \qquad (3.3.2)$$

In the method of weighted residual, a solution is sought in the form

$$u_n = \sum_{i=1}^{n} c_i \phi_i + \phi_0 \qquad (3.3.3)$$

and the parameters c_i are determined by requiring the residual of the approximation

$$E_n \equiv A\left(\sum_{j=1}^{n} c_j \phi_j + \phi_0\right) - f \neq 0 \qquad (3.3.4)$$

orthogonal to n linearly independent set of *weight functions* ψ_i:

$$\int_R \psi_i E_n(c_j, \phi_j, f) \, d\mathbf{x} = 0, \qquad i = 1, 2, \ldots, n \qquad (3.3.5)$$

The coordinate functions ϕ_0 and ϕ_i should satisfy the properties in Eq. (3.2.2), except that they should satisfy *all* of the boundary conditions:

ϕ_0 should satisfy all of the specified boundary conditions

ϕ_i should satisfy the homogeneous form of all specified boundary conditions

(3.3.6)

The variational principle referred to in Property 2 of Eq. (3.2.2) is now given by Eq. (3.3.5). Properties in Eq. (3.3.6) are required because the boundary conditions, both essential and natural, are not included in Eq. (3.3.5). Both properties now require ϕ_i to be of higher order than those used in the Ritz method. On the other hand, ψ_i can be any linearly independent set, such as $\{1, x, \dots\}$, and no continuity requirements are placed on ψ_i.

Equation (3.3.5) provides n linearly independent equations for the determination of the parameters c_i. If A is a nonlinear operator, the resulting algebraic equations will be nonlinear. Whenever A is linear, we have

$$A\left(\sum_{j=1}^{n} c_j \phi_j + \phi_0\right) = \sum_{j=1}^{n} c_j A(\phi_j) + A(\phi_0) \qquad (3.3.7)$$

and Eq. (3.3.5) becomes

$$\sum_{j=1}^{n}\left[\int_R \psi_i A(\phi_j)\, d\mathbf{x}\right] c_j - \int_R \psi_i [f - A(\phi_0)]\, d\mathbf{x} = 0$$

$$\sum_{j=1}^{n} a_{ij} c_j - q_i = 0, \qquad i = 1, 2, \dots, n \qquad (3.3.8a)$$

where

$$a_{ij} = \int_R \psi_i A(\phi_j)\, d\mathbf{x}, \qquad q_i = \int_R \psi_i [f - A(\phi_0)]\, d\mathbf{x} \qquad (3.3.8b)$$

Note that a_{ij} is not symmetric.

Returning to the question of selecting functions ϕ_i and ψ_i, in most problems of interest, the operator A permits the use of integration by parts (or gradient/divergence theorems) to transfer part of the differentiation to the weight functions ψ_i, provided that ψ_i are sufficiently differentiable. For example, if A is a $2m$th order linear differential operator, then the product $\psi_i A(\phi_j)$ can be integrated m times by parts to obtain an expression that involves mth order derivatives of both ψ_i and ϕ_j and the natural boundary conditions of the problem. As a specific example, consider the operator in

(3.3.2). The operator is a second-order (i.e., $m = 1$) operator. We have

$$
\int_R \psi_i A(\phi_j) \, d\mathbf{x} \equiv \int_R -\psi_i \mathrm{div}(k \, \mathrm{grad} \, \phi_j) \, d\mathbf{x}
$$

$$
= \int_R k \, \mathrm{grad} \, \psi_i \cdot \mathrm{grad} \, \phi_j \, d\mathbf{x} - \oint_S \psi_i k \frac{\partial \phi_j}{\partial n} \, dS
$$

$$
= \int_R k \, \mathrm{grad} \, \psi_i \cdot \mathrm{grad} \, \phi_j \, d\mathbf{x} - \int_{S_2} \psi_i \hat{g} \, dS - \int_{S_1} \psi_i k \frac{\partial \phi_j}{\partial n} \, dS
$$

$$(3.3.9)$$

where the divergence theorem is used in arriving at the second line and the natural boundary condition is included in the third line. If we select ψ_i such that $\psi_i = 0$ on S_1, the last term in Eq. (3.3.9) vanishes. Thus, both the coordinate functions ϕ_i and weight functions ψ_i should satisfy the requirements in Eq. (3.2.2).

In general, Eq. (3.3.8a) can be converted to a *weak form* if the operator A can be expressed as

$$
A = T^*(kT) \tag{3.3.10}
$$

where operator T^* is called the *adjoint* of T and is related to T by

$$
\int_R T(u) v \, d\mathbf{x} = \int_R u T^*(v) \, d\mathbf{x} + \int_{S_2} B_1 v B_2(u) \, dS \tag{3.3.11}
$$

for all u and v that satisfy the homogeneous form of essential boundary conditions associated with the operator A. For the operator in Eq. (3.3.2), T and T^* are given by

$$
T = \mathrm{grad}, \qquad T^* = -\mathrm{div}
$$

and Eq. (3.3.11) represents the Green–Gauss theorem. The weak form of Eq. (3.3.1), with A given by Eq. (3.3.10), is

$$
\sum_{j=1}^{n} \left[\int_R kT(\psi_i) T(\phi_j) \, d\mathbf{x} \right] c_j - \int_R [\psi_i f - kT(\psi_i) T(\phi_0)] \, d\mathbf{x} - \int_{S_2} \psi_i \hat{g} \, dS
$$

$$
- \sum_{j=1}^{n} c_j \int_{S_1} k \psi_i T(\phi_j) \, dS = 0
$$

$$
\sum_{j=1}^{n} b_{ij} c_j - f_i = 0 \tag{3.3.12a}
$$

where

$$
b_{ij} = \int_R kT(\psi_i) T(\phi_j) \, d\mathbf{x} - \int_{S_1} k \psi_i T(\phi_j) \, dS
$$

$$
f_i = \int_R [\psi_i f - kT(\psi_i) T(\phi_0)] \, d\mathbf{x} + \int_{S_2} \hat{g} \psi_i \, dS \tag{3.3.12b}
$$

The weighted residual method presented in this section is called the *Petrov–Galerkin* method when $\psi_i \neq \phi_i$. In problems for which a quadratic form or virtual work statement exists, the weighted residual method can be considered as a generalization of the Ritz method. The generalization involves using different coordinate functions for the dependent variable and the weight function:

$$u_n = \sum_{j=1}^{n} c_j \phi_j + \phi_0 \qquad \psi_i \neq \phi_i \tag{3.3.13}$$

Also, the method can be used to approximate the boundary conditions, especially the natural boundary conditions; see Section 3.3.4, Example 3.13. We look at other special cases of the weighted residual method in the sections ahead.

Example 3.7

Consider the following nondimensionalized form of the equation governing a variable cross section bar fixed at the left end and subjected to an axial force at the right end

$$-\frac{d}{dx}\left[(2-x)\frac{du}{dx}\right] = 0, \qquad 0 < x < 1$$

$$u(0) = 0, \qquad \frac{du}{dx}(1) = 1 \tag{3.3.14}$$

The virtual work statement associated with the equation is given by

$$0 = \int_0^1 (2-x)\frac{d\delta u}{dx}\frac{du}{dx}\,dx - \delta u(1) \tag{3.3.15}$$

Let us choose a two-parameter approximation of u,

$$u_2 = c_1\phi_1 + c_2\phi_2 + \phi_0 \tag{3.3.16}$$

We solve the problem using the Petrov–Galerkin method in two forms: using Eq. (3.3.8) and Eq. (3.3.12).

For the Petrov–Galerkin method based on Eq. (3.3.8), we use the coordinate functions ϕ_i and ϕ_0

$$\phi_0 = x, \qquad \phi_1 = 2x - x^2, \qquad \phi_2 = 3x - x^3 \tag{3.3.17}$$

which satisfy the required conditions: $\phi_0(0) = 0$, $\phi_0'(1) = 1$, $\phi_i(0) = 0$, $\phi_i'(1) = 0$. For the weight functions, we choose

$$\psi_1 = x, \qquad \psi_2 = x^2 \tag{3.3.18}$$

which are linearly independent. Substituting Eqs. (3.3.17) and (3.3.18) into Eq. (3.3.8b), we obtain

$$a_{ij} = \int_0^1 \psi_i \left\{ -\frac{d}{dx} \left[(2-x) \frac{d\phi_j}{dx} \right] \right\} dx, \qquad q_i = \int_0^1 \psi_i \frac{d}{dx} \left[(2-x) \frac{d\phi_0}{dx} \right] dx$$

or, in matrix form, we have

$$\begin{bmatrix} \frac{5}{3} & \frac{13}{4} \\ 1 & \frac{11}{5} \end{bmatrix} \begin{Bmatrix} c_1 \\ c_2 \end{Bmatrix} = \begin{Bmatrix} -\frac{1}{2} \\ -\frac{1}{3} \end{Bmatrix}, \qquad c_1 = -\frac{1}{25}, \qquad c_2 = -\frac{2}{15}$$

and the approximate solution is given by

$$u_2(x) = x - \tfrac{1}{25}(2x - x^2) - \tfrac{2}{15}(3x - x^3)$$

$$= 0.52x + 0.04x^2 + 0.1333x^3 \tag{3.3.19}$$

The exact solution of Eq. (3.3.14) is given by

$$u_0(x) = -\log\left(1 - \frac{x}{2}\right)$$

$$= 0.5x + 0.125x^2 + 0.04167x^3 + \cdots \tag{3.3.20}$$

For the Petrov–Galerkin method based on Eq. (3.3.12), we choose the coordinate functions and weight functions

$$\phi_0 = 0, \qquad \phi_1 = x, \qquad \phi_2 = x^2, \qquad \psi_1 = x^2, \qquad \psi_2 = x^3 \tag{3.3.21}$$

We obtain from Eq. (3.3.12a) [the boundary term in b_{ij} is zero because $\psi_i(0) = 0$]

$$\begin{bmatrix} \frac{4}{3} & \frac{5}{3} \\ \frac{5}{4} & \frac{9}{5} \end{bmatrix} \begin{Bmatrix} c_1 \\ c_2 \end{Bmatrix} = \begin{Bmatrix} 1 \\ 1 \end{Bmatrix}, \qquad c_1 = \tfrac{8}{19}, \qquad c_2 = \tfrac{5}{19}$$

The solution is given by

$$u_2(x) = \tfrac{8}{19}x + \tfrac{5}{19}x^2 = 0.421x + 0.263x^2 \tag{3.3.22}$$

If we solve the same problem by the Ritz method with the ϕ_i given in Eq. (3.3.21), we obtain

$$\begin{bmatrix} \frac{3}{2} & \frac{4}{3} \\ \frac{4}{3} & \frac{5}{3} \end{bmatrix} \begin{Bmatrix} c_1 \\ c_2 \end{Bmatrix} = \begin{Bmatrix} 1 \\ 1 \end{Bmatrix}, \qquad c_1 = \tfrac{6}{13}, \qquad c_2 = \tfrac{3}{13}$$

$$u_2(x) = \tfrac{6}{13}x + \tfrac{3}{13}x^2 = 0.462x + 0.231x^2 \tag{3.3.23}$$

For the choice of coordinate functions in Eq. (3.3.17), the Ritz method gives

$$\begin{bmatrix} \frac{7}{3} & \frac{43}{10} \\ \frac{43}{10} & \frac{81}{10} \end{bmatrix} \begin{Bmatrix} c_1 \\ c_2 \end{Bmatrix} = \begin{Bmatrix} -\frac{5}{3} + 1 \\ -\frac{13}{4} + 2 \end{Bmatrix}, \qquad c_1 = -\frac{5}{82}, \qquad c_2 = -\frac{5}{41}$$

$$u_2(x) = x - \frac{5}{82}(2x - x^2) - \frac{5}{41}(3x - x^3) = 0.5122x + 0.061x^2 + 0.122x^3$$

$$(3.3.24)$$

From the above calculations, it is clear that the Ritz method gives, for the linear problem at hand, more accurate results, than the weighted residual method. In general, for problems that permit the construction of a quadratic functional, the Ritz method gives the best results.

In the following sections, we consider some special cases of the weighted residual methods that differ from each other in the selection of the weight functions.

3.3.2 The Galerkin Method

The Galerkin method is a special case of the Petrov–Galerkin method (or, more exactly, the Petrov–Galerkin method is a generalization of the Galerkin method) and constitutes a generalization of the Ritz method. As we shall see, this method, when applied to problems with quadratic functionals (or virtual work principles), reduces to the Ritz method.

In the Galerkin method, we seek a solution of the form Eq. (3.3.3) such that the residual in Eq. (3.3.4) is orthogonal to each of the coordinate functions which satisfy the conditions in Eqs. (3.2.2) and (3.3.6):

$$\int_R E_n(c_j, \phi_j, f)\phi_i \, dx = 0, \qquad i = 1, 2, \ldots, n \qquad (3.3.25)$$

Equation (3.3.25) provides n linearly independent algebraic equations for the solution of the parameters c_i ($i = 1, 2, \ldots, n$) which, in conjunction with Eq. (3.3.3), gives the n-parameter Galerkin approximation of Eq. (3.3.1).

The Galerkin method is a special case of the Petrov–Galerkin method for the weight function $\psi_i = \phi_i$. It is more general than the Ritz method, because no quadratic functional or virtual work principle is necessary. If the governing equations are derivable from a variational principle, then the Galerkin method reduces to the Ritz method [and the conditions on ϕ_i are reduced to those in Eq. (3.2.2)]. This can be seen from Eq. (3.3.9). Also, for problems with all specified boundary conditions of the essential type, the Galerkin equations in Eqs. (3.3.8) and (3.3.12) give the same result, and the results coincide with those of the Ritz method for the same choice of coordinate functions. The following examples illustrate these ideas.

Example 3.8

Consider the differential equation

$$-\left(\frac{\partial^2 u}{\partial x^2} + \frac{\partial^2 u}{\partial y^2}\right) = f \quad \text{in } -a \leq x, y \leq a$$

$$u = 0 \quad \text{on the boundary} \qquad (3.3.26)$$

in a square region. The origin of the coordinate system is taken at the center of the domain, which is a square of length $2a$. The equation arises, among others (see Table 3.3), in connection with the transverse deflection of a membrane fixed on all sides and with the torsion of a square cross-section member (see Section 1.7.4). The function u denotes the deflection in the case of a membrane and the Prandtl stress function in the case of torsion problems; f denotes the transverse force on the membrane and $f = 2G\theta$, where θ is twist/unit length and G is the shear modulus, for the torsion problem.

The quadratic functional associated with Eq. (3.3.26) is given by

$$\Pi(u) = \int_{-a}^{a}\int_{-a}^{a} \frac{1}{2}\left[\left(\frac{\partial u}{\partial x}\right)^2 + \left(\frac{\partial u}{\partial y}\right)^2 - 2fu\right] dx\, dy \qquad (3.3.27)$$

which represents the total potential energy for the membrane and the total complementary energy for the torsion problem. The functional $\Pi(u)$ can be derived either from virtual work principle (see Example 2.11 for the procedure) or by the inverse procedure outlined in Section 3.2.5. The boundary condition in Eq. (3.3.26) is kinematic (the natural boundary condition for the equation involves the specification of $\partial u/\partial n$).

For the Galerkin method, we assume ($\phi_0 = 0$)

$$u_n = \sum_{j=1}^{n} c_j \phi_j(x, y), \qquad \delta u_n = \sum_{i=1}^{n} \delta c_i \phi_i(x, y) \qquad (3.3.28a)$$

where

$$\phi_1 = (x^2 - a^2)(y^2 - a^2), \qquad \phi_2 = (x^2 + y^2)\phi_1 \qquad (3.3.28b)$$

Clearly, ϕ_i meet the requirements. From Eq. (3.3.8), we have

$$a_{ij} = \int_{-a}^{a}\int_{-a}^{a} \phi_i A(\phi_j)\, dx\, dy, \qquad q_i = \int_{-a}^{a}\int_{-a}^{a} \phi_i f\, dx\, dy$$

For the two-parameter approximation, we have (for $f = $ constant, f_0)

$$a^8\begin{bmatrix} \frac{256}{45} & \frac{1024}{525}a^2 \\ \frac{1024}{525}a^2 & \frac{11264}{4725}a^4 \end{bmatrix}\begin{Bmatrix} c_1 \\ c_2 \end{Bmatrix} = f_0 a^6 \begin{Bmatrix} \frac{16}{9} \\ \frac{32}{45}a^2 \end{Bmatrix} \qquad (3.3.29a)$$

or

$$c_1 = \frac{1295f_0}{4432a^2}, \qquad c_2 = \frac{525f_0}{8864a^4} \qquad (3.3.29b)$$

and the solution is given by

$$u_2(x) = \frac{f_0 a^2}{8864}\left[2590 + 525(\bar{x}^2 + \bar{y}^2)\right](1 - \bar{x}^2)(1 - \bar{y}^2) \qquad (3.3.30)$$

where $\bar{x} = x/a$ and $\bar{y} = y/a$.

If we use Eq. (3.3.12), the coordinate functions that satisfy the required conditions are again given by Eq. (3.3.28b), and we have

$$b_{ij} = \int_{-a}^{a}\int_{-a}^{a}\left(\frac{\partial\phi_i}{\partial x}\frac{\partial\phi_j}{\partial x} + \frac{\partial\phi_i}{\partial y}\frac{\partial\phi_j}{\partial y}\right)dx\,dy, \qquad f_i = \int_{-a}^{a}\int_{-a}^{a}f_0\phi_i\,dx\,dy$$

$$(3.3.31)$$

Calculations will show that

$$b_{ij} = a_{ij} \quad \text{and} \quad f_i = q_i$$

and the solution coincides with that given in Eq. (3.3.30).

The exact solution to Eq. (3.3.26) can be obtained using the separation of variables method. We have

$$u_0(x, y) = \frac{16f_0 a^2}{\pi^3}\sum_{n=1,3,5,\ldots}^{\infty}\frac{1}{n^3}(-1)^{(n-1)/2}\left[1 - \frac{\cosh(n\pi y/2a)}{\cosh(n\pi/2)}\right]\cos\frac{n\pi x}{2a}$$

$$(3.3.32)$$

The value of u_0 at the center of the region is

$$u_0(0, 0) = 0.2942 f_0 a^2$$

whereas the Galerkin solution gives $0.2922 f_0 a^2$, which is only 0.68% in error. We consider this problem again in Sections 3.3.4 and 3.3.5.

The next example is concerned with the application of the Galerkin method to a problem in which both essential and natural boundary conditions are specified.

Example 3.9

Let us reconsider the differential equation of Example 3.8. This time we exploit the biaxial symmetry and model only a quadrant $(0 \le (x, y) \le a)$ of the domain. In that case, the boundary conditions along the symmetry lines, $x = 0$

and $y = 0$, are

$$\frac{\partial u}{\partial x} = 0 \text{ along } x = 0$$

$$\frac{\partial u}{\partial y} = 0 \text{ along } y = 0 \qquad (3.3.33)$$

These are the natural boundary conditions of the problem.

To solve the problem using Eq. (3.3.8), we must select ϕ_i that satisfy both the essential and natural boundary conditions of the problem. Two sets of such functions are given below ($\phi_0 = 0$):

$$\phi_i - \cos\frac{(2i-1)\pi x}{2a}\cos\frac{(2i-1)\pi y}{2a}, \qquad i - 1, 2, \ldots, n \qquad (3.3.34a)$$

$$\phi_i = (x^2 - a^2)(y^2 - a^2)x^{2i}y^{2i}, \qquad i = 0, 1, 2, \ldots, n \qquad (3.3.34b)$$

The one-parameter solutions corresponding to coordinate functions in Eq. (3.3.34a) and (3.3.34b) respectively, are given by

$$u_1(x, y) = \frac{32 f_0 a^2}{\pi^4}\cos\frac{\pi x}{2a}\cos\frac{\pi y}{2a}$$

$$u_1(x, y) = \frac{5 f_0 a^2}{16}\left[\left(\frac{x}{a}\right)^2 - 1\right]\left[\left(\frac{y}{a}\right)^2 - 1\right] \qquad (3.3.35)$$

The first one is 11.66% in error and the second one, which corresponds to the one-parameter approximation in Eq. (3.3.28), is 6.22% in error when compared to the exact solution.

To solve the problem using Eq. (3.3.12), we can select ϕ_i ($= \psi_i$) that satisfy only the essential boundary conditions of the problem. Of course, for the choice of coordinate functions in Eq. (3.3.34), one would get the same solution as in Eq. (3.3.35). Let us choose the following one-parameter approximation

$$u_1 = c_1(x - a)(y - a) \qquad (3.3.36)$$

Substituting Eq. (3.3.36) into Eq. (3.3.12) yields the solution

$$u_1 = \frac{3 f_0 a^2}{8}\left(\frac{x}{a} - 1\right)\left(\frac{y}{a} - 1\right)$$

This solution, when evaluated at $x = y = 0$, gives a value that is -27.46% in error, compared to the exact solution. The solution obtained by using the coordinate functions

$$\phi_i = (x^i - a^i)(y^i - a^i)$$

should give an increasingly better solution as the number of parameters is increased.

3.3.3   Least-Squares, Collocation, and Subdomain Methods

Least-Squares Method

The least-squares method is a variational method in which the integral of the square of the residual in the approximation of a given differential equation is minimized with respect to the parameters in the approximation:

$$\min\left\{ I(c_i) \equiv \int_R |E(c_j, \phi_j, f)|^2 \, d\mathbf{x} \right\} \tag{3.3.37a}$$

or

$$\int_R 2E(c_j, \phi_j, f) \frac{\partial E}{\partial c_i} \, d\mathbf{x} = 0, \qquad i = 1, 2, \dots, n \tag{3.3.37b}$$

where E is the residual defined in Eq. (3.3.4). Equation (3.3.37) provides n algebraic equations for the constants c_i, which then give the least-squares solution

$$u_n = \sum_{j=1}^{n} c_j \phi_j + \phi_0$$

First, we note that the least squares method is a special case of the weighted-residual method for the weight function

$$\psi_i = \frac{\partial E}{\partial c_i} \tag{3.3.38}$$

Therefore, the coordinate functions ϕ_i should satisfy the same conditions as in the case of the weighted-residual method. Next, if the operator A in the governing equation is linear, the weight function ψ_i becomes

$$\psi_i = \frac{\partial E}{\partial c_i} = A(\phi_i) \tag{3.3.39}$$

Then, from Eq. (3.3.8) we have

$$\sum_{j=1}^{n} \left[\int_R A(\phi_i) A(\phi_j) \, d\mathbf{x} \right] c_j - \int_R [A(\phi_i)f - A(\phi_i)A(\phi_0)] \, d\mathbf{x} = 0$$

or

$$\sum_{j=1}^{n} g_{ij}c_j - h_i = 0, \qquad i = 1, 2, \ldots, n \qquad (3.3.40a)$$

where

$$g_{ij} = \int_R A(\phi_i)A(\phi_j)\,dx, \qquad h_i = \int_R A(\phi_i)[f - A(\phi_0)]\,dx \quad (3.3.40b)$$

Note that the coefficient matrix is symmetric. The method requires higher-order coordinate functions than the Ritz and Galerkin methods, because the coefficient matrix g_{ij} involves the same operator as in the original differential equation and no trading of differentiation can be achieved. Such methods can be advantageous for first-order differential equations, because one obtains a symmetric coefficient matrix. Another advantage of the method, when applied to linear equations, is that not only u_n converges to the exact solution u_0, but Au_n also converges to f as n is increased.

Example 3.10

Consider the bending of a cantilever beam under uniformly distributed load, as discussed in Example 3.5. We wish to find the least-squares approximation to the fourth-order equation

$$\frac{d^2}{dx^2}\left(EI\frac{d^2w}{dx^2}\right) + f_0 = 0, \qquad 0 < x < L \qquad (3.3.41)$$

and the pair of equations (3.2.52). Obviously, Eq. (3.3.41) demands coordinate functions that are of degree four or higher, whereas the pair in Eq. (3.2.52) requires coordinate functions of degree two or higher.
From Eq. (3.3.41), we have the operator

$$A \equiv \frac{d^2}{dx^2}\left(EI\frac{d^2}{dx^2}\right)$$

and Eq. (3.3.40) becomes

$$g_{ij} = \int_0^L \frac{d^2}{dx^2}\left(EI\frac{d^2\phi_i}{dx^2}\right)\frac{d^2}{dx^2}\left(EI\frac{d^2\phi_j}{dx^2}\right)dx$$

$$h_i = \int_0^L \frac{d^2}{dx^2}\left(EI\frac{d^2\phi_i}{dx^2}\right)\left[-f_0 - \frac{d^2}{dx^2}\left(EI\frac{d^2\phi_0}{dx^2}\right)\right]dx \qquad (3.3.42)$$

For an n-parameter approximation, we must choose ϕ_i that satisfies all of the boundary conditions in Eq. (3.2.52c). Let us begin with a fourth-order polynomial for ϕ_1 and determine the constants in the polynomial to satisfy the boundary conditions:

$$\phi_1 = a_0 + a_1 x + a_2 x^2 + a_3 x^3 + a_4 x^4$$

$$\phi_1(0) = 0 \text{ gives } a_0 = 0, \qquad \phi_1'(0) = 0 \text{ gives } a_1 = 0$$

$$\phi_1''(L) = 0 \text{ gives } 2a_2 + 6a_3 L + 12a_4 L^2 = 0$$

$$\phi_1'''(L) = 0 \text{ gives } 6a_3 + 24a_4 L = 0 \tag{3.3.43a}$$

The last two equations have the nontrivial solution

$$a_3 = -\frac{2}{3L}a_2, \qquad a_4 = \frac{1}{6L^2}a_2, \qquad a_2 = \text{ arbitrary (nonzero)}$$

Setting $a_2 = 1$, we obtain

$$\phi_1 = \left(x^2 - \frac{2}{3L}x^3 + \frac{1}{6L^2}x^4\right) \tag{3.3.43b}$$

which is the desired function; other ϕ_i can be generated similarly with an increasing number of arbitrary constants in each ϕ_i, $i = 2, \ldots, n$.

A trigonometric function that satisfies all boundary conditions can be obtained by choosing

$$\phi_1''' = \cos\frac{\pi x}{2L} \tag{3.3.44a}$$

and integrating with respect to x and determining the constants with the use of the boundary conditions. We obtain, after lengthy algebra and setting arbitrary constants to unity,

$$\phi_1 = x^2 - \frac{4L}{\pi}\left(x - \frac{2L}{\pi}\sin\frac{\pi x}{2L}\right) \tag{3.3.44b}$$

For the choice of ϕ_1 from Eq. (3.3.43b), Eq. (3.3.42) gives

$$g_{11} = (EI)^2 L\left(\frac{4}{L^2}\right)^2 = \frac{16}{L^3}(EI)^2$$

$$h_1 = -\frac{4EI}{L}f_0, \qquad c_1 = -\frac{f_0 L^2}{4EI} \tag{3.3.45a}$$

The one-parameter least-squares solution becomes

$$w_1 = -\frac{f_0}{24EI}(x^4 - 4x^3 L + 6x^2 L^2) \tag{3.3.45b}$$

which coincides with the exact solution.

For the choice of ϕ_1 from Eq. (3.3.44b), we have

$$g_{11} = \left(\frac{EI\pi^2}{2L^2}\right)^2 \int_0^L \sin^2 \frac{\pi x}{2L} \, dx = \frac{(EI)^2 \pi^4}{8L^3}$$

$$h_1 = -\frac{EI\pi f_0}{L}, \qquad c_1 = -\frac{8f_0 L^2}{EI\pi^3} \qquad\qquad (3.3.46a)$$

and the solution is given by

$$w_1 - -\frac{8f_0 L^2}{EI\pi^3}\left[x^2 - \frac{4L}{\pi}\left(x - \frac{2L}{\pi}\sin\frac{\pi x}{2L}\right)\right] \qquad (3.3.46b)$$

The maximum deflection is $w_1(L) = -0.13864 \, f_0 L^4$, which is 10.9% in error, compared to the exact value.

Now we wish to solve the same problem using Eq. (3.2.52). Since we have two operators, we should begin with the basic equation (3.3.37):

$$\frac{\partial}{\partial c_i}\int_0^L \left(\sum_{j=1}^{n=1} b_j \frac{d^2\psi_j}{dx^2} + f\right)^2 dx = 0, \qquad i = 1, 2, 3$$

$$(3.3.47a)$$

$$\frac{\partial}{\partial c_i}\int_0^L \left(\sum_{j=1}^{n=2} a_j \frac{d^2\phi_j}{dx^2} - \frac{1}{EI}\sum_{j=1}^{m=1} b_j \psi_j\right)^2 dx = 0, \qquad i = 1, 2, 3$$

$$(3.3.47b)$$

where

$$w_2(x) = a_1\phi_1 + a_2\phi_2 = a_1 x^2 + a_2 x^3$$

$$M_2(x) = b_1\psi_1 = b_1(L - x)^2 \qquad\qquad (3.3.48)$$

which satisfy the boundary conditions in Eq. (3.2.52c). The parameters c_i have the meaning

$$c_1 = a_1, \qquad c_2 = a_2, \qquad c_3 = b_1$$

Clearly, we have one equation from Eq. (3.3.47a) (the other two are trivial) and three equations from Eq. (3.3.47b), whereas there are only three constants to be determined. To circumvent this difficulty, we minimize the error in Eq.

(3.3.47a) with respect to b_i and that in Eq. (3.3.47b) with respect to a_i and obtain the three equations needed:

$$\int_0^L \frac{d^2\psi_i}{dx^2} \left(\sum_{j=1}^m b_j \frac{d^2\psi_j}{dx^2} + f_0 \right) dx = 0$$

$$\int_0^L \frac{d^2\phi_i}{dx^2} \left(\sum_{j=1}^n a_j \frac{d^2\phi_j}{dx^2} - \frac{1}{EI} \sum_{j=1}^m b_j \psi_j \right) dx = 0 \qquad (3.3.49)$$

or

$$\sum_{j=1}^{m=1} M_{ij} b_j + t_i = 0, \qquad i = 1, 2, \ldots, m$$

$$\sum_{j=1}^{n=2} N_{ij} a_j - \sum_{j=1}^{m=1} P_{ij} b_j = 0, \qquad i = 1, 2, \ldots, n \qquad (3.3.50a)$$

where

$$M_{ij} = \int_0^L \frac{d^2\psi_i}{dx^2} \frac{d^2\psi_j}{dx^2} dx, \qquad N_{ij} = \int_0^L \frac{d^2\phi_i}{dx^2} \frac{d^2\phi_j}{dx^2} dx$$

$$P_{ij} = \int_0^L \frac{1}{EI} \frac{d^2\phi_i}{dx^2} \psi_j \, dx, \qquad t_i = \int_0^L \frac{d^2\psi_i}{dx^2} f_0 \, dx \qquad (3.3.50b)$$

Using the coordinate functions of Eq. (3.3.48) in Eq. (3.3.50), we obtain

$$M_{11} = 4L, \qquad N_{11} = 4L, \qquad N_{12} = N_{21} = 6L^2, \qquad N_{22} = 12L^3$$

$$P_{11} = \frac{2L^3}{3EI}, \qquad P_{21} = \frac{L^4}{2EI} \qquad t_1 = 2f_0 L$$

$$b_1 = -\frac{f_0}{2}, \qquad a_1 = -\frac{5f_0 L^2}{24EI}, \qquad a_2 = \frac{f_0 L}{12EI} \qquad (3.3.51a)$$

and the solution becomes

$$w_2(x) = -\frac{5f_0 L^2}{24EI} x^2 + \frac{f_0 L}{12EI} x^3$$

$$M_1(x) = -\frac{f_0}{2}(L - x)^2 \qquad (3.3.51b)$$

The results agree with those obtained in Example 3.5 by the Ritz approximation based on the total potential energy functional.

Collocation Method

In the collocation method, the parameters c_i are determined by forcing the residual in the approximation of the governing equation to vanish at n selected points $\mathbf{x}^i$ ($i = 1, 2, \ldots, n$) in the domain

$$E\left(\mathbf{x}^i, c_j, \phi_j, f\right) = 0 \qquad i = 1, 2, \ldots, n \qquad (3.3.52)$$

For linear operator equations, this yields

$$\sum_{j=1}^{n} A\left[\phi_j(\mathbf{x}^i)\right] c_j - f(\mathbf{x}^i) = 0 \qquad (3.3.53)$$

Equation (3.3.53) can be recast in a form that resembles the weighted-residual method. Let $\delta(\mathbf{x} - \mathbf{x}^i)$ denote the Dirac delta function

$$\int_R \delta(\mathbf{x} - \mathbf{x}^i) f(\mathbf{x}) \, d\mathbf{x} \equiv f(\mathbf{x}^i) \qquad (3.3.54)$$

Then Eq. (3.3.53) can be expressed in the alternative form

$$\sum_{j=1}^{n} \left\{ \int_R \delta(\mathbf{x} - \mathbf{x}^i) A\left[\phi_j(\mathbf{x})\right] d\mathbf{x} \right\} c_j - \int_R \delta(\mathbf{x} - \mathbf{x}^i) f(\mathbf{x}) \, d\mathbf{x} = 0 \quad (3.3.55)$$

A comparison of Eq. (3.3.54) with Eq. (3.3.5) shows that the weight functions ψ_i are given by

$$\psi_i(\mathbf{x}) = \delta(\mathbf{x} - \mathbf{x}^i). \qquad (3.3.56)$$

Thus, the collocation method is a special case of the weighted-residual method.

The selection of the *collocation points* $\mathbf{x}^i$ is crucial in obtaining a well-conditioned system of equations and a convergent solution. The collocation points should be located as evenly as possible to avoid ill-conditioning of the resulting equations.

Example 3.11

Consider a simply-supported beam under uniformly distributed load f_0. The governing differential equation is given by Eq. (3.3.41), and the boundary conditions are

$$w(0) = w(L) = 0, \qquad \frac{d^2 w}{dx^2}(0) = \frac{d^2 w}{dx^2}(L) = 0 \qquad (3.3.57)$$

We wish to find an approximation to the transverse deflection using the collocation method.

Let us consider a two-parameter approximation.

$$w_2 = c_1 \sin\frac{\pi x}{L} + c_2 \sin\frac{3\pi x}{L} \qquad (3.3.58)$$

with collocation points at $x = L/4$ and $x = L/2$. We obtain

$$EI\left[c_1 \left(\frac{\pi}{L}\right)^4 \sin\frac{\pi}{4} + c_2 \left(\frac{3\pi}{L}\right)^4 \sin\frac{3\pi}{4}\right] + f_0 = 0$$

$$EI\left[c_1 \left(\frac{\pi}{L}\right)^4 \sin\frac{\pi}{2} + c_2 \left(\frac{3\pi}{L}\right)^4 \sin\frac{3\pi}{2}\right] + f_0 = 0 \qquad (3.3.59)$$

which yields

$$EI\left(\frac{\pi}{L}\right)^4 \begin{bmatrix} \dfrac{1}{\sqrt{2}} & \dfrac{81}{\sqrt{2}} \\ 1 & -81 \end{bmatrix} \begin{Bmatrix} c_1 \\ c_2 \end{Bmatrix} = \begin{Bmatrix} -f_0 \\ -f_0 \end{Bmatrix}$$

or

$$c_1 = -\frac{(1+\sqrt{2})f_0 L^4}{2EI\pi^4}, \qquad c_2 = -\frac{(\sqrt{2}-1)f_0 L^4}{162EI\pi^4}$$

$$w_2(x) = -\frac{f_0 L^4}{162EI\pi^4}\left(195.55 \sin\frac{\pi x}{L} + 0.414 \sin\frac{3\pi x}{L}\right) \qquad (3.3.60)$$

Note that if we use the points $x = L/3$ and $x = 2L/3$ as collocation points, it would not be possible to compute c_2. Because of the symmetry, the use of points $x = L/4$ and $x = L/2$ in the half-beam is sufficient. The maximum deflection, $w_2(L/2) = 1.205 f_0 L^4/EI\pi^4 = -(f_0 L^4/80.87EI)$, is 5% in error compared to the exact value, $-(f_0 L^4/76.8EI)$.

Subdomain Method

In the subdomain method, the domain of the problem is subdivided into as many subdomains as there are adjustable parameters, and then the parameters are determined by making the residual orthogonal to a constant (say unity) in each domain:

$$\int_{R_i} E(c_j, \phi_j, f)\, d\mathbf{x} = 0, \qquad i = 1, 2, \ldots, n \qquad (3.3.61)$$

where R_i denotes the ith subdomain. Equation (3.3.61) states that the average value of the residual in each subdomain is zero. Obviously, in this method, negative errors can cancel positive errors to give least net error, although the sum of the absolute values of the errors is very large. Once again, we can interpret the subdomain method as a special case of the weighted-residual method for the weight function

$$\psi_i(x) = \begin{cases} 1, & \text{if } x \text{ is in } R_i \\ 0, & \text{otherwise} \end{cases} \qquad (3.3.62)$$

Example 3.12

Reconsider the simply supported beam of Example 3.11. Because of the symmetry, we can consider two subdomains in the first half of the beam: $R_1 = (0, L/4)$, $R_2 = (L/4, L/2)$. For the same choice of coordinate functions as in Example 3.11, we obtain

$$\int_0^{L/4} \left[c_1 EI \left(\frac{\pi}{L} \right)^4 \sin \frac{\pi x}{L} + c_2 EI \left(\frac{3\pi}{L} \right)^4 \sin \frac{3\pi x}{L} + f_0 \right] dx = 0$$

$$\int_{L/4}^{L/2} \left[c_1 EI \left(\frac{\pi}{L} \right)^4 \sin \frac{\pi x}{L} + c_2 EI \left(\frac{3\pi}{L} \right)^4 \sin \frac{3\pi x}{L} + f_0 \right] dx = 0$$

which result in

$$EI \left(\frac{\pi}{L} \right)^4 \frac{L}{\pi} \left(1 - \frac{1}{\sqrt{2}} \right) c_1 + EI \left(\frac{3\pi}{L} \right)^4 \frac{L}{3\pi} \left(1 + \frac{1}{\sqrt{2}} \right) c_2 + \frac{f_0 L}{4} = 0$$

$$EI \left(\frac{\pi}{L} \right)^4 \frac{L}{\pi} \frac{1}{\sqrt{2}} c_1 + EI \left(\frac{3\pi}{L} \right)^4 \frac{L}{3\pi} \left(-\frac{1}{\sqrt{2}} \right) c_2 + \frac{f_0 L}{4} = 0$$

$$(3.3.63)$$

whose solution is

$$c_1 = -\frac{(1 + \sqrt{2}) f_0 L^4}{4\sqrt{2} \, EI\pi^3}, \qquad c_2 = -\frac{(\sqrt{2} - 1) f_0 L^4}{108\sqrt{2} \, EI\pi^3}$$

$$w_2(x) = -\frac{f_0 L^4}{108\sqrt{2} \, EI\pi^3} \left(65.184 \sin \frac{\pi x}{L} + 0.414 \sin \frac{3\pi x}{L} \right) \qquad (3.3.64)$$

The center deflection obtained in the subdomain method, $w_2(L/2) = -(f_0 L^4 / 73.12 EI)$ is -5% in error, compared to the exact value.

3.3.4   The Kantorovich Method

In the Ritz and weighted residual methods, the solution of an ordinary or partial differential equation was reduced to the solution of a set of algebraic equations in terms of undetermined parameters. In the Kantorovich method, the problem of solving partial differential equations is reduced to the solution of ordinary differential equations in terms of undetermined functions. Hopefully, it is possible to solve the ordinary differential equations more precisely, perhaps in exact form, to achieve better accuracy than the Ritz approximation. Thus, the Kantorovich method is appropriate for problems with two or more independent variables. We describe the essence of the method, using a partial differential equation in two dimensions and the ideas introduced in connection with the Ritz method.

Consider the problem of determining the solution $u = u(x, y)$ to the variational problem

$$B(v, u) = l(v) \quad \text{for any admissible } v \tag{3.3.65}$$

where $B(v, u)$ and $l(v)$ are the bilinear and linear forms associated with the problem to be solved. We assume that the natural boundary conditions of the problem are already included in Eq. (3.3.65). Equation (3.3.65) is nothing but an alternative form of the principle of virtual displacements. It is the *weak form* (see Sec. 3.1.4) of the governing differential (i.e., Euler) equations

$$Au = f \quad \text{in } R$$

$$\mathscr{B}u = \hat{g} \quad \text{on } S_2 \tag{3.3.66}$$

In the Ritz method, we sought an n-parameter approximation of u in the form

$$u_n = \sum_{j=1}^{n} c_j \phi_j(x, y) + \phi_0(x, y) \tag{3.3.67}$$

where ϕ_0 and ϕ_j satisfy the conditions specified in Eq. (3.2.2), by requiring that Eq. (3.3.65) be satisfied for every $v \equiv \phi_i$, $i = 1, 2, \ldots, n$. Substituting Eq. (3.3.67) into Eq. (3.3.65) gives

$$B\left(\phi_i, \sum_{j=1}^{n} c_j \phi_j + \phi_0\right) = l(\phi_i)$$

$$\sum_{j=1}^{n} B(\phi_i, \phi_j) c_j = l(\phi_i) - B(\phi_i, \phi_0)$$

or

$$\sum_{j=1}^{n} b_{ij} c_j = f_i \tag{3.3.68a}$$

where

$$b_{ij} = B(\phi_i, \phi_j), \qquad f_i = l(\phi_i) - B(\phi_i, \phi_0) \qquad (3.3.68b)$$

Although Eqs. (3.3.65) and (3.3.68) are in slightly different forms compared to Eq. (3.2.8) and (3.2.10), they describe the Ritz method in general terms. For example, consider Poisson's equation in two dimensions

$$-\nabla^2 u = f \quad \text{in } R$$

$$u = \hat{u} \quad \text{on } S_1, \qquad \frac{\partial u}{\partial n} = \hat{g} \quad \text{on } S_2 \qquad (3.3.69)$$

The bilinear and linear forms for the equations are

$$B(v, u) = \int_R \text{grad } v \cdot \text{grad } u \, dx \, dy = \int_R \left(\frac{\partial v}{\partial x} \frac{\partial u}{\partial x} + \frac{\partial v}{\partial y} \frac{\partial u}{\partial y} \right) dx \, dy$$

$$l(v) = \int_R fv \, dx \, dy + \int_{S_2} \hat{g} v \, dS \qquad (3.3.70)$$

For the Ritz approximation, Eq. (3.3.70) requires ϕ_0 and ϕ_j to satisfy the following conditions:

$$\phi_0 = \hat{u} \quad \text{on } S_1, \qquad \phi_j = 0 \quad \text{on } S_1, \qquad j = 1, 2, \ldots, n \qquad (3.3.71a)$$

The associated algebraic equations become

$$b_{ij} = \int_R \left(\frac{\partial \phi_i}{\partial x} \frac{\partial \phi_j}{\partial x} + \frac{\partial \phi_i}{\partial y} \frac{\partial \phi_j}{\partial y} \right) dx \, dy$$

$$f_i = \int_R f\phi_i \, dx \, dy + \int_{S_2} \hat{g}\phi_i \, dS - \int_R \left(\frac{\partial \phi_i}{\partial x} \frac{\partial \phi_0}{\partial x} + \frac{\partial \phi_i}{\partial y} \frac{\partial \phi_0}{\partial y} \right) dx \, dy$$

$$(3.3.71b)$$

In the Kantorovich method, we seek an n-parameter solution of Eq. (3.3.65) in the form

$$u_n = \sum_{j=1}^{n} c_j(x)\phi_j(x, y) + \phi_0(x, y) \qquad (3.3.72)$$

where c_j are now undetermined *functions* of the coordinate x, and ϕ_0 and ϕ_j satisfy the following conditions:

$$\phi_0 = \hat{u} \quad \text{on } S_1 \qquad (3.3.73a)$$

and

$$\text{either } \phi_j = 0 \quad \text{on } S_1 \quad \text{or} \quad c_j = 0 \quad \text{on } S_1 \qquad (3.3.73b)$$

so that $c_j \phi_j = 0$ on S_1 for any $j = 1, 2, \ldots, n$. Thus, we have some flexibility in choosing functions ϕ_j. Substituting Eq. (3.3.72) for u and $\delta c_i \phi_i$ for v into Eq. (3.3.65) gives

$$B(\delta c_i \phi_i, u_n) = l(\delta c_i \phi_i) \quad (\text{sum on } i) \qquad (3.3.74)$$

where δc_i is the variation of c_i; δc_i are arbitrary, except that we require them to vanish on S_1. Since this is the variational form of Eq. (3.3.66), we can integrate by parts to obtain

$$\int_R \delta c_i \phi_i [A(u_n) - f] \, dx \, dy + \int_{S_1} \delta c_i \phi_i \mathscr{B}(u_n) \, dS + \int_{S_2} \delta c_i \phi_i [\mathscr{B}(u_n) - \hat{g}] \, dS = 0$$

$$(3.3.75)$$

The boundary terms drop out because $\delta c_i = 0$ on S_1 and $\mathscr{B}(u_n) = \hat{g}$ on S_2. Now we assume that the region R is such that we can separate the integration with respect to x and y:

$$\int_a^b \delta c_i \left\{ \int_P^Q [A(u_n) - f] \phi_i(x, y) \, dy \right\} dx = 0 \qquad (3.3.76)$$

Since $\delta c_i(x)$ are arbitrary in the interval (a, b), it follows from Eq. (3.3.76) that

$$\int_P^Q [A(u_n) - f] \phi_i \, dy = 0 \qquad (3.3.77a)$$

or

$$\sum_{j=1}^n \left[\int_P^Q \phi_i A(\phi_j) \, dy \right] c_j = \int_P^Q \phi_i [f - A(\phi_0)] \, dy \qquad (3.3.77b)$$

which gives a system of n ordinary differential equations in c_j.

When applied to Eq. (3.3.69), Eqs. (3.3.74)–(3.3.76) take the form (for $\phi_0 = \hat{u} = 0$)

$$\int_R \left[\frac{\partial}{\partial x}(\delta c_i \phi_i) \frac{\partial u_n}{\partial x} + \frac{\partial}{\partial y}(\delta c_i \phi_i) \frac{\partial u_n}{\partial y} \right] dx \, dy = \int_R \delta c_i \phi_i f \, dx \, dy + \int_{S_2} \hat{g} \delta c_i \phi_i \, dS$$

$$\int_R \left[-\delta c_i \phi_i \left(\frac{\partial^2 u_n}{\partial x^2} + \frac{\partial^2 u_n}{\partial y^2} \right) \right] dx \, dy + \oint_S \delta c_i \phi_i \frac{\partial u_n}{\partial n} \, dS = \int_R \delta c_i \phi_i f \, dx \, dy$$

$$+ \int_{S_2} \hat{g} \delta c_i \phi_i \, dS$$

Since $\delta c_i = 0$ on S_1 and $\partial u_{n}/\partial n = \hat{g}$ on S_2, the above equation simplifies to

$$\int_R \delta c_i \phi_i \left(- \nabla^2 u_n - f \right) dx\, dy = 0 \qquad (3.3.78)$$

which corresponds to Eq. (3.3.75) for $A = -\nabla^2$. Now, suppose that the region R is a rectangle: $0 \leq x \leq a$, $0 \leq y \leq b$. Then Eq. (3.3.78) can be expressed as

$$\int_0^a \delta c_i \left[\int_0^b \left(- \nabla^2 u_n - f \right) \phi_i\, dy \right] dx = 0$$

Since δc_i is an arbitrary function of x, we have

$$\int_0^b \left(- \nabla^2 u_n - f \right) \phi_i\, dy = 0, \qquad i = 1, 2, \ldots, n \qquad (3.3.79)$$

A comparison of Eq. (3.3.77a) with Eq. (3.3.25) shows that the Kantorovich method is a generalization of the Ritz method, and it is a special case of the weighted residual method. The main features are that the coordinate functions in the Kantorovich method are selected much like in the Galerkin method, and the residual in one coordinate direction is made orthogonal to the weight functions $\psi_i = \phi_i$.

We now illustrate the basic ideas of the Kantorovich method via the torsion problem for a square cross-section member.

Example 3.13

Consider the membrane/torsion problem considered in Example 3.8. Here we seek a one-parameter Kantorovich approximation of the form,

$$u_1 = c_1(x)\phi_1(y) = c_1(x)(y^2 - a^2) \qquad (3.3.80)$$

Since $u = 0$ on $x = \pm a$ and on $y = \pm a$, it follows that we must determine $c_1(x)$ such that it satisfies the conditions

$$c_1(-a) = c_1(a) = 0 \qquad (3.3.81)$$

From Eq. (3.3.77), we get

$$\int_{-a}^a \left(- \frac{d^2 c_1}{dx^2} \phi_1 - c_1 \frac{d^2 \phi_1}{dy^2} - f_0 \right) \phi_1\, dy = 0$$

$$- \int_{-a}^a \left[(y^2 - a^2) \frac{d^2 c_1}{dx^2} + 2c_1 + f_0 \right] (y^2 - a^2)\, dy = 0$$

Performing the integration and dividing throughout by the coefficient of

d^2c_1/dx^2 we get

$$\frac{d^2c_1}{dx^2} - \frac{5}{2a^2}c_1 - \frac{5f_0}{4a^2} = 0 \qquad (3.3.82)$$

This completes the application of the Kantorovich method. We can solve the ordinary differential equation (3.3.82) either exactly or by an approximate method. In the interest of improving the accuracy (over the Ritz solution), we consider the exact solution of Eq. (3.3.82) subjected to the boundary conditions in Eq. (3.3.81).

The exact solution of Eq. (3.3.82) is given by

$$c_1(x) = A \cosh kx + B \sinh kx - \frac{f_0}{2}, \qquad k = \sqrt{\frac{5}{2a^2}} \qquad (3.3.83)$$

Using the conditions in Eq. (3.3.81), the constants of integration, A and B, can be evaluated

$$A = \frac{f_0}{2 \cosh ka}, \qquad B = 0$$

The solution in Eq. (3.3.80) becomes

$$u_1 = -(y^2 - a^2)\left[1 - \frac{\cosh kx}{\cosh ka}\right]\frac{f_0}{2} \qquad (3.3.84)$$

A two-parameter Kantorovich approximation of the form

$$u_2(x, y) = (y^2 - a^2)\left[c_1(x) + c_2(x)y^2\right] \qquad (3.3.85)$$

gives the differential equations

$$\frac{8}{15}a^2\frac{d^2c_1}{dx^2} + \frac{8}{105}a^4\frac{d^2c_2}{dx^2} - \left(\frac{4}{3}c_1 + \frac{4}{15}a^2c_2\right) = \frac{2}{3}f_0$$

$$\frac{8}{105}a^4\frac{d^2c_1}{dx^2} + \frac{8}{315}a^6\frac{d^2c_2}{dx^2} - \left(\frac{4a^2}{15}c_1 + \frac{44}{105}a^4c_2\right) = \frac{2}{15}f_0a^2$$

$$(3.3.86a)$$

with the boundary conditions

$$c_1(-a) = c_1(a) = c_2(-a) = c_2(a) = 0 \qquad (3.3.86b)$$

The simultaneous differential equations in Eq. (3.3.86) can be solved as follows: let $D = d/dx$, $D^2 = d^2/dx^2$, and so on. Then we can write Eq.

(3.3.86a) in the operator form

$$
\begin{bmatrix}
\dfrac{8a^2}{15}D^2 - \dfrac{4}{3} & \dfrac{8a^4}{105}D^2 - \dfrac{4a^2}{15} \\[3mm]
\dfrac{8a^4}{105}D^2 - \dfrac{4a^2}{15} & \dfrac{8a^6}{315}D^2 - \dfrac{44a^4}{105}
\end{bmatrix}
\begin{Bmatrix} c_1 \\[3mm] c_2 \end{Bmatrix}
=
\begin{Bmatrix} \dfrac{2}{3}f_0 \\[3mm] \dfrac{2}{15}f_0 a^2 \end{Bmatrix}
\qquad (3.3.87)
$$

Using Cramer's rule, but keeping in mind that D's operate on the quantities in front of them, we obtain

$$
L(c_1) =
\begin{vmatrix}
\dfrac{2}{3}f_0 & \dfrac{8a^4}{105}D^2 - \dfrac{4a^2}{15} \\[3mm]
\dfrac{2}{15}f_0 a^2 & \dfrac{8a^6}{315}D^2 - \dfrac{44a^4}{105}
\end{vmatrix}
$$

$$
= \left(\dfrac{8}{315}a^6 D^2 - \dfrac{44}{105}a^4 \right) \dfrac{2}{3}f_0 - \left(\dfrac{8}{105}a^4 D^2 - \dfrac{4a^2}{15} \right) \dfrac{2}{15}f_0 a^2
$$

$$
= -\dfrac{384}{1575}f_0 a^4 \qquad (3.3.88a)
$$

$$
L(c_2) =
\begin{vmatrix}
\dfrac{8a^2}{15}D^2 - \dfrac{4}{3} & \dfrac{2}{3}f_0 \\[3mm]
\dfrac{8}{105}a^4 D^2 - \dfrac{4a^2}{15} & \dfrac{2}{15}f_0 a^2
\end{vmatrix}
= 0 \qquad (3.3.88b)
$$

where $L(\cdot)$ is the determinant of the operator matrix in Eq. (3.3.87),

$$
L = \dfrac{256a^8}{33075} \left(D^4 - \dfrac{28}{a^2}D^2 + \dfrac{63}{a^4} \right) \qquad (3.3.89)
$$

The general solutions of Eqs. (3.3.88a) and (3.3.88b) are

$$
c_1(x) = -\dfrac{f_0}{2} + A_1 \cosh k_1 x + A_2 \sinh k_1 x + A_3 \cosh k_2 x + A_4 \sinh k_2 x
$$

$$
c_2(x) = B_1 \cosh k_1 x + B_2 \sinh k_1 x + B_3 \cosh k_2 x + B_4 \sinh k_2 x \qquad (3.3.90)
$$

where A_i and B_i ($i = 1, 2, 3, 4$) are constants to be determined, and k_1^2 and k_2^2 are the roots of the quadratic equation

$$
k^4 - \dfrac{28}{a^2}k^2 + \dfrac{63}{a^4} = 0
$$

$$
k_1^2 = 14 - \sqrt{133}, \qquad k_2^2 = 14 + \sqrt{133} \qquad (3.3.91)
$$

Substituting Eq. (3.3.90) into the first equation in (3.3.86a), we obtain the relationships

$$\left(\frac{8a^2}{15}k_1^2 - \frac{4}{3}\right)A_i = \left(\frac{4a^2}{15} - \frac{8}{105}a^4k_1^2\right)B_i, \qquad i = 1, 2$$

$$\left(\frac{8a^2}{15}k_2^2 - \frac{4}{3}\right)A_i = \left(\frac{4a^2}{15} - \frac{8}{105}a^4k_2^2\right)B_i, \qquad i = 3, 4 \qquad (3.3.92)$$

Use of the boundary conditions in Eq. (3.3.86b) gives

$$A_2 = A_4 = B_2 = B_4 = 0$$

$$A_1\cosh k_1 a + A_3\cosh k_2 a = \frac{f_0}{2}$$

$$B_1\cosh k_1 a + B_3\cosh k_2 a = 0 \qquad (3.3.93)$$

which can be solved along with Eq. (3.3.92) for A_1, B_1, A_3, and B_3. Finally, we have

$$c_1(x) = -\frac{f_0}{2} + 0.5156\, f_0 \frac{\cosh k_1 x}{\cosh k_1 a} - 0.0156\, f_0 \frac{\cosh k_2 x}{\cosh k_2 a}$$

$$c_2(x) = -0.1138\, f_0 \frac{\cosh k_1 x}{\cosh k_1 a} + 0.1138\, f_0 \frac{\cosh k_2 x}{\cosh k_2 a} \qquad (3.3.94)$$

Equations (3.3.94) and (3.3.85) define the two-parameter Kantorovich solution of the torsion problem.

A comparison of the solutions obtained using the Galerkin method and the Kantorovich method (with one and two parameters) with the series solution of the torsion problem is presented in Table 3.4. The table also includes the shearing stress, $\sigma_{yz} = -\partial u/\partial x$, at $x = a$ and $y = 0$. The exact value of the maximum shear stress, which occurs at $x = a$ and $y = 0$, is $0.675 f_0$. It is clear from the results that the two-parameter solutions are more accurate than the one-parameter solutions, and that the Kantorovich method gives slightly more accurate solutions than the Galerkin method.

We remark that the Kantorovich method gives, in general, more accurate solutions, provided that one can solve the resulting differential equations exactly. The method has two limitations when compared to the Ritz method. First, it assumes that the solution is separable into two components; second, the resulting ordinary differential equations are difficult to solve for $n \geq 3$. The method can be used to reduce partial differential equations of time and space to ordinary differential equations in time, which can then be solved using the Laplace transforms. In the next example, we illustrate the use of the Kantorovich method with time-dependent problems.

Table 3.4 Comparison of the Solution $\bar{u}(x,0)$ of Eq. (3.3.26) Obtained by the Galerkin–Ritz and Kantorovich Methods with the Series Solution[a]

x $(y = 0)$	Series Solution Eq. (3.3.32)	Ritz–Galerkin Solution[b] Eq. (3.3.30)		Kantorovich Solution Eq. (3.3.85)	
		One Parameter	Two Parameter	One Parameter	Two Parameter
0.0	0.29445	0.31250	0.29219	0.30261	0.29473
0.10000	0.29194	0.30937	0.28986	0.30014	0.29222
0.20000	0.28437	0.30000	0.28278	0.29266	0.28462
0.30000	0.27158	0.28437	0.27075	0.27999	0.27177
0.40000	0.25328	0.26250	0.25340	0.26180	0.25339
0.50000	0.22909	0.23437	0.23025	0.23765	0.22909
0.60000	0.19854	0.20000	0.20065	0.20693	0.19839
0.70000	0.16104	0.15937	0.16382	0.16886	0.16072
0.80000	0.11591	0.11250	0.11884	0.12250	0.11546
0.90000	0.06239	0.05937	0.06463	0.06667	0.06203
1.00000	−0.00038	0.0	0.0	−0.00447	0.00000
		Shear Stress, $\bar{\sigma}_{yz}(a,0)$			
1.00000	0.67500	0.62500	0.70284	0.72636	0.66416

[a] $\bar{u} = u/f_0 a^2$; $\bar{\sigma}_{yz} = \sigma_{yz}/f_0 a$.
[b] The Ritz and Galerkin solutions are the same for the same coordinate functions.

Example 3.14

Consider the following nondimensionalized partial differential equation governing the transverse motion of a cable:

$$\frac{\partial u}{\partial t} - \frac{\partial^2 u}{\partial x^2} = 0, \quad 0 < x < 2$$

$$u(0,t) = u(2,t) = 0 \quad \text{for } t > 0$$

$$u(x,0) = 1.0 \quad \text{for } 0 < x < 2 \tag{3.3.95}$$

Owing to the symmetry about the $x = 1$ point, we solve the following equivalent problem

$$\frac{\partial u}{\partial t} - \frac{\partial^2 u}{\partial x^2} = 0, \quad 0 < x < 1 \tag{3.3.96}$$

$$u(0,t) = \frac{\partial u}{\partial x}(1,t) = 0 \quad \text{for } t > 0 \tag{3.3.97}$$

$$u(x,0) = 1.0 \quad \text{for } 0 < x < 1 \tag{3.3.98}$$

Let us consider the following form of a one-parameter Kantorovich approximation:

$$u_1(x, t) = c_1(t)\phi_1(x) = c_1(t)(2x - x^2) \qquad (3.3.99)$$

The function $\phi_1(x)$, which is a function of x only, satisfies the boundary conditions in Eq. (3.3.97). We should determine $c_1(t)$ such that the initial condition (3.3.98) is satisfied. This requires that

$$c_1(0) = 1 \qquad (3.3.100)$$

Substituting Eq. (3.3.99) into Eq. (3.3.77), we obtain

$$\int_0^1 \left(\frac{\partial u_1}{\partial t} - \frac{\partial^2 u_1}{\partial x^2} \right) \phi_1 \, dx = 0$$

$$\int_0^1 \left[\frac{dc_1}{dt}(2x - x^2) - c_1(-2) \right] (2x - x^2) \, dx = 0$$

or

$$\frac{dc_1}{dt} + \frac{5}{2}c_1 = 0 \qquad (3.3.101)$$

The solution of Eq. (3.3.101) is given by

$$c_1(t) = Ae^{-(5/2)t}, \qquad c_1(0) = 1 \to A = 1$$

and the complete solution becomes

$$u_1(x, t) = e^{-2.5t}(2x - x^2) \qquad (3.3.102)$$

The exact solution of Eqs. (3.3.96)–(3.3.98) is given by

$$u(x, t) = 2 \sum_{n=0}^{\infty} \frac{e^{-\lambda_n^2 t}\sin \lambda_n x}{\lambda_n}, \qquad \lambda_n = \frac{(2n + 1)\pi}{2} \qquad (3.3.103)$$

For a two-parameter approximation, we use the weak form

$$\int_0^1 \phi_i \left(\frac{\partial u}{\partial t} - \frac{\partial^2 u}{\partial x^2} \right) dx = 0$$

$$\int_0^1 \left(\phi_i \frac{\partial u}{\partial t} + \frac{\partial \phi_i}{\partial x} \frac{\partial u}{\partial x} \right) dx = 0 \qquad (3.3.104)$$

where ϕ_i satisfy only the essential boundary condition. It should be noted that such a weak formulation is not always possible with Eq. (3.3.77). For example,

Eq. (3.3.79) cannot be integrated by parts with respect to x and, therefore, we cannot obtain its weak form. For the problem at hand, we seek a solution in the form,

$$u_2(x,t) = c_1(t)\phi_1 + c_2(t)\phi_2 = c_1(t)x + c_2(t)x^2 \qquad (3.3.105)$$

where ϕ_1 and ϕ_2 satisfy the essential boundary condition, $u(0) = 0$. Substitution of Eq. (3.3.105) into Eq. (3.3.104) gives the ordinary differential equations in time,

$$\frac{1}{3}\frac{dc_1}{dt} + \frac{1}{4}\frac{dc_2}{dt} + c_1 + c_2 = 0$$

$$\frac{1}{4}\frac{dc_1}{dt} + \frac{1}{5}\frac{dc_2}{dt} + c_1 + \frac{4}{3}c_2 = 0 \qquad (3.3.106)$$

which must be solved subject to the initial condition in Eq. (3.3.98). From Eq. (3.3.105), it is clear that there is no way the initial condition will be satisfied exactly. Therefore, we satisfy the initial condition in the Galerkin sense:

$$\int_0^1 [u(x,0) - 1]\phi_i\, dx = 0, \qquad i = 1,2 \qquad (3.3.107a)$$

We obtain

$$\tfrac{1}{3}c_1(0) + \tfrac{1}{4}c_2(0) = \tfrac{1}{2}, \qquad \tfrac{1}{4}c_1(0) + \tfrac{1}{5}c_2(0) = \tfrac{1}{3} \qquad (3.3.107b)$$

Now we use the Laplace transform method to solve Eqs. (3.3.106) and (3.3.107b). Let $\bar{c}_i(s)$ denote the Laplace transform of $c_i(t)$:

$$L[c_i(t)] \equiv \int_{-\infty}^{\infty} c_i(t)e^{-st}\, dt \qquad (3.3.108)$$

The Laplace transform of the derivative of $c_i(t)$ is given by

$$L\left[\frac{dc_i}{dt}\right] = s\bar{c}_i(s) - c_i(0) \qquad (3.3.109)$$

Using Eqs. (3.3.109), Eq. (3.3.106) can be transformed to

$$\left(\tfrac{1}{3}s + 1\right)\bar{c}_1 + \left(\tfrac{1}{4}s + 1\right)\bar{c}_2 = \tfrac{1}{3}c_1(0) + \tfrac{1}{4}c_2(0)$$

$$\left(\tfrac{1}{4}s + 1\right)\bar{c}_1 + \left(\tfrac{1}{5}s + \tfrac{4}{3}\right)\bar{c}_2 = \tfrac{1}{4}c_1(0) + \tfrac{1}{5}c_2(0) \qquad (3.3.110)$$

The right-hand side can be substituted from Eq. (3.3.107b), and the resulting equations can be solved for $\bar{c}_1(s)$ and $\bar{c}_2(s)$

$$\bar{c}_1(s) = \frac{\frac{1}{2}(\frac{1}{5}s + \frac{4}{3}) - \frac{1}{3}(\frac{1}{4}s + 1)}{(\frac{1}{5}s + \frac{4}{3})(\frac{1}{3}s + 1) - (\frac{1}{4}s + 1)^2} = \frac{12s + 240}{3s^2 + 104s + 240}$$

$$\bar{c}_2(s) = \frac{\frac{1}{3}(\frac{1}{3}s + 1) - \frac{1}{2}(\frac{1}{4}s + 1)}{(\frac{1}{5}s + \frac{4}{3})(\frac{1}{3}s + 1) - (\frac{1}{4}s + 1)^2} = -\frac{10s + 120}{3s^2 + 104s + 240}$$

$$(3.3.111)$$

The inverse transform is given by

$$L^{-1}\left[\frac{s + a_1}{(s + a)(s + b)}\right] = \frac{a_1 - a}{b - a}e^{-at} + \frac{a_1 - b}{a - b}e^{-bt} \quad (3.3.112)$$

$$c_1(t) = 1.6408e^{-32.1807t} + 2.3592e^{-2.486t}$$

$$c_2(t) = -(2.265e^{-32.1807t} + 1.068e^{-2.486t}) \quad (3.3.113)$$

Equations (3.3.105) and (3.3.113) together give the two-parameter solution.

The two variational solutions in Eq. (3.3.102) and (3.3.105) are compared, for various values of t and x, with the series solution (3.3.103) in Table 3.5. The two-parameter solution is more accurate than the one-parameter solution and agrees more closely with the series solution for large values of time.

The last example illustrates the use of variational methods for the solution of time-dependent problems. The approximation is selected such that the parameters c_i are functions of time and ϕ_i are functions of spatial coordinates (of course, the original method of Kantorovich suggests that ϕ_i be a function of time also). The procedure used in Example 3.14 to obtain ordinary differential equations in time is known as the *semidiscrete* approximation.

3.3.5 The Trefftz Method

In all the variational methods discussed in this chapter, the coordinate functions were selected such that they satisfied the boundary conditions (most often, only the essential boundary conditions) of the problem, and the parameters were determined using a variational procedure, such as the minimization of a quadratic functional or setting the weighted-residual to zero. An alternative approach is to select the coordinate functions that satisfy the governing differential equation, but not the boundary conditions, and to determine the parameters, using a variational procedure that requires the satisfaction of the boundary conditions. The Trefftz method is one such method, and it provides an upper bound to the solution (i.e., the solution converges to the exact

Table 3.5 Comparison of the Variational Solution (by the Kantorovich Method) with the Series Solution of the Transient Motion of a Cable Subjected to Initial Displacement $u(x, 0) = 1$ [see Eqs. (3.3.95)–(3.3.98)]

t	x	Series Solution Eq. (3.3.103)	Variational Solution Eq. (3.3.102)	Eq. (3.3.105)
0.05	0.2	0.4727	0.3177	0.4265
	0.4	0.7938	0.5648	0.7413
	0.6	0.9418	0.7413	0.9443
	0.8	0.9880	0.8472	1.0357
	1.0	0.9965	0.8825	1.0154
0.25	0.2	0.2135	0.1927	0.2306
	0.4	0.4052	0.3426	0.4152
	0.6	0.5560	0.4496	0.5538
	0.8	0.6520	0.5139	0.6466
	1.0	0.6848	0.5353	0.6934
0.50	0.2	0.1145	0.1031	0.1238
	0.4	0.2177	0.1834	0.2230
	0.6	0.2996	0.2407	0.2975
	0.8	0.3522	0.2751	0.3473
	1.0	0.3703	0.2865	0.3725
0.75	0.2	0.0617	0.0552	0.0665
	0.4	0.1174	0.0981	0.1198
	0.6	0.1616	0.1288	0.1600
	0.8	0.1900	0.1472	0.1866
	1.0	0.1997	0.1534	0.2001
1.0	0.2	0.0333	0.0296	0.0357
	0.4	0.0633	0.0525	0.0643
	0.6	0.0872	0.0690	0.0858
	0.8	0.1025	0.0788	0.1002
	1.0	0.1077	0.0821	0.1075
1.50	0.2	0.0097	0.0085	0.0103
	0.4	0.0184	0.0151	0.0186
	0.6	0.0254	0.0198	0.0248
	0.8	0.0298	0.0226	0.0289
	1.0	0.0313	0.0235	0.0310

solution from above as the number of parameters is increased). Since the Ritz method provides a lower bound on the energy, a method that provides an upper bound is desirable (especially if the bounds are close) to assess the accuracy of the approximate solution. The Trefftz method is applicable to problems with essential boundary conditions. We describe the essence of the method below.

Consider the operator equation

$$A(u) - f = 0 \quad \text{in } R \tag{3.3.114}$$

$$\mathcal{B}(u) - \hat{u} = 0 \quad \text{on } S \tag{3.3.115}$$

where A and $\mathcal{B}$ are linear operators and f and $\hat{u}$ are given functions of position. In the Trefftz method, we seek an approximate solution of the form in Eq. (3.3.3), with ϕ_0 and ϕ_i satisfying the following properties:

ϕ_0 is a particular solution of Eq. (3.3.114)
ϕ_i are (nonconstant) solutions of the homogeneous equation,

$$A(u) = 0 \tag{3.3.116}$$

The parameters c_i are determined by setting the weighted-residual in the boundary condition (3.3.115) to zero:

$$\oint_S [\mathcal{B}(u_n) - \hat{u}] \frac{\partial \phi_i}{\partial n} \, dS = 0 \tag{3.3.117a}$$

or

$$\sum_{i=1}^{n} \left[\oint_S \frac{\partial \phi_i}{\partial n} \mathcal{B}(\phi_j) \, dS \right] c_j = \oint_S \frac{\partial \phi_i}{\partial n} \hat{u} \, dS$$

$$\sum_{j=1}^{n} \beta_{ij} c_j = \alpha_i \tag{3.117b}$$

where

$$\beta_{ij} = \oint_S \frac{\partial \phi_i}{\partial n} \mathcal{B}(\phi_j) \, dS, \qquad \alpha_i = \oint_S \frac{\partial \phi_i}{\partial n} \hat{u} \, dS \tag{3.117c}$$

Equation (3.3.117) gives n linear algebraic equations for the parameters c_j, $j = 1, 2, \ldots, n$. We illustrate the method via an example.

Example 3.15

The torsion of cylindrical members can be formulated alternatively in terms of the conjugate function Ψ, related to the warping function w (see Section 1.7.4), by the Cauchy-Riemann conditions

$$\frac{\partial w}{\partial x} = \frac{\partial \Psi}{\partial y}, \qquad \frac{\partial w}{\partial y} = -\frac{\partial \Psi}{\partial x} \tag{3.3.118}$$

From Eq. (3.3.118), it follows that

$$\nabla^2 \Psi = 0 \quad \text{in } -a < (x, y) < a \qquad (3.3.119a)$$

The boundary condition for Ψ can be expressed as ($f = 2G\theta$)

$$\Psi = \frac{f}{4}(x^2 + y^2) \quad \text{on } S \qquad (3.3.119b)$$

For the square section member of Example 3.13, we seek an approximate solution of Eq. (3.3.119) by the Trefftz method. Since the governing equation is homogeneous, we select $\phi_0 = 0$ and a harmonic function for ϕ_1. The lowest degree harmonic polynomial is given by $\phi_1 = (x^2 - y^2)$:

$$\Psi_1 = c_1(x^2 - y^2) \qquad (3.3.120)$$

We have (for $n = 1$ and $\mathcal{B} = 1$)

$$\beta_{ij} = \oint_S \frac{\partial \phi_i}{\partial n} \phi_j \, dS$$

$$\beta_{11} = \int_{-a}^{a} \left[\left(-\frac{\partial \phi_1}{\partial y} \right) \phi_1 \right]_{y=-a} dx + \int_{-a}^{a} \left(\frac{\partial \phi_1}{\partial x} \phi_1 \right)_{x=a} dy$$

$$+ \int_{a}^{-a} \left(\frac{\partial \phi_1}{\partial y} \phi_1 \right)_{y=a} (-dx) + \int_{a}^{-a} \left[\left(-\frac{\partial \phi_1}{\partial x} \right) \phi_1 \right]_{x=-a} (-dy)$$

$$= 2a \left[-\int_{-a}^{a} (x^2 - a^2) \, dx + \int_{-a}^{a} (a^2 - y^2) \, dy \right.$$

$$\left. - \int_{-a}^{a} (x^2 - a^2) \, dx + \int_{-a}^{a} (a^2 - y^2) \, dy \right]$$

$$= 8a \int_{-a}^{a} (a^2 - x^2) \, dx = \frac{32}{3} a^4$$

$$\alpha_1 = \oint_S \frac{f}{4} \frac{\partial \phi_1}{\partial n} (x^2 + y^2) \, dS$$

$$= \frac{af}{2} \left[-\int_{-a}^{a} (a^2 + x^2) \, dx + \int_{-a}^{a} (a^2 + y^2) \, dy \right.$$

$$\left. - \int_{-a}^{a} (x^2 + a^2) \, dx + \int_{-a}^{a} (a^2 + y^2) \, dy \right]$$

$$= 0 \qquad (3.3.121)$$

Thus, the choice $\phi_1 = x^2 - y^2$ yields the trivial solution for a square cross-section member. Therefore, we choose the next biharmonic polynomial,

$$\phi_1 = x^4 - 6x^2y^2 + y^4 \qquad (3.3.122)$$

which satisfies the equation $\nabla^2\phi_1 = 0$. We have

$$\beta_{11} = 4\int_{-a}^{a} (4a^3 - 12ax^2)(a^4 - 6a^2x^2 + x^4)\, dx = \frac{1536}{35}a^8$$

$$\alpha_1 = -\int_{-a}^{a} f(-12x^2a + 4a^3)(x^2 + a^2)\, dx = -\frac{64}{30}a^6f \qquad (3.3.123)$$

and the solution is given by (see Exercise 3.3.29)

$$\Psi_1 = -\frac{7f}{144a^2}\left(x^4 - 6x^2y^2 + y^4\right) \qquad (3.3.124a)$$

$$\Phi_1 = -\frac{7f}{144a^2}\left(x^4 - 6x^2y^2 + y^4\right) - \frac{f}{4}\left(x^2 + y^2\right) \qquad (3.3.124b)$$

The maximum shear stress, $\sigma_{yz}(a,0) = -(\partial\Phi_1/\partial x)(a,0)$ is given by

$$\sigma_{yz}(a,0) = \tfrac{5}{6}fa = 0.833fa$$

The maximum shear stress is about 23.4% greater than the exact solution. Of course, the solution can be improved by taking more parameters in the approximation.

The Trefftz method has limited use, because it can only be utilized for Dirichlet type boundary-value problems (i.e., boundary-value problems of the first type), and the specified boundary values should be nonzero [otherwise α_i in Eq. (3.3.117) will be zero]. Furthermore, it is not easy to find particular and complementary solutions of the problem.

3.3.6 Closing Remarks on Traditional Variational Methods

The single most important step in all the traditional variational methods presented in Sections 3.1–3.3 is the selection of the coordinate functions. The conditions on the coordinate functions provide the guidelines for their selection, but there exists no systematic procedure for their construction. In selecting the functions, one should give consideration (in addition to the properties discussed in various methods) to symmetry and other features of the particular problem. Begin with the lowest degree polynomials that are admissible, and then select higher degree polynomials without omitting intermediate ones. While trying to satisfy the essential boundary conditions on a portion of the boundary, one should not select coordinate functions that force zero essential boundary conditions on other parts of the boundary where it is not specified. For example, in the solution of the axial deformation of a bar fixed at the left end and subjected to an applied load at the right, use of $u = c_1 \sin(\pi x/L)$ will satisfy the essential boundary condition $u(0) = 0$; however, it also implies that $u(L) = 0$, which is not the right boundary condition at $x = L$. Of course, $\sin(\pi x/L)$ cannot represent any admissible configuration of the bar. Thus, without a judicious choice of the coordinate functions, the resulting equations can be trivial and the solutions can be wrong.

The selection of coordinate functions becomes more difficult for problems with irregular domains (i.e., noncircular and nonrectangular) or discontinuous data (i.e., loading or geometry). Further, the generation of coefficient matrices for the resulting algebraic equations cannot be automatized for a *class* of problems that differ from each other only in the geometry of the domain, boundary conditions, or loading. These limitations of the classical variational methods can be overcome by representing a given domain as a collection of geometrically simple subdomains for which we can generate the coordinate functions and determine the parameters by employing any of the variational methods over each subdomain. One such technique is provided by the finite element method, which is the topic of the next section.

Exercises 3.3

Determine the two-parameter Petrov–Galerkin solution for Exercises 1–5, using Eq. (3.3.8) of the following differential equations and compare, wherever possible, with the exact solutions.

1. $$-\frac{d^2u}{dx^2} = \frac{1}{1+x}, \qquad 0 < x < 1$$
    $$u(0) = u'(1) = 0$$

2. $$-\frac{d}{dx}\left[(1 + x^2)\frac{du}{dx}\right] + u = \sin \pi x + 3x - 1$$
    $$u(0) = u'(1) = 0$$

3. $$\left(\frac{d^2}{dr^2} + \frac{1}{r}\frac{d}{dr}\right)\left(\frac{d^2w}{dr^2} + \frac{1}{r}\frac{dw}{dr}\right) = f_0, \qquad 0 < r < a$$
    $$w = 0, \quad \frac{dw}{dr} = 0 \quad \text{at } r = a$$

4. $$\frac{d^2w}{dx^2} - \frac{M}{EI} = 0, \qquad \frac{d^2M}{dx^2} + f_0 = 0, \qquad 0 < x < L$$
    $$w(0) = \frac{dw}{dx}(0) = 0, \qquad M(L) = M_0, \qquad \frac{dM}{dx}(L) = 0$$

5. $$-\nabla^2 u = 0 \quad \text{in } 0 < (x, y) < 1$$
    $$u = \sin \pi x \quad \text{on } y = 0$$
    $$u = 0 \quad \text{on all other sides}$$

6–10. Repeat Exercises 1–5 using Eq. (3.3.12).

11. Solve the nonlinear differential equation

$$-\frac{d}{dx}\left(u\frac{du}{dx}\right) + 1 = 0, \qquad 0 < x < 1$$

$$u(1) = \sqrt{2}, \qquad u'(0) = 0$$

using Eq. (3.3.8). Use a one-parameter Petrov–Galerkin approximation. Compare the results with the exact solution, $u_0 = \sqrt{1 + x^2}$.

12. Repeat Exercise 11 using Eq. (3.3.12)

13. Find the first two eigenvalues associated with the differential equation

$$-\frac{d^2u}{dx^2} = \lambda u, \qquad 0 < x < 1$$

$$u(0) = 0, \qquad u(1) + u'(1) = 0$$

Use the Galerkin method.

14. Solve the Poisson equation

$$-\nabla^2 u = f_0 \quad \text{in a unit square}$$

$$u = 0 \quad \text{on the boundary}$$

using the following n-parameter Galerkin approximation

$$u_n = \sum_{i,j=1}^{n} c_{ij} \; \sin i\pi x \sin j\pi y$$

15. Solve the nonlinear equation in Exercise 11 by the Galerkin method.

16. Find a two-parameter Galerkin solution of a clamped (at both ends) beam under uniformly distributed load.

17. Solve the equation in Exercise 2 using the least-squares method.

18. Solve the problem of Exercise 13 using the least-squares method.

19. Solve the nonlinear equation in Exercise 11 by the least-squares method.

20. Solve the equation in Exercise 14 using the least-squares method.

21. Consider a cantilevered beam of variable flexural rigidity, $EI = a_0[2 - (x/L)^2]$ and carrying a distributed load, $f = f_0[1 - (x/L)]$. Find a three-parameter solution using the collocation method.

22. Repeat Exercise 13 using the collocation method.

23. Solve the nonlinear differential equation in Exercise 11 by the collocation method.

24. Solve the problem in Exercise 2 by the subdomain method.

25. Solve the problem in Exercise 14 using the one-point collocation method.

26. Use the Kantorovich method to find a two-parameter approximation of the form

$$u(x, y) = c_1(y)\cos\frac{\pi x}{2a} + c_2(y)\cos\frac{3\pi x}{2a}$$

to determine an approximation of the torsion problem in Example 3.8.

27. Use the Kantorovich method (i.e., semidiscrete method) to find a two-parameter approximation of the form

$$w(x, t) = c_1(t)(1 - \cos 2\pi x)$$

to determine an approximate solution of the equation

$$\frac{\partial^4 w}{\partial x^4} + \frac{\partial^2 w}{\partial t^2} = 0, \qquad 0 < x < 1, \qquad t > 0$$

$$w = \frac{\partial w}{\partial x} = 0 \quad \text{at } x = 0, 1 \quad \text{and} \quad t > 0$$

$$w = \sin \pi x - \pi x(1 - x), \qquad \frac{\partial w}{\partial t} = 0 \quad \text{at } t = 0, \qquad 0 < x < 1$$

Use the Galerkin method to satisfy the initial conditions.

28. Consider the Poisson equation

$$-\nabla^2 u = 0, \qquad 0 < x < 1, \qquad 0 < y < \infty$$

$$u(0, y) = u(1, y) = 0 \quad \text{for } y > 0$$

$$u(x, 0) = x(1 - x), \qquad u(x, \infty) = 0, \qquad 0 \le x \le 1$$

Assuming an approximation of the form

$$u(x, y) = c_1(y)x(1 - x)$$

find the differential equation for $c_1(y)$ and solve it exactly.

29. Use the Trefftz method to determine the Prandtl stress function $\Phi = \Psi - \frac{f}{4}(x^2 + y^2)$, by solving the differential equation

$$\nabla^2 \Psi = 0 \quad \text{in } -a < x < a, \qquad -a < y < a$$

$$\Psi = \frac{f}{4}(x^2 + y^2) \quad \text{on the boundary.}$$

Use two parameters with $\phi_1 = 1$ and ϕ_2 as in Eq. (3.3.122).

30. Recall from Chapter 2 that the problem of solving the equation

$$\nabla^2 u = 0 \quad \text{in } R$$

$$u = \hat{u} \quad \text{on } S$$

is equivalent to finding u such that $u = \hat{u}$ on S and

$$I(u) = \int_R \frac{1}{2}\left[\left(\frac{\partial u}{\partial x}\right)^2 + \left(\frac{\partial u}{\partial y}\right)^2\right] dx\, dy$$

is a minimum. Show that
(a) $I(v) = I(u) + I(u - v)$ for any v such that $v = \hat{u}$ on S
(b) $I(w) = I(w) - I(u - w)$ for any w such that $\nabla^2 w = 0$ on
R and $\oint_S (w - \hat{u})\dfrac{\partial w}{\partial n}\, dS = 0$

31. Show that the conditions in (b) of Exercise 30 are equivalent to the requirement that $I(u - w)$ be a minimum.

32. In the Trefftz method, show that if one of ϕ_i is a constant, the corresponding c_i are indeterminate.

33. Show that for a cantilever beam with end load P_0, the Ritz approximations based on the following functionals provide lower and upper bounds, respectively, for the exact solution:

$$\Pi(w) = \int_0^L \frac{EI}{2}\left(\frac{d^2 w}{dx^2}\right)^2 dx - Pw(L), \qquad w(0) = \frac{dw}{dx}(0) = 0$$

$$\Pi^*(M) = \int_0^L \frac{1}{2EI}\left(\frac{d^2 M}{dx^2}\right)^2 dx, \qquad M(L) = 0, \qquad \frac{dM}{dx}(L) = P_0$$

3.4   THE FINITE ELEMENT METHOD

3.4.1   Introduction

As noted in Section 3.3.6, the traditional variational methods are ineffective for solving problems that are geometrically complex, have discontinuous loads, or involve discontinuous material or geometric properties. In such cases, the selection of the coordinate functions is a formidable task. Even in cases where the coordinate functions are available, the computation of associated coefficient matrices cannot be automatized, because the coordinate functions are not always algebraic polynomials and they depend on the specific boundary conditions. Consequently, each time the essential boundary conditions are changed, for the same differential equation, the coordinate functions are also changed and the coefficient matrices have to be recalculated.

The finite element method uses the philosophy of the traditional variational methods to derive the algebraic equations relating undetermined coefficients. It differs in two ways from the traditional variational methods in generating the algebraic equations of the problem. First, the coordinate functions are always algebraic polynomials that are developed using ideas from the interpolation theory; second, the coordinate functions are developed for subdomains into which a given domain is divided. The subdomains, called *finite elements*, are geometrically simple shapes that permit a systematic construction of coordinate functions. Since the coordinate functions are algebraic polynomials, the computation of the coefficient matrices of the algebraic equations can be automatized on the computer. As we show, the construction of the coordinate functions employs concepts from interpolation theory, and the process is independent of the boundary conditions and the problem data. In short, the finite element method is a piecewise application of classical variational methods and it uses algebraic interpolation to represent the dependent variables to be determined. The undetermined parameters represent the values of the dependent variables at a finite number of preselected points (whose number and location dictates the degree and form of the coordinate functions) in the element. The method is well-suited for electronic computation and the development of general purpose computer programs.

The present study exposes the reader to some of the basic concepts of the finite element method. The material presented is introductory in nature and is confined to the basic finite elements of one- and two-dimensional problems in applied mechanics. The reader is asked to consult the references on the finite element method listed in the Bibliography for additional topics.

In this section, the method is described and illustrated for (i) one-dimensional second-order equations (e.g., bars, cables, etc.) (ii) one-dimensional fourth-order equations governing bending of beams, and (iii) two-dimensional second-order equations for one scalar variable (e.g., torsion, heat transfer, ground-water flow, and other problems) and elasticity problems. The general procedure is the same for all three classes of problems, but the details differ from each other and, therefore, it is best to consider the three classes of problems separately.

3.4.2 One-Dimensional Second-Order Equations

We use the following model equation to introduce the finite element method:

$$-\frac{d}{dx}\left(a\frac{du}{dx}\right) + cu = f, \qquad 0 < x < L \tag{3.4.1}$$

where $a = a(x)$, $c = c(x)$, and $f = f(x)$ are given functions of x. We are interested in determining an approximation of the true solution u that satisfies Eq. (3.4.1) and appropriate boundary conditions. It should be noted that Eq. (3.4.1) arises in many fields of engineering and applied science including solid mechanics.

To account for all possible cases of Eq. (3.4.1), we allow the data, a, c, and f, to take zero, continuous, or discontinuous values. For example, a tapered stepped shaft has discontinuities in its area of cross section at a finite number of points along the length,

$$a(x) = \begin{cases} a_1(x), & 0 < x < x_1 \\ a_2(x), & x_1 < x < x_2 \\ \vdots \\ a_n(x), & x_{n-1} < x < x_n = L \end{cases} \qquad (3.4.2)$$

Also, we must allow for various combinations of the two boundary conditions we derived for bar problems in Chapter 2,

$$u = \text{specified} \quad \text{or} \quad a\frac{du}{dx} = \text{specified} \qquad (3.4.3)$$

Thus, in a general case, we must seek an approximate solution to Eq. (3.4.1) in each of the subintervals of $(0, L)$ in which the equation has continuous coefficients (i.e., a, c, and f are zero, constant, or continuous functions of x). A step-by-step procedure for the finite element analysis of Eq. (3.4.1) for arbitrary values of a, c, and f, and arbitrary boundary conditions, is given below.

1. *Representation of a Domain by Elements.* A given domain should be divided into a number of subdomains, called *elements*. The collection of elements is called *finite-element mesh* (see Fig. 3.2). The number of divisions is analogous to the number of parameters in the variational methods. Therefore,

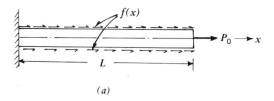

(a)

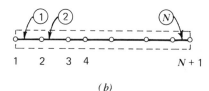

(b)

Figure 3.2 Finite-element discretization of an axially loaded member: (*a*) geometry and loading; (*b*) finite-element mesh.

the larger the number of elements, the more accurate the solution will be. The minimum number of subdivisions is equal to the number of subdivisions created by the discontinuities in the data of the problem. An element in one-dimensional equations is a line of finite length.

The elements of a domain are connected to neighboring elements at a finite number of points called *global nodes* (see Fig. 3.2). The word "global" refers to the whole problem, as opposed to an element of it. The end points of individual elements are called *element nodes*, and they match with a pair of global nodes. The relationship between the element nodes and global nodes is discussed in the following sections.

2. *Approximation Over an Element.* Consider an arbitrary element, $R_e = (x_A, x_B)$, located between points $x = x_A$ and x_B in the domain (see Fig. 3.3). We isolate the element and study a variational approximation of Eq. (3.4.1) using any one of the variational methods presented in Sections 3.2 and 3.3. The idea here is to find algebraic equations relating the nodal values of u, which take the place of the undetermined parameters in the traditional variational methods, to the nodal forces of the arbitrary (but typical of the collection) element and, thereby, to have a set of algebraic equations for every element in the mesh. We shall also see how to obtain the equations of the whole problem from the equations of individual elements.

For the element under consideration, we display the displacements and forces to complete the free-body-diagram of the element, as shown in Fig. 3.3. We label the left-end force, $[a(du/dx)]$ at $x = x_A$, as $-P_A$ and the right-end force, $[a(du/dx)]$ at $x = x_B$, as P_B, and the corresponding displacements by $u_A \equiv u(x_A)$ and $u_B \equiv u(x_B)$:

$$u(x_A) = u_A, \qquad u(x_B) = u_B$$

$$\left(a\frac{du}{dx}\right)\Bigg|_{x=x_A} = -P_A, \qquad \left(a\frac{du}{dx}\right)\Bigg|_{x=x_B} = P_B \qquad (3.4.4)$$

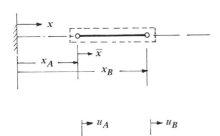

Figure 3.3 Geometry, displacements, and forces of a bar element.

Equations (3.4.4) can be thought of as the "boundary conditions" for the element R_e when the variational formulation is based on the virtual displacement (or total potential energy) principle. The variational statement for the bar element in Fig. 3.3, equivalently, for Eqs. (3.4.1) and (3.4.4), is given by

$$\Pi^e(u_e) = \int_{x_A}^{x_B} \left[\frac{a_e}{2} \left(\frac{du_e}{dx} \right)^2 + \frac{c_e}{2} (u_e)^2 - f_e u_e \right] dx - P_A u_A - P_B u_B$$

$$(3.4.5)$$

where the subscript e on the variables indicates that the variables are defined in element R_e. It is easy to verify that Eqs. (3.4.1) and (3.4.4) are the Euler equations of the total potential energy functional Π^e. If the element has any point forces at any point between the end points, its contribution to Π^e should be added.

As in any variational method, we seek an approximation of $u_e(x)$ in the form

$$u_e(x) = \sum_{j=1}^{n} u_j^e \psi_j^e(x) \qquad (3.4.6)$$

where ψ_j^e are the coordinate functions and u_j^e are parameters to be determined. The functions ψ_j^e are subjected to conditions similar to those in Eq. (3.2.2). In the present case, we require that

(i) ψ_j^e should be continuous as required by the functional $\Pi(u_e)$
(ii) ψ_j^e should be linearly independent and complete
(iii) ψ_j^e should be such that u_e satisfies the essential boundary conditions in Eq. (3.4.4)   (3.4.7)

Before discussing the generation of the coordinate functions ψ_j^e, we complete the Ritz approximation to derive the algebraic equations.

Substitution of Eq. (3.4.6) into Eq. (3.4.5) yields a function of u_j^e, $j = 1, 2, \ldots, n$. Then, by the total potential energy principle, Π^e should be a minimum with respect to u_i^e:

$$\frac{\partial \Pi^e}{\partial u_i^e} = 0, \qquad i = 1, 2, \ldots, n$$

$$= \int_{x_A}^{x_B} \left[a_e \frac{d\psi_i^e}{dx} \left(\sum_{j=1}^{n} u_j^e \frac{d\psi_j^e}{dx} \right) + c_e \psi_i^e \sum_{j=1}^{n} u_j^e \psi_j^e - \psi_i^e f_e \right] dx$$

$$- P_A \psi_i^e(x_A) - P_B \psi_i^e(x_B)$$

$$= \sum_{j=1}^{n} K_{ij}^e u_j^e - F_i^e \qquad (3.4.8a)$$

where

$$K_{ij}^e = \int_{x_A}^{x_B} \left(a_e \frac{d\psi_i^e}{dx} \frac{d\psi_j^e}{dx} + c_e \psi_i^e \psi_j^e \right) dx$$

$$F_i^e = \int_{x_A}^{x_B} \psi_i^e f_e \, dx + P_A \psi_i^e(x_A) + P_B \psi_i^e(x_B) \qquad (3.4.8b)$$

Equation (3.4.8a) is often referred to as the *finite-element model* of the differential equation (3.4.1).

Equation (3.4.8) provides n linear algebraic equation relating n parameters u_j^e ($j = 1, 2, \ldots, n$). The coefficient matrix $[K^e]$ is called the *stiffness matrix* and the column $\{F^e\}$ is called the generalized *force vector*. It should be noted that both u_i^e and (P_A, P_B), in general, are not known for an interior element. One cannot solve Eq. (3.4.8) because F_i^e is not completely known. The element equations (3.4.8) must be put together, *assembled*, appropriately to obtain the global equations, and specified boundary conditions on u and/or $[a(du/dx)]$ must be applied before solving the algebraic equations. Before we discuss the assembly of element equations, we next discuss the generation of coordinate functions.

3. *Derivation of the Coordinate Functions.* We derive the coordinate functions ψ_j^e using the properties in Eq. (3.4.7). First, we must select the value of n, which depends on the degree of the algebraic polynomials used to approximate u_e. The continuity requirement (i) places the minimum restriction that ψ_j^e be such that du_e/dx is nonzero. Let us begin with the linear polynomial, which meets the continuity requirement

$$u_e(x) = \alpha_0 + \alpha_1 x, \qquad x_A \leq x \leq x_B \qquad (3.4.9)$$

and derive ψ_j^e from it. Next, we choose α_0 and α_1 such that u_e satisfies the essential boundary conditions of the element:

$$u_A = u_e(x_A) = \alpha_0 + \alpha_1 x_A$$

$$u_B = u_e(x_B) = \alpha_0 + \alpha_1 x_B \qquad (3.4.10a)$$

These two equations can be solved for α_0 and α_1 to obtain

$$\alpha_0 = \frac{u_A x_B - u_B x_A}{x_B - x_A}, \qquad \alpha_1 = \frac{u_B - u_A}{x_B - x_A} \qquad (3.4.10b)$$

Substituting for α_0 and α_1 from Eq. (3.4.10b) into Eq. (3.4.9), we obtain

$$u_e(x) = \frac{(u_A x_B - u_B x_A) + (u_B - u_A)x}{x_B - x_A}$$

$$= u_A \frac{x_B - x}{x_B - x_A} + u_B \frac{x - x_A}{x_B - x_A} \qquad (3.4.11)$$

Note that u_A and u_B are the unknown nodal values at the end nodes of element R_e. For notational convenience, we denote u_A by u_1^e and u_B by u_2^e (i.e., u_j^e denotes the value of u_e at node j of element R_e). Then, Eq. (3.4.11) becomes

$$u_e(x) = u_1^e \psi_1^e + u_2^e \psi_2^e = \sum_{j=1}^{2} u_j^e \psi_j^e \qquad (3.4.12a)$$

where

$$\psi_1^e = \frac{x_B - x}{x_B - x_A}, \qquad \psi_2^e = \frac{x - x_A}{x_B - x_A} \qquad (3.4.12b)$$

Note that the value of n, implied by Eq. (3.4.12a), is equal to 2. The functions ψ_j^e ($j = 1, 2$) satisfy the properties

$$\psi_1^e + \psi_2^e = \sum_{j=1}^{2} \psi_j^e = 1$$

$$\psi_1^e(x_A) = 1, \qquad \psi_1^e(x_B) = 0$$

$$\psi_2^e(x_A) = 0, \qquad \psi_2^e(x_B) = 1 \qquad (3.4.13)$$

In view of these properties of ψ_j, the force vector in Eq. (3.4.8b) takes the form

$$F_i^e = \int_{x_A}^{x_B} f_e \psi_i^e \, dx + P_i^e = f_i^e + P_i^e \qquad (3.4.14)$$

where $P_1^e = P_A$ and $P_2^e = P_B$. Geometrically, ψ_j^e are the linear functions that vary from zero at one node to unity at the other (see Fig. 3.4). These functions are linearly independent because

$$\alpha \psi_1 + \beta \psi_2 = 0 \quad \text{implies} \quad \alpha = \beta = 0$$

The functions ψ_j^e are called the linear *interpolation functions*, because the sum in Eq. (3.4.12a) represents the linear interpolation of u_e between points $x = x_A$ and $x = x_B$ (see Fig. 3.4).

If we set up an element (or local) coordinate system $\bar{x}$ in the left node (i.e., node 1) of element R_e and denote the coordinate of the ith node of the element by $\bar{x}_i$, we can write Eqs. (3.4.12b) and (3.4.13) in the alternative form

$$\psi_1^e(\bar{x}) = 1 - \frac{\bar{x}}{h_e}, \qquad \psi_2^e(\bar{x}) = \frac{\bar{x}}{h_e}, \qquad h_e = x_B - x_A$$

$$\sum_{j=1}^{2} \psi_j^e(\bar{x}) = 1, \qquad \psi_j^e(\bar{x}_i) = \delta_{ij} \qquad (3.4.15)$$

where we use the following relationship between the global coordinate x and local coordinate $\bar{x}$ to rewrite ψ_j^e in terms of the local coordinate:

$$x = \bar{x} + x_A \qquad (3.4.16)$$

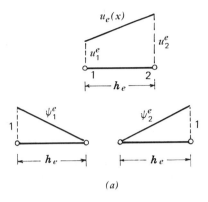

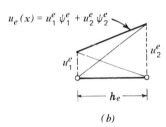

(b)

Figure 3.4 Linear approximation of the displacement of a bar: (*a*) linear interpolation functions; (*b*) finite-element approximation.

If a quadratic interpolation of u_e is desired, one can begin with the equation

$$u_e(x) = \alpha_0 + \alpha_1 x + \alpha_2 x^2, \qquad x_A \le x \le x_B \qquad (3.4.17)$$

and derive the corresponding ψ_j^e from it. In addition to the two boundary conditions, we need one more condition on u_e to eliminate α's from Eq. (3.4.17). This requires us to identify a third node inside the element. Let us select the third node in the middle of the element [i.e., at $x = (x_A + x_B)/2$]. In the interest of algebraic simplicity, we write Eq. (3.4.17) in terms of the local coordinate $\bar{x}$ and evaluate it at the three nodes:

$$u_e(\bar{x}) = \alpha_0 + \alpha_1(\bar{x} + x_A) + \alpha_2(\bar{x} + x_A)^2 = \bar{\alpha}_0 + \bar{\alpha}_1\bar{x} + \bar{\alpha}_2\bar{x}^2$$

$$u_1^e \equiv u_e(0) = \bar{\alpha}_0$$

$$u_2^e \equiv u_e\left(\frac{h_e}{2}\right) = \bar{\alpha}_0 + \bar{\alpha}_1\frac{h_e}{2} + \bar{\alpha}_2\frac{h_e^2}{4}$$

$$u_3^e \equiv u_e(h_e) = \bar{\alpha}_0 + \bar{\alpha}_1 h_e + \bar{\alpha}_2 h_e^2$$

Solving for $\bar{\alpha}_0$, $\bar{\alpha}_1$, and $\bar{\alpha}_2$, and substituting into Eq. (3.4.17), we get

$$u_e(x) = \sum_{j=1}^{n} u_j^e \psi_j^e(\bar{x}), \qquad 0 \le \bar{x} \le h_e, \qquad x_A \le x \le x_B \qquad (3.4.18a)$$

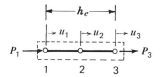

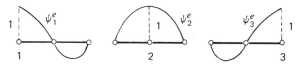

Figure 3.5 Quadratic element and its interpolation functions.

where

$$\psi_1^e(\bar{x}) = 1 - 3\frac{\bar{x}}{h_e} + 2\left(\frac{\bar{x}}{h_e}\right)^2, \qquad \psi_2^e(\bar{x}) = 4\frac{\bar{x}}{h_e}\left(1 - \frac{\bar{x}}{h_e}\right)$$

$$\psi_3^e(\bar{x}) = \frac{\bar{x}}{h_e}\left(\frac{2\bar{x}}{h_e} - 1\right) \qquad\qquad (3.4.18b)$$

Once again, we can verify that the ψ_j^e in Eq. (3.4.18) satisfy the properties in Eq. (3.4.15):

$$\sum_{j=1}^{n=3} \psi_j^e(\bar{x}) = 1, \qquad \psi_j^e(\bar{x}_i) = \delta_{ij}, \qquad j = 1, 2, 3 \qquad (3.4.18c)$$

where $\bar{x}_1 = x_A, \bar{x}_2 = x_A + (h_e/2), \bar{x}_3 = x_B$ (see Fig. 3.5).

The procedure described above can be used to derive interpolation functions of any degree. An alternative procedure that uses the interpolation properties (3.4.18c) to construct ψ_j^e is described in Exercise 3.4.3.

4. *Assembly (or Connectivity) of Elements.* The finite-element equations (3.4.8) can be specialized to each one of the elements in the mesh by assigning the values of x_A, x_B, a_e, c_e, and so on. Because each element in the mesh is connected to its neighboring elements at the global nodes, and the displacement is continuous from one element to the next, one can relate the nodal values of displacements at the interelement connecting nodes. To this end, let U_I denote the value of the displacement $u(x)$ at the Ith global node. Then we have the following correspondence between U_I and u_j^e (see Fig. 3.6) of a linear-element mesh:

$$u_1^1 = U_1, \qquad u_2^1 = u_1^2 \equiv U_2, \qquad \dots, \qquad u_2^e = u_1^{e+1} \equiv U_{e+1}, \qquad \dots, \qquad u_2^N \equiv U_{N+1}$$

$$(3.4.19)$$

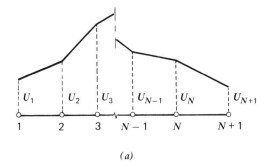

(a)

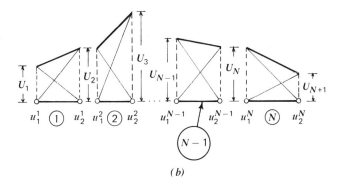

(b)

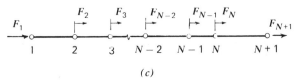

(c)

Figure 3.6 Connectivity (or assembly) of finite elements: (a) elementwise linear approximation of the displacement; (b) interelement continuity of the displacements; (c) connected mesh (or global mesh) of elements with global forces.

where N is the total number of linear elements. Equation (3.4.19) relates the global displacements to local displacements and enforces interelement continuity of the displacements (see Fig. 3.6a and b).

The assembly of element equations is based on the satisfaction of the total potential energy principle for the whole system:

$$\delta\Pi = \sum_{I=1}^{N+1} \frac{\partial\Pi}{\partial U_I}\delta U_I = 0 \quad \text{or} \quad \frac{\partial\Pi}{\partial U_I} = 0, \quad I = 1, 2, \dots, N+1$$

$$(3.4.20)$$

where Π is given in terms of Π^e by

$$\Pi = \sum_{e=1}^{N} \Pi^e = \sum_{e=1}^{N} \sum_{i=1}^{n=2} u_i^e \left(\frac{1}{2} \sum_{j=1}^{n=2} K_{ij}^e u_j^e - F_i^e \right)$$

$$= \frac{1}{2} \sum_{e=1}^{N} \left[u_1^e \left(K_{11}^e u_1^e + K_{12}^e u_2^e - 2F_1^e \right) + u_2^e \left(K_{21}^e u_1^e + K_{22}^e u_2^e - 2F_2^e \right) \right]$$

$$= \tfrac{1}{2} \left[u_1^1 \left(K_{11}^1 u_1^1 + K_{12}^1 u_2^1 - 2F_1^1 \right) + u_1^2 \left(K_{11}^2 u_1^2 + K_{12}^2 u_2^2 - 2F_1^2 \right) + \cdots \right.$$

$$\left. + u_2^1 \left(K_{21}^1 u_1^1 + K_{22}^1 u_2^1 - 2F_2^1 \right) + u_2^2 \left(K_{21}^2 u_1^2 + K_{22}^2 u_2^2 - 2F_2^2 \right) + \cdots \right]$$

$$= \tfrac{1}{2} \left[U_1 \left(K_{11}^1 U_1 + K_{12}^1 U_2 - 2F_1^1 \right) + U_2 \left(K_{11}^2 U_2 + K_{12}^2 U_3 - 2F_1^2 \right) + \cdots \right.$$

$$\left. + U_2 \left(K_{21}^1 U_1 + K_{22}^1 U_2 - 2F_2^1 \right) + U_3 \left(K_{21}^2 U_2 + K_{22}^2 U_3 - 2F_2^2 \right) + \cdots \right]$$

$$(3.4.21)$$

and, arriving at the last line, the correspondence in Eq. (3.4.19) is used to replace the element nodal displacements by the global nodal displacements. Using Eq. (3.4.20), we obtain

$$K_{11}^1 U_1 + \tfrac{1}{2} \left(K_{12}^1 + K_{21}^1 \right) U_2 - F_1^1 = 0$$

$$\tfrac{1}{2} \left(K_{12}^1 + K_{21}^1 \right) U_1 + \left(K_{22}^1 + K_{11}^2 \right) U_2 + \tfrac{1}{2} \left(K_{12}^2 + K_{21}^2 \right) U_3 + \cdots - F_2^1 - F_1^2 = 0$$

$$\tfrac{1}{2} \left(K_{12}^2 + K_{21}^2 \right) U_2 + \left(K_{22}^2 + K_{11}^3 \right) U_3 + \cdots - F_2^2 - F_1^3 = 0$$

$$\tfrac{1}{2} \left(K_{12}^{N-1} + K_{21}^{N-1} \right) U_{N-1} + \left(K_{22}^{N-1} + K_{11}^N \right) U_N$$

$$+ \tfrac{1}{2} \left(K_{12}^N + K_{21}^N \right) U_{N+1} - F_2^{N-1} - F_1^N = 0$$

$$\tfrac{1}{2} \left(K_{12}^N + K_{21}^N \right) U_N + K_{22}^N U_{N+1} - F_2^N = 0 \qquad (3.4.22)$$

Using the symmetry of the element stiffness matrices, $K_{ji}^e = K_{ij}^e$, we can write the above equations in the matrix form

$$
\begin{bmatrix}
K_{11}^1 & K_{12}^1 & 0 & & 0 & & 0 \\
K_{12}^1 & K_{22}^1 + K_{11}^2 & K_{12}^2 & & 0 & & 0 \\
0 & K_{12}^2 & K_{22}^2 + K_{11}^3 & & K_{12}^3 & & 0 \\
\vdots & \vdots & \vdots & \ddots & \vdots & & \vdots \\
0 & 0 & \cdots & & K_{22}^{N-1} + K_{11}^N & K_{12}^N \\
0 & 0 & 0 & \cdots & K_{12}^N & K_{22}^N
\end{bmatrix}
\begin{Bmatrix}
U_1 \\
U_2 \\
U_3 \\
\vdots \\
U_N \\
U_{N+1}
\end{Bmatrix}
$$

$$
=
\begin{Bmatrix}
F_1^1 \\
F_2^1 + F_1^2 \\
F_2^2 + F_1^3 \\
\vdots \\
F_2^{N-1} + F_1^N \\
F_2^N
\end{Bmatrix}
\tag{3.4.23}
$$

It is clear from Eq. (3.4.23) that the diagonal elements of stiffness matrices of elements R_e and R_{e+1} add up at global node $I = e + 1$, and the global stiffness element K_{IJ} is zero, if global nodes I and J do not belong to the same element. Thus, the resulting global stiffness matrix is not only symmetric, but is also banded, that is, all entries beyond a diagonal parallel to the main diagonal are zero. This is a feature of all finite element equations, irrespective of the differential equation being solved, and is a result of the piecewise definition of the coordinate functions. Keeping in mind the general pattern of the assembled stiffness matrix and force column, one can routinely assemble the element matrices for any number of elements.

   5. *Imposition of the Boundary Conditions.* The boundary conditions of the problem should be imposed on the assembled equations (3.4.23) before solving them. In all bar problems, we have either a displacement or a force (but *never* both) specified at a point. If the displacement is specified at a global node, the corresponding nodal displacement should be replaced by the specified value, the row and column corresponding to the node should be deleted, and the force column should be modified by subtracting the product of the stiffness coefficient in the column with the specified displacement.

   To impose a specified force, we must separate the contributions to F_i^e resulting from the distributed force f and internal force $a(du/dx)$ at the nodes:

$$
\begin{Bmatrix}
F_1^1 \\
F_2^1 + F_1^2 \\
\vdots \\
F_2^{N-1} + F_1^N \\
F_2^N
\end{Bmatrix}
=
\begin{Bmatrix}
f_1^1 \\
f_2^1 + f_1^2 \\
\vdots \\
f_2^{N-1} + f_1^N \\
f_2^N
\end{Bmatrix}
+
\begin{Bmatrix}
P_1^1 \\
P_2^1 + P_1^2 \\
\vdots \\
P_2^{N-1} + P_1^N \\
P_2^N
\end{Bmatrix}
\equiv \{F\} \quad (3.4.24)
$$

where f_i^e and P_i^e are the components defined in Eq. (3.4.14). At any node where the displacement is unknown, the externally applied point force there should equal the sum of internal forces. For instance, if U_I is unknown, then the force at node I is known (i.e., specified to be zero or some number, say P_0). Then, the equilibrium of forces at node I requires that

$$
P_2^{(I-1)} + P_1^{(I)} = P_0 \tag{3.4.25}
$$

To fix our thinking, suppose that the displacement of global node 1 is specified to be u_0 and the force at global node $N + 1$ is specified to be P_0 (e.g., a bar fixed at the left end, $u_0 = 0$, and subjected to distributed force f and point load P_0 at the right end). No point loads are specified at the intermediate to the end nodes, which implies that (see Fig. 3.6c)

$$
P_2^1 + P_1^2 = 0, \qquad P_2^2 + P_1^3 = 0, \quad \ldots \quad , P_2^{N-1} + P_1^N = 0 \tag{3.4.26}
$$

Then Eq. (3.4.23) can be reduced to

$$
\begin{bmatrix}
K_{22}^1 + K_{11}^2 & K_{12}^2 & 0 & \cdots & 0 & 0 \\
K_{12}^2 & K_{22}^2 + K_{11}^3 & K_{12}^3 & & 0 & 0 \\
\vdots & \vdots & \ddots & & \vdots & \vdots \\
0 & 0 \cdots & & & K_{22}^{N-1} + K_{11}^N & K_{12}^N \\
0 & 0 \cdots & & 0 & K_{12}^N & K_{22}^N
\end{bmatrix}
\begin{Bmatrix}
U_2 \\
U_3 \\
\vdots \\
U_N \\
U_{N+1}
\end{Bmatrix}
$$

$$
=
\begin{Bmatrix}
f_2^1 + f_1^2 \\
f_2^2 + f_1^3 \\
\vdots \\
f_2^{N-1} + f_1^N \\
f_2^N + P_0
\end{Bmatrix}
-
\begin{Bmatrix}
K_{12}^1 u_0 \\
0 \\
0 \\
\vdots \\
0
\end{Bmatrix}
\tag{3.4.27}
$$

which consists of N equations for the N unknowns, U_I, $I = 2, \ldots, N + 1$. Since the right-hand column is completely known, one can use any standard solution procedure for the solution of linear algebraic equations.

6. *Calculation of Reactions and Derivatives of Solution: Postprocessing.* The unknown reaction force at global node 1 can be computed from the first equation in Eq. (3.4.23) after the displacements U_I are computed:

$$P_1^1 = -f_1^1 + K_{11}^1 u_0 + K_{12}^1 U_2 \tag{3.4.28}$$

Note that P_1^1 is nothing but the value of $[-a(du/dx)]$ at $x = 0$.

The displacement at any point other than a node can be computed using the equation (see Fig. 3.6)

$$u(x) = \begin{cases} u_1(x), & 0 \le x \le h_1 \\ u_2(x), & h_1 \le x \le h_2 \\ \vdots \\ u_N(x), & h_{N-1} \le x \le h_N \end{cases} \tag{3.4.29}$$

where $u_e(x)$ is the finite-element solution in Eq. (3.4.6) in element R_e. For example, if $u(x_0)$, $h_1 \le x_0 \le h_2$, is desired, for a linear element we have

$$u(x_0) = u_2(x_0) = u_1^2 \psi_1(x_0) + u_2^2 \psi_2(x_0)$$

$$= U_2 \psi_1(x_0) + U_3 \psi_2(x_0) \tag{3.4.30}$$

Similarly, the derivatives of $u(x)$ can be computed from

$$\frac{du}{dx} = \begin{cases} \sum_{j=1}^{n} u_j^1 \dfrac{d\psi_j^1}{dx}, & 0 < x < h_1 \\ \sum_{j=1}^{n} u_j^2 \dfrac{d\psi_j^2}{dx}, & h_1 < x < h_2 \\ \vdots \\ \sum_{j=1}^{n} u_j^N \dfrac{d\psi_j^N}{dx}, & h_{N-1} < x < h_N \end{cases} \tag{3.4.31}$$

The nodal values of the derivative computed from the two elements sharing that node do not coincide, irrespective of the order of the elements, because the continuity of the derivative is not imposed in the procedure. Therefore, one should not use the nodal values of the derivatives without averaging them.

This completes the description of the basic steps in the finite-element analysis of the model problem. The derivation of the coordinate functions is not a usual step in the finite-element formulation, because these functions are already derived and are available in textbooks. One should only select appropriate coordinate functions for the problem at hand. The derivation is included here to illustrate the procedure. Although we used the terminology of solid mechanics in the development of the finite-element equations, all the concepts are applicable to any physical problem described by Eq. (3.4.1). In other words, if a computer program is developed to solve Eq. (3.4.1) for any boundary conditions, then the program can be used to analyze not only bars, but all other physical problems described by the equation (e.g., transverse deflection of cables, heat transfer in fins, flow through pipes, etc.). For additional details, the reader can consult Reddy (1984). In the following example, we illustrate the steps involved in the finite-element analysis of a composite bar.

Example 3.16

Consider the composite bar consisting of a tapered steel bar fastened to an aluminum rod of uniform cross section, and subjected to loads as shown in Fig. 3.7. We wish to determine the deformation in the bar using the linear elements.

The governing equations are given by

$$-\frac{d}{dx}\left(E_s A_s \frac{du}{dx}\right) = 0, \qquad 0 < x < h_1$$

$$-\frac{d}{dx}\left(E_a A_a \frac{du}{dx}\right) = 0, \qquad h_1 < x < h_1 + h_2 = L \qquad (3.4.32a)$$

where the subscript s refers to steel and a to aluminum. We need not worry

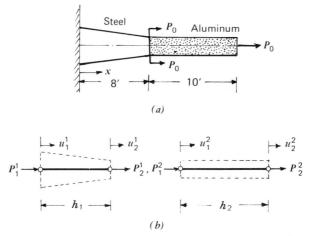

Figure 3.7 Axial deformation of a composite member: (a) geometry and loading; (b) finite-element representation.

about the boundary conditions until Step 5 of the procedure. We take the
following values of the data:

$$E_s = 30 \times 10^6 \text{ psi}, \qquad A_s = (c_1 + c_2 x)^2 \text{ in}^2., \qquad E_a = 10^7 \text{ psi}$$

$$A_a = 1 \text{ in}^2., \qquad h_1 = 96 \text{ in.}, \qquad L = 216 \text{ in.}, \qquad P_0 = 10,000 \text{ lb}$$

$$(3.4.32b)$$

We now follow the step by step procedure to develop the finite-element
equations. In actual practice, all of the steps can be carried out on a digital
computer.

 1. *Finite-Element Mesh.* From the discontinuity in the material property,
area of cross section, and loading at $x = h_1$, we are required to divide the
domain $R = (0, L)$ into at least two elements: $R_1 = (0, h_1)$ and $R_2 = (h_1, L)$.
For the mesh of two linear elements, there will be three global nodes; if
quadratic elements are used, there will be five global nodes.
 2. *Element Equations.* For each element, whether linear or quadratic, the
element stiffness matrix and force vector are given by Eq. (3.4.8), with

$$a_e = E_e \left(c_1^e + c_2^e x \right)^2, \qquad c_e = 0, \qquad f_e = 0$$

We now compute $[K^e]$ and $\{F^e\}$ using the linear interpolation functions.
 The element stiffness matrix and force vector can be computed using either
the local coordinate $\bar{x}$ or the global coordinate x. If we use Eq. (3.4.15), then
the transformation Eq. (3.4.16) from x to $\bar{x}$ should be used to convert all
functions of x into functions of $\bar{x}$. We have

$$K_{ij}^e = \int_{x_A = x_e}^{x_B = x_{e+1}} E_e \left(c_1^e + c_2^e x \right)^2 \frac{d\psi_i^e}{dx} \frac{d\psi_j^e}{dx} \, dx$$

$$= \int_0^{h_e} E_e \left[c_1^e + c_2^e (\bar{x} + x_e) \right]^2 \frac{d\psi_i^e}{d\bar{x}} \frac{d\psi_j^e}{d\bar{x}} \, d\bar{x}$$

$$K_{11}^e = \frac{E_e}{h_e^2} \int_{x_e}^{x_{e+1}} \left(c_1^e + c_2^e x \right)^2 dx$$

$$= \frac{E_e}{h_e^2} \left[\left(c_1^e \right)^2 h_e + \frac{\left(c_2^e \right)^2}{3} \left(x_{e+1}^3 - x_e^3 \right) + c_1^e c_2^e \left(x_{e+1}^2 - x_e^2 \right) \right]$$

$$K_{12}^e = K_{21}^e = -K_{11}^e, \qquad K_{22}^e = K_{11}^e$$

$$F_1^e = P_1^e, \qquad F_2^e = P_2^e$$

$$[K^e] = \frac{E_e A_e}{h_e} \begin{bmatrix} 1 & -1 \\ -1 & 1 \end{bmatrix}, \qquad \{F^e\} = \begin{Bmatrix} P_1^e \\ P_2^e \end{Bmatrix} \qquad (3.4.33)$$

where we have used the algebraic equalities, $a^2 - b^2 = (a - b)(a + b)$ and $a^3 - b^3 = (a^2 + b^2 + ab)(a - b)$

$$A_e = (c_1^e)^2 + \tfrac{1}{3}(c_2^e)^2(x_{e+1}^2 + x_e^2 + x_e x_{e+1}) + c_1^e c_2^e(x_{e+1} + x_e)$$

More specifically, we have ($c_1^1 = 1.5$, $c_2^1 = -0.5/96$, $c_1^2 = 1$, $c_2^2 = 0$, $h_1 = 96$, $h_2 = 120$)

$$[K^1] = \frac{4.75 \times 10^7}{96}\begin{bmatrix} 1 & -1 \\ -1 & 1 \end{bmatrix}, \quad \{F^1\} = \begin{Bmatrix} P_1^1 \\ P_2^1 \end{Bmatrix}$$

$$[K^2] = \tfrac{1}{120} \times 10^7\begin{bmatrix} 1 & -1 \\ -1 & 1 \end{bmatrix}, \quad \{F^2\} = \begin{Bmatrix} P_1^2 \\ P_2^2 \end{Bmatrix} \tag{3.4.34}$$

3. *Assembly of the Element Equations.* For the two-element mesh we have, from Eq. (3.4.23), the assembled set of equations

$$\begin{bmatrix} K_{11}^1 & K_{12}^1 & 0 \\ K_{21}^1 & K_{22}^1 + K_{11}^2 & K_{12}^2 \\ 0 & K_{21}^2 & K_{22}^2 \end{bmatrix}\begin{Bmatrix} U_1 \\ U_2 \\ U_3 \end{Bmatrix} = \begin{Bmatrix} P_1^1 \\ P_2^1 + P_1^2 \\ P_2^2 \end{Bmatrix} \tag{3.4.35}$$

For the problem at hand, these have the explicit form

$$10^7\begin{bmatrix} \dfrac{4.75}{96} & -\dfrac{4.75}{96} & 0 \\ -\dfrac{4.75}{96} & \dfrac{4.75}{96} + \dfrac{1}{120} & -\dfrac{1}{120} \\ 0 & -\dfrac{1}{120} & \dfrac{1}{120} \end{bmatrix}\begin{Bmatrix} U_1 \\ U_2 \\ U_3 \end{Bmatrix} = \begin{Bmatrix} P_1^1 \\ P_2^1 + P_1^2 \\ P_2^2 \end{Bmatrix}$$

$$\tag{3.4.36}$$

4. *Imposition of Boundary Conditions.* From Fig. 3.7, we have

$$u(0) \equiv U_1 = 0, \quad P_2^1 + P_1^2 = 2P_0, \quad P_2^2 = P_0 \tag{3.4.37a}$$

Using the boundary conditions, Eq. (3.4.36) can be written as

$$10^4\begin{bmatrix} 49.479 & -49.479 & 0 \\ -49.479 & 57.8125 & -8.333 \\ 0 & -8.333 & 8.333 \end{bmatrix}\begin{Bmatrix} 0 \\ U_2 \\ U_3 \end{Bmatrix} = \begin{Bmatrix} P_1^1 \\ 2P_0 \\ P_0 \end{Bmatrix} \tag{3.4.37b}$$

5. *Solution of the Equations.*   From Eq. (3.4.37b), we have (by deleting the first row and column)

$$
10^4 \begin{bmatrix} 57.8125 & -8.333 \\ -8.333 & 8.333 \end{bmatrix} \begin{Bmatrix} U_2 \\ U_3 \end{Bmatrix} = \begin{Bmatrix} 2P_0 \\ P_0 \end{Bmatrix} = \begin{Bmatrix} 20{,}000 \\ 10{,}000 \end{Bmatrix}
$$

or

$$
U_2 = 0.06063 \text{ in.,} \qquad U_3 = 0.18063 \text{ in.} \qquad\qquad (3.4.38a)
$$

6. *Postprocessing of the Results.*   The reaction force at the left end of the bar is given by

$$
P_1^1 \equiv 10^4 (49.479 U_1 - 49.479 U_2)
$$

$$
= -30{,}000 \text{ lb}
$$

The negative sign indicates that P_1^1 is acting opposite to the convention used in Fig. 3.7 and, therefore, the reaction is acting away from the end (i.e. tensile force). The magnitude of P_1^1 is consistent with the static equilibrium of the forces:

$$
P_1^1 + 2P_0 + P_0 = 0 \quad \text{or} \quad P_1^1 = -3P_0 = -30{,}000 \text{ lb}
$$

The axial displacement at any point x along the bar is given by

$$
u(x) = \begin{cases} u_1^1 \psi_1^1 + u_2^1 \psi_2^1 = 0.06063 \dfrac{x}{96}, & 0 \le x \le 96 \text{ in.} \\ u_1^2 \psi_1^2 + u_2^2 \psi_2^2 = -0.03537 + 0.001x, & 96 \text{ in.} \le x \le 216 \text{ in.} \end{cases}
$$

$$
(3.4.38b)
$$

and its first derivative is given by

$$
\frac{du}{dx}(x) = \begin{cases} \dfrac{0.06063}{96}, & 0 \le x \le 96 \text{ in.} \\ 0.001, & 96 \text{ in.} \le x \le 216 \text{ in.} \end{cases} \qquad (3.4.38c)
$$

The exact solution of Eq. (3.4.32a) subject to the boundary conditions

$$
u(0) = 0, \qquad \left(a \frac{du}{dx} \right)\Big|_{x=96^-} = \left(a \frac{du}{dx} \right)\Big|_{x=96^+} = 2P_0, \qquad \left(a \frac{du}{dx} \right)\Big|_{x=216} = P_0
$$

is given by

$$u_0(x) = \begin{cases} 0.128\dfrac{x}{288 - x}, & 0 \leq x \leq 96 \text{ in.} \\ 0.001\,(x - 32), & 96 \text{ in.} \leq x \leq 216 \text{ in.} \end{cases}$$

$$\frac{du_0}{dx}(x) = \begin{cases} \dfrac{36.864}{(288 - x)^2}, & 0 \leq x \leq 96 \text{ in.} \\ 0.001, & 96 \text{ in.} \leq x \leq 216 \text{ in.} \end{cases} \tag{3.4.39}$$

The exact solution at nodes 2 and 3 is given by

$$u_0(96) = 0.064 \text{ in.}, \qquad u_0(216) = 0.1840 \text{ in.}$$

Thus, the two-element solution is about 1.8% off from the maximum displacement.

Next, consider a two-element mesh of quadratic elements. After a lengthy algebraic manipulation, we obtain the following element matrix and force vector for Element 1:

$$[K^1] = 10^4 \begin{bmatrix} 142.19 & -159.37 & 17.18 \\ -159.37 & 266.67 & -107.29 \\ 17.18 & -107.29 & 90.10 \end{bmatrix}, \qquad \{F^1\} = \begin{Bmatrix} P_1^1 \\ 0 \\ P_3^1 \end{Bmatrix}$$

The assembled stiffness matrix is of order 5×5, and is of the form

$$[K] = \begin{bmatrix} K_{11}^1 & K_{12}^1 & K_{13}^1 & 0 & 0 \\ K_{21}^1 & K_{22}^1 & K_{23}^1 & 0 & 0 \\ K_{31}^1 & K_{32}^1 & K_{33}^1 + K_{11}^2 & K_{12}^2 & K_{13}^2 \\ 0 & 0 & K_{21}^2 & K_{22}^2 & K_{23}^2 \\ 0 & 0 & K_{31}^2 & K_{32}^2 & K_{33}^2 \end{bmatrix} \tag{3.4.40}$$

After imposing boundary conditions ($U_1 = 0$, $P_3^1 + P_1^2 = 20{,}000$, $P_3^2 = 10{,}000$), and solving the resulting 4×4 equations, we obtain

$$U_2 = 0.02572 \text{ in.}, \quad U_3 = 0.06392 \text{ in.}, \quad U_4 = 0.12392 \text{ in.}, \quad U_5 = 0.18392 \text{ in.}$$

Obviously, the two-element solution obtained using the quadratic element is very accurate.

A comparison of the finite-element solution obtained by various meshes of linear and quadratic elements with the exact solution is presented in Table 3.6.

Table 3.6 Comparison of the Finite-Element Solutions with the Exact Solution of the Bar Problem Considered in Example 3.16

x	Exact Solution	Linear Elements					Quadratic Elements	
		2 Elements[a]	3 Elements[b]	5 Elements[c]	6 Elements[d]	8 Elements[e]	2 Elements[a]	3 Elements[b]
16	0.00753	—	—	—	—	0.00752	—	—
32	0.01600	—	—	0.01593	—	0.01598	—	—
48	0.02560	—	0.02532	—	0.02553	0.02557	—	0.02560
64	0.03657	—	—	0.03638	—	0.03652	—	—
72	0.04267	—	—	—	0.04253	—	—	—
80	0.04923	—	—	—	—	0.04916	—	—
96	0.06400	0.06063	0.06309	0.06359	0.06377	0.06390	0.06392	0.06399
156	0.12400	—	—	0.12359	0.12377	0.1239	—	—
216	0.18400	0.18063	0.18309	0.18359	0.18377	0.1840	0.18392	0.18399

[a] $h_1 = 96,\ h_2 = 120.$
[b] $h_1 = h_2 = 48,\ h_3 = 120.$
[c] $h_1 = h_2 = h_3 = 32,\ h_4 = h_5 = 60.$
[d] $h_1 = h_2 = h_3 = h_4 = 24,\ h_5 = h_6 = 60.$
[e] $h_1 = \cdots = h_6 = 16,\ h_7 = h_8 = 60.$

3.4.3 One-Dimensional Fourth-Order Equations

Here we consider the finite-element formulation of the fourth-order equation governing the bending of elastic beams:

$$\frac{d^2}{dx^2}\left(EI\frac{d^2w}{dx^2}\right) + f = 0, \qquad 0 < x < L \qquad (3.4.41)$$

The basic steps in the formulation are the same as those discussed for bars, but the specific mathematical details differ from those of second-order equations. Since the basic terminology of the finite-element method is introduced in the preceding section, we go straight to the approximation over an element.

Approximation Over an Element

A finite-element for a beam is again a free-body-diagram of a portion, $R_e = (x_A, x_B)$, of a typical beam, as shown in Fig. 3.8. From the total potential energy principle for beams, we know that the essential boundary conditions involve the specification of w and dw/dx,

$$w(x_A) \equiv u_1^e, \qquad \frac{dw}{dx}(x_A) \equiv -u_2^e, \qquad w(x_B) \equiv u_3^e, \qquad \frac{dw}{dx}(x_B) \equiv -u_4^e$$

$$(3.4.42)$$

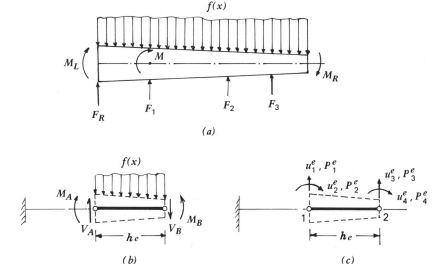

Figure 3.8 Finite-element representation of a beam: (*a*) geometry and loading on a beam; (*b*) free-body diagram of an element; (*c*) beam element with primary (displacement) and secondary (force) degrees of freedom.

and the natural boundary conditions involve the specification of the second-
and third-order derivatives of w,

$$\left[\frac{d}{dx}\left(EI\frac{d^2w}{dx^2}\right)\right]_{x=x_A} \equiv P_1^e, \qquad \left(EI\frac{d^2w}{dx^2}\right)\Big|_{x=x_A} \equiv P_2^e$$

$$\left[\frac{d}{dx}\left(EI\frac{d^2w}{dx^2}\right)\right]_{x=x_B} \equiv -P_3^e, \qquad \left(EI\frac{d^2w}{dx^2}\right)\Big|_{x=x_B} \equiv -P_4^e$$

$$(3.4.43)$$

Obviously, inclusion of the minus sign in Eqs. (3.4.42) and (3.4.43) has a
purpose, as will be apparent; also, the quantities defined in these equations are
consistent with the notation in Fig. 3.8.

The total potential energy functional for the beam element is given by

$$\Pi^e(w_e) = \int_{x_A}^{x_B}\left[\frac{E_eI_e}{2}\left(\frac{d^2w_e}{dx^2}\right)^2 + w_ef_e\right]dx - P_1^eu_1^e - P_2^eu_2^e - P_3^eu_3^e - P_4^eu_4^e$$

$$(3.4.44)$$

where the sum $\sum_{i=1}^{4}P_i^eu_i^e$ represents the work done by the forces P_1^e and P_3^e
and moments P_2^e and P_4^e in moving and rotating the ends of the element
through displacements u_1^e and u_3^e and slopes u_2^e and u_4^e.

To derive the coordinate functions for the Ritz approximation of the
function w_e, we must first select a polynomial that is continuous and complete
up to the degree required. It is clear from the functional Π^e that the function
we select should be twice-differentiable to yield nonzero strain energy, and it
should be cubic to yield nonzero shear forces P_1^e and P_3^e. Another reason for
selecting w_e to be a cubic polynomial is that we need four parameters in the
polynomial to satisfy all four essential boundary conditions in Eq. (3.4.42).
Thus, a minimum degree polynomial for the fourth-order equation is cubic:

$$w_e = \alpha_0 + \alpha_1\bar{x} + \alpha_2\bar{x}^2 + \alpha_3\bar{x}^3 \qquad (3.4.45)$$

where $\bar{x}$ denotes the local coordinate. Next, select the parameters α_0, α_1, α_2,
and α_3 such that w_e satisfies the essential boundary conditions Eq. (3.4.42):

$$u_1^e \equiv w_e(0) = \alpha_0$$

$$u_2^e \equiv -\frac{dw_e}{dx}(0) = -\alpha_1$$

$$u_3^e \equiv w_e(h_e) = \alpha_0 + \alpha_1h_e + \alpha_2h_e^2 + \alpha_3h_e^3$$

$$u_4^e \equiv -\frac{dw_e}{dx}(h_e) = -\alpha_1 - 2\alpha_2h_e - 3\alpha_3h_e^2 \qquad (3.4.46)$$

where $h_e = x_B - x_A$. Solving Eq. (3.4.46) for α_0, α_1, α_2, and α_3 in terms of u_i^e

$(i = 1, 2, 3, 4)$ and substituting the result into Eq. (3.4.45), we obtain

$$w_e(x) = \sum_{j=1}^{4} u_j^e \phi_j^e(x) \tag{3.4.47}$$

where the coordinate functions ϕ_j^e for the beam element are given by

$$\phi_1^e = 1 - 3\left(\frac{\bar{x}}{h_e}\right)^2 + 2\left(\frac{\bar{x}}{h_e}\right)^3$$

$$\phi_2^e = -\bar{x}\left[1 - \left(\frac{\bar{x}}{h_e}\right)\right]^2$$

$$\phi_3^e = 3\left(\frac{\bar{x}}{h_e}\right)^2 - 2\left(\frac{\bar{x}}{h_e}\right)^3$$

$$\phi_4^e = -\bar{x}\left[\left(\frac{\bar{x}}{h_e}\right)^2 - \left(\frac{\bar{x}}{h_e}\right)\right] \tag{3.4.48}$$

The coordinate functions ϕ_i^e are called the *Hermite cubics*. We note the following properties of ϕ_i^e, $i = 1, 2, 3, 4$ (see Fig. 3.9):

$$\phi_{2\alpha-1}^e(\bar{x}_\beta) = \delta_{\alpha\beta}, \qquad \phi_{2\alpha}^e(\bar{x}_\beta) = 0, \qquad \sum_{\alpha=1}^{2}\phi_{2\alpha-1}^e = 1$$

$$\frac{d\phi_{2\alpha-1}^e}{dx}(\bar{x}_\beta) = 0, \qquad -\frac{d\phi_{2\alpha}^e}{dx}(\bar{x}_\beta) = \delta_{\alpha\beta} \qquad (\alpha, \beta = 1, 2) \tag{3.4.49}$$

where $\bar{x}_1 = 0$ and $\bar{x}_2 = h_e$ are the local coordinates of nodes 1 and 2 of element $R_e = (x_e, x_{e+1})$, and x_e is the global coordinate of global node e.

By including the slopes u_2^e and u_4^e as nodal quantities, we will be able to enforce the slope continuity at the interelement boundaries during the assembly process. Of course, the total potential energy formulation of the governing equation yields slopes as part of the essential boundary conditions. All dependent variables and their derivatives that enter the specification of the

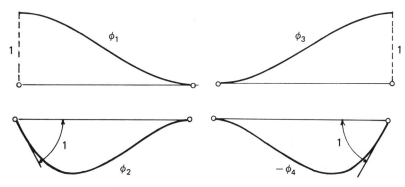

Figure 3.9 Hermite cubic interpolation functions for the beam element.

essential boundary conditions of a problem always end up as the nodal variables, and they take the role of the undetermined coefficients of the Ritz method.

To derive the Ritz equations corresponding to the functional in Eq. (3.4.44), we substitute Eq. (3.4.47) for w_e into Π^e, and set its derivatives with respect to each u_j^e, $j = 1, 2, 3, 4$, to zero:

$$\Pi^e(u_1^e, u_2^e, u_3^e, u_4^e) = \int_{x_A = x_e}^{x_B = x_{e+1}} \left[\frac{E_e I_e}{2} \left(\sum_{j=1}^{4} u_j^e \frac{d^2 \phi_j^e}{dx^2} \right)^2 + \sum_{i=1}^{4} u_i^e \phi_i^e f_e \right] dx$$

$$- \sum_{i=1}^{4} P_i^e u_i^e$$

$$= \frac{1}{2} \sum_{i=1}^{4} \sum_{j=1}^{4} u_i^e \left(\int_{x_e}^{x_{e+1}} E_e I_e \frac{d^2 \phi_i^e}{dx^2} \frac{d^2 \phi_j^e}{dx^2} dx \right) u_j^e$$

$$+ \sum_{i=1}^{4} u_i^e \left(\int_{x_e}^{x_{e+1}} f_e \phi_i^e \, dx - P_i^e \right)$$

$$\equiv \frac{1}{2} \sum_{i=1}^{4} \left(\sum_{j=1}^{4} u_i^e K_{ij}^e u_j^e - u_i^e F_i^e \right) \tag{3.4.50a}$$

$$\frac{\partial \Pi^e}{\partial u_i^e} = 0 = \sum_{j=1}^{4} K_{ij}^e u_j^e - F_i^e \tag{3.4.50b}$$

where

$$K_{ij}^e = \int_{x_e}^{x_{e+1}} E_e I_e \frac{d^2 \phi_i^e}{dx^2} \frac{d^2 \phi_j^e}{dx^2} \, dx, \qquad F_i^e = -\int_{x_e}^{x_{e+1}} f_e \phi_i^e \, dx + P_i^e$$

$$\tag{3.4.50c}$$

In Eq. (3.4.50), it is understood that ϕ_i^e are expressed in terms of the global coordinate x. In the element coordinate $\bar{x}$, Eq. (3.4.50c) takes the simple form,

$$K_{ij}^e = \int_0^{h_e} \bar{E}_e \bar{I}_e \frac{d^2 \phi_i^e}{d\bar{x}^2} \frac{d^2 \phi_j^e}{d\bar{x}^2} \, d\bar{x}, \qquad F_i^e = -\int_0^{h_e} \bar{f}_e \phi_i^e \, d\bar{x} + P_i^e \tag{3.4.50d}$$

where $\bar{E}_e$, $\bar{I}_e$, and $\bar{f}_e$ are the transformed functions,

$$\bar{E}_e = E_e(\bar{x}), \qquad \bar{I}_e = I_e(\bar{x}), \qquad \bar{f}_e = f(\bar{x})$$

For the case in which E_e, I_e, and f_e are constant over the element R_e, the stiffness matrix $[K^e]$ and force vector $\{F^e\}$ are given by

$$[K] = \frac{2EI}{h^3} \begin{bmatrix} 6 & -3h & -6 & -3h \\ -3h & 2h^2 & 3h & h^2 \\ -6 & 3h & 6 & 3h \\ -3h & h^2 & 3h & 2h^2 \end{bmatrix}, \quad \{F\} = -\frac{fh}{12} \begin{Bmatrix} 6 \\ -h \\ 6 \\ h \end{Bmatrix} + \begin{Bmatrix} P_1 \\ P_2 \\ P_3 \\ P_4 \end{Bmatrix}$$

$$\tag{3.4.51}$$

where the label e on the variables is omitted in the interest of brevity. One can show that the first column of the force vector represents the "statically equivalent" forces and moments at the nodes caused by the uniformly distributed load f over the elements.

When E, I, and f are functions of x, the hand computation becomes tedious; however, for any algebraic polynomial variations of E, I, and f, the stiffness and force coefficients can be accurately evaluated (numerically) on a computer.

Assembly of the Element Equations

The assembly procedure for the beam element is analogous to that described for the bar element. The difference is that in the beam element there are two unknowns, called *degrees of freedom*, per node. Consequently, at every global node that is shared by two elements, the elements in the last two rows and columns of the Ith element overlap with the elements in the first two rows and columns of the $(I + 1)$th element. For a three-element case, the assembled force vector and stiffness matrix are shown below (see Fig. 3.10):

$$\{F\} = \begin{Bmatrix} f_1^1 \\ f_2^1 \\ f_3^1 + f_1^2 \\ f_4^1 + f_2^2 \\ f_3^2 + f_1^3 \\ f_4^2 + f_2^3 \\ f_3^3 \\ f_4^3 \end{Bmatrix} + \begin{Bmatrix} P_1^1 \\ P_2^1 \\ P_3^1 + P_1^2 \\ P_4^1 + P_2^2 \\ P_3^2 + P_1^3 \\ P_4^2 + P_2^3 \\ P_3^3 \\ P_4^3 \end{Bmatrix} \qquad (3.4.52a)$$

$$[K] = \begin{bmatrix} K_{11}^1 & K_{12}^1 & K_{13}^1 & K_{14}^1 & 0 & 0 & 0 & 0 \\ K_{21}^1 & K_{22}^1 & K_{23}^1 & K_{24}^1 & 0 & 0 & 0 & 0 \\ K_{31}^1 & K_{32}^1 & K_{33}^1 + K_{11}^2 & K_{34}^1 + K_{12}^2 & K_{13}^2 & K_{14}^2 & 0 & 0 \\ K_{41}^1 & K_{42}^1 & K_{43}^1 + K_{21}^2 & K_{44}^1 + K_{22}^2 & K_{23}^2 & K_{24}^2 & 0 & 0 \\ 0 & 0 & K_{31}^2 & K_{32}^2 & K_{33}^2 + K_{11}^3 & K_{34}^2 + K_{12}^3 & K_{13}^3 & K_{14}^3 \\ 0 & 0 & K_{41}^2 & K_{42}^2 & K_{43}^2 + K_{21}^3 & K_{44}^2 + K_{22}^3 & K_{23}^3 & K_{24}^3 \\ 0 & 0 & 0 & 0 & K_{31}^3 & K_{32}^3 & K_{33}^3 & K_{34}^3 \\ 0 & 0 & 0 & 0 & K_{41}^3 & K_{42}^3 & K_{43}^3 & K_{44}^3 \end{bmatrix}$$

$$(3.4.52b)$$

Imposition of Boundary Conditions

The boundary conditions on the generalized displacements can be imposed in the manner described earlier for bars. The boundary (or equilibrium) condi-

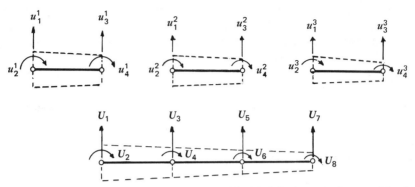

Figure 3.10 Assembled mesh of beam elements with global displacement degrees of freedom.

tions on forces are imposcd by modifying the second column of the assembled force vector [see Eq. (3.4.52a)]. If, for example, the force at the Ith global node is specified to be F_0, and the moment at the Kth global node is specified to be M_0, then the following assembled coefficients are modified:

$$P_3^{I-1} + P_1^I = F_0, \qquad P_4^{K-1} + P_2^K = M_0 \qquad (3.4.53)$$

If no force or moment is specified at a nodal point that is unconstrained, it is understood that the force and moment are zero there.

The solution and postcomputation of the displacements and their derivatives at various points of the beam follow along the same lines as those described for bar problems. We now consider a specific example to illustrate the steps involved in the finite-element analysis of beams.

Example 3.17

Consider the indeterminate beam shown in Fig. 3.11. The beam is made of steel ($E = 30 \times 10^6$ psi) and the cross-sectional dimensions are 2×3 in.^{2} ($I = 4.5$ in.^{4}). We are interested in finding the deflections and stresses in the beam.

Because of the discontinuity in loading, the beam should be divided into three elements: $R_1 = (0, 16)$, $R_2 = (16, 36)$, and $R_3 = (36, 48)$: the element lengths are: $h_1 = 16$ in., $h_2 = 20$ in., and $h_3 = 12$ in.

The element equations for all three elements are readily available from Eq. (3.4.51). The only exception is that the load vector for Element 1, which has linearly varying distributed load, must be computed. The load variation is given by

$$f^1 = \left(30 - \tfrac{10}{16}x\right), \qquad f^2 = 20, \qquad f^3 = 0$$

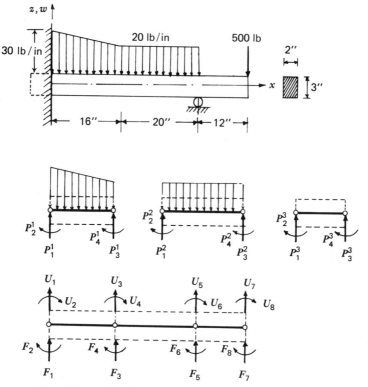

Figure 3.11 Finite-element modeling of an indeterminate beam.

and their contribution to the element nodes can be computed using Eq. (3.4.50d). For Element 1, we have

$$f_i^1 = -\int_0^{16}\left(30 - \tfrac{10}{16}x\right)\phi_i\,dx, \qquad i = 1, 2, 3, 4$$

We have

$$\{f^1\} = -\begin{Bmatrix} 216.00 \\ -554.67 \\ 184.00 \\ 512.00 \end{Bmatrix}, \qquad \{f^2\} = -\begin{Bmatrix} 200.00 \\ -666.67 \\ 200.00 \\ 666.67 \end{Bmatrix}, \qquad \{f^3\} = \{0\}$$

$$(3.4.54)$$

The element stiffness matrices can be computed from Eq. (3.4.51) by substituting appropriate values of h, E, and I. For example, for Element 1, we have

$(h_1 = 16$ in., $EI = 135 \times 10^6$ lb–in.$^2)$,

$$[K^1] = 10^6 \begin{bmatrix} 0.3955 & -3.1641 & -0.3955 & -3.1641 \\ -3.1641 & 33.750 & 3.1641 & 16.875 \\ -0.3955 & 3.1641 & 0.3955 & 3.1641 \\ -3.1641 & 16.875 & 3.1641 & 33.750 \end{bmatrix}$$

The assembled matrix is of the same form as that given in Eq. (3.4.52b). The boundary conditions for the problem are

$$w(0) = 0 \rightarrow U_1 = 0, \qquad \frac{dw}{dx}(0) = 0 \rightarrow U_2 = 0, \qquad w(36) = 0 \rightarrow U_5 = 0$$

$$P_3^1 + P_1^2 = 0, \qquad P_4^1 + P_2^2 = 0, \qquad P_4^2 + P_2^3 = 0, \qquad P_3^3 = -500,$$

$$P_4^3 = 0 \tag{3.4.55}$$

Note that P_1^1, P_2^1, and $P_3^2 + P_1^3$ are the reactions that are not known, and are to be calculated in the post computation. Since the specified essential boundary conditions are homogeneous, one can delete the rows and columns corresponding to the specified displacements (i.e., delete rows and columns 1, 2, and 5) and solve the remaining five equations for U_3, U_4, U_6, U_7, and U_8:

$$U_3 = 0.000322 \text{ in.}, \qquad U_4 = -0.00005935 \text{ rad}, \qquad U_6 = 0.0002513 \text{ rad},$$

$$U_7 = -0.00515 \text{ in.}, \qquad U_8 = 0.000518 \text{ rad}$$

The exact solution to the problem can be obtained (using the classical beam theory) by integrating the moment expression, which can be readily obtained (in terms of the reactions). We have

$$w_0(x) = \begin{cases} \left(\frac{1}{192}x^5 - \frac{15}{12}x^4 + \frac{691}{15}x^3 - 268.543x^2\right)/EI, & 0 \le x \le 16 \text{ in.} \\ \left(-\frac{5}{6}x^4 + 32.7333x^3 - 55.20967x^2 - 537.086x\right)/EI, \\ & 16 \text{ in.} \le x \le 36 \text{ in.} \\ \left(83.333x^3 - 12000.0x^2 + 506066.1x - 6554372.27\right)/EI, \\ & 36 \text{ in.} \le x \le 48 \text{ in.} \end{cases}$$

$$\tag{3.4.56}$$

The exact solution w_0 and $(-dw_0/dx)$ at the nodal points is given by

$$0.000322 \text{ in.}, \qquad -0.00005935 \text{ rad}, \qquad 0.0 \text{ in.}, \qquad 0.00025136 \text{ rad},$$

$$-0.0051496 \text{ in.}, \qquad 0.000518 \text{ rad}$$

Thus, the finite-element solution at the nodes is in excellent agreement with the exact solution. The finite-element results for deflections, slopes, and bending moments (calculated in the postcomputation) at points other than the nodes are compared with the exact values in Table 3.7 for three different meshes. The nodal values of deflections and slopes do not change with the mesh, because they have already converged to the exact solution. The values of deflections and slopes at points other than nodes, and the bending moment, were computed using the finite-element equation (3.4.47) and its derivatives.

We close this section by remarking that one can use Reissner's variational principle [see Eq. (3.2.58) in Example 3.5] to derive a finite-element model that includes deflection and bending moment as degrees of freedom at the nodes. The resulting finite element equations resemble those in Eq. (3.2.61). Indeed, by choosing $\phi_i = \psi_i$ to be the linear interpolation functions in Eq. (3.4.15), one can obtain the element matrices for Reissner's finite-element model directly

Table 3.7 Comparison of the Finite-Element Solution with the Exact Solution of the Beam Problem Considered in Example 3.17

		$-w \times 10^6$		$(-dw/dx) \times 10^6$		$(d^2w/dx^2) \times 10^6$	
x	Mesh[a]	FEM	Exact	FEM	Exact	FEM	Exact
2.0	3-element	-0.8570		-1.1553		-1.025	
	5-element	4.1405	5.3739	3.3386	4.1552	0.4663	0.3219
	6-element	5.2319		4.1558		0.4638	
6.0	3-element	-18.451		-8.8348		-2.8147	
	5-element	8.3936	9.6048	-4.4199	-5.2328	-4.3456	-4.4727
	6-element	9.4751		-5.2323		-4.3431	
12.0	3-element	-138.23		-33.776		-5.4992	
	5-element	-122.59	-120.81	-39.663	-39.672	-5.4794	-5.9238
	6-element	-122.59		39.663		-5.4794	
21.0	3-element	-674.36		-71.562		3.5507	
	5-element	-643.49	-639.63	-62.303	-62.304	3.5507	2.9335
	6-element	-643.49		-62.303		3.5507	
31.0	3-element	-812.93		83.797		27.521	
	5-element	-782.07	-778.19	74.537	74.532	27.521	26.904
	6-element	-782.07		74.537		27.521	
42.0	3-element	2174.9		451.36		22.222	
	5-element	2174.9	2174.8	451.36	451.36	22.222	22.222
	6-element	2174.9		451.36		22.222	

[a]3-Element mesh: $h_1 = 16$, $h_2 = 20$, $h_3 = 12$.
5-Element mesh: $h_1 = 8$, $h_2 = 8$, $h_3 = 10$, $h_4 = 10$, $h_5 = 12$.
6-Element mesh: $h_1 = 4$, $h_2 = 4$, $h_3 = 8$, $h_4 = 10$, $h_5 = 10$, $h_6 = 12$.

from Eq. (3.2.61). The reader is asked to compute these matrices and use them to solve the problem of Example 3.17 (see Exercise 3.4.10). One can use Eqs. (3.2.52) and the least-squares method to derive a least-squares finite-element model for beams (see Exercise 3.4.11).

3.4.4 Two-Dimensional Second-Order Equations

One-Variable Problems: Torsion

Here we consider the finite-element analysis of a partial differential equation governing a scalar function. As a model equation, we consider the following general second-order equation in two dimensions:

$$-\frac{\partial}{\partial x}\left(c_{11}\frac{\partial u}{\partial x} + c_{12}\frac{\partial u}{\partial y}\right) - \frac{\partial}{\partial y}\left(c_{21}\frac{\partial u}{\partial x} + c_{22}\frac{\partial u}{\partial y}\right) + c_0 u = f \quad \text{in } R$$

$$(3.4.57)$$

The coefficients c_{ij} $(i, j = 1, 2)$ and c_0, and the source term f are given functions of position (x, y) in the domain R. Equation (3.4.57) arises in the study of a number of engineering problems, including torsion of constant cross-section members, heat transfer, irrotational flow of a fluid, transverse deflection of a membrane, and so on. The constants c_{ij} are generally material property constants for an orthotropic medium. When $c_{11} = c_{22}$, $c_0 = c_{12} = c_{21} = 0$, and $f = 2G\theta$, we obtain the Poisson equation governing the torsion of isotropic prismatic members. Thus, the finite-element procedure to be described for Eq. (3.4.57) is applicable to *any* problem that can be formulated as one of solving Eq. (3.4.57). We now describe a step-by-step procedure for the finite-element formulation of Eq. (3.4.57).

1. *Representation of a Domain by Elements.* Two-dimensional domains, contrary to one-dimensional problems, can be represented by more than one type of geometric shape. For example, a rectangular domain can be repre-

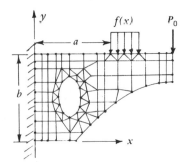

Figure 3.12 Representation of a plane elastic body by triangular, rectangular, and quadrilateral elements.

sented by rectangular elements, triangular elements, or quadrilateral elements (see Fig. 3.12). Without reference to a specific geometric shape, we simply denote a typical element by R_e and proceed to discuss the approximation of Eq. (3.4.57). It should be apparent to the reader by now that the finite-element formulation of a given differential equation can be discussed without the knowledge of (or reference to) the specific form of the interpolation functions. The specific form, which depends on the geometry of the element, is needed to compute the coefficients when we actually solve the problem.

2. *Finite-Element Formulation.* To develop a finite-element model of Eq. (3.4.57) using the Ritz–Galerkin methods, we need a variational formulation, either in the Galerkin integral form or in a functional form, of Eq. (3.4.57). If one traces back to the origin of Eq. (3.4.57), one finds that it is either derived from vectorial mechanics or a variational principle. If the variational statement is available, we simply proceed, as we did in one-dimensional problems, to develop the Ritz approximation of the equation. If such a variational problem is not readily available, we can develop one from the given differential equation by following a procedure that is precisely the inverse of that used in deriving the Euler equations from a functional (or, in general, variational statement). Here we demonstrate the inverse procedure for the equation at hand (see Sec. 3.2.5).

We begin with the last step in the derivation of the Euler equations (which can also be interpreted as the first step in the Galerkin method). We take the source term to the left side of Eq. (3.4.57), multiply it with an arbitrary variation δu, and integrate the product over the area of the *element*, R_e:

$$0 = \int_{R_e} \delta u \left[-\frac{\partial}{\partial x} \left(c_{11} \frac{\partial u}{\partial x} + c_{12} \frac{\partial u}{\partial y} \right) \right.$$
$$\left. -\frac{\partial}{\partial y} \left(c_{21} \frac{\partial u}{\partial x} + c_{22} \frac{\partial u}{\partial y} \right) + c_0 u - f \right] dx\, dy \qquad (3.4.58)$$

Since Eq. (3.4.58) contains even-order derivatives of u, it is possible for us to trade half the differentiation to δu by using integration by parts (or the component form of the Green–Gauss theorem). We have

$$0 = \int_{R_e} \left(c_{11} \frac{\partial \delta u}{\partial x} \frac{\partial u}{\partial x} + c_{12} \frac{\partial \delta u}{\partial x} \frac{\partial u}{\partial y} + c_{21} \frac{\partial \delta u}{\partial y} \frac{\partial u}{\partial x} \right.$$
$$\left. + c_{22} \frac{\partial \delta u}{\partial y} \frac{\partial u}{\partial y} + c_0 \delta u\, u - \delta u f \right) dx\, dy$$
$$-\oint_{S_e} \delta u \left[\left(c_{11} \frac{\partial u}{\partial x} + c_{12} \frac{\partial u}{\partial y} \right) n_x + \left(c_{21} \frac{\partial u}{\partial x} + c_{22} \frac{\partial u}{\partial y} \right) n_y \right] dS \qquad (3.4.59)$$

The quantity in the square brackets of the boundary integral will be denoted by t_n^e. We know from Chapter 2 that the boundary integral in a variational formulation provides the classification of the boundary conditions into *essential* and *natural* type. For the problem, we recognize that specification of u

constitutes the essential boundary condition and specification of t_n^e constitutes the natural boundary condition. From now on, t_n^e takes the role of $P \equiv [a(du/dx)]$ in one-dimensional problems. Equation (3.4.59) is the variational form of Eq. (3.4.57), and it forms the basis for the finite-element model. One more step in the inverse procedure will take us to the functional, if it exists. In the present case, it is obvious that the functional exists only if $c_{12} = c_{21}$. Using the properties of the variational operator, we obtain

$$0 = \delta \int_{R_e} \left\{ \frac{1}{2} \left[c_{11} \left(\frac{\partial u}{\partial x} \right)^2 + 2c_{12} \frac{\partial u}{\partial x} \frac{\partial u}{\partial y} + c_{22} \left(\frac{\partial u}{\partial y} \right)^2 + c_0 u^2 \right] - fu \right\} dx\, dy$$

$$- \delta \oint_{S_e} u t_n^e\, dS$$

$$\equiv \delta \Pi^e(u) \tag{3.4.60}$$

Note that in writing the functional Π^e, it is assumed that t_n^e is a specified quantity, and that is how it is treated in the finite-element method.

Now we seek an approximate solution of u over element R_e. Once again, we seek an approximation of the form

$$u = \sum_{j=1}^{n} u_j^e \psi_j^e(x, y) \tag{3.4.61}$$

where u_j^e are the values of u at element node j, and ψ_j^e are the interpolation functions. These are derived later for two different elements. Substituting Eq. (3.4.61) for u and ψ_i for δu into Eq. (3.4.59), we obtain

$$0 = \sum_{j=1}^{n} \left[\int_{R_e} \left(c_{11} \frac{\partial \psi_i}{\partial x} \frac{\partial \psi_j}{\partial x} + c_{12} \frac{\partial \psi_i}{\partial x} \frac{\partial \psi_j}{\partial y} + c_{21} \frac{\partial \psi_i}{\partial y} \frac{\partial \psi_j}{\partial x} \right. \right.$$

$$\left. \left. + c_{22} \frac{\partial \psi_i}{\partial y} \frac{\partial \psi_j}{\partial y} + c_0 \psi_i \psi_j \right) dx\, dy \right] u_j^e - \int_{R_e} \psi_i f\, dx\, dy - \oint_{S_e} \psi_i t_n^e\, dS$$

$$= \sum_{j=1}^{n} K_{ij}^e u_j^e - f_i^e - P_i^e \tag{3.4.62a}$$

where

$$K_{ij}^e = \int_{R_e} \left(c_{11} \frac{\partial \psi_i}{\partial x} \frac{\partial \psi_j}{\partial x} + c_{12} \frac{\partial \psi_i}{\partial x} \frac{\partial \psi_j}{\partial y} + c_{21} \frac{\partial \psi_i}{\partial y} \frac{\partial \psi_j}{\partial x} \right.$$

$$\left. + c_{22} \frac{\partial \psi_i}{\partial y} \frac{\partial \psi_j}{\partial y} + c_0 \psi_i \psi_j \right) dx\, dy$$

$$f_i^e = \int_{R_e} \psi_i f\, dx\, dy, \qquad P_i^e = \oint_{S_e} \psi_i t_n^e\, dS \tag{3.4.62b}$$

This completes the development of the finite element model of Eq. (3.4.57).

3. *Derivation of Interpolation Functions for Linear Triangular and Rectangular Elements.*

TRIANGULAR ELEMENT.   Since a triangle is defined uniquely by three points that form its vertices, we choose the vertex points as the nodes. These nodes are connected to the nodes of adjoining elements in a finite-element mesh (see Fig. 3.13). The variational form in Eq. (3.4.59) requires us to choose ψ_i^e that have nonzero derivatives with respect to x and y. In other words, ψ_i^e should be at least a linear polynomial in x and y. Suppose that we use a linear polynomial of the form,

$$u_e = c_0 + c_1 x + c_2 y + c_3 xy \qquad (3.4.63)$$

The polynomial has four constants and, therefore, four nodes must be iden-

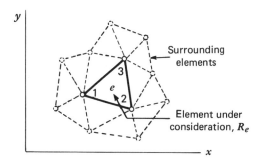

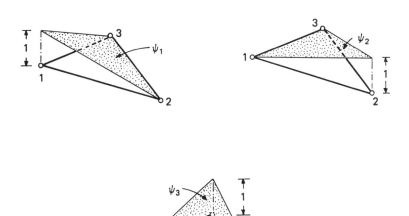

Figure 3.13   Interpolation functions of a linear triangular element.

tified (one can retain c_3 as a parameter that has no obvious physical meaning —called a *nodeless variable*) to write Eq. (3.4.63) in the form of Eq. (3.4.61). We have a choice: we can either identify a fourth node, say at the centroid of the element, or omit the fourth term involving c_3. Note that we cannot omit any other term in the polynomial, because we will spoil the completeness property. We choose to omit the c_3 term.

Proceeding as in the case of one-dimensional elements, we obtain

$$u_i^e \equiv u_e(x_i, y_i) = c_0 + c_1 x_i + c_2 y_i, \qquad i = 1, 2, 3$$

$$\begin{Bmatrix} u_1^e \\ u_2^e \\ u_3^e \end{Bmatrix} = \begin{bmatrix} 1 & x_1 & y_1 \\ 1 & x_2 & y_2 \\ 1 & x_3 & y_3 \end{bmatrix} \begin{Bmatrix} c_0 \\ c_1 \\ c_2 \end{Bmatrix} \tag{3.4.64}$$

and

$$u_e = \sum_{i=1}^{3} u_i^e \psi_i^e(x, y), \qquad \psi_i^e = \frac{1}{2A_e}(\alpha_i^e + \beta_i^e x + \gamma_i^e y) \tag{3.4.65a}$$

where $2A_e$ represents the value of the determinant of the matrix in Eq. (3.4.64) (or A_e represents the area of the triangle), and

$$\alpha_i^e = x_j y_k - x_k y_j, \qquad \beta_i^e = y_j - y_k, \qquad \gamma_i^e = x_k - x_j,$$

$$i \neq j \neq k, \qquad i, j, k = 1, 2, 3 \tag{3.4.65b}$$

and the indices on α_i, β_i, and γ_i permute in a natural order. For example, α_1 is given by setting $i = 1$, $j = 2$, and $k = 3$:

$$\alpha_1 = x_2 y_3 - x_3 y_2$$

Here x_i and y_i denote the x- and y-coordinates (global) of node i of the element. The interpolation functions ψ_i^e once again satisfy the properties listed in Eq. (3.4.15). The shapes of these functions are shown in Fig. 3.13.

RECTANGULAR ELEMENT. A rectangular element is uniquely defined by the four corner points (see Fig. 3.14). Therefore, we can use the four-term polynomial in Eq. (3.4.63) to derive the interpolation functions. We have

$$\begin{Bmatrix} u_1^e \\ u_2^e \\ u_3^e \\ u_4^e \end{Bmatrix} = \begin{bmatrix} 1 & x_1 & y_1 & x_1 y_1 \\ 1 & x_2 & y_2 & x_2 y_2 \\ 1 & x_3 & y_3 & x_3 y_3 \\ 1 & x_4 & y_4 & x_4 y_4 \end{bmatrix} \begin{Bmatrix} c_0 \\ c_1 \\ c_2 \\ c_3 \end{Bmatrix}$$

By inverting the equations and substituting the result for c's into Eq. (3.4.63), we obtain

$$\psi_1^e = \left(1 - \frac{\bar{x}}{a}\right)\left(1 - \frac{\bar{y}}{b}\right)$$

$$\psi_2^e = \frac{\bar{x}}{a}\left(1 - \frac{\bar{y}}{b}\right)$$

$$\psi_3^e = \frac{\bar{x}}{a}\frac{\bar{y}}{b}$$

$$\psi_4^e = \left(1 - \frac{\bar{x}}{a}\right)\frac{\bar{y}}{b}$$

$$\bar{x} = x - x_1, \qquad \bar{y} = y - y_1 \qquad\qquad (3.4.66)$$

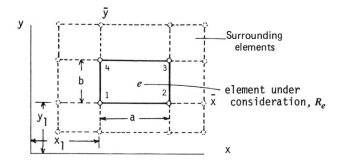

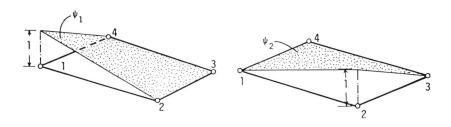

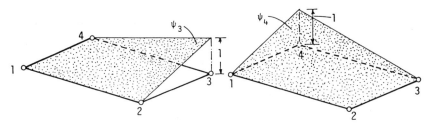

Figure 3.14 Interpolation functions of a linear rectangular element.

The functions are geometrically represented in Fig. 3.14. In calculating element matrices, one finds that the use of the local coordinate system $(\bar{x}, \bar{y})$ is more convenient than using the global coordinates (x, y).

EXPLICIT FORM OF THE FINITE-ELEMENT MATRICES FOR CONSTANT COEFFICIENTS. Here we give the values of $[K^e]$ and $\{f^e\}$, as defined in Eq. (3.4.62b), for constant values of c_{ij}, c_0, and f. We have

$$[K^e] = c_{11}[S^x] + c_{12}[S^{xy}] + c_{21}[S^{xy}]^T + c_{22}[S^y] + c_0[S]$$

$$(3.4.67a)$$

where S_{ij}^x, S_{ij}^{xy}, S_{ij}^y, and S_{ij} are defined by

$$S_{ij}^x = \int_{R_e} \frac{\partial \psi_i}{\partial x} \frac{\partial \psi_j}{\partial x} \, dx \, dy, \qquad S_{ij}^y = \int_{R_e} \frac{\partial \psi_i}{\partial y} \frac{\partial \psi_j}{\partial y} \, dx \, dy$$

$$S_{ij}^{xy} = \int_{R_e} \frac{\partial \psi_i}{\partial x} \frac{\partial \psi_j}{\partial y} \, dx \, dy, \qquad S_{ij} = \int_{R_e} \psi_i \psi_j \, dx \, dy \qquad (3.4.67b)$$

For the linear triangular and rectangular elements, these matrices are given below:

Triangular Element

$$S_{ij}^x = \frac{I_{00}}{4A} \beta_i \beta_j, \qquad S_{ij}^y = \frac{I_{00}}{4A} \gamma_i \gamma_j, \qquad S_{ij}^{xy} = \frac{I_{00}}{4A} \beta_i \gamma_j$$

$$S_{ij} = \frac{I}{4A} \alpha_i \alpha_j, \qquad f_i = \frac{f}{2}(\alpha_i I_{00} + \beta_i I_{10} + \gamma_i I_{01}) \qquad (3.4.68)$$

where

$$I_{ij} = \frac{1}{A} \int_{R_e} x^i y^j \, dx \, dy$$

$$I = I_{00} \alpha_i \alpha_j + I_{10}(\alpha_i \beta_j + \alpha_j \beta_i) + I_{01}(\alpha_i \gamma_j + \alpha_j \gamma_i)$$

$$+ I_{20} \beta_i \beta_j + I_{02} \gamma_i \gamma_j + I_{11}(\beta_i \gamma_j + \beta_j \gamma_i)$$

$$I_{00} = 1, \qquad I_{10} = x_0, \qquad I_{01} = y_0, \qquad I_{11} = \frac{1}{12}\left(\sum_{i=1}^{3} x_i y_i + 9x_0 y_0 \right)$$

$$I_{20} = \frac{1}{12}\left(\sum_{i=1}^{3} x_i^2 + 9x_0^2 \right), \qquad I_{02} = \frac{1}{12}\left(\sum_{i=1}^{3} y_i^2 + 9y_0^2 \right)$$

$$x_0 = \frac{1}{3} \sum_{i=1}^{3} x_i, \qquad y_0 = \frac{1}{3} \sum_{i=1}^{3} y_i \qquad (3.4.69)$$

Rectangular Element

$$[S^x] = \frac{b}{6a} \begin{bmatrix} 2 & -2 & -1 & 1 \\ -2 & 2 & 1 & -1 \\ -1 & 1 & 2 & -2 \\ 1 & -1 & -2 & 2 \end{bmatrix}, \quad [S^y] = \frac{a}{6b} \begin{bmatrix} 2 & 1 & -1 & -2 \\ 1 & 2 & -2 & -1 \\ -1 & -2 & 2 & 1 \\ -2 & -1 & 1 & 2 \end{bmatrix},$$

$$[S^{xy}] = \frac{1}{4} \begin{bmatrix} 1 & 1 & -1 & -1 \\ -1 & -1 & 1 & 1 \\ -1 & -1 & 1 & 1 \\ 1 & 1 & -1 & -1 \end{bmatrix}, \quad [S] = \frac{ab}{36} \begin{bmatrix} 4 & 2 & 1 & 2 \\ 2 & 4 & 2 & 1 \\ 1 & 2 & 4 & 2 \\ 2 & 1 & 2 & 4 \end{bmatrix}$$

$$\{f\} = f\frac{ab}{4}\{1\ 1\ 1\ 1\}^T \tag{3.4.70}$$

4. *Assembly of Element Equations.* The assembly of finite-element equations of two-dimensional elements is based on the same principle as used for one-dimensional elements. The proof is algebraically complicated; therefore, it is not repeated here for the two-dimensional elements. Instead, we give a simple procedure for assembling the element equations.

Consider the finite-element mesh of linear triangular and rectangular elements shown in Fig. 3.15. We illustrate the procedure by considering a couple of elements of the mesh. First, we label the elements and global nodes and then label the element nodes. We find the following correspondence between the global and local nodes:

Element Number	Global Node Numbers Corresponding to Element Node Numbers			
1	1	2	5	4
2	2	3	6	5
3	4	5	7	0
4	5	6	8	7
5	7	8	9	0

or

$$[B] = \begin{bmatrix} 1 & 2 & 5 & 4 \\ 2 & 3 & 6 & 5 \\ 4 & 5 & 7 & 0 \\ 5 & 6 & 8 & 7 \\ 7 & 8 & 9 & 0 \end{bmatrix}$$

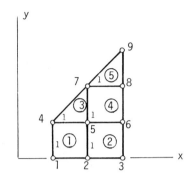

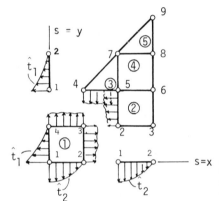

Figure 3.15 A finite-element mesh of triangular and rectangular elements and interelement force computation.

where zero indicates no correspondence between the global and element nodes. Array $[B]$ is called a *correspondence* or *connectivity matrix*. The correspondence tells us how individual elements fit into their places in the finite element mesh. If a global node number is shared by several elements, it is obvious that the stiffness and force coefficients corresponding to that node are contributed to all elements sharing that node. Using this idea, we now describe the assembly procedure for problems with one degree of freedom (i.e., one primary unknown) per node.

The assembly is based on the following relation:

$$K_{IJ} = K_{ij}^{\alpha} + K_{kl}^{\beta} + K_{pq}^{\gamma} + \cdots$$

$$F_{I} = F_{i}^{\alpha} + F_{k}^{\beta} + F_{p}^{\gamma} + \cdots \tag{3.4.71}$$

Here $\alpha, \beta, \gamma, \ldots$, denote the labels of elements that share the global nodes I and J, and $(i, j), (k, l), (p, q), \ldots$, are the pairs of element nodes that have one-to-one correspondence with the global nodes. For example, pair $(5, 7)$ of the global nodes is shared by two elements: element 3 and element 4; the corresponding element pairs are $(2, 3)$ and $(1, 4)$. Therefore, the global stiffness coefficient K_{57} (i.e., stiffness coefficient in row 5 and column 7 of the assembled stiffness matrix) is given by

$$K_{57} = K_{23}^3 + K_{14}^4$$

Similarly, global node 5 is shared by elements 1, 2, 3, and 4; the element nodes corresponding to global node 5 are 3, 4, 2, and 1, respectively. Therefore, the global stiffness coefficient in row and column 5, and the force coefficient in row 5 are given by

$$K_{55} = K_{33}^1 + K_{44}^2 + K_{22}^3 + K_{11}^4$$

$$F_5 = F_3^1 + F_4^2 + F_2^3 + F_1^4$$

The stiffness coefficient in row 5 and column 9 of the global stiffness matrix have no contribution from any element because both nodes do not belong to any single element:

$$K_{59} = 0.0$$

Thus, one can use the global-local node correspondence to assemble the element stiffness matrices and force vectors. This procedure can be automatized on a computer, by generating the connectivity matrix (of integers) for a given finite element mesh, and then used to implement Eq. (3.4.71). For additional details, the reader can consult Reddy (1984).

The global stiffness matrix and force vector for the mesh shown in Fig. 3.15 are given in Eq. (3.4.72) on the next page.

5. *Imposition of Boundary Conditions.* The implementation of essential boundary conditions is the same as usual. The imposition of natural boundary conditions involves some additional calculations, because the boundaries of two-dimensional elements involve one-dimensional (line) elements. When the element under consideration is an element R_e that does not have any of its sides on the boundary of the domain, no calculations are necessary, because all of its element-boundary contributions to P_i^e are assumed to cancel with similar contributions from the adjoining elements (i.e., Newton's third law concerning action and reaction is assumed to hold). Of course, any contributions from the body forces (i.e., the source term f) are already accounted for in f_i^e, which add to similar contributions from the adjoining elements (see Fig. 3.15).

When an element R_e has one or more sides (not just a node) on the boundary of the domain, then we must be given, as part of the problem statement, either u (primary unknown) or t_n (secondary unknown) on the boundary. When u is given, we impose it directly by giving the value of u at a

$$
[K] =
\begin{array}{c c}
 & \begin{array}{ccccccccc} 1 & \ \ 2 & \ \ 3 & \ \ 4 & \ \ 5 & \ \ 6 & \ \ 7 & \ \ 8 & \ \ 9 \end{array} \\
\begin{array}{c} 1 \\ 2 \\ 3 \\ 4 \\ 5 \\ 6 \\ 7 \\ 8 \\ 9 \end{array} &
\left[
\begin{array}{ccccccccc}
K_{11}^1 & & & & & & & & \\
K_{21}^1 & K_{22}^1 + K_{11}^2 & & & & & & & \\
0 & K_{21}^2 & K_{22}^2 & & & \multicolumn{2}{c}{\text{Transpose of}} & & \\
K_{41}^1 & K_{42}^1 & 0 & K_{44}^1 + K_{11}^3 & & \multicolumn{2}{c}{\text{the entries from}} & & \\
K_{31}^1 & K_{32}^1 + K_{41}^2 & K_{42}^2 & K_{34}^1 + K_{21}^3 & K_{33}^1 + K_{44}^2 + K_{22}^3 + K_{11}^4 & \multicolumn{2}{c}{\text{the lower-diagonal}} & & \\
0 & K_{31}^2 & K_{32}^2 & 0 & K_{34}^2 + K_{21}^4 & K_{33}^2 + K_{22}^4 & & & \\
0 & 0 & 0 & K_{31}^3 & K_{32}^3 + K_{41}^4 & K_{42}^4 & K_{33}^3 + K_{44}^4 + K_{11}^5 & & \\
0 & 0 & 0 & 0 & K_{31}^4 & K_{32}^4 & K_{34}^4 + K_{21}^5 & K_{33}^4 + K_{22}^5 & \\
0 & 0 & 0 & 0 & 0 & 0 & K_{31}^5 & K_{32}^5 & K_{33}^5
\end{array}
\right]
\end{array}
\tag{3.4.72}
$$

$$
\{F\}^{T} =
\left\{
\underset{1}{F_1^1}\ \
\underset{2}{\left(F_2^1 + F_1^2\right)}\ \
\underset{3}{F_2^2}\ \
\underset{4}{\left(F_4^1 + F_1^3\right)}\ \
\underset{5}{\left(F_3^1 + F_4^2 + F_2^3 + F_1^4\right)}\ \
\underset{6}{\left(F_3^2 + F_2^4\right)}\ \
\underset{7}{\left(F_3^3 + F_4^4 + F_1^5\right)}\ \
\underset{8}{\left(F_3^4 + F_2^5\right)}\ \
\underset{9}{F_3^5}
\right\}
$$

global node to the global variable at that node. When t_n is given, in general as a function of position on the boundary, we must calculate P_i^e using Eq. (3.4.62b). It should be noted that the normal derivative of the dependent variable of the problem is zero (i.e., $t_n = 0$) along the lines of symmetry. For all linear elements, the evaluation of P_i involves integration over a side of the element. It turns out that when the linear interpolation functions $\psi_i(x, y)$ of two-dimensional elements are evaluated on a line, we get the linear interpolation functions $\psi(s)$ of a one-dimensional (bar) element. Therefore, the boundary integral in Eq. (3.4.62b) can be easily evaluated over a side of the element. As an example, consider element 1 of the finite-element mesh shown in Fig. 3.15; we have

$$\left(P_i^1\right)_{1-4} = \int_0^{h_{14}} \psi_i^1(s)|_{1-4}\hat{t}_1(s)\, ds, \qquad \left(P_i^1\right)_{1-2} = \int_0^{h_{12}} \psi_i^1(s)|_{1-2}\hat{t}_2(s)\, ds,$$

$$(i = 1, 2)$$

$$\psi_1 = 1 - (s/h_\alpha), \qquad \psi_2 = (s/h_\alpha), \qquad h_1 = h_{14}, \qquad h_2 = h_{12}$$

$$(3.4.73)$$

where s denotes the local coordinate used to describe $\hat{t}_i$ and ψ_i on the sides of the elements, and h_{ij} denotes the length of side $i–j$. The contributions from the other two sides of element 1 cancel, as pointed out earlier, with similar contributions from elements 2 and 3. Thus, at node 1 we have

$$P_1^1 = \int_0^{h_{14}} \psi_1^1(s)|_{1-4}\hat{t}_1(s)\, ds + \int_0^{h_{12}} \psi_1^1(s)|_{1-2}\hat{t}_2(s)\, ds \qquad (3.4.74)$$

where $\psi_1^1(s)|_{1-4} = 1 - (s/h_{14})$ and $\psi_1^1(s)|_{1-2} = 1 - (s/h_{12})$. If $\hat{t}_1$ and $\hat{t}_2$ are constants, we obtain,

$$P_1^1 = \frac{\hat{t}_1 h_{14}}{2} + \frac{\hat{t}_2 h_{12}}{2}$$

This is the value we use to replace P_1^1 in the global force vector.

The solution of equations and postprocessing are common steps that are involved in any numerical method and we do not discuss them here. This completes the general introduction to the finite-element analysis of two-dimensional problems involving a scalar quantity. The following example illustrates the procedure for a specific problem.

Example 3.18

Here we consider the torsion of the hollow square cross-section prismatic member shown in Fig. 3.16. For hollow cross sections, the stress function

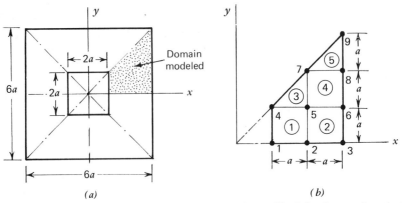

Figure 3.16 Configuration of a hollow cross-section member and its finite element discretization: (*a*) domain modelled; (*b*) finite-element mesh.

$u = \Phi$ must not only satisfy Eq. (1.7.39a) inside the region and Eq. (1.7.39b) on the outer boundary, but along the boundary of the hole we must specify that $u = \hat{u}$, where $\hat{u}$ is a nonzero constant. Owing to the biaxial and diagonal symmetry, one can use the shaded portion shown in Fig. 3.16*a*. As a first choice, we use a mesh of two triangular elements and three rectangular elements, as shown in Fig. 3.16*b*. If we were to use a mesh of rectangles only (or, if the cross section is a rectangle), then we must use a quadrant of the domain to model the problem.

The element matrices are given by Eq. (3.4.67) with $c_{11} = c_{22} = 1$, $c_{12} = c_{21} = c_0 = 0$, and $f = 2G\theta$. We have

$$[K^1] = [K^2] = [K^4], \qquad \{f^1\} = \{f^2\} = \{f^4\}, \qquad [K^3] = [K^5]$$

$$\{f^3\} = \{f^5\}$$

where

$$[K^1] = \frac{1}{6}\begin{bmatrix} 4 & -1 & -2 & -1 \\ -1 & 4 & -1 & -2 \\ -2 & -1 & 4 & -1 \\ -1 & -2 & -1 & 4 \end{bmatrix}, \qquad \{f^1\} = \frac{fa^2}{4}\begin{Bmatrix} 1 \\ 1 \\ 1 \\ 1 \end{Bmatrix}$$

$$[K^3] = \frac{1}{2}\begin{bmatrix} 2 & -1 & -1 \\ -1 & 1 & 0 \\ -1 & 0 & 1 \end{bmatrix}, \qquad \{f^3\} = \frac{fa^2}{6}\begin{Bmatrix} 1 \\ 1 \\ 1 \end{Bmatrix}$$

Note that, although the two right-angle triangle elements are geometrically equal to one rectangular element, their element matrices are not equal (therefore, the solution, in general, will not be equal).

The assembled system of equations is given by (see Eq. 3.4.72),

$$
\frac{1}{6}
\begin{bmatrix}
4 \\
-1 & \text{⑧} & & & & \text{symmetric} \\
0 & -1 & 4 \\
-1 & -2 & 0 & 10 \\
-2 & \text{(-2)} & -2 & -4 & \text{⑮} \\
0 & -2 & -1 & 0 & -2 & 8 \\
0 & \text{⓪} & 0 & -3 & \text{(-1)} & -2 & \text{⑪} \\
0 & 0 & 0 & 0 & -2 & -1 & -4 & 7 \\
0 & 0 & 0 & 0 & 0 & 0 & -3 & 0 & 3
\end{bmatrix}
\begin{Bmatrix}
U_1 \\ U_2 \\ U_3 \\ U_4 \\ U_5 \\ U_6 \\ U_7 \\ U_8 \\ U_9
\end{Bmatrix}
$$

$$
= \frac{fa^2}{12}
\begin{Bmatrix}
3 \\ 6 \\ 3 \\ 5 \\ 11 \\ 6 \\ 7 \\ 5 \\ 2
\end{Bmatrix}
+
\begin{Bmatrix}
P_1^1 \\
P_2^1 + P_1^2 \\
P_2^2 \\
P_4^1 + P_1^3 \\
P_3^1 + P_4^2 + P_2^3 + P_1^4 \\
P_3^2 + P_2^4 \\
P_3^3 + P_4^4 + P_1^5 \\
P_3^4 + P_2^5 \\
P_3^5
\end{Bmatrix}
\tag{3.4.75}
$$

The boundary conditions of the problem require us to set

$$U_1 = U_4 = \hat{u}, \qquad U_3 = U_6 = U_8 = U_9 = 0$$

$$P_2^1 + P_1^2 = 0, \qquad P_3^1 + P_4^2 + P_2^3 + P_1^4 = 0, \qquad P_3^3 + P_4^4 + P_1^5 = 0$$

$$\tag{3.4.76}$$

Deleting the rows and columns corresponding to the specified degrees of freedom, we get the *condensed set of equations* [see the circled entries in Eq.

3.4.75)]

$$\frac{1}{6}\begin{bmatrix} 8 & -2 & 0 \\ -2 & 15 & -1 \\ 0 & -1 & 11 \end{bmatrix}\begin{Bmatrix} U_2 \\ U_5 \\ U_7 \end{Bmatrix} = \frac{fa^2}{12}\begin{Bmatrix} 6 \\ 11 \\ 7 \end{Bmatrix} - \frac{1}{6}\begin{Bmatrix} -U_1 - 2U_4 \\ -2U_1 - 4U_4 \\ 0 - 3U_4 \end{Bmatrix}$$

$$(3.4.77a)$$

whose solution is given by (using Cramer's rule)

$$U_2 = \frac{1}{1268}\left(620fa^2 + 630\hat{u}\right), \qquad U_5 = \frac{1}{1268}\left(578fa^2 + 618\right)$$

$$U_7 = \frac{1}{1268}\left(456fa^2 + 402\right) \qquad (3.4.77b)$$

For $3a = 1$, $f = 2$, and $\hat{u} = 1$, we get

$$U_2 = 0.6055, \qquad U_5 = 0.5887, \qquad U_7 = 0.3970$$

Table 3.8 Comparison of the Finite Element Solution (FEM) with the Finite Difference (FDM), Finite Difference with Relaxation (FDMR), and Exact Solutions of the Torsion of a Solid Square Cross-Section Member

		$u/G\theta a^2$			$\sigma_{yz}/G\theta a$		
Source	Mesh[a]	$x = y = 0$	$x = a/2, y = 0$	$x = y = a/2$	Value	x/a	y/a
FEM	1 × 1	0.7500	—	—	0.5625	0.5	0.5
(rectangles)	2 × 2	0.6214	0.4821	0.3857	0.9161	0.75	0.25
	4 × 4	0.5968	0.4644	0.3676	1.1182	0.875	0.125
	8 × 8	0.5912	0.4601	0.3636	1.2301	0.9375	0.0625
FEM	1 × 1	0.6667	—	—	0.9167	0.6667	0.3333[b]
(triangles)	2 × 2	0.625	0.4583	0.3542	0.9167	0.8333	0.1667
	4 × 4	0.6026	0.4583	0.3601	1.1139	0.8333	0.1667
	8 × 8	0.5938	0.4586	0.3617	1.2288	0.9167	0.0833
FDM[c]	4 × 4	0.5625	0.4375	0.3438	1.2920	—	—
FDMR[c]	4 × 4	0.5822	0.4531	0.3573	—	—	—
Exact Solution		0.5889	0.4582	—	1.356	1.0000	0.0

[a]Uniform meshes.
[b]Centroid location; however, they are the same on the entire element.
[c]See Crandall (1956), pp. 243–261.

The corresponding values obtained using rectangles in a quadrant and triangles in an octant of the domain are, respectively:

$$U_2 = 0.5865, \qquad U_5 = 0.5128, \qquad U_7 = 0.3365$$

$$U_2 = 0.5594, \qquad U_5 = 0.5278, \qquad U_7 = 0.3194$$

Since the exact solution for the problem is not available, we cannot assess the accuracy of the finite-element solution. However, we can compare the results for a solid cross-section member. Table 3.8 contains a comparison of the finite-element solution with the exact solution in Eq. (3.3.32) and the finite difference solution (see Crandall, 1956). The finite-element solutions obtained using both meshes converge to the exact stress function values and the maximum shear stress value. This completes the example.

Two-Variable Problems: Plane Elasticity

Here we study a finite-element formulation based on the total potential energy functional of the plane elasticity (see Section 1.7.5). Once again we begin with the governing equations (1.7.49) and develop the variational statement of the problem. We note that the displacement vector in plane elasticity has two components, u and v. Therefore, we must approximate both displacement components independently; an element will have two degrees of freedom (u_i, v_i) at each node (see Fig. 3.17). We begin with the variational formulation and associated finite-element model of Eq. (1.7.49).

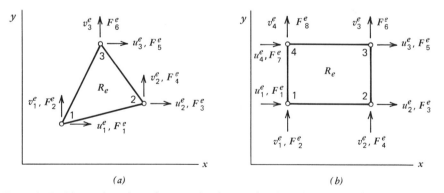

(a)   (b)

Figure 3.17 Linear triangular and rectangular elements for plane elasticity problems: (a) triangular (CST) element; (b) rectangular element.

The variational form of Eq. (1.7.49) over an element R_e is obtained by multiplying the first equation with variation δu and the second one with variation δv and integrating the results by parts to trade the second differential to δu and δv. We have (with $c_{66} = c_{33}$)

$$0 = \int_{R_e} \left[\frac{\partial \delta u}{\partial x} \left(c_{11} \frac{\partial u}{\partial x} + c_{12} \frac{\partial v}{\partial y} \right) + c_{33} \frac{\partial \delta u}{\partial y} \left(\frac{\partial u}{\partial y} + \frac{\partial v}{\partial x} \right) - \delta u f_x \right] dx \, dy$$

$$- \oint_{S_e} \delta u t_x \, dS$$

$$0 = \int_{R_e} \left[c_{33} \frac{\partial \delta v}{\partial x} \left(\frac{\partial u}{\partial y} + \frac{\partial v}{\partial x} \right) + \frac{\partial \delta v}{\partial y} \left(c_{12} \frac{\partial u}{\partial x} + c_{22} \frac{\partial v}{\partial y} \right) - \delta v f_y \right] dx \, dy$$

$$- \oint_{S_e} \delta v t_y \, dS \qquad\qquad (3.4.78)$$

where t_x and t_y denote the boundary forces defined in Eq. (1.7.50). Note that the plane stress or plane strain assumption enters the equations via the specification of $c_{ij}(i, j = 1, 2, 3)$.

Substituting

$$u = \sum_{j=1}^{n} u_j^e \psi_j^e(x, y), \qquad v = \sum_{j=1}^{n} v_j^e \psi_j^e(x, y) \qquad (3.4.79)$$

for u and v, and ψ_i for δu and δv into Eq. (3.4.78), we obtain

$$0 = \sum_{j=1}^{n} K_{ij}^{11} u_j + \sum_{j=1}^{n} K_{ij}^{12} v_j - F_i^1$$

$$0 = \sum_{j=1}^{n} K_{ij}^{21} u_j + \sum_{j=1}^{n} K_{ij}^{22} v_j - F_i^2$$

or

$$[K^{11}]\{u\} + [K^{12}]\{v\} = \{F^1\}$$

$$[K^{21}]\{u\} + [K^{22}]\{v\} = \{F^2\} \qquad\qquad (3.4.80a)$$

where

$$K_{ij}^{11} = \int_{R_e} \left(c_{11} \frac{\partial \psi_i}{\partial x} \frac{\partial \psi_j}{\partial x} + c_{33} \frac{\partial \psi_i}{\partial y} \frac{\partial \psi_j}{\partial y} \right) dx\, dy$$

$$K_{ij}^{12} = K_{ji}^{21} = \int_{R_e} \left(c_{12} \frac{\partial \psi_i}{\partial x} \frac{\partial \psi_j}{\partial y} + c_{33} \frac{\partial \psi_i}{\partial y} \frac{\partial \psi_j}{\partial x} \right) dx\, dy$$

$$K_{ij}^{22} = \int_{R_e} \left(c_{33} \frac{\partial \psi_i}{\partial x} \frac{\partial \psi_j}{\partial x} + c_{22} \frac{\partial \psi_i}{\partial y} \frac{\partial \psi_j}{\partial y} \right) dx\, dy$$

$$F_i^1 = \int_{R_e} \psi_i f_x \, dx\, dy + \oint_{S_e} \psi_i \hat{t}_x \, dS$$

$$F_i^2 = \int_{R_e} \psi_i f_y \, dx\, dy + \oint_{S_e} \psi_i \hat{t}_y \, dS \tag{3.4.80b}$$

These element coefficients can be represented in terms of $[S^x]$, $[S^y]$, and $[S^{xy}]$ defined in Eq. (3.4.67b), and they can be computed for the linear triangular element ($n = 3$) or rectangular element ($n = 4$).

The assembly procedure discussed previously for the torsion problem applies to each of the coefficient matrices in Eq. (3.4.80a). Note that there are two components of forces on the boundary. When an element has its sides on the boundary of the domain, the force contribution to its nodes follows the same procedure as described in Eq. (3.4.73). But these contributions should be divided into components along the global x and y coordinates, and only like components should be added at the global nodes. Now we consider an example problem.

Example 3.19

Consider the steel plate ($E = 30 \times 10^6$ psi, $\nu = 0.25$) shown in Fig. 3.18. We can use the symmetry about the x-axis and model only the upper half of the plate. We shall solve the problem using a mesh of triangles and a mesh of rectangles. The plate constants c_{ij} are given by

$$c_{11} = \frac{E_1 h}{1 - \nu_{12}\nu_{21}}, \qquad c_{22} = \frac{E_2 h}{1 - \nu_{12}\nu_{21}}$$

$$c_{12} = c_{21} = c_{11}\nu_{12}, \qquad c_{33} = G_{12} h \tag{3.4.81}$$

where h is the thickness of the plate.

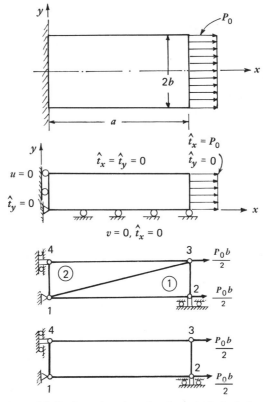

Figure 3.18 Geometry and finite-element meshes of a plane elastic body (a = 16 in., b = 2 in.).

Triangles. Using a two-element mesh, we obtain (after lengthy algebra), $f_i^1 = 0$, $f_i^2 = 0$, and

$$
[K^1] = c
\begin{bmatrix}
b^2 & & & & & \\
0 & \beta & & & \text{symmetric} & \\
-b^2 & \gamma & b^2 + \alpha & & & \\
\nu ab & -\beta & -\gamma & a^2 + \beta & & \\
0 & -\gamma & -\alpha & \gamma & \alpha & \\
-\nu ab & 0 & \nu ab & -a^2 & 0 & a^2
\end{bmatrix}
$$

$$
[K^2] = c
\begin{bmatrix}
\alpha & & & & & \\
0 & a^2 & & & \text{symmetric} & \\
0 & -\nu ab & b^2 & & & \\
-\gamma & 0 & 0 & \beta & & \\
-\alpha & -\nu ab & -b^2 & -\gamma & b^2 + \alpha & \\
\gamma & -a^2 & \nu ab & -\beta & -\gamma & a^2 + \beta
\end{bmatrix}
$$

$$(3.4.82a)$$

where a and b are the sides of the element, and

$$c = \frac{Eh}{2ab(1 - v^2)}, \qquad \alpha = \frac{1 - v}{2}a^2, \qquad \beta = \frac{1 - v}{2}b^2, \qquad \gamma = \frac{1 - v}{2}ab$$

$$(3.4.82b)$$

Note that the element matrices in Eq. (3.4.82a) are in a rearranged form of Eq. (3.4.80a); the arrangement involved writing the nodal displacement vector in the form (see Fig. 3.17a),

$$\{u_1 \quad v_1 \quad u_2 \quad v_2 \quad u_3 \quad v_3\}$$

and moving their coefficients accordingly, to retain the symmetry of the stiffness matrix. The reader is asked to verify this.

The assembly of the newly arranged element stiffness is a bit more involved. One has to keep in mind that each node has two degrees of freedom. As a result, two columns and rows of the stiffness matrices correspond to a single node. For example, consider node 1 of the mesh. We have

$K_{11} = K_{11}^1 + K_{11}^2$ (corresponding to u at node 1)

$K_{12} = K_{12}^1 + K_{12}^2$ (corresponding to the coupling between u and v at node 1)

$K_{22} = K_{22}^1 + K_{22}^2$ (corresponding to v at node 1)

$K_{15} = K_{15}^1 + K_{13}^2$ (corresponding to the coupling between u at node 1 and u at node 3)

$K_{16} = K_{16}^1 + K_{14}^2$ (corresponding to the coupling between u at node 1 and v at node 3)

Thus, the assembled stiffness matrix is given by

$$[K] = \begin{bmatrix} K_{11}^1 + K_{11}^2 & K_{12}^1 + K_{12}^2 & K_{13}^1 & K_{14}^1 & K_{15}^1 + K_{13}^2 & K_{16}^1 + K_{14}^2 & K_{15}^2 & K_{16}^2 \\ & K_{22}^1 + K_{22}^2 & K_{23}^1 & K_{24}^1 & K_{25}^1 + K_{23}^2 & K_{26}^1 + K_{24}^2 & K_{25}^2 & K_{26}^2 \\ & & K_{33}^1 & K_{34}^1 & K_{35}^1 & K_{36}^1 & 0 & 0 \\ & & & K_{44}^1 & K_{45}^1 & K_{46}^1 & 0 & 0 \\ & & & & K_{55}^1 + K_{33}^2 & K_{56}^1 + K_{34}^2 & K_{35}^2 & K_{36}^2 \\ & \text{symmetric} & & & & K_{66}^1 + K_{44}^2 & K_{45}^2 & K_{46}^2 \\ & & & & & & K_{55}^2 & K_{56}^2 \\ & & & & & & & K_{66}^2 \end{bmatrix}$$

$$(3.4.83a)$$

and the assembled force vector is

$$\{F\} = \begin{Bmatrix} F_1^1 + F_1^2 \\ F_2^1 + F_2^2 \\ F_3^1 \\ F_4^1 \\ F_5^1 + F_3^2 \\ F_6^1 + F_4^2 \\ F_5^2 \\ F_6^2 \end{Bmatrix} \qquad (3.4.83b)$$

The boundary conditions are (see Fig. 3.18),

$$U_1 = V_1 = V_2 = U_4 = 0$$

$$P_3^1 = \frac{bP_0}{2}, \qquad P_5^1 + P_3^2 = \frac{bP_0}{2}, \qquad P_6^1 + P_4^2 = 0, \qquad P_6^2 = 0 \quad (3.4.84)$$

Use of these conditions in Eqs. (3.4.83a and b) results in the following condensed set of equations,

$$10^6 h \begin{bmatrix} 97.00 & -96.00 & 4.00 & 0.00 \\ -96.00 & 97.00 & 0.00 & 4.00 \\ 4.00 & 0.00 & 256.37 & -0.375 \\ 0.00 & 4.00 & -0.375 & 256.37 \end{bmatrix} \begin{Bmatrix} U_2 \\ U_3 \\ V_3 \\ V_4 \end{Bmatrix} = P_0 \begin{Bmatrix} 0.5 \\ 0.5 \\ 0.0 \\ 0.0 \end{Bmatrix}$$

The solution is given by

$$U_2 = U_3 = 0.5333 \times 10^{-6} \frac{P_0}{h} \quad \text{(in.)}$$

$$V_3 = V_4 = -0.00833 \times 10^{-6} \frac{P_0}{h} \quad \text{(in.)} \qquad (3.4.85a)$$

The stresses in each of the elements can be calculated using Eq. (1.7.45). We obtain

$$\sigma_x^1 = \sigma_x^2 = 1.0667 \frac{P_0}{h} \quad \text{(psi)}, \quad \sigma_{xy}^1 = \sigma_{xy}^2 = 0.0, \quad \sigma_y^1 = \sigma_y^2 = 0.2667 \frac{P_0}{h} \quad \text{(psi)}$$

$$(3.4.85b)$$

Rectangle. The one-element model is equivalent to the two-element model discussed above. The element matrix is given by Eq. (3.4.86a), where

$$[K] = \begin{bmatrix}
2(\alpha+\lambda) & & & & & & & \\[4pt]
\frac{1}{4}(c_{33}+c_{12}) & 2(\beta+\mu) & & & & & \text{symmetric} & \\[4pt]
-(2\alpha-\lambda) & \frac{1}{4}(c_{33}-c_{12}) & 2(\alpha+\lambda) & & & & & \\[4pt]
\frac{1}{4}(c_{12}-c_{33}) & \beta-2\mu & -\frac{1}{4}(c_{12}+c_{33}) & 2(\beta+\mu) & & & & \\[4pt]
-(\alpha+\lambda) & -\frac{1}{4}(c_{12}+c_{33}) & \alpha-2\lambda & \frac{1}{4}(c_{33}-c_{12}) & 2(\alpha+\lambda) & & & \\[4pt]
-\frac{1}{4}(c_{12}+c_{33}) & -(\beta+\mu) & \frac{1}{4}(c_{12}-c_{33}) & -2\beta+\mu & \frac{1}{4}(c_{12}+c_{33}) & 2(\beta+\mu) & & \\[4pt]
-(\alpha+2\lambda) & \frac{1}{4}(c_{12}-c_{33}) & -(\alpha+\lambda) & \frac{1}{4}(c_{12}+c_{33}) & -2\alpha+\lambda & \frac{1}{4}(c_{12}+c_{33}) & 2(\alpha+\lambda) & \\[4pt]
\frac{1}{4}(c_{33}-c_{12}) & -(2\beta+\mu) & \frac{1}{4}(c_{33}+c_{12}) & -(\beta+\mu) & \frac{1}{4}(c_{12}-c_{33}) & \beta-\mu & -\frac{1}{4}(c_{12}+c_{33}) & 2(\beta+\mu)
\end{bmatrix}$$

$$(3.4.86a)$$

$$\alpha = \frac{c_{11}b}{6a}, \qquad \beta = \frac{c_{22}a}{6b}, \qquad \mu = \frac{c_{33}b}{6a}, \qquad \lambda = \frac{c_{33}a}{6b} \qquad (3.4.86b)$$

After imposing the boundary conditions of Eq. (3.4.84), the condensed set of equations for the rectangular element is given by [see the boxed coefficients in Eq. (3.4.86a)]

$$10^6 h \begin{bmatrix} 64.667 & -63.667 & -1.000 & 5.000 \\ -63.667 & 64.667 & 5.000 & -1.000 \\ -1.000 & 5.000 & 170.920 & 85.083 \\ 5.000 & -1.000 & 85.083 & 170.920 \end{bmatrix} \begin{Bmatrix} U_2 \\ U_3 \\ V_3 \\ V_4 \end{Bmatrix} = P_0 \begin{Bmatrix} 0.5 \\ 0.5 \\ 0.0 \\ 0.0 \end{Bmatrix}$$

The displacements and stresses are given by

$$U_2 = U_3 = 0.5333 \times 10^{-6} \frac{P_0}{h} \quad \text{(in.)},$$

$$V_3 = V_4 = -0.00833 \times 10^{-6} \frac{P_0}{h} \quad \text{(in.)}$$

$$\sigma_x^1 = \frac{P_0}{h} \quad \text{(psi)}, \qquad \sigma_y^1 = 0.0, \qquad \sigma_{xy}^1 = 0.0 \qquad (3.4.87)$$

Note that the axial displacements predicted by the finite element model, which is based on the plane stress equations, are equal to those given by the theory of strength of materials,

$$u_{\text{axial}} = \frac{(P_0 a)(b)}{(ah)E} = 0.5333 \times 10^{-6} \frac{P_0}{h}$$

This is not too surprising, because the plate dimensions are such that $(a/b = 8)$ it can be treated as a uniaxially loaded member. When $b/a = 1$ ($b = 16$), displacements, under the same total load, are given by

$$U_2 = U_3 = 0.06667 \times 10^{-6} \frac{P_0}{h}$$

Thus, the plane elasticity solution is different from a uniaxial bar solution.

The finite-element formulation of two-dimensional problems presented in this chapter is limited to neither linear nor rectangular elements. One can use quadrilateral and higher-order elements, which yield different stiffness coefficients than the triangular and rectangular elements discussed here. In general, the exact evaluation of the coefficients of the stiffness matrices and force vectors are algebraically complicated for quadrilateral and higher-order elements. Therefore, numerical integration is used to calculate the coefficients. When the domain under consideration is curved, it is also possible to approximate the domain, much the same way we approximated the dependent variables of the problem. When the same interpolation functions are used to approximate both geometry and dependent unknowns, the resulting formulations are called *isoparametric* formulations. These ideas and other additional details are too numerous to include in this study. We close this section with an example that illustrates the use of quadrilateral elements for the stress analysis of a plate with a circular hole. The problem is solved using the FEM2D computer program from the author's finite element text (see Reddy, 1984).

Example 3.20

Consider a thin steel plate with a circular hole, as shown in Fig. 3.19a. We analyze the problem for normal stress across the section AB. Owing to the abrupt change in the cross section along AB, a steep gradient in stress and strain is expected. The maximum normal stress σ_x along AB is known to be three times that of the applied uniform stress σ_0 at the edges. To capture the maximum stress, a finer mesh near the hole is required. To illustrate the effect of mesh size, a series of gradually finer meshes near the hole are used to determine the maximum stress.

Exploiting the biaxial symmetry, only one quadrant of the plate is modeled. The boundary conditions on the displacements are shown in Fig. 3.19b. The applied distributed load at the edge should be integrated along the boundary $x = 0$, in each element, and should be used as applied forces at nodes 1–6. We have $f_i = 0$ for all elements, and all other boundary forces, except for the reactions and known axial forces along $x = 0$, are zero. Along $x = 0$, the forces are

$$P_1^1 = -\int_0^1 \sigma_0 \psi_1^1 \, dy, \qquad P_7^1 + P_1^2 = -\int_0^1 \sigma_0 \psi_2^1 \, dy - \int_0^1 \sigma_0 \psi_1^2 \, dy$$

$$P_7^2 + P_1^3 = -\int_0^1 \sigma_0 \psi_2^2 \, dy - \int_0^1 \sigma_0 \psi_1^3 \, dy,$$

$$P_7^3 + P_1^4 = -\int_0^1 \sigma_0 \psi_2^3 \, dy - \int_0^{0.5} \sigma_0 \psi_1^4 \, dy$$

$$P_7^4 + P_1^5 = -\int_0^{0.5} \sigma_0 \psi_2^4 \, dy - \int_0^{0.5} \sigma_0 \psi_1^5 \, dy, \qquad P_7^5 = -\int_0^{0.5} \sigma_0 \psi_2^5 \, dy$$

where $\psi_1^e = (1 - y)/h_e$, $\psi_2^e = y/h_e$, and h_e is the length of the line elements along the boundary $x = 0$. The evaluation of the integrals gives

$$P_1^1 = -500, \qquad P_7^1 + P_1^2 = -1000, \qquad P_7^2 + P_1^3 = -1000,$$

$$P_7^3 + P_1^4 = -750, \qquad P_7^4 + P_1^5 = -500, \qquad P_7^5 = -250$$

Figure 3.19b contains the set of four meshes of quadrilateral elements used in the present study. Table 3.9 contains a comparison of stresses and displacements obtained by various meshes. As can be seen from Fig. 3.20, the maximum normal stress along AB varies from 1000 psi at the edge to about 3000 psi near the hole. This is consistent with experimental and other numerical results.

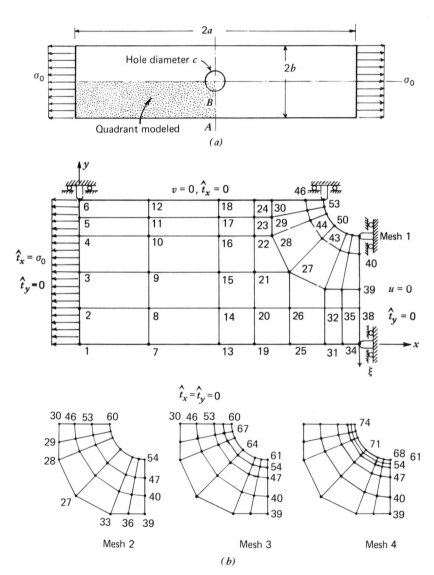

Figure 3.19 Geometry and finite-element meshes of a plate with a circular hole: (a) geometry of the plate; (b) finite-element meshes (a = 8 in., b = 4 in.).

Table 3.9 Comparison of Maximum Normal Stress and Horizontal Displacement[a] Along the Hole Boundary Obtained by Various Meshes of the Plate with a Circular Hole (see Example 3.20)

	Stress [$\sigma_x(8, y)$]				$-u \times 10^6$			
y	Mesh 1	Mesh 2	Mesh 3	Mesh 4	Mesh 1	Mesh 2	Mesh 3	Mesh 4
0.50	1020.05	1025.74	1025.46	1025.37	19.657	19.891	20.847	20.976
1.25	1129.30	1135.03	1138.85	1139.37	37.502	37.943	39.857	40.229
1.75	—	1209.53	1217.81	1218.89	70.343	71.186	73.259	73.499
2.00	1304.42	—	—	—	93.798	94.828	96.828	97.169
2.25	—	1387.88	1407.75	1410.43	99.649	100.68	102.48	102.68
2.65	—	—	1840.00	1845.10	101.84	102.87	104.59	105.04
2.75	2116.83	2136.31	—	—				
2.85	—	—	—	2350.79				
2.90	—	—	2598.3	—				
2.95	—	—	—	2834.77				

[a]The results are given along the hole boundary.

3.4.5 Closure

The brief introduction of the finite-element method given in the preceding sections should provide the reader with the basic idea of the method and its usefulness in solving practical engineering problems. Advanced topics, such as isoparametric elements and numerical integration, time-dependent problems, nonlinear problems, three-dimensional elements, and so on, are dealt with in text books on the subject, and interested readers should consult the Bibliography. In Chapter 4, we consider briefly the finite-element analysis of plates and shells.

The analysis of time-dependent problems by the finite-element method proceeds along the lines discussed for traditional variational methods in

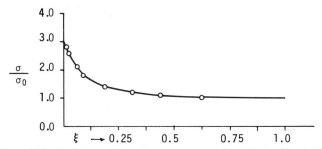

Figure 3.20 Variation of normal stress across section AB (see Fig. 3.19).

Section 3.3. The interpolation of the dependent variable $u(x, y, t)$ at time t is assumed to be of the form

$$u(x, y, t) = \sum_{i=1}^{n} u_i(t)\psi_i(x, y) \qquad (3.4.88)$$

where u_i are the nodal values of u at time t. Substituting Eq. (3.4.88) into a variational form (e.g., Hamilton's principle) of the equation of motion for the problem, we obtain the finite-element model whose general form is

$$[M^e]\{\ddot{u}\} + [K^e]\{u\} = \{F^e\} \qquad (3.4.89)$$

where $[M^e]$ is the mass matrix. To complete the solution, we can solve Eq. (3.4.89), for simple cases, by the Laplace transform method or use finite-difference methods to approximate the time derivatives. For additional details the reader is referred to the finite element text by Reddy (1984).

The generality of the finite-element method, coupled with its adaptability to digital computers, has led to the development of a number of general purpose computer programs that are extensively used in the analysis and design of many engineering structures. Many general purpose programs are cited in the survey articles listed in the Bibliography.

Exercises 3.4

1. Derive the finite-element model of the equation,

$$-\frac{d}{dx}\left[(1 + x)\frac{du}{dx}\right] = f, \qquad 0 < x < 1$$

   for the boundary conditions

   $$u(0) = \hat{u}, \qquad \left[(1 + x)\frac{du}{dx} + u\right]_{x=1} = \hat{q}$$

2. Solve the problem in Exercise 1 for $f = 0$, $\hat{u} = 1$, and $\hat{q} = 0$. Use two linear elements.

3. Use the following procedure to derive the cubic interpolation functions for a line element with equally-spaced (four) nodes. Set up the local coordinate $\bar{x}$ at the left-end node (i.e., node 1), and then use the interpolation property of ψ_i, $\psi_i(\bar{x}_j) = \delta_{ij}$, $i, j = 1, 2, 3, 4$ to write ψ_i as a polynomial that vanishes at all $\bar{x}_j, \bar{x}_j \neq \bar{x}_i$,

   $$\psi_1(\bar{x}) = c_1(\bar{x} - \bar{x}_2)(\bar{x} - \bar{x}_3)(\bar{x} - \bar{x}_4)$$

   $$\psi_2(\bar{x}) = c_2(\bar{x} - \bar{x}_1)(\bar{x} - \bar{x}_3)(\bar{x} - \bar{x}_4), \quad \text{and so on}$$

and determine $c_1, c_2, \ldots$ such that $\psi_1(\bar{x}_1) = 1$, $\psi_2(\bar{x}_2) = 1$, and so on.

4. A steel plate ($E = 20.7 \times 10^6$ N/cm², $\nu = 0.3$) 1 cm thick is loaded as shown in Fig. E4. Use two- and four-element meshes of linear elements to determine the deflections at the nodes and stresses in the elements. Compare the finite-element solution with the exact solution of the problem.

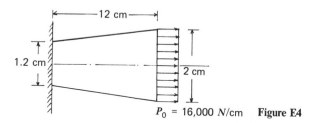

$P_0 = 16{,}000$ N/cm   Figure E4

5. Assuming elementwise constant area of cross section, determine the finite-element solution of the problem in Exercise 4 and compare the two solutions. Assume that the area of each element is the average of the areas at node 1 and node 2 of that element.

6. Find the stress and compression in each section of the composite member shown in Fig. E6. Use $E_s = 30 \times 10^6$ psi, $E_a = 10 \times 10^6$ psi, and $E_b = 15 \times 10^6$ psi.

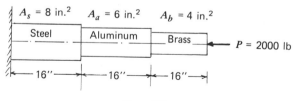

Figure E6

7. Determine the stresses in each member of the composite member in Exercise 6 if the total elongation is 0.02 in.

8. The two members in Fig. E8 are fastened together and to rigid walls. If the members are stress free before they are loaded, what will be the stresses and deformations in each after the two 50,000-lb

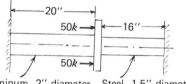

Aluminum, 2″ diameter   Steel, 1.5″ diameter   Figure E8

loads are applied? Use $E_s = 30 \times 10^6$ psi and $E_a = 10 \times 10^6$ psi; the aluminum rod is 2 in. in diameter and the steel rod is 1.5 in. in diameter.

9. Consider the following pair of differential equations [see Eq. (3.2.52)] associated with the bending of beams,

$$\frac{d^2 w}{dx^2} - \frac{M}{EI} = 0, \qquad \frac{d^2 M}{dx^2} + f_0 = 0$$

Using the Reissner variational principle [see Eq. (3.2.58)], derive a finite-element model of the equations.

10. Using the linear interpolation functions in Eq. (3.4.12) compute the coefficients in the finite-element equations of Exercise 9.

11. Use the least-squares method to construct the finite-element model of the equations in Exercise 9.

12–16. Use the minimum possible number of elements to analyze (i.e., determine deflections and slopes at the nodes, and reactions at the supports of) the following beam structures (see Figs. E12–E16).

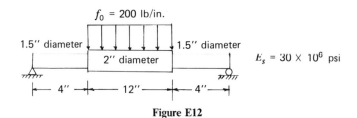

Figure E12

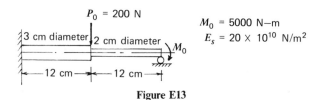

Figure E13

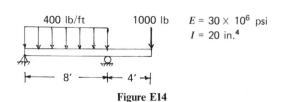

Figure E14

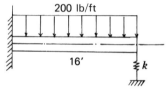

Figure E15 $E = 30 \times 10^6$ psi; $k = 1.5 \times 10^6$ lb/in; $I = 40$ in^4.

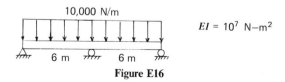

Figure E16

17. Construct the finite-element model, based on the total potential energy functional, for the following differential equation, which arises in connection with the buckling of columns,

$$\frac{d^2}{dx^2}\left(b\frac{d^2w}{dx^2}\right) + P\frac{d^2w}{dx^2} = f, \qquad 0 < x < L$$

where P is a constant, and f and b are given functions of x.

18. Compute the stiffness coefficients of the finite-element equations in Exercise 17. Make use of the results already available in the text. Take b and f as constants.

19. Use the finite-element model of Exercises 17 and 18 to determine the critical buckling load P of a simply-supported beam (when $f = 0$). Use one element in half beam. This exercise requires the solution of an eigenvalue problem.

20. The axisymmetric bending of a thin, uniform thickness, circular plate can be described by a one-dimensional differential equation. For orthotropic material, the differential equation is given by

$$\frac{1}{r}\frac{d}{dr}\left[-D_{11}\left(r\frac{d^3w}{dr^3} + \frac{d^2w}{dr^2}\right) + \frac{1}{r}D_{22}\frac{dw}{dr}\right] + f(r) = 0$$

where r is the radial coordinate, and D_{11} and D_{22} are the material constants. Find the variational formulation and associated finite-element model of the equation over a typical element.

21. Solve the torsion problem for the stress function associated with the triangular section member. Use the mesh shown in Fig. E21.

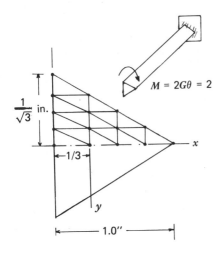

Figure E21

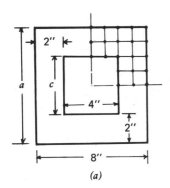

(a)

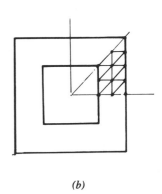

(b)

Figure E22

22. Consider the torsion of a hollow square cross section. The stress function Φ is required to satisfy Eq. (1.7.39a) in the domain, and

$$\Phi = 0 \quad \text{on outer boundary}$$

$$\Phi = 2G\theta c^2 \quad \text{on inner boundary}$$

The hole dimension c is $a/c = 2$, where a is the outside dimension of the square. Solve the problem using the meshes shown in Fig. E22a and b.

23. Give the variational formulation and associated finite element model of the equation

$$-\frac{\partial}{\partial x}\left(K_1 \frac{\partial u}{\partial x}\right) - \frac{\partial}{\partial y}\left(K_2 \frac{\partial u}{\partial y}\right) = f \quad \text{in } R$$

where K_1, K_2, and f are given functions of x and y in a two-dimensional domain R, and account for the following boundary condition in the model:

$$\left(K_1 \frac{\partial u}{\partial x} n_x + K_2 \frac{\partial u}{\partial y} n_y \right) + h(u - u_0) + q = 0$$

Here h and u_0 are known constants, and q is a given function on boundary of the element. This equation arises in connection with convective heat transfer in two dimensions.

24. Evaluate the finite-element coefficient matrices of a linear rectangular element for the model in Exercise 23. Use the results already available in the text.

25. Use a procedure similar to that suggested in Exercise 3 to develop the interpolation functions of a square element with eight nodes: four at the corners and four at the midsides. Set up a local coordinate system at the center of the element (See Fig. E25). For example, $\psi_1(\bar{x}, \bar{y})$ is a quadratic function of the form

$$\psi_1(\bar{x}, \bar{y}) = c_1(a - \bar{x})(a - \bar{y})(\bar{x} + \bar{y} + a)$$

26. Solve the plate problem in Exercise 4 with a 1×2 mesh of triangular elements in a symmetric half plate and compare the results with those in Exercise 4. Use the plane stress assumption. You can use a calculator or a computer to assemble and solve the resulting finite-element equations.

27. Solve the problem in Exercise 5 using 1×2 mesh of rectangular elements.

28. Solve the plane strain problem of Exercise 1.7.17 using the finite-element shown in Fig. E28. Take $r_0 = 8$ in., $r_i = 2$ in., $p_0 = 0$, $p_i = 100$ psi, $E = 30 \times 10^6$ psi, $\nu = 0.3$, and $\theta = 22.5°$.

29. The nodal values of a function u at the nodes of a triangular element are given in Fig. E29. Determine the value of u, $\partial u/\partial x$, and $\partial u/\partial y$ at the point (x_0, y_0).

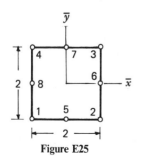

Figure E25

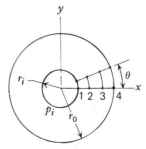

Figure E28

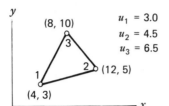

$$u_1 = 3.0$$
$$u_2 = 4.5$$
$$u_3 = 6.5$$

Figure E29

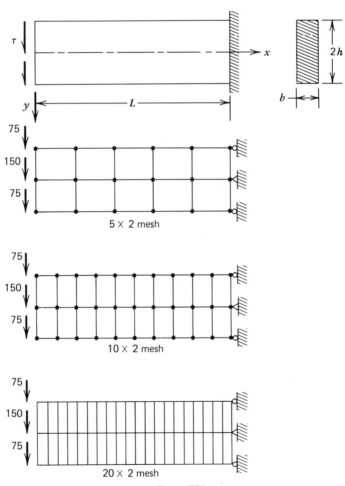

Figure E30

30. Model the problem of the bending of a narrow cantilevered beam (see Example 1.5) bent by an end force by the finite-element method (Fig. E30). The height to length ratio is 0.2, and the end shear stress is $\tau = 150$ psi. Use rectangular elements to study the convergence, and take $E = 30 \times 10^6$ psi, $\nu = 0.25$, $h = 1$ in., $L = 10$ in., and $b = 1$ in.

31. Solve the time-dependent problem of Example 3.14 using a two-element mesh for the spatial approximation and the Galerkin method for the time approximation.

4

THEORY AND ANALYSIS OF PLATES AND SHELLS

4.1 CLASSICAL THEORY OF PLATES

4.1.1 Governing Equations

A plate is a solid body, bounded by two parallel flat surfaces, whose lateral dimensions (the width and length of a rectangular plate or the diameter of a circular plate) are large compared to the distance (the thickness of the plate) between the flat surfaces. The theory of elastic plates is an approximation of three-dimensional elasticity theory to two dimensions, and it can be used to determine the state of stress and deformation in plates. The approximations are simplifications that allow us only to consider the deformation of the midplane of the plate.

Plates can be classified into two groups: *thin* plates and *thick* plates. A plate is said to be thin when the ratio of the thickness to the smaller span length is less than $1/20$ (a commonly accepted ratio). The small-deflection theory of bending of thin plates is known as the *classical plate theory* and is based on assumptions similar to those used in the Euler–Bernoulli theory (or the classical beam theory). The fundamental assumptions of the Kirchhoff–Love or classical plate theory are:

1. The displacements of the midsurface are small compared with the thickness of the plate and, therefore, the slope of the deflected surface is very small and the square of the slope is negligible compared to unity.

2. The midplane of the plate remains unstrained, that is, the midplane of the plate is the neutral plane, after the bending.

3. *The Kirchhoff–Love Assumption.* Plane sections initially normal to the midsurface remain *plane* and *normal* to the midsurface after bending. Analogous to beams, this assumption implies that the transverse shear strains e_{xz} and e_{yz} are negligible. The deflection of the plate is thus associated principally with bending strains. Consequently, the transverse normal strain e_z resulting from transverse loading can be neglected.

4. The transverse normal stress σ_z is small compared with the other stress components of the plate and, therefore, can be neglected.

These assumptions give results that do not differ significantly from those obtained using the three-dimensional theory of elasticity for a vast majority of thin plate problems. When the transverse deflection is not small (but comparable to the thickness) and inplane displacement gradients are small, assumptions 1 and 2 are replaced by the single assumption that the products and squares of the slopes of inplane displacements are small compared to unity and can be neglected. This leads to the *von Karman theory of plates* (see von Karman, 1910, and von Karman et al., 1932).

In view of assumptions 3 and 4, the plate problem can be classified as a plane stress problem. Using the assumption 3 and Eq. (1.3.11) we can write

$$e_x = \frac{\partial u}{\partial x}, \qquad e_y = \frac{\partial v}{\partial y}, \qquad e_z = \frac{\partial w}{\partial z} = 0$$

$$2e_{xy} = \frac{\partial u}{\partial y} + \frac{\partial v}{\partial x}, \qquad 2e_{xz} = \frac{\partial u}{\partial z} + \frac{\partial w}{\partial x} = 0, \qquad 2e_{yz} = \frac{\partial v}{\partial z} + \frac{\partial w}{\partial y} = 0$$

$$(4.1.1)$$

The relation $\partial w/\partial z = 0$ implies that w is a function of x and y only:

$$w = w(x, y) \qquad (4.1.2)$$

In other words, the transverse deflection does not vary over the plate thickness. Integration of $2e_{xz}$ and $2e_{yz}$ with respect to z gives

$$u = -\frac{\partial w}{\partial x} z + f(x, y), \qquad v = -\frac{\partial w}{\partial y} z + g(x, y) \qquad (4.1.3)$$

where f and g denote the functions of integration. Clearly, for $z = 0$, we obtain

$$f(x, y) = u(x, y, 0), \qquad g(x, y) = v(x, y, 0) \qquad (4.1.4)$$

indicating that f and g are the inplane displacements along the x- and y-directions of the midplane (i.e., $z = 0$ plane). We denote these displacements by u_0 and v_0, respectively. We have, from Eq. (4.1.3),

$$u(x, y, z) = u_0(x, y) - z\frac{\partial w}{\partial x}$$

$$v(x, y, z) = v_0(x, y) - z\frac{\partial w}{\partial y} \qquad (4.1.5)$$

The displacement expressions in Eq. (4.1.5) are analogous to the displacement derived in the classical beam theory. Figure 4.1 illustrates the relationship for the u displacement.

Substituting the displacement field from Eq. (4.1.5) into Eq. (4.1.1), we obtain

$$e_x = \frac{\partial u_0}{\partial x} - z\frac{\partial^2 w}{\partial x^2}, \qquad e_y = \frac{\partial v_0}{\partial y} - z\frac{\partial^2 w}{\partial y^2}, \qquad 2e_{xy} = \frac{\partial u_0}{\partial y} + \frac{\partial v_0}{\partial x} - 2z\frac{\partial^2 w}{\partial x\,\partial y}$$

$$(4.1.6)$$

These strains are sometimes referred to as the *Kirchhoff strains*.

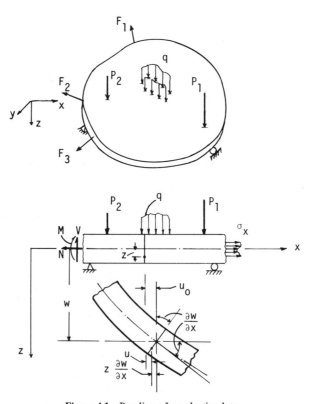

Figure 4.1 Bending of an elastic plate.

To derive the governing equilibrium equations of the classical plate theory, we use the principle of virtual displacements. From Eq. (2.3.4) we have

$$0 = \int_V \left(\sigma_{ij}\, \delta e_{ij} - f_i\, \delta u_i \right) dV - \oint_S \hat{t}_i\, \delta u_i\, dS \qquad (4.1.7)$$

where V is the volume of the plate and S is the boundary of V on which the tractions $\hat{t}_i$ are specified. In Eq. (4.1.7), the sum on repeated indices is implied. The volume V of the plate can be represented by

$$V = R \times \left(\frac{-h}{2}, \frac{h}{2} \right) \qquad (4.1.8)$$

where R is the midplane of the plate and h is the total thickness of the plate. The surface S of the volume V consists of three parts: the top surface ($z = -h/2$), the bottom surface ($z = h/2$), and the edge C of the plate (see

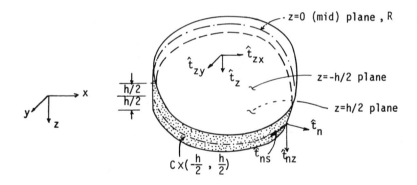

(a) Boundary stresses

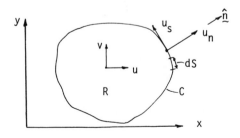

(b) Displacements

Figure 4.2   Boundary stresses and displacements in a plate.

Fig. 4.2). Consequently, the boundary integral in Eq. (4.1.7) can be written as

$$\oint_S \hat{t}_i \,\delta u_i \, dS = \int_{-h/2}^{h/2} \oint_C (\hat{t}_n \,\delta u_n + \hat{t}_{ns} \,\delta u_s + \hat{t}_{nz} \,\delta w) \, dS \, dz$$

$$+ \int_R (\hat{t}_{zx} \,\delta u + \hat{t}_{zy} \,\delta v + \hat{t}_z \,\delta w)\Big|_{-h/2}^{h/2} dx \, dy \qquad (4.1.9)$$

where $\hat{t}_n$ and $(\hat{t}_{ns}, \hat{t}_{nz})$ are the specified normal and tangential components, measured per unit area, and δu_n and $(\delta u_s, \delta w)$ are components of the normal and tangential displacements on the edge C. The second integral on the right-hand side in Eq. (4.1.9) can be simplified using the following conditions (hereafter, we use dS to denote the arc length along contour C as well as a surface element, as will be apparent in the context):

On Face $z = \dfrac{-h}{2}$

$\hat{t}_z = q(x, y)$ (specified distributed transverse load)

$\hat{t}_{zx} = 0, \qquad \hat{t}_{zy} = 0$ (no tangential surface forces are specified)

On Face $z = h / 2$

$\hat{t}_z = \hat{t}_{zx} = \hat{t}_{zy} = 0$ (normal and tangential forces are specified to be zero)   (4.1.10)

The first integral on the right-hand side of Eq. (4.1.9) can be simplified using the relations

$$u_n = u_{n0} - z \frac{\partial w}{\partial n}, \qquad u_s = u_{s0} - z \frac{\partial w}{\partial s} \qquad (4.1.11)$$

We have

$$\int_{-h/2}^{h/2} \oint_C (\hat{t}_n \,\delta u_n + \hat{t}_{ns} \,\delta u_s + \hat{t}_{nz} \,\delta w) \, dS \, dz$$

$$= \oint_C \left(\hat{N}_n \,\delta u_{n0} - \hat{M}_n \frac{\partial \delta w}{\partial n} + \hat{N}_s \,\delta u_{s0} - \hat{M}_s \frac{\partial \delta w}{\partial s} + \hat{Q}_n \,\delta w \right) dS$$

$$(4.1.12)$$

where $\hat{N}_n$, $\hat{N}_s$, $\hat{M}_n$, $\hat{M}_s$, and $\hat{Q}_n$ are the resultants defined by

$$\hat{N}_n = \int_{-h/2}^{h/2} \hat{t}_n \, dz, \qquad \hat{N}_s = \int_{-h/2}^{h/2} \hat{t}_{ns} \, dz$$

$$\hat{M}_n = \int_{-h/2}^{h/2} z\hat{t}_n \, dz, \qquad \hat{M}_s = \int_{-h/2}^{h/2} z\hat{t}_{ns} \, dz$$

$$\hat{Q}_n = \int_{-h/2}^{h/2} \hat{t}_{nz} \, dz \qquad (4.1.13)$$

Thus, Eq. (4.1.7) takes the form ($f_i = 0$)

$$0 = \int_{-h/2}^{h/2} \left[\int_R (\sigma_x \delta e_x + \sigma_y \delta e_y + \sigma_{xy} 2 \delta e_{xy}) \, dx \, dy \right] dz - \int_R q \delta w \, dx \, dy$$

$$- \int_{C_1} \hat{N}_n \delta u_{n0} \, dS - \int_{C_2} \hat{N}_s \delta u_{s0} \, dS + \int_{C_3} \hat{M}_n \frac{\partial \delta w}{\partial n} \, dS + \int_{C_4} \hat{M}_s \frac{\partial \delta w}{\partial s} \, dS$$

$$- \int_{C_5} \hat{Q}_n \delta w \, dS \qquad (4.1.14)$$

where C_i ($i = 1, 2, \ldots, 5$) are the portions of the boundary of the midplane R on which $\hat{N}_n$, $\hat{N}_s$, $\hat{M}_n$, $\hat{M}_s$, and $\hat{Q}_n$, respectively, are specified. In general, the boundary segments C_1, C_2, C_3, C_4 and C_5 are not all the same; they complement, respectively, the portions of the boundary on which u_n, u_s, $\partial w/\partial n$, $\partial w/\partial s$, and w are specified. From an examination of the boundary terms in Eq. (4.1.14), it is clear that the specifications of the inplane normal and tangential displacements, the transverse deflection, and the normal and tangential derivatives of the transverse deflection constitute the essential boundary conditions. We return to this point shortly.

To derive the equilibrium equations of the classical plate theory, we now introduce the following resultants:

Inplane Stress Resultants

$$N_x = \int_{-h/2}^{h/2} \sigma_x \, dz, \qquad N_y = \int_{-h/2}^{h/2} \sigma_y \, dz, \qquad N_{xy} = \int_{-h/2}^{h/2} \sigma_{xy} \, dz$$

$$(4.1.15a)$$

Bending Moment Resultants

$$M_x = \int_{-h/2}^{h/2} z\sigma_x \, dz, \qquad M_y = \int_{-h/2}^{h/2} z\sigma_y \, dz, \qquad M_{xy} = \int_{-h/2}^{h/2} z\sigma_{xy} \, dz$$

$$(4.1.15b)$$

Shear Force Resultants

$$Q_x = \int_{-h/2}^{h/2} \sigma_{xz}\,dz, \qquad Q_y = \int_{-h/2}^{h/2} \sigma_{yz}\,dz \qquad (4.1.15c)$$

These resultants are shown on a plate element in Fig. 4.3. From the figure, it is clear that N_{yx}, for example, signifies the shear force acting on the edge perpendicular to the y-axis, and that it is measured per unit length. On an edge that is not parallel to any of the coordinate axes, the stress and moment resultants can be related to the components in the x- and y-coordinates by equations of the form,

$$N_n = N_x n_x^2 + N_y n_y^2 + 2n_x n_y N_{xy}$$

$$N_s = \left(N_y - N_x\right)n_x n_y + N_{xy}\left(n_x^2 - n_y^2\right)$$

$$M_n = M_x n_x^2 + M_y n_y^2 + 2n_x n_y M_{xy}$$

$$M_s = \left(M_y - M_x\right)n_x n_y + M_{xy}\left(n_x^2 - n_y^2\right) \qquad (4.1.16)$$

Returning to the virtual work statement, we substitute for strains from Eq. (4.1.6) and the definitions in Eqs. (4.1.15) into Eq. (4.1.14) to obtain

$$0 = \int_R \left[N_x \frac{\partial \delta u_0}{\partial x} - M_x \frac{\partial^2 \delta w}{\partial x^2} + N_y \frac{\partial \delta v_0}{\partial y} - M_y \frac{\partial^2 \delta w}{\partial y^2} + N_{xy}\left(\frac{\partial \delta u_0}{\partial y} + \frac{\partial \delta v_0}{\partial x} \right) \right.$$

$$\left. - 2M_{xy}\frac{\partial^2 \delta w}{\partial x\,\partial y} - q\,\delta w \right] dx\,dy - \int_{C_1} \hat{N}_n \delta u_{n0}\,dS$$

$$- \int_{C_2} \hat{N}_s \delta u_{s0}\,dS + \int_{C_3} \hat{M}_n \frac{\partial \delta w}{\partial n}\,dS + \int_{C_4} \hat{M}_s \frac{\partial \delta w}{\partial s}\,dS - \int_{C_5} \hat{Q}_n \delta w\,dS$$

$$(4.1.17a)$$

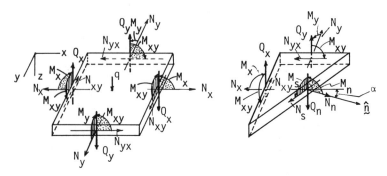

Figure 4.3 Stress resultants and moment resultants at a point in the plate and on the boundary.

Integrating the expressions under the integral over R by parts, we obtain

$$0 = \int_R \left[-\delta u_0 \left(\frac{\partial N_x}{\partial x} + \frac{\partial N_{xy}}{\partial y} \right) - \delta v_0 \left(\frac{\partial N_{xy}}{\partial x} + \frac{\partial N_y}{\partial y} \right) \right.$$

$$\left. - \delta w \left(\frac{\partial^2 M_x}{\partial x^2} + 2\frac{\partial^2 M_{xy}}{\partial x\,\partial y} + \frac{\partial^2 M_y}{\partial y^2} + q \right) \right] dx\,dy$$

$$+ \oint_C \left[\delta u_0 \left(N_x n_x + N_{xy} n_y \right) + \delta v_0 \left(N_{xy} n_x + N_y n_y \right) \right.$$

$$- \frac{\partial \delta w}{\partial x} \left(M_x n_x + M_{xy} n_y \right) - \frac{\partial \delta w}{\partial y} \left(M_{xy} n_x + M_y n_y \right)$$

$$\left. + \delta w \left(\frac{\partial M_x}{\partial x} n_x + \frac{\partial M_{xy}}{\partial x} n_y + \frac{\partial M_{xy}}{\partial y} n_x + \frac{\partial M_y}{\partial y} n_y \right) \right] dS$$

$$- \int_{C_1} \hat{N}_n \delta u_{n0}\, dS - \int_{C_2} \hat{N}_s \delta u_{s0}\, dS + \int_{C_3} \hat{M}_n \frac{\partial \delta w}{\partial n}\, dS + \int_{C_4} \hat{M}_s \frac{\partial \delta w}{\partial s}\, dS$$

$$- \int_{C_5} \hat{Q}_n \delta w\, dS \tag{4.1.17b}$$

where we used the component form of the gradient theorem, Eq. (1.1.15), to transfer the differentiation from δu_0, δv_0, and δw to their coefficients.

To simplify the boundary terms, we convert all of the x and y components to the normal and tangential components using the following relations:

$$u_{n0} = u_0 n_x + v_0 n_y, \qquad u_{s0} = -u_0 n_y + v_0 n_x$$

$$u_0 = u_{n0} n_x - u_{s0} n_y, \qquad v_0 = u_{n0} n_y + u_{s0} n_x$$

$$\frac{\partial}{\partial n} = n_x \frac{\partial}{\partial x} + n_y \frac{\partial}{\partial y}, \qquad \frac{\partial}{\partial s} = n_x \frac{\partial}{\partial y} - n_y \frac{\partial}{\partial x}$$

$$\frac{\partial}{\partial x} = n_x \frac{\partial}{\partial n} - n_y \frac{\partial}{\partial s}, \qquad \frac{\partial}{\partial y} = n_y \frac{\partial}{\partial n} + n_x \frac{\partial}{\partial s} \tag{4.1.18}$$

Substituting Eq. (4.1.18) for δu_0, δv_0, $\partial \delta w/\partial x$, and $\partial \delta w/\partial y$ into the first boundary integral, and rearranging the terms in Eq. (4.1.17b), we obtain

$$
0 = \int_R \left[-\delta u_0 \left(\frac{\partial N_x}{\partial x} + \frac{\partial N_{xy}}{\partial y} \right) - \delta v_0 \left(\frac{\partial N_{xy}}{\partial x} + \frac{\partial N_y}{\partial y} \right) \right.
$$

$$
\left. -\delta w \left(\frac{\partial^2 M_x}{\partial x^2} + 2 \frac{\partial^2 M_{xy}}{\partial x\, \partial y} + \frac{\partial^2 M_y}{\partial y^2} + q \right) \right] dx\, dy
$$

$$
+ \oint_C \left(N_n \delta u_{n0} + N_s \delta u_{s0} - M_n \frac{\partial \delta w}{\partial n} - M_s \frac{\partial \delta w}{\partial s} + Q_n \delta w \right) dS
$$

$$
- \int_{C_1} \hat{N}_n \delta u_{n0}\, dS - \int_{C_2} \hat{N}_s \delta u_{s0}\, dS + \int_{C_3} \hat{M}_n \frac{\partial \delta w}{\partial n}\, dS + \int_{C_4} \hat{M}_s \frac{\partial \delta w}{\partial s}\, dS
$$

$$
- \int_{C_5} \hat{Q}_n \delta w\, dS \tag{4.1.19}
$$

where N_n, N_s, M_n, and M_s are given by Eq. (4.1.16), and

$$
Q_n = \left(\frac{\partial M_x}{\partial x} + \frac{\partial M_{xy}}{\partial y} \right) n_x + \left(\frac{\partial M_y}{\partial y} + \frac{\partial M_{xy}}{\partial x} \right) n_y \tag{4.1.20}
$$

Next, we assume that the variations δu_{n0}, δu_{s0}, δw, $\partial \delta w / \partial n$, and $\partial \delta w / \partial s$ are zero, respectively, on boundary segments $C - C_1$, $C - C_2$, $C - C_3$, $C - C_4$, and $C - C_5$, where C is the total boundary. Now, using the usual argument that the variations $(\delta u_0, \delta v_0, \delta w)$ are arbitrary in R and $(\delta u_{n0}, \delta u_{s0}, \delta w, \partial \delta w / \partial n, \partial \delta w / \partial s)$ are arbitrary on boundary segments C_1 thru C_5 respectively, we obtain the equilibrium differential equations and natural boundary conditions of a plate:

$$
\delta u_0: \quad \frac{\partial N_x}{\partial x} + \frac{\partial N_{xy}}{\partial y} = 0 \tag{4.1.21a}
$$

$$
\delta v_0: \quad \frac{\partial N_{xy}}{\partial x} + \frac{\partial N_y}{\partial y} = 0 \tag{4.1.21b}
$$

$$
\delta w: \quad \left(\frac{\partial^2 M_x}{\partial x^2} + 2 \frac{\partial^2 M_{xy}}{\partial x\, \partial y} + \frac{\partial^2 M_y}{\partial y^2} \right) + q = 0 \tag{4.1.21c}
$$

$$
\delta u_{n0}: \quad N_n - \hat{N}_n = 0 \quad \text{on } C_1 \tag{4.1.22a}
$$

$$
\delta u_{s0}: \quad N_s - \hat{N}_s = 0 \quad \text{on } C_2 \tag{4.1.22b}
$$

$$
\delta \left(\frac{\partial w}{\partial n} \right): \quad M_n - \hat{M}_n = 0 \quad \text{on } C_3 \tag{4.1.22c}
$$

$$
\delta \left(\frac{\partial w}{\partial s} \right): \quad M_s - \hat{M}_s = 0 \quad \text{on } C_4 \tag{4.1.22d}
$$

$$
\delta w: \quad Q_n - \hat{Q}_n = 0 \quad \text{on } C_5 \tag{4.1.22e}
$$

The essential boundary conditions of plates are given by

$$u_{n0} = \hat{u}_{n0}, \qquad u_{s0} = \hat{u}_{s0}, \qquad w = \hat{w}$$

$$\frac{\partial w}{\partial n} = \hat{\theta}_n, \qquad \frac{\partial w}{\partial s} = \hat{\theta}_s \qquad (4.1.23)$$

Various commonly used boundary conditions of plates are listed in Fig. 4.4.

Equations (4.1.21)–(4.1.23) completely describe the equilibrium of a plate satisfying the assumptions of the classical plate theory. The first two equations in Eq. (4.1.21) represent the equilibrium of forces along the x and y directions, and the third is along the z direction. The first two equilibrium equations can be viewed as the integral (through the thickness) form of the equilibrium equations (1.7.43).

A comment concerning the boundary conditions is in order. The last two boundary conditions in Eq. (4.1.22) are usually combined, based on physical considerations, into a single boundary condition. This is done by integrating the following term in Eq. (4.1.12),

$$-\oint_C M_s \frac{\partial \delta w}{\partial s}\, dS = \oint_C \frac{\partial M_s}{\partial s} \delta w\, dS - [M_s \delta w]_C$$

The term in the square brackets is zero at the end points of the boundary curve (because the end points coincide). Now the integral contribution on the right-hand side of the above expression can be added to Q_n in Eq. (4.1.12). Thus, the boundary condition resulting from the arbitrariness of δw on $C_4 = C_5$ in Eq. (4.1.19) is given by

$$V_n \equiv Q_n + \frac{\partial M_s}{\partial s} = \hat{V}_n \quad \text{on } C_4 = C_5 \qquad (4.1.24)$$

This boundary condition is known as the *Kirchhoff free-edge condition*.

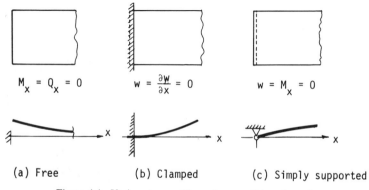

(a) Free $\qquad$ (b) Clamped $\qquad$ (c) Simply supported

Figure 4.4 Various types of boundary conditions in a plate.

The equilibrium differential equations (4.1.21) can also be derived by applying Newton's laws to an element of the plate. However, the variational approach not only gives the governing differential equations, but it also gives the natural boundary conditions and the form of the dependent variables in the essential boundary conditions.

To express the equilibrium equations in terms of the displacements, we use Eq. (1.7.45), (4.1.6), (4.1.14), and (4.1.15) to relate the stress and moment resultants to the displacements. We have

$$N_x = A_{11}\frac{\partial u_0}{\partial x} + A_{12}\frac{\partial v_0}{\partial y}$$

$$N_y = A_{12}\frac{\partial u_0}{\partial x} + A_{22}\frac{\partial v_0}{\partial y}$$

$$N_{xy} = A_{66}\left(\frac{\partial u_0}{\partial y} + \frac{\partial v_0}{\partial x}\right) \tag{4.1.25a}$$

$$M_x = -\left(D_{11}\frac{\partial^2 w}{\partial x^2} + D_{12}\frac{\partial^2 w}{\partial y^2}\right)$$

$$M_y = -\left(D_{12}\frac{\partial^2 w}{\partial x^2} + D_{22}\frac{\partial^2 w}{\partial y^2}\right)$$

$$M_{xy} = -2D_{66}\frac{\partial^2 w}{\partial x \partial y} \tag{4.1.25b}$$

where A_{ij} and D_{ij} are given by

$$A_{ij} = \int_{-h/2}^{h/2} c_{ij}\, dz, \qquad D_{ij} = \int_{-h/2}^{h/2} z^2 c_{ij}\, dz, \qquad (i,j = 1,2,6) \tag{4.1.25c}$$

and c_{ij} are given by Eq. (1.7.46a). Substituting Eqs. (4.1.25a and b) into Eq. (4.1.21), we obtain

$$\frac{\partial}{\partial x}\left(A_{11}\frac{\partial u_0}{\partial x} + A_{12}\frac{\partial v_0}{\partial y}\right) + A_{66}\frac{\partial}{\partial y}\left(\frac{\partial u_0}{\partial y} + \frac{\partial v_0}{\partial x}\right) = 0$$

$$A_{66}\frac{\partial}{\partial x}\left(\frac{\partial u_0}{\partial y} + \frac{\partial v_0}{\partial x}\right) + \frac{\partial}{\partial y}\left(A_{12}\frac{\partial u_0}{\partial x} + A_{22}\frac{\partial v_0}{\partial y}\right) = 0 \tag{4.1.26a}$$

$$D_{11}\frac{\partial^4 w}{\partial x^4} + 2(D_{12} + 2D_{66})\frac{\partial^4 w}{\partial x^2 \partial y^2} + D_{22}\frac{\partial^4 w}{\partial y^4} = q \tag{4.1.26b}$$

Note that Eqs. (4.1.26a) are uncoupled from Eq. (4.1.26b). Thus, the inplane displacements u_0 and v_0 can be determined from Eqs. (4.1.26a) and the transverse deflection from Eq. (4.1.26b). For an isotropic elastic body, the last equation reduces to the well-known biharmonic equation for w:

$$\nabla^4 w = \frac{q}{D}$$

$$\nabla^4 = \nabla^2 \nabla^2 = \frac{\partial^4}{\partial x^4} + 2\frac{\partial^4}{\partial x^2 \partial y^2} + \frac{\partial^4}{\partial y^4}, \qquad D = \frac{Eh^3}{12(1 - \nu^2)} \qquad (4.1.27)$$

Equation (4.1.27) is valid in all coordinate systems, including the polar coordinates.

Example 4.1

Many structural members that carry transverse loads, such as end plates, pump diaphragms, and turbine disks are circular in shape. For circular plates, it is convenient to use the polar coordinates (r, θ) to describe the governing equations. We use the principle of total potential energy to obtain the governing differential equation. For an isotropic plate, the total potential energy in polar coordinates is given by (see Exercise 4.1.3)

$$\Pi = \frac{D}{2}\int_R \left[\left(\frac{\partial^2 w}{\partial r^2} + \frac{1}{r}\frac{\partial w}{\partial r} + \frac{1}{r^2}\frac{\partial^2 w}{\partial \theta^2} \right)^2 - 2(1 - \nu)\frac{\partial^2 w}{\partial r^2}\left(\frac{1}{r}\frac{\partial w}{\partial r} + \frac{1}{r^2}\frac{\partial^2 w}{\partial \theta^2} \right) \right.$$

$$\left. + 2(1 - \nu)\left(\frac{1}{r}\frac{\partial^2 w}{\partial r \partial \theta} - \frac{1}{r^2}\frac{\partial w}{\partial \theta} \right)^2 - qw \right] r\, dr\, d\theta$$

$$- \int_{C_1} \hat{V}_r w\, dS - \int_{C_2} \hat{V}_\theta w\, dr + \int_{C_3} \hat{M}_r \frac{\partial w}{\partial r}\, dS + \int_{C_4} \hat{M}_\theta \left(\frac{1}{r}\frac{\partial w}{\partial \theta} \right) dr \qquad (4.1.28)$$

where C_1 thru C_4 denote portions of the boundary C on which the shear forces and the bending moments are specified, $dS = r\, d\theta$, and the quantities $\hat{M}_r$, $\hat{M}_\theta$, $\hat{V}_r$, and $\hat{V}_\theta$ are the specified bending moments and shear forces.

Using the total potential energy principle, we set $\delta\Pi = 0$ and obtain

$$0 = D\int_R \left[\left(\frac{\partial^2 w}{\partial r^2} + \frac{1}{r}\frac{\partial w}{\partial r} + \frac{1}{r^2}\frac{\partial^2 w}{\partial \theta^2} \right)\left(\frac{\partial^2 \delta w}{\partial r^2} + \frac{1}{r}\frac{\partial \delta w}{\partial r} + \frac{1}{r^2}\frac{\partial^2 \delta w}{\partial \theta^2} \right) \right.$$

$$- (1 - \nu)\frac{\partial^2 w}{\partial r^2}\left(\frac{1}{r}\frac{\partial \delta w}{\partial r} + \frac{1}{r^2}\frac{\partial^2 \delta w}{\partial \theta^2} \right)$$

$$- (1 - \nu)\frac{\partial^2 \delta w}{\partial r^2}\left(\frac{1}{r}\frac{\partial w}{\partial r} + \frac{1}{r^2}\frac{\partial^2 w}{\partial \theta^2} \right)$$

$$+ 2(1 - \nu)\left(\frac{1}{r}\frac{\partial^2 w}{\partial r \partial \theta} - \frac{1}{r^2}\frac{\partial w}{\partial \theta} \right)$$

$$\left. \times \left(\frac{1}{r}\frac{\partial^2 \delta w}{\partial r \partial \theta} - \frac{1}{r^2}\frac{\partial \delta w}{\partial \theta} \right) - q\delta w \right] r\, dr\, d\theta$$

$$- \int_{C_1} \hat{V}_r \delta w r\, d\theta - \int_{C_2} \hat{V}_\theta \delta w\, dr + \int_{C_3} \hat{M}_r \frac{\partial \delta w}{\partial r} r\, d\theta + \int_{C_4} \hat{M}_\theta \frac{1}{r}\frac{\partial \delta w}{\partial \theta}\, dr$$

Integration by parts of the terms over the domain and algebraic simplification
of the resulting expressions yield,

$$0 = D \int_R \delta w \left[\left(\frac{\partial^2}{\partial r^2} + \frac{1}{r}\frac{\partial}{\partial r} + \frac{1}{r^2}\frac{\partial^2}{\partial \theta^2} \right) \left(\frac{\partial^2 w}{\partial r^2} + \frac{1}{r}\frac{\partial w}{\partial r} + \frac{1}{r^2}\frac{\partial^2 w}{\partial \theta^2} \right) - \frac{q}{D} \right] r\, dr\, d\theta$$

$$+ D \oint_C \left\{ -r \left[\frac{\partial}{\partial r}(\nabla^2 w) + \frac{1-\nu}{r}\frac{\partial}{\partial \theta}\left(\frac{1}{r}\frac{\partial^2 w}{\partial r\, \partial \theta} - \frac{1}{r^2}\frac{\partial w}{\partial \theta} \right) \right] \right.$$

$$\left. + (1-\nu)\frac{\partial}{\partial r}\left(\frac{1}{r}\frac{\partial^2 w}{\partial \theta^2} \right) \right\} \delta w\, d\theta$$

$$+ D \oint_C \left\{ -\left[\frac{1}{r}\frac{\partial}{\partial \theta}(\nabla^2 w) + (1-\nu)\frac{\partial}{\partial r}\left(\frac{1}{r}\frac{\partial^2 w}{\partial r\, \partial \theta} - \frac{1}{r^2}\frac{\partial w}{\partial \theta} \right) \right] \right.$$

$$\left. + \frac{1-\nu}{r}\frac{\partial^3 w}{\partial r^2\, \partial \theta} - \frac{2(1-\nu)}{r}\left(\frac{1}{r}\frac{\partial^2 w}{\partial r\, \partial \theta} - \frac{1}{r^2}\frac{\partial w}{\partial \theta} \right) \right\} \delta w\, dr$$

$$+ D \oint_C r \left[\frac{\partial^2 w}{\partial r^2} + \nu\left(\frac{1}{r}\frac{\partial w}{\partial r} + \frac{1}{r^2}\frac{\partial^2 w}{\partial \theta^2} \right) \right] \delta\left(\frac{\partial w}{\partial r} \right) d\theta$$

$$+ D \oint_C \left[\frac{1}{r}\frac{\partial w}{\partial r} + \frac{1}{r^2}\frac{\partial^2 w}{\partial \theta^2} + \nu\frac{\partial^2 w}{\partial r^2} \right] \delta\left(\frac{1}{r}\frac{\partial w}{\partial \theta} \right) dr$$

$$+ D \oint_C (1-\nu)\left(\frac{1}{r}\frac{\partial^2 w}{\partial r\, \partial \theta} - \frac{1}{r^2}\frac{\partial w}{\partial \theta} \right)\underline{\underline{\left[\frac{\partial}{\partial r}(\delta w)\, dr + \frac{\partial}{\partial \theta}(\delta w)\, d\theta \right]}}$$

$$- \int_{C_1} \hat{V}_r \delta w\, dS - \int_{C_2} \hat{V}_\theta \delta w\, dr + \int_{C_3} \hat{M}_r \delta\left(\frac{\partial w}{\partial r} \right) dS + \int_{C_4} \hat{M}_\theta \delta\left(\frac{1}{r}\frac{\partial w}{\partial \theta} \right) dr$$

$$(4.1.29)$$

where

$$\nabla^2 = \frac{\partial^2}{\partial r^2} + \frac{1}{r}\frac{\partial}{\partial r} + \frac{1}{r^2}\frac{\partial^2}{\partial \theta^2}$$

Next we integrate the terms that are double underscored by parts. Let

$$F = (1-\nu)\left(\frac{1}{r}\frac{\partial^2 w}{\partial r\, \partial \theta} - \frac{1}{r^2}\frac{\partial w}{\partial \theta} \right)$$

and integrate the expression

$$D \oint_C F(r,\theta)\left[\frac{\partial}{\partial r}(\delta w)\, dr + \frac{\partial}{\partial \theta}(\delta w)\, d\theta \right] = -D \oint_C \left(\frac{\partial F}{\partial r}\delta w\, dr + \frac{\partial F}{\partial \theta}\delta w\, d\theta \right)$$

$$(4.1.30)$$

where the boundary terms are zero, because they are evaluated at the ends of a closed contour C. The two terms in Eq. (4.1.30) are precisely the negative of the underlined terms in Eq. (4.1.29) and, therefore, they are cancelled. For arbitrary variations of w, $\partial w/\partial r$, $\partial w/\partial \theta$ in R and/or on portions of the boundary C, Eq. (4.1.29) implies that

$$\delta w: \quad \left(\frac{\partial^2}{\partial r^2} + \frac{1}{r}\frac{\partial}{\partial r} + \frac{1}{r^2}\frac{\partial^2}{\partial \theta^2}\right)\left(\frac{\partial^2 w}{\partial r^2} + \frac{1}{r}\frac{\partial w}{\partial r} + \frac{1}{r^2}\frac{\partial^2 w}{\partial \theta^2}\right)$$

$$-\frac{q}{D} = 0 \quad \text{in } R \tag{4.1.31}$$

$$\delta w: \quad V_r - \hat{V}_r = 0 \qquad \text{in the } r\text{-direction on } C_1$$

$$\delta w: \quad V_\theta - \hat{V}_\theta = 0 \qquad \text{in the } \theta\text{-direction on } C_2$$

$$\delta\left(\frac{\partial w}{\partial r}\right): \quad -(M_r - \hat{M}_r) = 0 \qquad \text{in the } r\text{-direction on } C_3$$

$$\delta\left(\frac{\partial w}{\partial \theta}\right): \quad -(M_\theta - \hat{M}_\theta) = 0 \qquad \text{in the } \theta\text{-direction on } C_4 \tag{4.1.32}$$

where

$$V_r = Q_r + \frac{1}{r}\frac{\partial M_{r\theta}}{\partial \theta}, \qquad V_\theta = Q_\theta + \frac{\partial M_{r\theta}}{\partial r}$$

$$Q_r = -D\frac{\partial}{\partial r}(\nabla^2 w), \qquad Q_\theta = -D\frac{1}{r}\frac{\partial}{\partial \theta}(\nabla^2 w)$$

$$M_{r\theta} = -(1-\nu)D\left(\frac{1}{r}\frac{\partial^2 w}{\partial r\,\partial \theta} - \frac{1}{r^2}\frac{\partial w}{\partial \theta}\right)$$

$$M_r = -D\left[\frac{\partial^2 w}{\partial r^2} + \nu\left(\frac{1}{r}\frac{\partial w}{\partial r} + \frac{1}{r^2}\frac{\partial^2 w}{\partial \theta^2}\right)\right]$$

$$M_\theta = -D\left[\frac{1}{r}\frac{\partial w}{\partial r} + \frac{1}{r^2}\frac{\partial^2 w}{\partial \theta^2} + \nu\frac{\partial^2 w}{\partial r^2}\right] \tag{4.1.33}$$

These equations can also be derived from Eq. (4.1.25b) using the transformation relation between the Cartesian coordinates and polar coordinates

$$x = r\cos\theta, \qquad y = r\sin\theta \tag{4.1.34}$$

and relations

$$\frac{\partial w}{\partial x} = \frac{\partial w}{\partial r}\cos\theta - \frac{1}{r}\frac{\partial w}{\partial \theta}\sin\theta$$

$$\frac{\partial w}{\partial y} = \frac{\partial w}{\partial r}\sin\theta + \frac{1}{r}\frac{\partial w}{\partial \theta}\cos\theta \tag{4.1.35}$$

4.1.2 Exact Solutions

Solving the partial differential equations in Eqs. (4.1.26) for complex geometries and boundary conditions is a formidable task. For certain boundary and loading conditions, and circular and rectangular geometries, exact solutions of Eqs. (4.1.27) or (4.1.26a) can be obtained. In this section, we consider the exact solutions of Eq. (4.1.27) for some specific plate problems for which the inplane displacements are zero. More specifically, we consider the axisymmetric bending of circular plates with both simply supported and clamped boundary conditions, and the bending of simply supported—on all edges or two parallel edges—rectangular plates subjected to sinusoidal, uniform, or point loads.

Axisymmetric Bending of Circular Plates

The deflection w of a circular plate is a function of the radial coordinate r only when the applied load q and the boundary conditions are independent of the angular coordinate θ (see Fig. 4.5). Such plates are said to undergo axisymmetric bending. For this case, the partial differential equation (4.1.31) becomes an ordinary differential equation in r. The Laplace operator ∇^2 in polar coordinates is given by

$$\nabla^2 = \frac{\partial^2}{\partial r^2} + \frac{1}{r}\frac{\partial}{\partial r} + \frac{1}{r^2}\frac{\partial^2}{\partial \theta^2} \tag{4.1.36}$$

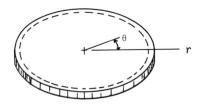

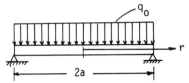

Figure 4.5 An axisymmetrically loaded circular plate.

For the axisymmetric case, we have $w = w(r)$, so that

$$\nabla^4 w = \left(\frac{d^2}{dr^2} + \frac{1}{r}\frac{d}{dr}\right)\left(\frac{d^2 w}{dr^2} + \frac{1}{r}\frac{dw}{dr}\right) \qquad (4.1.37)$$

and Eq. (4.1.31) becomes

$$\frac{1}{r}\frac{d}{dr}\left\{r\frac{d}{dr}\left[\frac{1}{r}\frac{d}{dr}\left(r\frac{dw}{dr}\right)\right]\right\} = \frac{q}{D} \qquad (4.1.38)$$

For a uniformly distributed load, $q = q_0$, the general solution of Eq. (4.1.38) is given by

$$w = \left(c_1 \ln r + c_2 r^2 \ln r + c_3 r^2 + c_4\right) + \left(\frac{q_0 r^4}{64D}\right) \qquad (4.1.39)$$

The first expression in Eq. (4.1.39) corresponds to the solution of the homogeneous equation, w_h, and the last term is the particular solution, w_p. The constants of integration, c_1 thru c_4, can be evaluated using the boundary conditions. It should be noted that the boundary in the axisymmetric case involves, analogous to a beam, only two points. We now consider two cases of boundary conditions:

(i) Simply supported: $w = 0, \qquad M_r = 0 \quad$ at $r = a$

(ii) Clamped: $w = 0, \qquad \dfrac{dw}{dr} = 0 \quad$ at $r = a$ $\qquad (4.1.40)$

If the deflections at the center of the plate are finite (which is the case for all solid plates), c_1 and c_2 must be zero.

For the simply supported boundary conditions, the constants c_1 thru c_4 in Eq. (4.1.39) are given by

$$c_1 = c_2 = 0, \qquad c_3 = -\frac{q_0 a^2}{32D}\left(\frac{3+\nu}{1+\nu}\right), \qquad c_4 = \frac{q_0 a^4}{64D}\left(\frac{5+\nu}{1+\nu}\right)$$

The plate deflection becomes

$$w = \frac{q_0 a^4}{64D}\left(\xi^4 - 2\frac{3+\nu}{1+\nu}\xi^2 + \frac{5+\nu}{1+\nu}\right), \qquad \xi = \frac{r}{a} \qquad (4.1.41)$$

The maximum deflection occurs at the center of the plate,

$$w_{max} = w(0) = \frac{q_0 a^4}{64D}\left(\frac{5+\nu}{1+\nu}\right) \qquad (4.1.42)$$

The stresses and bending moments are given by [see Eq. (4.1.33)]

$$\sigma_r = \frac{3q_0 z}{4h^3}(3+\nu)(a^2 - r^2)$$

$$\sigma_\theta = \frac{3q_0 z}{4h^3}\left[(3+\nu)a^2 - (1+3\nu)r^2\right] \qquad (4.1.43)$$

$$M_r = \frac{q_0}{16}(3+\nu)(a^2 - r^2)$$

$$M_\theta = \frac{q_0}{16}\left[(3+\nu)a^2 - (1+3\nu)r^2\right]$$

The maximum stress occurs at the center of the plate,

$$(\sigma_r)_{\text{max}} = (\sigma_\theta)_{\text{max}} = \frac{3}{8}(3+\nu)q_0\left(\frac{a}{h}\right)^2 \qquad (4.1.44)$$

For the clamped boundary conditions, the constants of integration are given by

$$c_1 = c_2 = 0, \qquad c_3 = -\frac{q_0 a^2}{32 D}, \qquad c_4 = \frac{q_0 a^4}{64 D}$$

and the deflection becomes

$$w = \frac{q_0}{64 D}(a^2 - r^2)^2 \qquad (4.1.45)$$

The maximum displacement occurs at the center of the plate

$$w_{\text{max}} = \frac{q_0 a^4}{64 D} \qquad (4.1.46)$$

which is about one-fourth of the deflection of the simply supported plate (for $\nu = \frac{1}{3}$). The expressions for the stresses and bending moments are given by

$$\sigma_r = \frac{3q_0 z}{4h^3}\left[(1+\nu)a^2 - (3+\nu)r^2\right]$$

$$\sigma_\theta = \frac{3q_0 z}{4h^3}\left[(1+\nu)a^2 - (1+3\nu)r^2\right]$$

$$M_r = \frac{q_0}{16}\left[(1+\nu)a^2 - (3+\nu)r^2\right]$$

$$M_\theta = \frac{q_0}{16}\left[(1+\nu)a^2 - (1+3\nu)r^2\right] \qquad (4.1.47)$$

The extreme value of stress occurs at the edge, for $\nu = \frac{1}{3}$,

$$(\sigma_r)_{\max} = \frac{6}{h^2}(M_r)_{\max} = -\frac{3q_0}{4}\left(\frac{a}{h}\right)^2 \qquad (4.1.48)$$

Note that the extreme value of stress in the clamped plate is only 60% of that in the simply supported plate (for $\nu = \frac{1}{3}$).

Example 4.2

Consider the axisymmetric bending of a simply supported circular plate of isotropic material, subjected to linearly varying axisymmetric loading that varies from zero at the center to q_0 at the outer edge. We have

$$q(r) = \frac{r}{a}q_0$$

For this case, the particular solution of Eq. (4.1.38) is given by

$$w_p = -\frac{1}{D}\int\frac{dr}{r}\int M(r)r\,dr$$

$$M(r) = -\int\frac{dr}{r}\int q(r)r\,dr \qquad (4.1.49)$$

We have

$$M(r) = -\frac{q_0 r^3}{9a}, \qquad w_p = \frac{q_0 r^5}{225Da} \qquad (4.1.50)$$

and the general solution for the problem becomes,

$$w = w_h + w_p = c_1\ln r + c_2 r^2\ln r + c_3 r^2 + c_4 + \frac{q_0 r^5}{225Da} \qquad (4.1.51)$$

For solid circular plates, $c_1 = c_2 = 0$, and the remaining two constants are determined from

$$w(a) = 0, \qquad M_r(a) = 0$$

We have

$$c_3 = -\frac{q_0 a^2}{90D}\left(\frac{4+\nu}{1+\nu}\right), \qquad c_4 = \frac{q_0 a^4}{150D}\left(\frac{6+\nu}{1+\nu}\right) \qquad (4.1.52)$$

Equation (4.1.51) defines the solution with $c_1 = c_2 = 0$ and c_3 and c_4 given by Eq. (4.1.52).

Axisymmetric bending of circular plates with point loads and annular plates with various types of loadings and boundary conditions can be analyzed using the methods presented here. Some of the problems that admit exact solutions are included in the exercises. We return to circular plate problems in the sections on approximate methods.

Navier's Solution for Simply Supported Rectangular Plates

Consider a rectangular plate of planeform dimensions a and b, which is simply supported on all edges and subjected to a transverse distributed load $q(x, y)$, as shown in Fig. 4.6. The origin of the coordinate system is taken at the lower left corner of the plate. The boundary conditions are

$$w = 0, \qquad M_x = 0 \quad \text{at } x = 0 \quad \text{and} \quad x = a$$

$$w = 0, \qquad M_y = 0 \quad \text{at } y = 0 \quad \text{and} \quad y = b \qquad (4.1.53)$$

Navier's solution, sometimes called the *forced* solution of the differential equation, consists of seeking the solution of Eq. (4.1.26b) by a double trigonometric series that satisfies the boundary conditions in Eq. (4.1.53). The method can be summarized by the following three steps:

1. Express the deflection in double Fourier sine series,

$$w(x, y) = \sum_{m=1}^{\infty} \sum_{n=1}^{\infty} W_{mn} \sin \alpha x \sin \beta y \qquad (4.1.54)$$

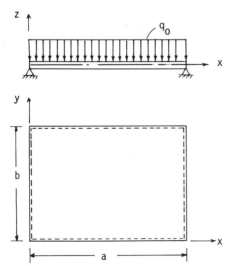

Figure 4.6 A uniformly loaded, simply supported plate.

where $\alpha = m\pi/a$ and $\beta = n\pi/b$, and W_{mn} are parameters to be determined.

2. Expand the transverse load into double sine series,

$$q(x, y) = \sum_{m=1}^{\infty} \sum_{n=1}^{\infty} q_{mn} \sin \alpha x \sin \beta y \qquad (4.1.55)$$

where q_{mn} are given by

$$q_{mn} = \frac{4}{ab} \int_0^a \int_0^b q \sin \alpha x \sin \beta y \, dx \, dy \qquad (4.1.56)$$

3. Substitute Eqs. (4.1.54) and (4.1.55) into Eq. (4.1.26b) and solve for W_{mn} in terms of q_{mn}:

$$\sum_{m=1}^{\infty} \sum_{n=1}^{\infty} W_{mn} \left(D_{11}\alpha^4 + 2H\alpha^2\beta^2 + D_{22}\beta^4 \right) \sin \alpha x \sin \beta y$$

$$= \sum_{m=1}^{\infty} \sum_{n=1}^{\infty} q_{mn} \sin \alpha x \sin \beta y$$

or, for fixed values of m and n, we have

$$W_{mn} = \frac{q_{mn}}{D_{11}\alpha^4 + 2H\alpha^2\beta^2 + D_{22}\beta^4}, \qquad H = D_{12} + 2D_{66} \qquad (4.1.57)$$

The solution w is given by Eq. (4.1.54) with the known coefficients W_{mn}. The stresses are given by

$$\sigma_x = -\left(c_{11}\frac{\partial^2 w}{\partial x^2} + c_{12}\frac{\partial^2 w}{\partial y^2} \right) z$$

$$\sigma_y = -\left(c_{12}\frac{\partial^2 w}{\partial x^2} + c_{22}\frac{\partial^2 w}{\partial y^2} \right) z$$

$$\sigma_{xy} = -2c_{66}\frac{\partial^2 w}{\partial x \partial y} z \qquad (4.1.58)$$

and the bending moments are given by Eq. (4.1.25b). For isotropic plates, Eq. (4.1.57) becomes

$$W_{mn} = \frac{q_{mn}}{D(\alpha^2 + \beta^2)^2} \qquad (4.1.59)$$

We now compute the coefficients q_{mn} and W_{mn} for various loadings.

Sinusoidal Loading.   For $q = q_0\sin(\pi x/a)\sin(\pi y/b)$, Eq. (4.1.56) gives q_{11} $= q_0$ and $q_{mn} = 0$ for all m, $n \neq 1$. We have

$$w(x, y) = \frac{q_0 a^4 b^4}{\pi^4 \left(D_{11} b^4 + 2Ha^2 b^2 + D_{22} a^4 \right)} \sin\frac{\pi x}{a} \sin\frac{\pi y}{b} \quad (4.1.60)$$

Uniform Loading on the Entire Plate.   For $q = q_0$, Eq. (4.1.56) gives

$$q_{mn} = \begin{cases} \dfrac{16 q_0}{\pi^2 mn}, & m, n = 1, 3, 5, \dots \\ 0, & m, n = 2, 4, 6, \dots \end{cases} \quad (4.1.61a)$$

and

$$w(x, y) = \frac{16 q_0 a^4 b^4}{\pi^6} \sum_{m,n=1,3,\dots}^{\infty} \frac{1}{mn D_{mn}} \sin\frac{m\pi x}{a} \sin\frac{n\pi y}{b} \quad (4.1.61b)$$

where

$$D_{mn} = \left(D_{11} m^4 b^4 + 2Hm^2 n^2 a^2 b^2 + D_{22} n^4 a^4 \right) \quad (4.1.61c)$$

Concentrated Load P_0 at Point (x_0, y_0).   For this case, Eq. (4.1.56) gives

$$q_{mn} = \begin{cases} \dfrac{4P_0}{ab} \sin\dfrac{m\pi x_0}{a} \sin\dfrac{n\pi y_0}{b}, & m, n \text{ odd} \\ 0, & m, n \text{ even} \end{cases} \quad (4.1.62a)$$

and the deflection becomes

$$w = \frac{4P_0 a^3 b^3}{\pi^4} \sum_{m,n=1,3,\dots}^{\infty} \frac{1}{D_{mn}} \sin\frac{m\pi x_0}{a} \sin\frac{n\pi y_0}{b} \sin\frac{m\pi x}{a} \sin\frac{n\pi y}{b}$$

$$(4.1.62b)$$

For $x_0 = a/2$ and $y_0 = b/2$, Eq. (4.1.62b) simplifies to

$$w = \frac{4P_0 a^3 b^3}{\pi^4} \sum_{m,n=1,3,\dots}^{\infty} \frac{(-1)^{(m+n)/2-1}}{D_{mn}} \sin\frac{m\pi x}{a} \sin\frac{n\pi y}{b} \quad (4.1.63)$$

Equations (4.1.54)–(4.1.57) can be used to compute the solution for a simply supported plate under any load that can be represented by a double trigonometric sine series. An example is given below.

Example 4.3

Suppose that a simply supported rectangular plate is subjected to a hydrostatic pressure that varies linearly along the x-axis, as shown in Fig. 4.7. We wish to determine the deflection and maximum bending moment.

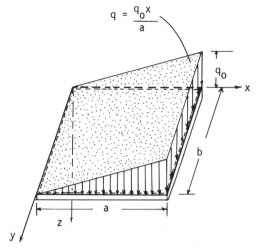

Figure 4.7 A hydrostatically loaded, simply supported rectangular plate.

The distributed load is given by

$$q(x, y) = q_0 \frac{x}{a}$$

and the coefficients q_{mn} of the double trigonometric series are given by

$$q_{mn} = \frac{4}{ab} \int_0^a \int_0^b q_0 \frac{x}{a} \sin \frac{m\pi x}{a} \sin \frac{n\pi y}{b} \, dx \, dy$$

$$= \frac{4q_0}{a^2 b} \frac{b}{n\pi} \left(-\cos \frac{n\pi y}{b} \right) \Big|_0^b \left[-\frac{a}{m\pi} x \cos \frac{m\pi x}{a} + \left(\frac{a}{m\pi} \right)^2 \sin \frac{m\pi x}{a} \right]_0^a$$

$$= \frac{4q_0}{a^2 b} \frac{b}{n\pi} [1 - \cos n\pi] \left[-\frac{a^2}{m\pi} \cos m\pi \right]$$

$$= -\frac{8q_0 \cos m\pi}{mn\pi^2} \quad \text{for all } m, \quad \text{and} \quad n \text{ odd} \qquad (4.1.64)$$

The solution becomes

$$w = \frac{8q_0 a^4 b^4}{\pi^6 D} \sum_{m,n=1,3,\dots}^{\infty} \frac{(-1)^{m+1} \sin \frac{m\pi x}{a} \sin \frac{n\pi y}{b}}{mn(m^2 b^2 + n^2 a^2)^2} \qquad (4.1.65)$$

The bending moment M_x is given by

$$M_x = \frac{8q_0 a^2 b^2}{\pi^4} \sum_{m,n=1,3,\dots}^{\infty} (-1)^{m+1} \frac{m^2 b^2 + \nu n^2 a^2}{mn(m^2 b^2 + n^2 a^2)^2} \sin \frac{m\pi x}{a} \sin \frac{n\pi y}{b}$$

$$(4.1.66)$$

The maximum deflection and bending moment for the square plate ($b = a$) is given by

$$w\left(\frac{a}{2}, \frac{a}{2}\right) = \frac{8q_0 a^4}{\pi^6 D} \sum_{m,n=1,3,\ldots}^{\infty} \frac{(-1)^{(m+n)/2-1}}{mn(m^2 + n^2)^2} \approx 0.00203 \frac{q_0 a^4}{D}$$

$$M_x\left(\frac{a}{2}, \frac{a}{2}\right) = \frac{8q_0 a^2}{\pi^4} \sum_{m,n=1,3,\ldots}^{\infty} \frac{(-1)^{(m+n)/2-1}(m^2 + \nu n^2)}{mn(m^2 + n^2)^2} \approx 0.0235 q_0 a^2$$

$$(4.1.67)$$

The Navier method is limited to simply supported (on all four edges), rectangular plates, because both boundary conditions and loading can be represented by double trigonometric series. Next, we consider Levy's method for rectangular plates, which uses a single Fourier series (in one coordinate direction) to reduce the governing partial differential equation to an ordinary differential equation with respect to the other coordinate. The ordinary differential equation can then be solved exactly. The method resembles the Kantorovich method discussed in Section 3.3.4.

Levy's Solution for Rectangular Plates

The method of Levy (1899) can be used to solve rectangular plate problems for which the boundary conditions along two opposite edges are such that they can be satisfied by trigonometric functions. The other two edges can each have arbitrary boundary conditions (see Fig. 4.8). Thus, Levy's method is restricted to rectangular plates with any two opposite edges simply supported. The

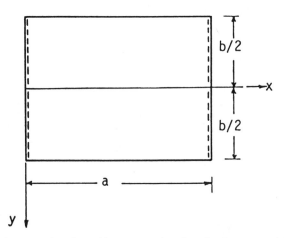

Figure 4.8 A rectangular plate with two opposite edges simply supported ($x = 0, a$).

method seeks the solution of Eq. (4.1.27) as a sum of the homogeneous solution w_h and the particular solution w_p,

$$w = w_h + w_p \tag{4.1.68}$$

The homogeneous solution is selected in the general form

$$w_h = \sum_{m=1}^{\infty} W_m(y)\sin\frac{m\pi x}{a} \tag{4.1.69}$$

where the function W_m must be determined such that it satisfies both the homogeneous form of Eq. (4.1.27),

$$\nabla^4 w = 0 \tag{4.1.70}$$

and the boundary conditions on edges $y = \pm b/2$ (see Fig. 4.8). Substituting Eq. (4.1.69) into Eq. (4.1.70), we obtain

$$\sum_{m=1}^{\infty}\left[\frac{d^4W_m}{dy^4} - 2\left(\frac{m\pi}{a}\right)^2\frac{d^2W_m}{dy^2} +\left(\frac{m\pi}{a}\right)^4 W_m\right]\sin\frac{m\pi x}{a} = 0$$

which implies, because $\sin(m\pi x/a)$ are a linearly independent set of functions, that

$$\frac{d^4W_m}{dy^4} - 2\left(\frac{m\pi}{a}\right)^2\frac{d^2W_m}{dy^2} +\left(\frac{m\pi}{a}\right)^4 W_m = 0 \quad \text{for all } m = 1,2,\ldots$$

$$\tag{4.1.71}$$

The general solution of Eq. (4.1.71) is given by

$$W_m = A_m\sinh\frac{m\pi y}{a} + B_m\cosh\frac{m\pi y}{a} + C_m y \sinh\frac{m\pi y}{a}$$

$$+ D_m y\cosh\frac{m\pi y}{a} \tag{4.1.72}$$

where A_m, B_m, C_m, and D_m are constants to be determined using the boundary conditions on edges $y = \pm b/2$.

The particular solution can be expressed in the form,

$$w_p = \sum_{m=1}^{\infty} Y_m(y)\sin\frac{m\pi x}{a} \tag{4.1.73}$$

where Y_m are functions to be determined. The load $q(x, y)$ should be such that we can expand it in the same form as the particular solution,

$$q(x, y) = \sum_{m=1}^{\infty} q_m(y)\sin\frac{m\pi x}{a} \qquad (4.1.74)$$

where

$$q_m(y) = \frac{2}{a}\int_0^a q(x, y)\sin\frac{m\pi x}{a}\, dx \qquad (4.1.75)$$

Substituting Eqs. (4.1.73) and (4.1.74) into Eq. (4.1.27), we obtain

$$\frac{d^4 Y_m}{dy^4} - 2\left(\frac{m\pi}{a}\right)^2\frac{d^2 Y_m}{dy^2} + \left(\frac{m\pi}{a}\right)^4 Y_m = \frac{q_m}{D} \qquad (4.1.76)$$

The solution of Eq. (4.1.71) and the particular solution of Eq. (4.1.76) define, respectively, the homogeneous solution in Eq. (4.1.69) and the particular solution in Eq. (4.1.73) and, therefore, we obtain the total solution from Eq. (4.1.68). We illustrate the method via an example.

Example 4.4

Consider a rectangular plate, clamped at $y = 0$, simply supported on edges $x = 0$, $x = a$, and $y = b$ (see Fig. 4.9), and subjected to uniformly distributed load q_0.

Because of symmetry about the line $x = a/2$, the expressions in Eqs. (4.1.69) and (4.1.73) can be summed only for odd integers of m. The homogeneous solution is given by Eq.(4.1.72). To determine the particular solution, we

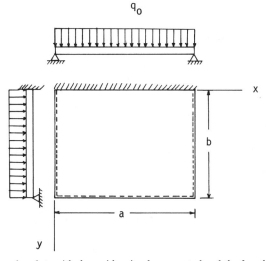

Figure 4.9 A rectangular plate with three sides simply supported and the fourth ($y = 0$) clamped.

must solve Eq. (4.1.76) with q_m given from Eq. (4.1.75),

$$q_m = \frac{4q_0}{m\pi}, \qquad m \text{ odd}$$

The particular solution of Eq. (4.1.76) is given by

$$Y_m = \frac{4q_0 a^4}{m^5 \pi^5 D}$$

so that Eq. (4.1.73) becomes

$$w_p = \frac{4q_0 a^4}{\pi^5 D} \sum_{m=1}^{\infty} \frac{1}{m^5} \sin\frac{m\pi x}{a}, \qquad m \text{ odd} \qquad (4.1.77)$$

The general solution is given by

$$w(x, y) = \sum_{m=1,3,\ldots}^{\infty} \left(A_m\sinh\frac{m\pi y}{a} + B_m\cosh\frac{m\pi y}{a} + C_m y \sinh\frac{m\pi y}{a} \right.$$

$$\left. + D_m y\cosh\frac{m\pi y}{a} + \frac{4q_0 a^4}{m^5\pi^5 D} \right)\sin\frac{m\pi x}{a} \qquad (4.1.78)$$

The boundary conditions of the plate are given by

$$w = 0, \qquad \frac{\partial w}{\partial y} = 0 \quad \text{at } y = 0$$

$$w = 0, \qquad \frac{\partial^2 w}{\partial y^2} = 0 \quad \text{at } y = b \qquad (4.1.79)$$

Use of these boundary conditions in Eq. (4.1.78) leads to the following expressions for A_m, B_m, C_m, and D_m

$$A_m = \frac{2q_0 a^4}{m^5\pi^5 D}\left(\frac{2\cosh^2 k_m - 2\cosh k_m - k_m\sinh k_m}{\cosh k_m\sinh k_m - k_m} \right)$$

$$B_m = -\frac{4q_0 a^4}{m^5\pi^5 D}$$

$$C_m = \frac{2q_0 a^4}{m^5\pi^5 D}\left(\frac{\dfrac{2m\pi}{a}\sinh k_m\cosh k_m - \dfrac{m\pi}{a}\sinh k_m - \left(\dfrac{m\pi}{a}\right)^2 b\cosh k_m}{\cosh k_m\sinh k_m - k_m} \right)$$

$$D_m = -\frac{m\pi}{a}A_m, \qquad k_m = \frac{m\pi b}{a} \qquad (4.1.80)$$

Now the solution of the problem is given by Eq. (4.1.78) with the constants A_m, B_m, C_m, and D_m defined by Eq. (4.1.80).

For a square plate, $a = b$, the deflection at $x = y = a/2$ and the bending moment at $x = a/2$ and $y = 0$ are given by

$$w\left(\frac{a}{2},\frac{a}{2}\right) = 0.0028\frac{q_0 a^4}{D}$$

$$M_y\left(\frac{a}{2},0\right) = 0.08 q_0 a^2 = M_{max}$$

This completes the example.

4.1.3 Variational Solutions

In this section, we utilize the traditional variational methods of Sections 3.2 and 3.3 to determine approximate solutions of certain plate problems that either do not admit exact solutions or the exact solutions are algebraically too complicated to obtain. From the discussions of Sections 3.2 and 3.3, we know that the variational methods reduce the governing partial differential equations to algebraic equations for the parameters c_i in the series approximation

$$w(x, y) \approx w_n = \sum_{i=1}^{n} c_i \phi_i(x, y) + \phi_0(x, y) \tag{4.1.81}$$

where ϕ_i and ϕ_0 are suitably chosen functions. Although the variational solution w_n does not satisfy Eq. (4.1.27), it represents a best possible approximation, in a variational sense, of the deflection surface. The key to obtaining a good approximation of the actual deflection is the choice of the coordinate functions. Trigonometric functions and beam-deflection formulas provide suitable coordinate functions for plates. In the following examples, we consider approximate solutions of several plate problems.

Example 4.5 (Circular Plates on Elastic Foundation)

Here we apply the Ritz method to determine an approximate solution of an orthotropic circular plate (of radius a) resting freely (i.e., without edge constraints) on an elastic foundation (of modulus k) and subjected to a point load P_0 at the center of the plate.

The total potential energy functional for the axisymmetric problem is given by

$$\Pi(w) = \pi \int_0^a \left[D_{11}\left(\frac{d^2 w}{dr^2}\right)^2 + 2D_{12}\frac{1}{r}\frac{dw}{dr}\frac{d^2 w}{dr^2} + D_{22}\left(\frac{1}{r}\frac{dw}{dr}\right)^2 + kw^2 \right] r\,dr$$
$$- P_0 w(0) \tag{4.1.82}$$

where D_{11}, D_{12}, and D_{22} are the bending stiffnesses of the orthotropic plate. We seek the Ritz solution of the form,

$$w_n = c_1\phi_1 + c_2\phi_2 + \cdots + c_n\phi_n \qquad (4.1.83)$$

where ϕ_i $(i = 1, 2, \ldots, n)$ are linearly independent functions, that are *not* subjected to any boundary conditions, because there are no specified essential boundary conditions. We select ϕ_i to be

$$\phi_1 = 1, \qquad \phi_2 = r^2, \qquad \phi_3 = r^3, \ldots, \qquad \phi_n = r^n \qquad (4.1.84)$$

Note that the choice $\phi_1 = 1$ is admissible (and necessary) because it contributes to the strain energy stored in the elastic foundation and to the potential of the applied load P_0. We omitted $\phi_2 = r$, because it contributes to the coefficient $1/r$ that, when integrated, results in an indeterminate value. Substituting Eq. (4.1.83) into Eq. (4.1.82), we get

$$\Pi(c_1, c_2, \ldots, c_n) = \pi \int_0^a \left[D_{11}\left(\sum_{i=1}^n c_i \frac{d^2\phi_i}{dr^2} \right)^2 + 2D_{12}\frac{1}{r}\left(\sum_{i=1}^n c_i \frac{d\phi_i}{dr} \right)\left(\sum_{j=1}^n c_j \frac{d^2\phi_j}{dr^2} \right) \right.$$

$$\left. + D_{22}\left(\frac{1}{r}\sum_{i=1}^n c_i \frac{d\phi_i}{dr} \right)^2 + k\left(\sum_{i=1}^n c_i\phi_i \right)^2 \right] r\, dr - P_0 c_1$$

Using the total potential energy principle $\delta\Pi = 0$, we obtain

$$0 = \frac{\partial\Pi}{\partial c_m} = 2\pi \int_0^a \left\{ D_{11}\frac{d^2\phi_m}{dr^2}\left(\sum_{i=1}^n c_i \frac{d^2\phi_i}{dr^2} \right) \right.$$

$$+ \frac{1}{r}D_{12}\left[\frac{d\phi_m}{dr}\left(\sum_{i=1}^n c_i \frac{d^2\phi_i}{dr^2} \right) + \frac{d^2\phi_m}{dr^2}\left(\sum_{i=1}^n c_i \frac{d\phi_i}{dr} \right) \right]$$

$$+ D_{22}\frac{1}{r^2}\frac{d\phi_m}{dr}\left(\sum_{i=1}^n c_i \frac{d\phi_i}{dr} \right)$$

$$\left. + k\phi_m\left(\sum_{i=1}^n c_i\phi_i \right) \right\} r\, dr - P_0 \delta_{1m}$$

for $m = 1, 2, \ldots, n$, and $n \geq 2$. In matrix form, we get

$$\sum_{i=1}^n A_{mi}c_i = P_0\delta_{1m} \qquad (4.1.85a)$$

where

$$A_{mi} = 2\pi \int_0^a \left[D_{11} \frac{d^2\phi_m}{dr^2} \frac{d^2\phi_i}{dr^2} + D_{12} \frac{1}{r}\left(\frac{d\phi_m}{dr} \frac{d^2\phi_i}{dr^2} + \frac{d^2\phi_m}{dr^2} \frac{d\phi_i}{dr} \right) \right.$$

$$\left. + D_{22} \frac{1}{r^2} \frac{d\phi_m}{dr} \frac{d\phi_i}{dr} + k\phi_m\phi_i \right] r \, dr \qquad (4.1.85b)$$

For $n = 2$, we have

$$A_{11} = 2\pi \int_0^a (0 + kr)\, dr = k\pi a^2$$

$$A_{12} = A_{21} = 2\pi \int_0^a (0 + kr^3)\, dr = \tfrac{1}{2}k\pi a^4$$

$$A_{22} = 2\pi \int_0^a \left[D_{11}4r + D_{12}(4r + 4r) + D_{22}4r + kr^5 \right] dr$$

$$= 2\pi \left[2(D_{11} + 2D_{12} + D_{22})a^2 + \frac{k}{6}a^6 \right]$$

and

$$\frac{\pi a^2}{2} \left[\begin{array}{c|c} 2k & ka^2 \\ \hline ka^2 & 8(D_{11} + 2D_{12} + D_{22}) + \tfrac{2}{3}ka^4 \end{array} \right] \left\{ \begin{array}{c} c_1 \\ c_2 \end{array} \right\} = \left\{ \begin{array}{c} P_0 \\ 0 \end{array} \right\}$$

The solution is given by

$$c_1 = \frac{P_0}{\pi a^2 k}\left[1 + \left(\frac{16\overline{D}}{ka^4} + \frac{1}{3} \right)^{-1} \right], \qquad c_2 = -\frac{P_0}{\pi ka^4}\left(\frac{1}{6} + \frac{8\overline{D}}{ka^4} \right)^{-1}$$

$$(4.1.86)$$

where $\overline{D} = D_{11} + 2D_{12} + D_{22}$. The two-parameter Ritz solution is given by

$$w_2(r) = c_1 + c_2 r^2 \qquad (4.1.87)$$

with c_1 and c_2 given by Eq. (4.1.86). The maximum deflection is given by $w(0) = c_1$.

The next example is concerned with the determination of the exact center deflection of the asymmetrical bending of a clamped circular plate by Betti's reciprocity theorem (see Section 2.6.4). The exact solution for any r and θ can be obtained by the methods presented in Section 4.1.2 (see Exercise 4.1.7).

Example 4.6

Consider an isotropic clamped circular plate of radius a and subjected to a linearly varying load of the form (see Fig. 4.10)

$$q(r,\theta) = q_0 + q_1 \frac{r}{a}\cos\theta \qquad (4.1.88)$$

According to the reciprocity theorem of Betti, Eq. (2.6.22), we have

$$P_0 w_c = \int_0^{2\pi}\int_0^a q(r,\theta)w(r)r\,dr\,d\theta \qquad (4.1.89)$$

where P_0 is the force at the center of the plate, $w(r)$ is the deflection at point r (for any θ) caused by P_0, $q(r,\theta)$ is the asymmetric load in Eq. (4.1.88), and w_c is the deflection at the center resulting from $q(r,\theta)$. Thus, if we know $w(r)$ caused by a point load at the center, then we can determine w_c from Eq. (4.1.89)

$$w_c = \frac{1}{P_0}\int_0^{2\pi}\int_0^a q(r,\theta)w(r)r\,dr\,d\theta \qquad (4.1.90)$$

From Exercise 4.1.5, the deflection at r owing to a point load P_0 at the center of a clamped circular plate is given by

$$w(r) = \frac{P_0}{16\pi D}\left(2r^2\ln\frac{r}{a} + a^2 - r^2\right) \qquad (4.1.91)$$

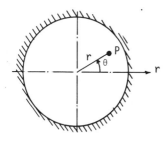

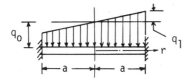

Figure 4.10 An asymmetrically loaded circular plate.

Substituting Eqs. (4.1.88) and (4.1.91) into Eq. (4.1.90), and performing the integration, we obtain

$$w_c = \frac{1}{16\pi D} \int_0^{2\pi} \int_0^a \left(q_0 + q_1 \frac{r}{a} \cos\theta \right) \left(2r^2 \ln\frac{r}{a} + a^2 - r^2 \right) r \, dr \, d\theta$$

$$= \frac{q_0 a^4}{64D}$$

The exact solution of the problem at any point (r, θ) is given by

$$w(r, \theta) = \frac{q_0}{64D}(a^2 - r^2)^2 + \frac{q_1}{192D}\frac{r}{a}(a^2 - r^2)^2 \cos\theta \quad (4.1.92)$$

We now turn to rectangular plates. The first example of the variational solution of rectangular plates is concerned with the bending of a clamped rectangular plate under uniformly distributed transverse load or a point load at the center.

Example 4.7

Consider a clamped orthotropic rectangular plate of dimensions a and b, which is subjected to uniformly distributed transverse load q. The total potential energy of the plate is given by

$$\Pi(w) = \int_0^a \int_0^b \left\{ \frac{1}{2}\left[D_{11}\left(\frac{\partial^2 w}{\partial x^2}\right)^2 + 2D_{12}\frac{\partial^2 w}{\partial y^2}\frac{\partial^2 w}{\partial x^2} + D_{22}\left(\frac{\partial^2 w}{\partial y^2}\right)^2 \right. \right.$$

$$\left. \left. + 4D_{66}\left(\frac{\partial^2 w}{\partial x \partial y}\right)^2 \right] - qw \right\} dx \, dy \quad (4.1.93)$$

where w and its first derivatives are subjected to the following boundary conditions

$$w = \frac{\partial w}{\partial x} = 0 \quad \text{at } x = 0, a$$

$$w = \frac{\partial w}{\partial y} = 0 \quad \text{at } y = 0, b \quad (4.1.94)$$

Since all of the boundary conditions are of the essential type, the Ritz and Galerkin methods require the same approximation functions. We seek a Ritz approximation of the form,

$$w(x, y) \approx w_{mn} = \sum_{i=1}^m \sum_{j=1}^n c_{ij}\left(1 - \cos\frac{2i\pi x}{a}\right)\left(1 - \cos\frac{2j\pi y}{b}\right) \quad (4.1.95)$$

where $\phi_{ij} = [1 - \cos(2i\pi x/a)][1 - \cos(2j\pi y/b)]$ satisfy the boundary conditions in Eq. (4.1.94). Introducing Eq. (4.1.95) into Eq. (4.1.93), and differentiating the result with respect to c_{rs}, we obtain

$$0 = \frac{\partial \Pi}{\partial c_{rs}} = \sum_{i=1}^{m} \sum_{j=1}^{n} c_{ij} \int_0^a \int_0^b \left\{ D_{11} \left(\frac{2i\pi}{a} \right)^4 \cos\frac{2i\pi x}{a} \cos\frac{2r\pi x}{a} \left(1 - \cos\frac{2j\pi y}{b} \right) \right.$$

$$\times \left(1 - \cos\frac{2s\pi y}{b} \right) + D_{12} \left(\frac{2j\pi}{b} \right)^2 \left(\frac{2i\pi}{a} \right)^2$$

$$\times \left[\left(1 - \cos\frac{2i\pi x}{a} \right) \cos\frac{2r\pi x}{a} \cos\frac{2j\pi y}{b} \left(1 - \cos\frac{2s\pi y}{b} \right) \right.$$

$$\left. + \cos\frac{2i\pi x}{a} \left(1 - \cos\frac{2r\pi x}{a} \right) \left(1 - \cos\frac{2j\pi y}{b} \right) \cos\frac{2s\pi y}{b} \right]$$

$$+ D_{22} \left(\frac{2j\pi}{b} \right)^4 \left(1 - \cos\frac{2i\pi x}{a} \right) \cos\frac{2j\pi y}{b} \left(1 - \cos\frac{2r\pi x}{a} \right) \cos\frac{2s\pi y}{b}$$

$$+ 4D_{66} \left(\frac{2i\pi}{a} \right)^2 \left(\frac{2j\pi}{b} \right)^2 \sin\frac{2i\pi x}{a} \sin\frac{2j\pi y}{b} \sin\frac{2r\pi x}{a} \sin\frac{2s\pi y}{b} \right\} dx\, dy$$

$$- \int_0^a \int_0^b q_0 \left(1 - \cos\frac{2r\pi x}{a} \right) \left(1 - \cos\frac{2s\pi y}{b} \right) dx\, dy \qquad (4.1.96)$$

Using the identities,

$$\int_0^a \sin\frac{2i\pi x}{a} \sin\frac{2j\pi x}{a}\, dx = \begin{cases} 0, & i \neq j \\ \frac{a}{2}, & i = j \end{cases}$$

$$\int_0^a \cos\frac{2i\pi x}{a} \sin\frac{2j\pi x}{a}\, dx = \begin{cases} 0, & i \neq j \\ \frac{a}{2}, & i = j \end{cases}$$

$$\int_0^a \left(1 - \cos\frac{2i\pi x}{a} \right) \left(1 - \cos\frac{2j\pi x}{a} \right) dx = \begin{cases} a, & i \neq j \\ \frac{3a}{2}, & i = j \end{cases} \qquad (4.1.97)$$

we can simplify the expression in Eq. (4.1.96),

$$
4ab\pi^4\Bigg\{ \sum_{r=1}^{m}\sum_{s=1}^{n} c_{rs}\bigg[3D_{11}\Big(\frac{r}{a}\Big)^4 + 2D_{12}\Big(\frac{r}{a}\Big)^2\Big(\frac{s}{b}\Big)^2
$$

$$
+ 3D_{22}\Big(\frac{s}{b}\Big)^4 + 4D_{66}\Big(\frac{r}{a}\Big)^2\Big(\frac{s}{b}\Big)^2\bigg]
$$

$$
+ \sum_{r=1}^{m}\sum_{j=1}^{n} 2D_{11}\Big(\frac{r}{a}\Big)^4 c_{rj} + \sum_{s=1}^{n}\sum_{i=1}^{m} 2D_{22}\Big(\frac{s}{b}\Big)^4 c_{si}\Bigg\} - q_0 ab = 0,
$$

$$
(r \neq j \quad \text{and} \quad s \neq i) \quad (4.1.98)
$$

For a one-parameter approximation, we obtain

$$
4ab\pi^4 c_{11}\left(\frac{3D_{11}}{a^4} + \frac{2D_{12}}{a^2 b^2} + \frac{3D_{22}}{b^4} + \frac{4D_{66}}{a^2 b^2}\right) = q_0 ab
$$

$$
c_{11} = \frac{q_0 a^4}{4\pi^4}\left(3D_{11} + 2D_{12}\alpha^2 + 3D_{22}\alpha^4 + 4D_{66}\alpha^2\right)^{-1} \quad (4.1.99)
$$

where $\alpha = a/b$. For $a = b$, we have

$$
c_{11} = \frac{q_0 a^4}{4\pi^4}\left[3D_{11} + 2(D_{12} + 2D_{66}) + 3D_{22}\right]^{-1} \quad (4.1.100)
$$

For $D_{11} = D_0$, $D_{22} = D_0/2$, and $D_{12} + 2D_{66} = 1.248 D_0$, we obtain

$$
c_{11} = \frac{q_0 a^4}{27.984\pi^4 D_0} \quad (4.1.101)
$$

For isotropic plates, we get $c_{11} = q_0 a^4/32\pi^4 D$. The solution is given by

$$
w_{11} = c_{11}\left(1 - \cos\frac{2\pi x}{a}\right)\left(1 - \cos\frac{2\pi y}{b}\right) \quad (4.1.102)
$$

The maximum deflections of orthotropic and isotropic square plates are given, respectively, by

$$
w_{\max} = w_{11}\left(\frac{a}{2}, \frac{b}{2}\right) = \frac{q_0 a^4}{6.996\pi^4 D_0} = 0.0014674\frac{q_0 a^4}{D_0}
$$

$$
w_{\max} = \frac{q_0 a^4}{8\pi^4 D} = 0.00128\frac{q_0 a^4}{D} \quad (4.1.103)
$$

For a point load P_0 at the center, Eqs. (4.1.99)–(4.1.103) are valid when q_0 is replaced by

$$q_0 = \frac{4P_0}{ab} \qquad (4.1.104)$$

because the work done by the point load P_0 is given by $P_0 w(a/2, b/2)$, which is equal to $4P_0$ for $m = n = 1$. The maximum deflection of a clamped square plate subjected to a point load P_0 at the center is given by

$$w_{max} = \frac{P_0 a^2}{2\pi^4 D} = 0.005133 \frac{P_0 a^2}{D} \qquad (4.1.105)$$

In the next example, we reconsider a clamped plate problem and seek its approximate solution by the Kantorovich method. One should note the similarity between the method of Levy and the method of Kantorovich.

Example 4.8

Consider an isotropic, clamped rectangular plate under uniformly distributed load q_0. We seek a one-parameter Kantorovich solution of the form

$$w_1 = c_1(x)\phi(y) \qquad (4.1.106)$$

For convenience, the origin of the coordinate system is taken at the center of the plate. We select the coordinate function ϕ_1 to be equal to $\phi_1 = (y^2 - b^2)^2$, which satisfies the boundary conditions $w = \partial w/\partial y = 0$ at $y = \pm b$. The function $c_1(x)$ is required to satisfy the boundary conditions

$$c_1 = \frac{dc_1}{dx} = 0 \quad \text{at } x = \mp a \qquad (4.1.107)$$

Note that the plate is of dimensions $2a \times 2b$. The Kantorovich integral from Eq. (3.3.77a) becomes

$$\int_{-b}^{b} \left(\nabla^4 w_1 - \frac{q_0}{D} \right) \phi_1 \, dy = 0$$

$$\int_{-b}^{b} \left[24c_1 + 2(12y^2 - 4b^2)\frac{d^2 c_1}{dx^2} + (y^2 - b^2)^2 \frac{d^4 c_1}{dx^4} - \frac{q_0}{D} \right] (y^2 - b^2)^2 \, dy = 0$$

$$(4.1.108)$$

and, after performing the integration, we get

$$\frac{256b^9}{315} \frac{d^4 c_1}{dx^4} - \frac{512b^7}{105} \frac{d^2 c_1}{dx^2} + \frac{384}{15} b^5 c_1 = \frac{16}{15} b^5 \frac{q_0}{D}$$

or

$$\frac{d^4 c_1}{d\xi^4} - 6\frac{d^2 c_1}{d\xi^2} + \frac{63}{2}c_1 = \frac{21}{16}\frac{q_0}{D} \qquad (4.1.109)$$

where $\xi = x/b$. The homogeneous solution of Eq. (4.1.109) can be calculated using the roots of the characteristic equation associated with Eq. (4.1.109).

$$d^4 - 6d^2 + \tfrac{63}{2} = 0$$

$$(d^2)_{1,2} = 3 \pm \sqrt{9 - \tfrac{63}{2}} = 3 \pm i4.7434 \qquad (4.1.110a)$$

where $i = \sqrt{-1}$. To determine the roots d_1 thru d_4, we use the polar form. We have

$$(d^2)_{1,2} = x \pm iy = r(\cos\theta \pm i\sin\theta) = re^{\pm i\theta}$$

where $x = 3$, $y = 4.7434$, $r = \sqrt{x^2 + y^2} = 5.6125$, and $\theta = \tan^{-1}(y/x) = 57.69°$. Therefore, we have

$$(d)_{1-4} = \pm\sqrt{r}\,e^{\pm i\theta/2} = \pm\sqrt{r}\left(\cos\frac{\theta}{2} \pm i\sin\frac{\theta}{2}\right)$$

$$= \pm(2.075 \pm i1.143)$$

or

$$d_1 = 2.075 + 1.143i, \qquad d_2 = 2.075 - 1.143i,$$

$$d_3 = -(2.075 + 1.143i), \qquad d_4 = -(2.075 - 1.143i) \qquad (4.1.110b)$$

Therefore, the homogeneous solution is given by

$$c_{1h} = \bar{A}_1 e^{(\alpha+i\beta)\xi} + \bar{A}_2 e^{(\alpha-i\beta)\xi} + \bar{A}_3 e^{-(\alpha+i\beta)\xi} + \bar{A}_e^{-(\alpha-i\beta)\xi}$$

$$= A_1\cosh\alpha\xi\cos\beta\xi + A_2\cosh\alpha\xi\sin\beta\xi + A_3\sinh\alpha\xi\sin\beta\xi$$

$$+ A_4\sinh\alpha\xi\cos\beta\xi \qquad (4.1.111)$$

where $\alpha = 2.075$, $\beta = 1.143$, and $\xi = x/b$. The particular solution is given by

$$c_{1p} = \frac{2}{63}\left(\frac{21}{16}\frac{q_0}{D}\right) = \frac{q_0}{24D} \qquad (4.1.112)$$

The general solution of Eq. (4.1.109) becomes

$$c_1(x) = c_{1h} + c_{1p} \qquad (4.1.113)$$

Because of the symmetry of the expected solution about the y-axis, it follows that $A_2 = A_4 = 0$. The remaining constants A_1 and A_3 can be determined using boundary conditions in Eq. (4.1.107) at $x = \pm a$. We obtain

$$A_1 \cosh k_1 \cos k_2 + A_3 \sinh k_1 \sin k_2 = -\frac{q_0}{24D}$$

$$A_1\left(\frac{\alpha}{b}\sinh k_1 \cos k_2 - \frac{\beta}{b}\cosh k_1 \sin k_2\right)$$

$$+ A_3\left(\frac{\alpha}{b}\cosh k_1 \sin k_2 + \frac{\beta}{b}\sinh k_1 \cos k_2\right) = 0 \qquad (4.1.114a)$$

where $k_1 = \alpha a/b$ and $k_2 = \beta a/b$. Solving these equations, we get

$$A_1 = \frac{\mu_1}{\mu_0}\left(\frac{q_0}{24D}\right), \qquad A_3 = \frac{\mu_2}{\mu_0}\left(\frac{q_0}{24D}\right) \qquad (4.1.114b)$$

where

$$\mu_0 = \beta \sinh k_1 \cosh k_1 + \alpha \sin k_2 \cos k_2$$

$$\mu_1 = -(\alpha \cosh k_1 \sin k_2 + \beta \sinh k_1 \cos k_2)$$

$$\mu_2 = \alpha \sinh k_1 \cos k_2 - \beta \cosh k_1 \sin k_2 \qquad (4.1.114c)$$

Thus, the one-parameter Kantorovich solution becomes

$$w_1 = \frac{q_0}{24D}\left[\left(\frac{\mu_1}{\mu_0}\cosh\frac{\alpha x}{b}\cos\frac{\beta x}{b} + \frac{\mu_2}{\mu_0}\sinh\frac{\alpha x}{b}\sin\frac{\beta x}{b}\right) + 1\right](y^2 - b^2)^2$$

$$(4.1.115)$$

The maximum deflection occurs at the center of the plate; for a square plate of dimensions $2a \times 2a$, the maximum deflection is given by

$$w_{max} = w_1(0,0) = \frac{q_0 a^4}{24D}\left(\frac{\mu_1}{\mu_0} + 1\right) = 0.02074\frac{q_0 a^4}{D} \qquad (4.1.116a)$$

For a square plate of dimension $a \times a$, this reduces to

$$w_{max} = 0.001296\frac{q_0 a^4}{D} \qquad (4.1.116b)$$

Much of the discussion presented above is also valid for a rectangular plate with edges $y = \pm b$ clamped and $x = \pm a$ simply supported, and subjected to a uniformly distributed load. The general solution in Eq. (4.1.113) is valid with the following boundary conditions on c_1:

$$c_1(\pm a) = \frac{d^2 c_1}{dx^2}(\pm a) = 0 \qquad (4.1.117)$$

These conditions give the first equation in Eq. (4.1.114a) and

$$A_1 \sinh k_1 \sin k_2 - A_3 \cosh k_1 \cos k_2 = \frac{q_0}{24D}\left(\frac{\beta^2 - \alpha^2}{2\alpha\beta}\right) \qquad (4.1.118)$$

The solution of these equations is given by

$$A_1 = -\frac{q_0}{24D}\frac{\cosh k_1 \cos k_2 + a_0 \sinh k_1 \sin k_2}{\cosh^2 k_1 \cos^2 k_2 + \sinh^2 k_1 \sin^2 k_2}$$

$$A_3 = \frac{q_0}{24D}\frac{a_0 \cosh k_1 \cos k_2 + \sinh k_1 \sin k_2}{\cosh^2 k_1 \cos^2 k_2 + \sinh^2 k_1 \sin^2 k_2} \qquad (4.1.119)$$

where $a_0 = (\beta^2 - \alpha^2)/2\alpha\beta$. For a square plate, the maximum deflection is given by

$$w_{\max} = w(0,0) = 0.0248\frac{q_0 a^4}{D} \qquad (4.1.120)$$

This solution is, after dividing by 16, on the lower side of that in Example 4.4. This completes the example.

The last example of this section is concerned with a rectangular plate that has mixed boundary conditions (i.e., boundary conditions of both natural and essential types). The problem can be solved exactly using Levy's method (see Exercise 4.1.12).

Example 4.9

Consider a uniformly loaded isotropic rectangular plate that has as its edges $x = 0$ and $x = a$ simply supported, the side $y = 0$ clamped, and the side $y = b$ free. We wish to determine a one-parameter Ritz solution to the problem.

We choose an approximation of the form

$$w_1 = c_1\phi_1, \qquad \phi_1 = \left(\frac{y}{b}\right)^2 \sin\frac{\pi x}{a} \qquad (4.1.121)$$

The solution satisfies the essential boundary conditions of the problem:

$$w_1(0, y) = w_1(a, y) = w_1(x, 0) = \frac{\partial w_1}{\partial y}(x, 0) = 0 \qquad (4.1.122)$$

Substituting Eq. (4.1.121) into the total potential energy expression in Eq. (4.1.93), we obtain

$$\Pi(c_1) = \frac{Dc_1^2}{2} \int_0^a \int_0^b \left\{ \left[-\left(\frac{\pi}{a}\right)^2 \left(\frac{y}{b}\right)^2 \sin\frac{\pi x}{a} + \frac{2}{b^2} \sin\frac{\pi x}{a} \right]^2 \right.$$

$$- 2(1 - \nu) \left[-\frac{2}{b^2} \left(\frac{\pi}{a}\right)^2 \left(\frac{y}{b}\right)^2 \sin^2\frac{\pi x}{a} \right.$$

$$\left. \left. - \left(2\frac{\pi}{ab}\frac{y}{b}\sin\frac{\pi x}{a} \right)^2 \right] \right\} dx\, dy$$

$$- c_1 \int_0^a \int_0^b q_0 \left(\frac{y}{b}\right)^2 \sin\frac{\pi x}{a} \, dx\, dy$$

$$= \frac{D}{2} c_1^2 \int_0^a \sin^2\frac{\pi x}{a} \, dx \int_0^b \left[\left(\frac{2}{b^2} - \frac{\pi^2}{a^2 b^2}y^2\right)^2 + \frac{12(1 - \nu)\pi^2}{a^2 b^4}y^2 \right] dy$$

$$- c_1 q_0 \int_0^a \sin\frac{\pi x}{a} \, dx \int_0^b \frac{y^2}{b^2} \, dy$$

$$= \frac{Dc_1^2 a}{4} \left(\frac{4}{b^3} + \frac{\pi^4 b}{5a^4} - \frac{4\pi^2}{3a^2 b} + \frac{4(1 - \nu)\pi^2}{a^2 b} \right) - \frac{2abc_1 q_0}{3\pi} \qquad (4.1.123)$$

Using the total potential energy principle, $\delta\Pi = d\Pi/dc_1 = 0$, we get

$$c_1 = \frac{20 q_0 b^4}{D\pi \left[60 + 3\pi^4 \alpha^4 + 20\pi^2 \alpha^2 (2 - 3\nu) \right]}, \qquad \alpha = a/b \qquad (4.1.124)$$

and the solution is given by Eq. (4.1.121) with the known value of c_1. The maximum deflection is given by

$$w_{\max} = w_1\left(\frac{a}{2}, b\right) = c_1 \qquad (4.1.125)$$

which, for a square plate with $\nu = 0.3$, is equal to $0.01118 q_0 a^4/D$.

Exercises 4.1

1. Using the rectangular plate element shown in Fig. E1, equilibrium of forces in the z-direction, and moments about the x- and y-axes, derive

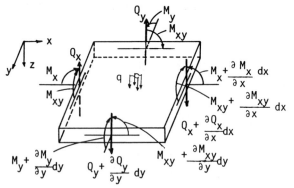

Figure E1

the equilibrium equations

$$\frac{\partial Q_x}{\partial x} + \frac{\partial Q_y}{\partial y} + q = 0$$

$$\frac{\partial M_{xy}}{\partial x} + \frac{\partial M_y}{\partial y} - Q_y = 0$$

$$\frac{\partial M_x}{\partial x} + \frac{\partial M_{xy}}{\partial y} - Q_x = 0$$

Eliminate Q_x and Q_y to obtain Eq. (4.1.21c).

2. Using the relations in Eq. (4.1.25), express the variation of the total potential energy expression in Eq. (4.1.17a) in terms of the displacements u_0, v_0, and w (and their variations). Obtain the total potential energy functional from the resulting expression.

3. Using the transformation equations (4.1.35), rewrite the total potential energy expression, obtained in Exercise 2 above, in terms of the polar coordinates. Assume that u_0 and v_0 are zero. Specialize the result for isotropic plates.

4. Use Eq. (4.1.35) to transform Eq. (4.1.25b) in Cartesian coordinates to those in polar coordinates.

5. Assuming deflection of the form

$$w = c_1 r^2 \ln r + c_2 r^2 + c_3$$

show that the expressions for the deflections of the axisymmetric bending of an isotropic circular plate with a concentrated load P_0 at the center are given by

(a) Simply supported: $w = \dfrac{P_0}{16\pi D}\left[2r^2\ln\dfrac{r}{a} + \dfrac{3 + \nu}{1 + \nu}(a^2 - r^2)\right]$

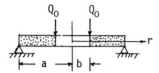

Figure E6

(b) Clamped: $w = \dfrac{P_0}{16\pi D}\left(2r^2\ln\dfrac{r}{a} + a^2 - r^2\right)$

where a is the radius and D is the flexural rigidity of the plate. *Hint:* Use the additional condition that the vertical shear force Q_r should be equal to $-P_0/2\pi r$ (for any r).

6. Determine the deflection of the axisymmetric bending of a simply supported isotropic circular plate with a concentric circular hole (see Fig. E6) when the plate is under vertical shear force Q_0 per unit circumferential length, uniformly distributed over the inner edge. Take the inner and outer radii of the plate to be b and a, respectively. *Hint:* Integrate the differential equation

$$\frac{d}{dr}\left[\frac{1}{r}\frac{d}{dr}\left(r\frac{dw}{dr}\right)\right] = \frac{Q_0 b}{Dr}$$

and determine the three constants of integration, using the boundary conditions at the inner and outer edges.

7. Consider a clamped isotropic circular plate under linearly varying load,

$$q(r,\theta) = q_0 + q_1\frac{r}{a}\cos\theta$$

where (r,θ) denote the coordinates, with the origin of the coordinate system being at the center of the plate. The deflection is given by

$$w = w_h + w_p$$

where w_h is the homogeneous solution

$$w_h = c_1 + c_2 r^2 + \left(c_3 r + c_4 r^3\right)\cos\theta$$

and w_p is the particular solution,

$$w_p = \frac{q_0 r^4}{64D} + \frac{q_1 r^5\cos\theta}{192 aD}$$

Determine the constants c_1 thru c_4 and obtain the deflection

$$w(r,\theta) = \frac{q_0}{64D}\left(a^2 - r^2\right)^2 + \frac{q_1}{192D}\frac{r}{a}\left(a^2 - r^2\right)^2\cos\theta$$

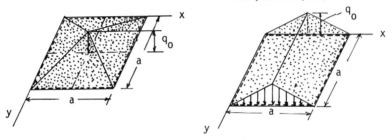

Figure E8   Figure E10

8. Use Navier's method to determine the deflection $w(x, y)$ of a simply supported isotropic square plate loaded with the pyramid-type loading shown in Fig. E.8.

9. Show that the solution in Eq. (4.1.61b) for a simply supported rectangular plate under uniformly distributed load degenerates to that of a simply supported beam as $b/a \to 0$ and $v = 0$:

$$w(y) = \frac{4q_0 b^4}{EI\pi^4} \sum_{n=1,3,\ldots}^{\infty} \frac{1}{n^5} \sin \frac{n\pi y}{b}$$

where I is the moment of inertia. Use the identity

$$4 \sum_{m=1,3,\ldots}^{\infty} \frac{\sin(m\pi x/a)}{m\pi} = 1$$

10. Repeat Exercise 8 for the triangular load shown in Fig. E10.

11. Use Levy's method to show that the transverse deflection of the simply supported isotropic plate (see Fig. E11) under uniform load is given by

$$w = \frac{4q_0 a^4}{\pi^5 D} \sum_{m=1,3,\ldots}^{\infty} \frac{1}{m^5} \left(1 - \frac{2 + \alpha_m \tanh \alpha_m}{2 \cosh \alpha_m} \cosh \frac{2\alpha_m y}{b}\right.$$

$$\left. + \frac{\alpha_m y}{b \cosh \alpha_m} \sinh \frac{2\alpha_m y}{b}\right) \sin \frac{m\pi x}{a}$$

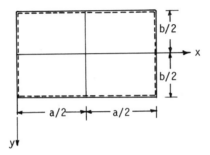

Figure E11

where $\alpha_m = m\pi b/2a$. *Hint:* First obtain $q_m = 4q_0/\pi m$ and then the particular solution of Eq. (4.1.76),

$$w_p = \frac{4q_0 a^4}{\pi^5 D} \sum_{m=1}^{\infty} \frac{1}{m^5} \sin\frac{m\pi x}{a}$$

12. Use Levy's method to determine the deflection of an isotropic rectangular plate clamped at $y = -(b/2)$, and $y = b/2$, simply supported on edges $x = 0$ and a, and subjected to uniformly distributed load q_0 (see Fig. E11).

13. Use Levy's method to show that the deflection of an isotropic, simply supported rectangular plate under hydrostatic loading, $q(x, y) = q_0(x/a)$, is given by

$$w = \frac{2q_0 a^4}{\pi^5 D} \sum_{m=1,2,\ldots}^{\infty} \frac{(-1)^{m+1}}{m^5}\left(1 - \frac{2 + \alpha_m\tanh\alpha_m}{2\cosh\alpha_m}\cosh\frac{2\alpha_m y}{b}\right.$$

$$\left. + \frac{\alpha_m y}{b\cosh\alpha_m}\sinh\frac{2\alpha_m y}{b}\right)\sin\frac{m\pi x}{a}$$

where $\alpha_m = m\pi b/2a$ (see Exercise 10 and Example 4.3).

14. An $a \times a$ square plate, with two opposite edges simply supported ($x = 0, a$) and the other two edges clamped ($y = \pm b/2$, $b = a$; see Fig. E11), is subjected to sinusoidal load,

$$q = q_0\sin\frac{\pi x}{a}\sin\frac{\pi y}{a}$$

Determine the deflection using Levy's method. Show that the maximum deflection and maximum bending moment are given by

$$w_{max} = 0.00154\frac{q_0 a^4}{D}$$

$$M_{max} = \left(M_y\right)_{max} = 0.0268 q_0 a^2$$

15. Use Levy's method to determine the deflection of an isotropic, simply supported rectangular plate under uniform load over half of the plate (see Fig. E.11 for the geometry and Fig. E15 for the loading).

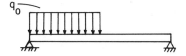

Figure E15   (see Fig. E11 for the geometry).

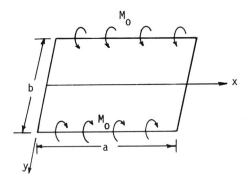

Figure E16

Hint: First determine $q_m = (2q_0/m\pi)[1 - \cos(m\pi/2)]$, and then the particular solution

$$w_p = \frac{2q_0 a^4}{\pi^5 D}\left(\sum_{m=1,3,\dots}^{\infty}\frac{1}{m^5}\sin\frac{m\pi x}{a} + 2\sum_{m=2,6,10\dots}^{\infty}\frac{1}{m^5}\sin\frac{m\pi x}{a}\right)$$

16. Consider an isotropic, simply supported rectangular plate subjected to symmetrically distributed edge moment M_0 (see Fig. E16) at $y = \pm b/2$. Assume deflection in the form

$$w = \sum_{m=1}^{\infty}\left(A_m\cosh\frac{m\pi y}{a} + B_m y\sinh\frac{m\pi y}{a}\right)\sin\frac{m\pi x}{a}$$

and determine the constants A_m and B_m using the boundary conditions

$$w = 0 \quad \text{at } y = \pm b/2$$

$$-D\frac{\partial^2 w}{\partial y^2} = M_0 \quad \text{at } y = \pm b/2$$

Note that the boundary conditions $w = \partial^2 w/\partial x^2 = 0$ at $x = 0$ and a are already satisfied by the assumed deflection.

17. A simply supported isotropic circular plate is subjected to hydrostatic loading, $q = q_0(r/a)$ (see Fig. E17). Use the reciprocity theorem to determine the center deflection.

18. Determine the center deflection owing to a point load Q_0 applied at a radial distance b from the center of a clamped, isotropic circular plate.

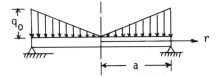

Figure E17

19. The strain energy of a circular plate is given by

$$U = \frac{D}{2} \int_R \left[\left(\frac{\partial^2 w}{\partial r^2} + \frac{1}{r} \frac{\partial w}{\partial r} + \frac{1}{r^2} \frac{\partial^2 w}{\partial \theta^2} \right)^2 \right.$$

$$- 2(1 - \nu) \frac{\partial^2 w}{\partial r^2} \left(\frac{1}{r} \frac{\partial w}{\partial r} + \frac{1}{r^2} \frac{\partial^2 w}{\partial \theta^2} \right)$$

$$\left. + 2(1 - \nu) \left(\frac{1}{r} \frac{\partial^2 w}{\partial r \partial \theta} - \frac{1}{r^2} \frac{\partial w}{\partial \theta} \right)^2 \right] r \, dr \, d\theta$$

For axisymmetric bending, the transverse deflection is only a function of the radial coordinate. Use the total energy principle to determine a one-parameter Ritz solution of a simply supported, isotropic circular plate under uniformly distributed load. What is the one-parameter Galerkin solution?

20. Rework the problem in Example 4.5 when the plate is annular (instead of solid), with the inner and outer radii b and a, respectively, and subjected to uniformly distributed load.

21. Find a one-parameter Galerkin solution of an isotropic annular plate, clamped at outer edge, and subjected to vertical shear force Q_0 at the inner edge.

22. Find a one-parameter Galerkin solution to the clamped, square plate problem in Example 4.7 using an algebraic (instead of a trigonometric) coordinate function. Assume that the plate is made of an isotropic material (see also Exercise 27 below).

23. Rework the clamped, square plate problem in Example 4.7 when the distributed load is replaced by a point load P_0 acting at a location $x = x_0$ and $y = y_0$.

24. Find a one-parameter Ritz solution of an orthotropic rectangular plate, simply supported at the edges and resting on an elastic foundation, and subjected to uniform load.

25. Rework Exercise 24 for the case in which the transverse load is a point load P_0 applied at the center of the plate.

26. Find a one-parameter Ritz approximation of the deflection of the plate in Example 4.4.

27. Solve the clamped plate problem of Example 4.7 using

$$\phi_{ij} = (x^2 - ax)^2 (y^2 - by)^2 x^{i-1} y^{j-1}$$

Show that the one-parameter solution is given by

$$w_{11} = \frac{6.125 q_0 (x^2 - ax)^2 (y^2 - by)^2}{7 D_{11} b^4 + 4(D_{12} + 2 D_{66}) a^2 b^2 + 7 D_{22} a^4}$$

4.2 SHEAR DEFORMATION THEORIES OF PLATES

4.2.1 Introduction

It is well-known from experimental observations that the Poisson–Kirchhoff theory of plates—in which it is assumed that normals to the midplane before deformation remain straight and normal to the plane after deformation—underpredicts deflections and overpredicts natural frequencies. These results are due to the neglect of transverse shear stains in the classical plate theory (CPT). From the equilibrium considerations, one finds that the transverse shear stresses, although small, are nonzero. Indeed, we can calculate these stresses by substituting σ_x, σ_y, and σ_{xy} into the equilibrium equations. Substitution of Eq. (4.1.6) into Hooke's law for plane stress, we obtain ($u_0, v_0 = 0$)

$$\sigma_x = -\frac{Ez}{1-v^2}\left(\frac{\partial^2 w}{\partial x^2} + v\frac{\partial^2 w}{\partial y^2}\right) = \frac{12 M_x z}{h^3}$$

$$\sigma_y = -\frac{Ez}{1-v^2}\left(\frac{\partial^2 w}{\partial y^2} + v\frac{\partial^2 w}{\partial x^2}\right) = \frac{12 M_y z}{h^3}$$

$$\sigma_{xy} = -\frac{Ez}{1+v}\frac{\partial^2 w}{\partial x\,\partial y} = \frac{12 M_{xy} z}{h^3} \qquad (4.2.1)$$

where M_x, M_y, and M_{xy} are substituted from Eqs. (4.1.25b) for an isotropic body. Substituting Eq. (4.2.1) into the equilibrium equations (1.2.40), we obtain

$$\frac{\partial \sigma_x}{\partial x} + \frac{\partial \sigma_{xy}}{\partial y} + \frac{\partial \sigma_{xz}}{\partial z} = 0 \rightarrow \frac{\partial \sigma_{xz}}{\partial z} = -\frac{12}{h^3}z\left(\frac{\partial M_x}{\partial x} + \frac{\partial M_{xy}}{\partial y}\right)$$

Integrating with respect to z, we get

$$\sigma_{xz} = -\frac{6z^2}{h^3}\left(\frac{\partial M_x}{\partial x} + \frac{\partial M_{xy}}{\partial y}\right) + f(x, y)$$

where $f(x, y)$ is a function to be determined, subjected to the boundary conditions $\sigma_{xz} = 0$ at $z = \pm h/2$. We obtain

$$f(x, y) = \frac{3}{2h}\left(\frac{\partial M_x}{\partial x} + \frac{\partial M_{xy}}{\partial y}\right)$$

$$\sigma_{xz} = \frac{3}{2h}\left(\frac{\partial M_x}{\partial x} + \frac{\partial M_{xy}}{\partial y}\right)\left[1 - 4\left(\frac{z}{h}\right)^2\right] \qquad (4.2.2a)$$

Similarly, we obtain

$$\sigma_{yz} = \frac{3}{2h}\left(\frac{\partial M_{xy}}{\partial x} + \frac{\partial M_y}{\partial y}\right)\left[1 - 4\left(\frac{z}{h}\right)^2\right] \qquad (4.2.2b)$$

The transverse normal stress can be computed from the third equilibrium equation, $\sigma_{xz,x} + \sigma_{yz,y} + \sigma_{z,z} = 0$ [see Eq. (1.2.40)] by integrating with respect to z and using the boundary condition $\sigma_z(x, y, h/2) = 0$,

$$\sigma_z = -\frac{q}{2}\left[1 - 3\frac{z}{h} + 4\left(\frac{z}{h}\right)^3\right] \qquad (4.2.2c)$$

where, in arriving at the last equation, we have used Eqs. (4.2.2a and b) and (4.1.21c). Thus, if the stresses σ_z, σ_{xz}, and σ_{yz} are small compared to σ_x, σ_y and σ_{xy}, the classical plate theory is consistent with assumptions 3 and 4. A nondimensional analysis shows that σ_x, σ_y, and σ_{xy} are of the order qa^2/h^2, whereas σ_{xz} and σ_{yz} are of order qa/h, and σ_z is of order q, where a is the characteristic length of the plate. Therefore, for plates with side-to-thickness ratios greater than 10, the transverse normal and shear stresses are negligible when compared to the remaining stresses. For thick plates ($a/h \leq 10$), especially in the vicinity of concentrated loads, the contribution of transverse shear stresses can be large. The errors in deflections, stresses, natural frequencies, and buckling loads are even higher for plates and shells made of composite materials like graphite–epoxy and boron–epoxy, whose elastic modulus to shear modulus ratios are very large (e.g., of the order of 25–40, instead of 2.6 for typical isotropic materials). These high ratios render classical theories inadequate for the analysis of composite plates and shells. An adequate theory must account for an accurate distribution of transverse shear stresses.

Refined plate theories—mainly those of Levy (1877), Reissner (1944, 1945), Hencky (1947), Mindlin (1951), and Kromm (1953, 1955)—are improvements of the classical plate theory in that they include the effect of transverse shear deformation and/or stresses. The Levy–Reissner–Hencky–Mindlin type theories (see Nelson and Lorch, 1974, Librescu, 1975, and Lo, Christensen, and Wu, 1977), do not satisfy the stress-free boundary conditions

$$\sigma_{xz}\left(x, y, \pm\frac{h}{2}\right) = \sigma_{yz}\left(x, y, \pm\frac{h}{2}\right) = 0 \qquad (4.2.3)$$

Kromm's theory (see Panc, 1975) satisfies these boundary conditions, but it does not account for the rotation of normals resulting from shear deformation. Recently, Reddy (1984a) presented a refined higher-order theory that satisfies the stress-free boundary conditions, and accounts for shear rotations and parabolic variation of the transverse shear stresses. In the following sections, we study the Hencky–Mindlin first-order shear deformation plate theory (FSDPT) and the Kromm–Reddy higher-order shear deformation plate theory (HSDPT).

4.2.2 A First-Order Theory

Analogous to the Timoshenko beam theory, in the shear deformation plate theory, we assume that plane sections originally perpendicular to the longitudinal plane of the plate remain plane, but not necessarily perpendicular to the longitudinal plane (see Fig. 1.14). We begin with the displacement field, which can be derived using the procedure described in Section 1.7

$$u(x, y, z) = u_0(x, y) + z\psi_x(x, y)$$

$$v(x, y, z) = v_0(x, y) + z\psi_y(x, y)$$

$$w(x, y, z) = w_0(x, y) \tag{4.2.4}$$

where (u_0, v_0, w_0) denote displacement components of a point along the (x, y, z) coordinates, and ψ_x and ψ_y denote the rotations of a line element, originally perpendicular to the longitudinal plane, about the y and x axes, respectively. The strains associated with the displacement field (4.2.4) are given by

$$e_x = \frac{\partial u_0}{\partial x} + z\frac{\partial \psi_x}{\partial x}, \qquad e_y = \frac{\partial v_0}{\partial y} + z\frac{\partial \psi_y}{\partial y},$$

$$2e_{xy} = \frac{\partial u_0}{\partial y} + \frac{\partial v_0}{\partial x} + z\left(\frac{\partial \psi_x}{\partial y} + \frac{\partial \psi_y}{\partial x}\right)$$

$$2e_{xz} = \psi_x + \frac{\partial w_0}{\partial x}, \qquad 2e_{yz} = \psi_y + \frac{\partial w_0}{\partial y}, \qquad e_z = 0 \tag{4.2.5}$$

Note that the transverse shear strains are nonzero. We regain the classical plate theory by setting

$$\psi_x = -\frac{\partial w_0}{\partial x}, \qquad \psi_y = -\frac{\partial w_0}{\partial y} \tag{4.2.6}$$

The equilibrium equations of the theory can be obtained from the total potential energy principle

$$\delta \Pi = 0$$

where Π is the total potential energy,

$$\Pi\left(u_0, v_0, w_0, \psi_x, \psi_y\right)$$

$$= \frac{1}{2}\int_V \left(\sigma_x e_x + \sigma_y e_y + 2\sigma_{xy} e_{xy} + 2\sigma_{xz} e_{xz} + 2\sigma_{yz} e_{yz}\right) dx\, dy\, dz$$

$$- \int_R q w_0 \, dx\, dy - \int_{C_1} \hat{N}_n u_{n0}\, dS - \int_{C_2} \hat{N}_s u_{s0}\, dS$$

$$- \int_{C_3} \hat{M}_n \psi_n\, dS - \int_{C_4} \hat{M}_s \psi_s\, dS - \int_{C_5} \hat{Q}_n w_0\, dS$$

$$= \frac{1}{2}\int_R \left[N_x \frac{\partial u_0}{\partial x} + M_x \frac{\partial \psi_x}{\partial x} + N_y \frac{\partial v_0}{\partial y} + M_y \frac{\partial \psi_y}{\partial y} \right.$$

$$+ N_{xy}\left(\frac{\partial u_0}{\partial y} + \frac{\partial v_0}{\partial x}\right) + M_{xy}\left(\frac{\partial \psi_x}{\partial y} + \frac{\partial \psi_y}{\partial x}\right)$$

$$\left. + Q_x\left(\psi_x + \frac{\partial w_0}{\partial x}\right) + Q_y\left(\psi_y + \frac{\partial w_0}{\partial y}\right) \right] dx\, dy$$

$$- \int_R q w_0 \, dx\, dy - \int_{C_1} \hat{N}_n u_{n0}\, dS - \int_{C_2} \hat{N}_s u_{s0}\, dS$$

$$- \int_{C_3} \hat{M}_n \psi_n\, dS - \int_{C_4} \hat{M}_s \psi_s\, dS - \int_{C_5} \hat{Q}_n w_0\, dS \qquad (4.2.7)$$

where R denotes the midplane of the plate, and N_x, M_x, and so on, are the stress and moment resultants defined in Eqs. (4.1.15a–c), ψ_n and ψ_s are normal and tangential rotations [defined analogous to u_{n0} and u_{s0}, respectively; see Eq. (4.1.18)].

Note that the total potential energy functional Π is a function of the five displacement functions u_0, v_0, w_0, ψ_x, and ψ_y. Therefore, the application of the principle of total potential energy gives the following five equations:

$$\delta u_0: \quad \frac{\partial N_x}{\partial x} + \frac{\partial N_{xy}}{\partial y} = 0$$

$$\delta v_0: \quad \frac{\partial N_{xy}}{\partial x} + \frac{\partial N_y}{\partial y} = 0$$

$$\delta w_0: \quad \frac{\partial Q_x}{\partial x} + \frac{\partial Q_y}{\partial y} + q = 0$$

$$\delta \psi_x: \quad \frac{\partial M_x}{\partial x} + \frac{\partial M_{xy}}{\partial y} - Q_x = 0$$

$$\delta \psi_y: \quad \frac{\partial M_{xy}}{\partial x} + \frac{\partial M_y}{\partial y} - Q_y - 0 \qquad (4.2.8)$$

The natural boundary conditions of the plate are given by

$$\delta u_{n0}: \quad N_n - \hat{N}_n = 0 \text{ on } C_1$$

$$\delta u_{s0}: \quad N_s - \hat{N}_s = 0 \text{ on } C_2$$

$$\delta \psi_n: \quad M_n - \hat{M}_n = 0 \text{ on } C_3$$

$$\delta \psi_s: \quad M_s - \hat{M}_s = 0 \text{ on } C_4$$

$$\delta w_0: \quad Q_n - \hat{Q}_n = 0 \text{ on } C_5 \qquad (4.2.9a)$$

Recall that C_1 thru C_5 denote the segments of the total boundary C such that on $C - C_1$, $C - C_2$, $C - C_3$, $C - C_4$, and $C - C_5$, essential boundary conditions are specified:

$$u_{n0} = \hat{u}_{n0} \text{ on } C - C_1$$

$$u_{s0} = \hat{u}_{s0} \text{ on } C - C_2$$

$$\psi_n = \hat{\psi}_n \text{ on } C - C_3$$

$$\psi_s = \hat{\psi}_s \text{ on } C - C_4$$

$$w_0 = \hat{w}_0 \text{ on } C - C_5 \qquad (4.2.9b)$$

For a rectangular plate, with coordinate axes parallel to the edges of the plate, the natural and essential boundary conditions take the following simple form:

	Natural		Essential
Specify	N_x	or	u_0
	N_y	or	v_0
	M_x	or	ψ_x
	M_y	or	ψ_y
	Q_x (or Q_y)	or	w_0 (4.2.10)

In using the total potential energy principle, we assumed that Hooke's law holds. In other words, the stress and moment resultants are related to the displacement gradients by

$$\left\{ \begin{array}{c} N_x \\ N_y \\ N_{xy} \end{array} \right\} = \left[\begin{array}{ccc} A_{11} & A_{12} & 0 \\ A_{12} & A_{22} & 0 \\ 0 & 0 & A_{66} \end{array} \right] \left\{ \begin{array}{c} \dfrac{\partial u_0}{\partial x} \\[2mm] \dfrac{\partial v_0}{\partial y} \\[2mm] \dfrac{\partial u_0}{\partial y} + \dfrac{\partial v_0}{\partial x} \end{array} \right\} \tag{4.2.11a}$$

$$\left\{ \begin{array}{c} M_x \\ M_y \\ M_{xy} \end{array} \right\} = \left[\begin{array}{ccc} D_{11} & D_{12} & 0 \\ D_{12} & D_{22} & 0 \\ 0 & 0 & D_{66} \end{array} \right] \left\{ \begin{array}{c} \dfrac{\partial \psi_x}{\partial x} \\[2mm] \dfrac{\partial \psi_y}{\partial y} \\[2mm] \dfrac{\partial \psi_x}{\partial y} + \dfrac{\partial \psi_y}{\partial x} \end{array} \right\} \tag{4.2.11b}$$

$$\left\{ \begin{array}{c} Q_y \\ Q_x \end{array} \right\} = \left[\begin{array}{cc} A_{44} & 0 \\ 0 & A_{55} \end{array} \right] \left\{ \begin{array}{c} \psi_y + \dfrac{\partial w_0}{\partial y} \\[2mm] \psi_x + \dfrac{\partial w_0}{\partial x} \end{array} \right\} \tag{4.2.11c}$$

where A_{ij} and D_{ij} are the plate stiffnesses defined in Eq. (4.1.25c), and

$$A_{ij} = K_i K_j \int_{-h/2}^{h/2} c_{ij} \, dz, \qquad i, j = 4, 5 \tag{4.2.12}$$

where K_i are the shear correction coefficients. The material stiffness coefficients c_{44}, c_{45}, and c_{55} for an orthotropic material are given by

$$c_{44} = G_{23}, \qquad c_{45} = 0, \qquad c_{55} = G_{13} \tag{4.2.13}$$

Using Eqs. (4.2.11), Eqs. (4.2.8) can be expressed in terms of the displacements. We obtain

$$\frac{\partial}{\partial x}\left(A_{11}\frac{\partial u_0}{\partial x} + A_{12}\frac{\partial v_0}{\partial y}\right) + A_{66}\frac{\partial}{\partial y}\left(\frac{\partial u_0}{\partial y} + \frac{\partial v_0}{\partial x}\right) = 0$$

$$A_{66}\frac{\partial}{\partial x}\left(\frac{\partial u_0}{\partial y} + \frac{\partial v_0}{\partial x}\right) + \frac{\partial}{\partial y}\left(A_{12}\frac{\partial u_0}{\partial x} + A_{22}\frac{\partial v_0}{\partial y}\right) = 0$$

$$A_{55}\frac{\partial}{\partial x}\left(\psi_x + \frac{\partial w_0}{\partial x}\right) + A_{44}\frac{\partial}{\partial y}\left(\psi_y + \frac{\partial w_0}{\partial y}\right) + q = 0$$

$$\frac{\partial}{\partial x}\left(D_{11}\frac{\partial \psi_x}{\partial x} + D_{12}\frac{\partial \psi_y}{\partial y}\right) + D_{66}\frac{\partial}{\partial y}\left(\frac{\partial \psi_x}{\partial y} + \frac{\partial \psi_y}{\partial x}\right) - A_{55}\left(\psi_x + \frac{\partial w_0}{\partial x}\right) = 0$$

$$D_{66}\frac{\partial}{\partial x}\left(\frac{\partial \psi_x}{\partial y} + \frac{\partial \psi_y}{\partial x}\right) + \frac{\partial}{\partial y}\left(D_{12}\frac{\partial \psi_x}{\partial x} + D_{22}\frac{\partial \psi_y}{\partial y}\right) - A_{44}\left(\psi_y + \frac{\partial w_0}{\partial y}\right) = 0$$

$$(4.2.14)$$

Note that the inplane displacements are *not* coupled to w_0, ψ_x, and ψ_y.

Equations (4.2.14), in general, cannot be solved exactly. However, for simply-supported boundary conditions (to be defined shortly) and rectangular geometries, one can develop exact solutions much the same way as was done for the classical plate theory.

Here we present an exact solution of Eq. (4.2.14) for two special cases: (1) axisymmetric bending of a simply supported isotropic circular plate under uniform load and (2) bending of a simply supported rectangular plate under sinusoidal or uniform load. We begin with the circular plate.

Axisymmetric Bending of a Simply Supported Circular Plate

Since the inplane displacements are not coupled to w_0, ψ_x, and ψ_y, they are identically zero (for the linear theory used here) in the absence of any inplane loads. The last three equations of Eq. (4.2.14) reduce to the following two equations, in cylindrical coordinates, for the axisymmetric case:

$$\frac{1}{r}\frac{d}{dr}(rQ) + q = 0$$

$$\frac{d}{dr}(rM_r) - M_\theta - rQ = 0 \tag{4.2.15}$$

where M_r, M_θ, and Q are related to the transverse deflection w and rotation

ψ by

$$M_r = D_{11}\frac{d\psi}{dr} + D_{12}\frac{\psi}{r}$$

$$M_\theta = D_{12}\frac{d\psi}{dr} + D_{22}\frac{\psi}{r}$$

$$Q = A_{55}\left(\psi + \frac{dw_0}{dr}\right), \qquad A_{55} = K^2 G_{13} h \qquad (4.2.16)$$

The boundary conditions of the problem are

$$w_0 = M_r = 0 \quad \text{at } r = a \qquad (4.2.17)$$

where a is the radius of the plate.

Next, we seek solutions of Eqs. (4.2.15)–(4.2.17) for the case $q = q_0 = $ constant, $D_{11} = D_{22} = D$, and $D_{12} = \nu D$. The total deflection w_0 of the plate can be represented, for the linear case, as the sum of deflection w_b owing to bending alone (i.e., without shear deformation) and deflection w_s due to shear only. Since we already have w_b from Eq. (4.1.41), we concentrate on obtaining w_s.

The deflection w_s of the midsurface caused by the shear alone is related to the shear force Q at the midsurface by $[\psi = -(dw_s/dr)]$

$$\frac{dw_s}{dr} = -\frac{3}{2}\frac{Q}{Gh} \qquad (4.2.18)$$

where we have used $\sigma_{xz} = 2Ge_{xz} = -G(dw_s/dr)$, and then Eq. (4.2.2a) for $z = 0$ and the forth equation in Eq. (4.2.8) to write $\sigma_{xz} = 3Q/2h$. The vertical shear force Q at a distance r from the center is given (from the equilibrium of an element, we have $2\pi r Q = \pi r^2 q_0$) by

$$Q = \frac{q_0 r}{2} \qquad (4.2.19)$$

Integrating Eq. (4.2.18) with Q given by Eq. (4.2.19), we obtain

$$w_s = -\frac{3}{2Gh}\int_a^r Q\, dr$$

$$= -\frac{3q_0}{8Gh}(r^2 - a^2)$$

$$= \frac{q_0 a^2 h^2}{16(1 - \nu)D}\left[1 - \left(\frac{r}{a}\right)^2\right]$$

The total deflection is given by

$$w_0 = \frac{q_0 a^4}{64D}\left[\frac{r^4}{a^4} - \frac{6+2\nu}{1+\nu}\frac{r^2}{a^2} + \frac{5+\nu}{1+\nu} + \frac{4h^2}{a^2(1-\nu)}\left(1 - \frac{r^2}{a^2}\right)\right] \qquad (4.2.20)$$

The ratio of the shear deflection to the bending deflection at $r = 0$ is given by

$$\left(\frac{w_s}{w_b}\right)_{r=0} = \frac{4h^2}{(1-\nu)a^2}\frac{1+\nu}{5+\nu} = \frac{4(1+\nu)}{(1-\nu)(5+\nu)}\left(\frac{h}{a}\right)^2 \qquad (4.2.21)$$

For plates with radius-to-thickness ratios greater than or equal to 10, the ratio w_s/w_b is less than 0.015 (for $\nu = 1/3$) and, therefore, the transverse shear strains can be neglected (i.e., the classical plate theory is adequate for $a/h \geq 10$).

Bending of Simply Supported Rectangular Plates

Consider a rectangular plate of planeform dimensions a and b and thickness h. Suppose that the plate is subjected to a distributed transverse load q, and that the edges are simply supported:

$$u_0 - w_0 = \psi_x = N_y = M_y = 0 \quad \text{at } y = 0 \text{ and } b$$

$$v_0 = w_0 = \psi_y = N_x = M_x = 0 \quad \text{at } x = 0 \text{ and } a \qquad (4.2.22)$$

Under these conditions, we can obtain the exact solution of Eqs. (4.2.14) and (4.2.22). The procedure is the same as that described in Eqs. (4.1.54)–(4.1.57). The solution of the last three equations in Eq. (4.2.14) is of the form

$$w_0 = \sum_{m,n=1}^{\infty} W_{mn} \sin \alpha x \sin \beta y$$

$$\psi_x = \sum_{m,n=1}^{\infty} X_{mn} \cos \alpha x \sin \beta y$$

$$\psi_y = \sum_{m,n=1}^{\infty} Y_{mn} \sin \alpha x \cos \beta y \qquad (4.2.23)$$

where $\alpha = m\pi/a$ and $\beta = n\pi/b$, and W_{mn}, X_{mn}, and Y_{mn} are the coefficients to be determined by substituting Eq. (4.2.23) into Eq. (4.2.14). We obtain, for any fixed m and n, the following equations

$$\begin{bmatrix} K_{11} & K_{12} & K_{13} \\ K_{12} & K_{22} & K_{23} \\ K_{13} & K_{23} & K_{33} \end{bmatrix} \begin{Bmatrix} W_{mn} \\ X_{mn} \\ Y_{mn} \end{Bmatrix} = \begin{Bmatrix} -Q_{mn} \\ 0 \\ 0 \end{Bmatrix} \qquad (4.2.24)$$

where Q_{mn} are the Fourier coefficients in the representation of the load q,

$$q = \sum_{m,n=1}^{\infty} Q_{mn} \sin \alpha x \sin \beta y \tag{4.2.25}$$

and K_{ij} are the coefficients

$$K_{11} = \alpha^2 A_{55} + \beta^2 A_{44}, \qquad K_{12} = \alpha A_{55}, \qquad K_{13} = \beta A_{44}$$

$$K_{22} = D_{11}\alpha^2 + D_{66}\beta^2 + A_{55}, \qquad K_{23} = (D_{12} + D_{66})\alpha\beta$$

$$K_{33} = D_{66}\alpha^2 + D_{22}\beta^2 + A_{44} \tag{4.2.26}$$

We now consider an example.

Example 4.10

Consider a simply supported, square plate under sinusoidal loading.

$$q = q_0 \sin\frac{\pi x}{a} \sin\frac{\pi y}{a} \tag{4.2.27}$$

The exact solution of the classical plate theory is given by Eq. (4.1.60). The exact solution of the shear deformation theory is given by

$$w_0 = -\frac{q_0(K_{22}K_{33} - K_{12}K_{12})}{K_0} \tag{4.2.28}$$

where K_{ij} are the coefficients defined in Eq. (4.2.26) and K_0 is the determinant of the coefficient matrix in Eq. (4.2.24). Table 4.1 contains the nondimensionalized deflection, $\bar{w} = -w(h^3 E_2/q_0 a^4)10^2$, for isotropic ($\nu = 0.3$) and orthotropic ($E_1 = 25E_2$, $G_{23} = 0.2E_2$, $G_{13} = G_{12} = 0.5E_2$, $\nu_{12} = 0.25$) square plates. For $a/h = 10$, the shear deformation theory predicts the center deflection about 5.64% larger for isotropic plates and 48% larger for orthotropic plates. Thus, the effect of shear deformation is more pronounced in orthotropic plates than in isotropic plates.

Table 4.1 also contains the nondimensionalized center deflections of isotropic and orthotropic plates under uniform loading. The solutions were obtained using $m, n = 1, 3, \ldots, 19$ in the series (4.2.23).

We close this section by commenting on the shear deformation theory described here and related higher-order shear deformation theories (see Lo, Christensen, and Wu, 1977 for a review). The higher-order theories are based

Table 4.1 Nondimensionalized Center Deflections of Square, Simply Supported
Plates under Sinusoidal and Uniform Loadings

Load[a]	a/h	\multicolumn{6}{c}{Shear Deformation Theory[b]}	Classical Plate Theory
		5	10	12.5	20	50	100	
SSL	Isotropic	3.4349	2.9607	2.9038	2.8422	2.8090	2.8042	2.8027
	Orthotropic	1.2327	0.6383	0.5644	0.4836	0.4397	0.4334	0.4313
UDL	Isotropic	5.3556	4.6660	4.5832	4.4936	4.4453	4.4438	4.4436
	Orthotropic	1.8159	0.9519	0.8442	0.7262	0.6620	0.6528	0.6497

[a] SSL = Sinusoidal load; UDL = uniform load.
[b] $K_4^2 = K_5^2 = 5/6$.

on the kinematic assumption (on the displacement field)

$$u(x, y, z, t)$$

$$= u_0(x, y, t) + z\psi_x(x, y, t) + z^2\xi_x(x, y, t) + z^3\zeta_x(x, y, t) + \cdots$$

$$v(x, y, z, t)$$

$$= v_0(x, y, t) + z\psi_y(x, y, t) + z^2\xi_y(x, y, t) + z^3\zeta_y(x, y, t) + \cdots$$

$$w(x, y, z, t)$$

$$= w_0(x, y, t) + z\psi_z(x, y, t) + z^2\xi_z(x, y, t) + z^3\zeta_z(x, y, t) + \cdots$$

$$(4.2.29)$$

where t denotes time. Note that with each power of the thickness coordinate, an additional dependent unknown is introduced into each component of the total displacement. Therefore, the equations of motion derived from Hamilton's principle for higher-order theories would be cumbersome and would involve a large number of dependent unknowns, most of which cannot be interpreted physically. Further, these theories do not satisfy the stress-free boundary conditions in Eq. (4.2.3). In the next section, we study a refined theory that satisfies the stress-free boundary conditions and contains the same dependent unknowns as in the first-order shear deformation theory described in this section.

4.2.3 A Refined Higher-Order Theory

The present theory is based on a displacement assumption of the type in Eq. (4.2.29). However, the specific form chosen satisfies the boundary conditions of Eq. (4.2.3). This particular choice eliminates the need for the specification of the shear correction factors, and allows for distortion (of third order) of normals to the midplane of the undeformed plate. As the following paragraphs show, this refined theory predicts displacements and natural frequencies very accurately.

We begin with the displacement field in which the displacements along the x- and y-directions are expanded as cubic functions of the thickness coordinate, and the transverse deflection is assumed to be constant through thickness (see Reddy, 1984a):

$$u_1(x, y, z, t) = u(x, y, t) + z\psi_x(x, y, t) + z^2\xi_x(x, y, t) + z^3\zeta_x(x, y, t)$$

$$u_2(x, y, z, t) = v(x, y, t) + z\psi_y(x, y, t) + z^2\xi_y(x, y, t) + z^3\zeta_y(x, y, t)$$

$$u_3(x, y, z, t) = w(x, y, t) \tag{4.2.30}$$

Here u, v, and w denote the displacements of a point (x, y) on the midplane, and ψ_x and ψ_y are the rotations of normals to the midplane about the y and x axes, respectively. The functions ξ_x, ζ_x, ξ_y, and ζ_y will be determined using the condition that the transverse shear stresses, $\sigma_{xz} = \sigma_5$ and $\sigma_{yz} = \sigma_4$ vanish on the plate top and bottom surfaces:

$$\sigma_5\left(x, y, \pm\frac{h}{2}, t\right) = 0, \qquad \sigma_4\left(x, y, \pm\frac{h}{2}, t\right) = 0 \tag{4.2.31}$$

These conditions are equivalent to, for orthotropic plates, the requirement that the corresponding strains be zero on these surfaces. We have

$$e_5 = \frac{\partial u_1}{\partial z} + \frac{\partial u_3}{\partial x} = \psi_x + 2z\xi_x + 3z^2\zeta_x + \frac{\partial w}{\partial x}$$

$$e_4 = \frac{\partial u_2}{\partial z} + \frac{\partial u_3}{\partial y} = \psi_y + 2z\xi_y + 3z^2\zeta_y + \frac{\partial w}{\partial y} \tag{4.2.32}$$

Setting $e_5(x, y, \pm h/2, t)$ and $e_4(x, y, \pm h/2, t)$ to zero, we obtain

$$\xi_x = 0, \qquad \xi_y = 0$$

$$\zeta_x = -\frac{4}{3h^2}\left(\frac{\partial w}{\partial x} + \psi_x\right), \qquad \zeta_y = -\frac{4}{3h^2}\left(\frac{\partial w}{\partial y} + \psi_y\right) \tag{4.2.33}$$

The displacement field in Eq. (4.2.30) becomes

$$u_1 = u + z\left[\psi_x - \frac{4}{3}\left(\frac{z}{h}\right)^2\left(\psi_x + \frac{\partial w}{\partial x}\right)\right]$$

$$u_2 = v + z\left[\psi_y - \frac{4}{3}\left(\frac{z}{h}\right)^2\left(\psi_y + \frac{\partial w}{\partial y}\right)\right]$$

$$u_3 = w \tag{4.2.34}$$

Although cubic variation of the in-plane displacements through thickness is taken into account, the displacement field in Eq. (4.2.34) contains the same number of dependent variables as in the first-order shear deformation theory.

The linear strains associated with the displacement field in Eq. (4.2.34), in the notation of Eq. (1.5.4), are

$$e_1 = e_1^0 + z\left(\kappa_1^0 + z^2\kappa_1^2\right)$$

$$e_2 = e_2^0 + z\left(\kappa_2^0 + z^2\kappa_2^2\right)$$

$$e_3 = 0$$

$$e_4 = e_4^0 + z^2\kappa_4^2$$

$$e_5 = e_5^0 + z^2\kappa_5^2$$

$$e_6 = e_6^0 + z\left(\kappa_6^0 + z^2\kappa_6^2\right) \tag{4.2.35}$$

where

$$e_1^0 = \frac{\partial u}{\partial x}, \qquad \kappa_1^0 = \frac{\partial \psi_x}{\partial x}, \qquad \kappa_1^2 = -\frac{4}{3h^2}\left(\frac{\partial \psi_x}{\partial x} + \frac{\partial^2 w}{\partial x^2}\right)$$

$$e_2^0 = \frac{\partial v}{\partial y}, \qquad \kappa_2^0 = \frac{\partial \psi_y}{\partial y}, \qquad \kappa_2^2 = -\frac{4}{3h^2}\left(\frac{\partial \psi_y}{\partial y} + \frac{\partial^2 w}{\partial y^2}\right)$$

$$e_4^0 = \psi_y + \frac{\partial w}{\partial y}, \qquad \kappa_4^2 = -\frac{4}{h^2}\left(\psi_y + \frac{\partial w}{\partial y}\right)$$

$$e_5^0 = \psi_x + \frac{\partial w}{\partial x}, \qquad \kappa_5^2 = -\frac{4}{h^2}\left(\psi_x + \frac{\partial w}{\partial x}\right)$$

$$e_6^0 = \frac{\partial u}{\partial y} + \frac{\partial v}{\partial x}, \qquad \kappa_6^0 = \frac{\partial \psi_x}{\partial y} + \frac{\partial \psi_y}{\partial x}, \qquad \kappa_6^2 = -\frac{4}{3h^2}\left(\frac{\partial \psi_x}{\partial y} + \frac{\partial \psi_y}{\partial x} + 2\frac{\partial^2 w}{\partial x \partial y}\right)$$

$$\tag{4.2.36}$$

It is interesting to note that the new strain components contain higher-order derivatives of the transverse deflection.

For a plate of constant thickness h and made of an orthotropic material (i.e., the plate possesses a plane of elastic symmetry parallel to the x-y plane) the constitutive equations can be written as

$$
\begin{Bmatrix} \sigma_1 \\ \sigma_2 \\ \sigma_6 \end{Bmatrix} = \begin{bmatrix} Q_{11} & Q_{12} & 0 \\ Q_{12} & Q_{22} & 0 \\ 0 & 0 & Q_{66} \end{bmatrix} \begin{Bmatrix} e_1 \\ e_2 \\ e_6 \end{Bmatrix}, \qquad \begin{Bmatrix} \sigma_4 \\ \sigma_5 \end{Bmatrix} = \begin{bmatrix} Q_{44} & 0 \\ 0 & Q_{55} \end{bmatrix} \begin{Bmatrix} e_4 \\ e_5 \end{Bmatrix}
$$

$$(4.2.37)$$

where Q_{ij} are the plane-stress reduced elastic constants in the material axes of the plate:

$$
Q_{11} = \frac{E_1}{1 - \nu_{12}\nu_{21}}, \qquad Q_{12} = \frac{\nu_{12}E_2}{1 - \nu_{12}\nu_{21}}, \qquad Q_{22} = \frac{E_2}{1 - \nu_{12}\nu_{21}}
$$

$$
Q_{44} = G_{23}, \qquad Q_{55} = G_{13}, \qquad Q_{66} = G_{12} \qquad (4.2.38)
$$

Here we use Hamilton's principle to derive the equations of motion appropriate for the displacement field (4.2.34) and constitutive equations (4.2.37). We have

$$
0 = -\int_0^t \left[\int_{-h/2}^{h/2} \int_R (\sigma_1 \delta e_1 + \sigma_2 \delta e_2 + \sigma_6 \delta e_6 + \sigma_4 \delta e_4 + \sigma_5 \delta e_5)\, dA\, dz \right.
$$

$$
\left. - \int_R q\,\delta w\, dA \right] dt + \frac{\delta}{2}\int_0^t \int_{-h/2}^{h/2}\int_R \rho \left[(\dot{u}_1)^2 + (\dot{u}_2)^2 + (\dot{u}_3)^2 \right] dA\, dz\, dt
$$

$$
= -\int_0^t \int_R \left\{ N_1 \frac{\partial \delta u}{\partial x} + M_1 \frac{\partial \delta \psi_x}{\partial x} + P_1 \left[-\frac{4}{3h^2}\left(\frac{\partial \delta \psi_x}{\partial x} + \frac{\partial^2 \delta w}{\partial x^2} \right) \right] \right.
$$

$$
+ N_2 \frac{\partial \delta v}{\partial y} + M_2 \frac{\partial \delta \psi_y}{\partial y} + P_2 \left[-\frac{4}{3h^2}\left(\frac{\partial \delta \psi_y}{\partial y} + \frac{\partial^2 \delta w}{\partial y^2} \right) \right]
$$

$$
+ N_6 \left(\frac{\partial \delta u}{\partial y} + \frac{\partial \delta v}{\partial x} \right) + M_6 \left(\frac{\partial \delta \psi_x}{\partial y} + \frac{\partial \delta \psi_y}{\partial x} \right)
$$

$$
+ P_6 \left[-\frac{4}{3h^2}\left(\frac{\partial \delta \psi_x}{\partial y} + \frac{\partial \delta \psi_y}{\partial x} + 2\frac{\partial^2 \delta w}{\partial x\, \partial y} \right) \right]
$$

$$+ Q_2 \left(\delta \psi_y + \frac{\partial \delta w}{\partial y} \right) + R_2 \left[-\frac{4}{h^2} \left(\delta \psi_y + \frac{\partial \delta w}{\partial y} \right) \right]$$

$$+ Q_1 \left(\delta \psi_x + \frac{\partial \delta w}{\partial x} \right) + R_1 \left[-\frac{4}{h^2} \left(\delta \psi_x + \frac{\partial \delta w}{\partial x} \right) \right] - q \delta w \Bigg\} \, dx \, dy \, dt$$

$$- \int_0^t \int_R \Bigg\{ \delta u \left[I_1 \ddot{u} + \left(I_2 - \frac{4}{3h^2} I_4 \right) \ddot{\psi}_x + \left(-\frac{4}{3h^2} \right) I_4 \frac{\partial \ddot{w}}{\partial x} \right]$$

$$+ \delta v \left[I_1 \ddot{v} + \left(I_2 - \frac{4}{3h^2} I_4 \right) \ddot{\psi}_y - \frac{4}{3h^2} I_4 \frac{\partial \ddot{w}}{\partial y} \right]$$

$$+ \delta w \left[I_1 \ddot{w} - \left(\frac{4}{3h^2} \right)^2 I_7 \left(\frac{\partial^2 \ddot{w}}{\partial x^2} + \frac{\partial^2 \ddot{w}}{\partial y^2} \right) \right.$$

$$\left. + \frac{4}{3h^2} I_4 \left(\frac{\partial \ddot{u}}{\partial x} + \frac{\partial \ddot{v}}{\partial y} \right) + \frac{4}{3h^2} \left(I_5 - \frac{4}{3h^2} I_7 \right) \left(\frac{\partial \ddot{\psi}_x}{\partial x} + \frac{\partial \ddot{\psi}_y}{\partial y} \right) \right]$$

$$+ \delta \psi_x \left[\left(I_2 - \frac{4}{3h^2} I_4 \right) \ddot{u} + \left(I_3 - \frac{8}{3h^2} I_5 + \frac{16}{9h^4} I_7 \right) \ddot{\psi}_x \right.$$

$$\left. - \frac{4}{3h^2} \left(I_5 - \frac{4}{3h^2} I_7 \right) \frac{\partial \ddot{w}}{\partial x} \right]$$

$$+ \delta \psi_y \left[\left(I_2 - \frac{4}{3h^2} I_4 \right) \ddot{v} + \left(I_3 - \frac{8}{3h^2} I_5 + \frac{16}{9h^4} I_7 \right) \ddot{\psi}_y \right.$$

$$\left. - \frac{4}{3h^2} \left(I_5 - \frac{4}{3h^2} I_7 \right) \frac{\partial \ddot{w}}{\partial y} \right] \Bigg\} \, dx \, dy \, dt \qquad (4.2.39)$$

where the stress resultants N_i, M_i, P_i, Q_i, and R_i are defined by

$$(N_i, M_i, P_i) = \int_{-h/2}^{h/2} \sigma_i (1, z, z^3) \, dz \qquad (i = 1, 2, 6)$$

$$(Q_2, R_2) = \int_{-h/2}^{h/2} \sigma_4 (1, z^2) \, dz$$

$$(Q_1, R_1) = \int_{-h/2}^{h/2} \sigma_5 (1, z^2) \, dz \qquad (4.2.40)$$

and the inertias I_i ($i = 1, 2, 3, 4, 5, 7$) are defined by

$$(I_1, I_2, I_3, I_4, I_5, I_7) = \int_{-h/2}^{h/2} \rho (1, z, z^2, z^3, z^4, z^6) \, dz \qquad (4.2.41)$$

Integrating the expression in Eq. (4.2.39) by parts, and collecting the coefficients of δu, δv, δw, $\delta \psi_x$, and $\delta \psi_y$, we obtain the following equations of motion:

$$\delta u: \quad \frac{\partial N_1}{\partial x} + \frac{\partial N_6}{\partial y} = I_1 \ddot{u} + \bar{I}_2 \ddot{\psi}_x - \frac{4}{3h^2} I_4 \frac{\partial \ddot{w}}{\partial x}$$

$$\delta v: \quad \frac{\partial N_6}{\partial x} + \frac{\partial N_2}{\partial y} = I_1 \ddot{v} + \bar{I}_2 \ddot{\psi}_y - \frac{4}{3h^2} I_4 \frac{\partial \ddot{w}}{\partial y}$$

$$\delta w: \quad \frac{\partial Q_1}{\partial x} + \frac{\partial Q_2}{\partial y} + q - \frac{4}{h^2} \left(\frac{\partial R_1}{\partial x} + \frac{\partial R_2}{\partial y} \right) + \frac{4}{3h^2} \left(\frac{\partial^2 P_1}{\partial x^2} + 2 \frac{\partial^2 P_6}{\partial x \partial y} + \frac{\partial^2 P_2}{\partial y^2} \right)$$

$$= I_1 \ddot{w} - \left(\frac{4}{3h^2} \right)^2 I_7 \left(\frac{\partial^2 \ddot{w}}{\partial x^2} + \frac{\partial^2 \ddot{w}}{\partial y^2} \right) + \frac{4}{3h^2} I_4 \left(\frac{\partial \ddot{u}}{\partial x} + \frac{\partial \ddot{v}}{\partial y} \right)$$

$$+ \frac{4}{3h^2} \bar{I}_5 \left(\frac{\partial \ddot{\psi}_x}{\partial x} + \frac{\partial \ddot{\psi}_y}{\partial y} \right)$$

$$\delta \psi_x: \quad \frac{\partial M_1}{\partial x} + \frac{\partial M_6}{\partial y} - Q_1 + \frac{4}{h^2} R_1 - \frac{4}{3h^2} \left(\frac{\partial P_1}{\partial x} + \frac{\partial P_6}{\partial y} \right)$$

$$= \bar{I}_2 \ddot{u} + \bar{I}_3 \ddot{\psi}_x - \frac{4}{3h^2} \bar{I}_5 \frac{\partial \ddot{w}}{\partial x}$$

$$\delta \psi_y: \quad \frac{\partial M_6}{\partial x} + \frac{\partial M_2}{\partial y} - Q_2 + \frac{4}{h^2} R_2 - \frac{4}{3h^2} \left(\frac{\partial P_6}{\partial x} + \frac{\partial P_2}{\partial y} \right)$$

$$= \bar{I}_2 \ddot{v} + \bar{I}_3 \ddot{\psi}_y - \frac{4}{3h^2} \bar{I}_5 \frac{\partial \ddot{w}}{\partial y} \tag{4.2.42}$$

where

$$\bar{I}_2 = I_2 - \frac{4}{3h^2} I_4, \quad \bar{I}_5 = I_5 - \frac{4}{3h^2} I_7, \quad \bar{I}_3 = I_3 - \frac{8}{3h^2} I_5 + \frac{16}{9h^4} I_7$$

$$\tag{4.2.43}$$

The boundary conditions are of the form: specify

$$u_n \text{ or } N_n$$

$$u_s \text{ or } N_s$$

$$w \text{ or } Q_n$$

$$\frac{\partial w}{\partial n} \text{ or } P_n \quad \text{on } C$$

$$\psi_n \text{ or } M_n$$

$$\psi_s \text{ or } M_s \tag{4.2.44}$$

where C is the boundary of the midplane R of the plate, and

$$u_n = u n_x + v n_y, \qquad u_{ns} = - u n_y + v n_x$$

$$N_n = N_1 n_x^2 + N_2 n_y^2 + 2 N_6 n_x n_y$$

$$N_s = (N_2 - N_1) n_x n_y + N_6 \left(n_x^2 - n_y^2 \right)$$

$$M_n = \hat{M}_1 n_x^2 + \hat{M}_2 n_y^2 + 2 \hat{M}_6 n_x n_y$$

$$M_s = (\hat{M}_2 - \hat{M}_1) n_x n_y + \hat{M}_6 \left(n_x^2 - n_y^2 \right)$$

$$Q_n = \hat{Q}_1 n_x + \hat{Q}_2 n_y - \frac{4}{3h^2} \frac{\partial P_s}{\partial s}$$

$$\hat{M}_i = M_i - \frac{4}{3h^2} P_i \qquad (i = 1, 2, 6)$$

$$\hat{Q}_i = Q_i - \frac{4}{h^2} R_i \qquad (i = 1, 2)$$

$$\frac{\partial}{\partial n} = n_x \frac{\partial}{\partial x} + n_y \frac{\partial}{\partial y}, \qquad \frac{\partial}{\partial s} = n_x \frac{\partial}{\partial y} - n_y \frac{\partial}{\partial x} \qquad (4.2.45)$$

and P_n and P_s are defined by expressions analogous to N_n and N_s, respectively. This completes the derivation of the governing equations.

The resultants defined in Eq. (4.2.40) can be related to the strains in Eq. (4.2.36) by the plate constitutive equations:

$$\begin{Bmatrix} N_1 \\ N_2 \\ N_6 \end{Bmatrix} = \begin{bmatrix} A_{11} & A_{12} & 0 \\ & A_{22} & 0 \\ \text{symmetric} & & A_{66} \end{bmatrix} \begin{Bmatrix} e_1^0 \\ e_2^0 \\ e_6^0 \end{Bmatrix}$$

$$\begin{Bmatrix} \begin{Bmatrix} M_1 \\ M_2 \\ M_6 \end{Bmatrix} \\ \begin{Bmatrix} P_1 \\ P_2 \\ P_6 \end{Bmatrix} \end{Bmatrix} = \begin{bmatrix} \begin{bmatrix} D_{11} & D_{12} & 0 \\ & D_{22} & 0 \\ \text{symmetric} & & D_{66} \end{bmatrix} & \begin{bmatrix} F_{11} & F_{12} & 0 \\ & F_{22} & 0 \\ \text{symmetric} & & F_{66} \end{bmatrix} \\ & \begin{bmatrix} H_{11} & H_{12} & 0 \\ & H_{22} & 0 \\ \text{symmetric} & & H_{66} \end{bmatrix} \end{bmatrix} \begin{Bmatrix} \kappa_1^0 \\ \kappa_2^0 \\ \kappa_6^0 \\ \kappa_1^2 \\ \kappa_2^2 \\ \kappa_6^2 \end{Bmatrix}$$

$$\begin{Bmatrix} Q_2 \\ Q_1 \\ R_2 \\ R_1 \end{Bmatrix} = \begin{bmatrix} A_{44} & 0 & D_{44} & 0 \\ 0 & A_{55} & 0 & D_{55} \\ D_{44} & 0 & F_{44} & 0 \\ 0 & D_{55} & 0 & F_{55} \end{bmatrix} \begin{Bmatrix} e_4^0 \\ e_5^0 \\ \kappa_4^2 \\ \kappa_5^2 \end{Bmatrix} \qquad (4.2.46)$$

where A_{ij}, D_{ij}, and so on, are the plate stiffnesses, defined by

$$\left(A_{ij}, D_{ij}, F_{ij}, H_{ij}\right) = \int_{-h/2}^{h/2} Q_{ij}\left(1, z^2, z^4, z^6\right) dz \qquad (i, j = 1, 2, 6)$$

$$\left(A_{ij}, D_{ij}, F_{ij}\right) = \int_{-h/2}^{h/2} Q_{ij}\left(1, z^2, z^4\right) dz \qquad (i, j = 4, 5) \qquad (4.2.47)$$

or

$$A_{ij} = Q_{ij}h, \qquad D_{ij} = Q_{ij}\frac{h^3}{12}$$

$$ \qquad\qquad\qquad\qquad (i, j = 1, 2, 6)$$

$$F_{ij} = Q_{ij}\frac{h^5}{80}, \qquad H_{ij} = Q_{ij}\frac{h^7}{448}$$

$$A_{44} = G_{23}h, \qquad A_{55} = G_{13}h$$

$$D_{44} = G_{23}\frac{h^3}{12}, \qquad D_{55} = G_{13}\frac{h^3}{12}$$

$$F_{44} = G_{23}\frac{h^5}{80}, \qquad F_{55} = G_{13}\frac{h^5}{80} \qquad (4.2.48)$$

Exact Solution of Simply Supported, Rectangular Plates

Here we consider the Navier solutions of the equations of the refined theory presented above. Since the coupling between stretching and bending is zero for linear theory of orthotropic plates, we consider only the flexural displacements. The following simply supported boundary conditions are used (see Fig. 4.6 for the geometry and the coordinate system):

$$w(x, 0) = w(x, b) = w(0, y) = w(a, y) = 0$$

$$P_2(x, 0) = P_2(x, b) = P_1(0, y) = P_1(a, y) = 0$$

$$M_2(x, 0) = M_2(x, b) = M_1(0, y) = M_1(a, y) = 0$$

$$\psi_x(x, 0) = \psi_x(x, b) = \psi_y(0, y) = \psi_y(a, y) = 0 \qquad (4.2.49)$$

To recast the equations of motion, Eqs. (4.2.42), in terms of the generalized displacements, we first express the resultants in terms of the displacements and then substitute the resultants into Eqs. (4.2.42). We have

$$N_1 = A_{11}\frac{\partial u}{\partial x} + A_{12}\frac{\partial v}{\partial y}$$

$$N_2 = A_{12}\frac{\partial u}{\partial x} + A_{22}\frac{\partial v}{\partial y}$$

$$N_6 = A_{66}\left(\frac{\partial u}{\partial y} + \frac{\partial v}{\partial x}\right)$$

$$M_1 = D_{11}\frac{\partial \psi_x}{\partial x} + D_{12}\frac{\partial \psi_y}{\partial y} + F_{11}\left(-\frac{4}{3h^2}\right)\left(\frac{\partial \psi_x}{\partial x} + \frac{\partial^2 w}{\partial x^2}\right)$$

$$+ F_{12}\left(-\frac{4}{3h^2}\right)\left(\frac{\partial \psi_y}{\partial y} + \frac{\partial^2 w}{\partial y^2}\right)$$

$$M_2 = D_{12}\frac{\partial \psi_x}{\partial x} + D_{22}\frac{\partial \psi_y}{\partial y} + F_{12}\left(-\frac{4}{3h^2}\right)\left(\frac{\partial \psi_x}{\partial x} + \frac{\partial^2 w}{\partial x^2}\right)$$

$$+ F_{22}\left(-\frac{4}{3h^2}\right)\left(\frac{\partial \psi_y}{\partial y} + \frac{\partial^2 w}{\partial y^2}\right)$$

$$M_6 = D_{66}\left(\frac{\partial \psi_x}{\partial y} + \frac{\partial \psi_y}{\partial x}\right) + F_{66}\left(-\frac{4}{3h^2}\right)\left(\frac{\partial \psi_x}{\partial y} + \frac{\partial \psi_y}{\partial x} + 2\frac{\partial^2 w}{\partial x\,\partial y}\right)$$

$$Q_2 = A_{44}\left(\psi_y + \frac{\partial w}{\partial y}\right) + D_{44}\left(-\frac{4}{h^2}\right)\left(\psi_y + \frac{\partial w}{\partial y}\right)$$

$$Q_1 = A_{55}\left(\psi_x + \frac{\partial w}{\partial x}\right) + D_{55}\left(-\frac{4}{h^2}\right)\left(\psi_x + \frac{\partial w}{\partial x}\right)$$

$$P_1 = F_{11}\frac{\partial \psi_x}{\partial x} + F_{12}\frac{\partial \psi_y}{\partial y} + H_{11}\left(-\frac{4}{3h^2}\right)\left(\frac{\partial \psi_x}{\partial x} + \frac{\partial^2 w}{\partial x^2}\right)$$

$$+ H_{12}\left(-\frac{4}{3h^2}\right)\left(\frac{\partial \psi_y}{\partial y} + \frac{\partial^2 w}{\partial y^2}\right)$$

$$P_2 = F_{12}\frac{\partial \psi_x}{\partial x} + F_{22}\frac{\partial \psi_y}{\partial y} + H_{12}\left(-\frac{4}{3h^2}\right)\left(\frac{\partial \psi_x}{\partial x} + \frac{\partial^2 w}{\partial x^2}\right)$$

$$+ H_{22}\left(-\frac{4}{3h^2}\right)\left(\frac{\partial \psi_y}{\partial y} + \frac{\partial^2 w}{\partial y^2}\right)$$

$$P_6 = F_{66}\left(\frac{\partial \psi_y}{\partial x} + \frac{\partial \psi_x}{\partial y}\right) + \left(-\frac{4}{3h^2}\right)H_{66}\left(\frac{\partial \psi_x}{\partial y} + \frac{\partial \psi_y}{\partial x} + 2\frac{\partial^2 w}{\partial x\,\partial y}\right)$$

$$R_2 = D_{44}\left(\frac{\partial w}{\partial y} + \psi_y\right) + F_{44}\left(-\frac{4}{h^2}\right)\left(\psi_y + \frac{\partial w}{\partial y}\right)$$

$$R_1 = D_{55}\left(\frac{\partial w}{\partial x} + \psi_x\right) + F_{55}\left(-\frac{4}{h^2}\right)\left(\psi_x + \frac{\partial w}{\partial x}\right) \tag{4.2.50}$$

The last three equations in Eq. (4.2.42) now can be expressed in terms of the displacements (note that $I_2 = I_4 = 0$):

$$\frac{4}{3h^2}\left[F_{11}\frac{\partial^3\psi_x}{\partial x^3} + H_{11}\left(-\frac{4}{3h^2}\right)\left(\frac{\partial^3\psi_x}{\partial x^3} + \frac{\partial^4 w}{\partial x^4}\right) + F_{12}\frac{\partial^3\psi_y}{\partial x^2\partial y}\right.$$

$$+ H_{12}\left(-\frac{4}{3h^2}\right)\left(\frac{\partial^3\psi_y}{\partial x^2\partial y} + \frac{\partial^4 w}{\partial x^2\partial y^2}\right) + F_{12}\frac{\partial^3\psi_x}{\partial y^2\partial x}$$

$$+ H_{12}\left(-\frac{4}{3h^2}\right)\left(\frac{\partial^3\psi_x}{\partial y^2\partial x} + \frac{\partial^4 w}{\partial x^2\partial y^2}\right) + F_{22}\frac{\partial^3\psi_y}{\partial y^3}$$

$$+ H_{22}\left(-\frac{4}{3h^2}\right)\left(\frac{\partial^3\psi_y}{\partial y^3} + \frac{\partial^4 w}{\partial y^4}\right) + 2F_{66}\left(\frac{\partial^3\psi_y}{\partial x^2\partial y} + \frac{\partial^3\psi_x}{\partial y^2\partial x}\right)$$

$$\left.+ 2H_{66}\left(-\frac{4}{3h^2}\right)\left(\frac{\partial^3\psi_x}{\partial y^2\partial x} + \frac{\partial^3\psi_y}{\partial x^2\partial y} + 2\frac{\partial^4 w}{\partial x^2\partial y^2}\right)\right]$$

$$-\frac{4}{h^2}\left[D_{55}\left(\frac{\partial^2 w}{\partial x^2} + \frac{\partial\psi_x}{\partial x}\right) + F_{55}\left(-\frac{4}{h^2}\right)\left(\frac{\partial\psi_x}{\partial x} + \frac{\partial^2 w}{\partial x^2}\right)\right.$$

$$\left.+ D_{44}\left(\frac{\partial^2 w}{\partial y^2} + \frac{\partial\psi_y}{\partial y}\right) + F_{44}\left(-\frac{4}{h^2}\right)\left(\frac{\partial\psi_y}{\partial y} + \frac{\partial^2 w}{\partial y^2}\right)\right]$$

$$+ \left[A_{55}\left(\frac{\partial\psi_x}{\partial x} + \frac{\partial^2 w}{\partial x^2}\right) + D_{55}\left(-\frac{4}{h^2}\right)\left(\frac{\partial\psi_y}{\partial x} + \frac{\partial^2 w}{\partial x^2}\right)\right.$$

$$\left.+ A_{44}\left(\frac{\partial\psi_y}{\partial y} + \frac{\partial^2 w}{\partial y^2}\right) + D_{44}\left(-\frac{4}{h^2}\right)\left(\frac{\partial\psi_y}{\partial y} + \frac{\partial^2 w}{\partial y^2}\right)\right] + q$$

$$= I_1\frac{\partial^2 w}{\partial t^2} - \left(\frac{4}{3h^2}\right)^2 I_7\left(\frac{\partial^4 w}{\partial x^2\partial t^2} + \frac{\partial^4 w}{\partial y^2\partial t^2}\right) + \frac{4}{3h^2}\bar{I}_5\left(\frac{\partial^3\psi_x}{\partial x\partial t^2} + \frac{\partial^3\psi_y}{\partial y\partial t^2}\right)$$

$$(4.2.51a)$$

$$D_{11}\frac{\partial^2\psi_x}{\partial x^2} + D_{12}\frac{\partial^2\psi_y}{\partial x\partial y} + F_{11}\left(-\frac{4}{3h^2}\right)\left(\frac{\partial^2\psi_x}{\partial x^2} + \frac{\partial^3 w}{\partial x^3}\right)$$

$$+ F_{12}\left(-\frac{4}{3h^2}\right)\left(\frac{\partial^2\psi_y}{\partial x\partial y} + \frac{\partial^3 w}{\partial x\partial y^2}\right) + D_{66}\left(\frac{\partial^2\psi_x}{\partial y^2} + \frac{\partial^2\psi_y}{\partial x\partial y}\right)$$

$$+ F_{66}\left(-\frac{4}{3h^2}\right)\left(\frac{\partial^2\psi_x}{\partial y^2} + \frac{\partial^2\psi_y}{\partial x\partial y} + 2\frac{\partial^3 w}{\partial x\partial y^2}\right)$$

$$-\left[A_{55}\left(\psi_x + \frac{\partial w}{\partial x}\right) + D_{55}\left(-\frac{4}{h^2}\right)\left(\psi_x + \frac{\partial w}{\partial x}\right)\right]$$

$$
-\frac{4}{3h^2}\left[F_{11}\frac{\partial^2 \psi_x}{\partial x^2} + H_{11}\left(-\frac{4}{3h^2}\right)\left(\frac{\partial^2 \psi_x}{\partial x^2} + \frac{\partial^3 w}{\partial x^3}\right) + F_{12}\frac{\partial^2 \psi_y}{\partial x \, \partial y} \right.
$$

$$
+ H_{12}\left(-\frac{4}{3h^2}\right)\left(\frac{\partial^2 \psi_y}{\partial x \, \partial y} + \frac{\partial^3 w}{\partial x \, \partial y^2}\right) + F_{66}\left(\frac{\partial^2 \psi_y}{\partial x \, \partial y} + \frac{\partial^2 \psi_x}{\partial y^2}\right)
$$

$$
\left. + H_{66}\left(-\frac{4}{3h^2}\right)\left(\frac{\partial^2 \psi_x}{\partial y^2} + \frac{\partial^2 \psi_y}{\partial x \, \partial y} + 2\frac{\partial^3 w}{\partial x \, \partial y^2}\right) \right]
$$

$$
+ \frac{4}{h^2}\left[D_{55}\left(\frac{\partial w}{\partial x} + \psi_x\right) + F_{55}\left(-\frac{4}{h^2}\right)\left(\psi_x + \frac{\partial w}{\partial x}\right) \right]
$$

$$
= \bar{I}_3 \frac{\partial^2 \psi_x}{\partial t^2} - \frac{4}{3h^2}\bar{I}_5 \frac{\partial^3 w}{\partial x \, \partial t^2}
$$

(4.2.51b)

$$
D_{66}\left(\frac{\partial^2 \psi_x}{\partial x \, \partial y} + \frac{\partial^2 \psi_y}{\partial x^2}\right) + F_{66}\left(-\frac{4}{3h^2}\right)\left(\frac{\partial^2 \psi_x}{\partial x \, \partial y} + \frac{\partial^2 \psi_y}{\partial x^2} + 2\frac{\partial^3 w}{\partial x^2 \, \partial y}\right)
$$

$$
+ D_{12}\frac{\partial^2 \psi_x}{\partial x \, \partial y} + D_{22}\frac{\partial^2 \psi_y}{\partial y^2} + F_{12}\left(-\frac{4}{3h^2}\right)\left(\frac{\partial^2 \psi_x}{\partial x \, \partial y} + \frac{\partial^3 w}{\partial x^2 \, \partial y}\right)
$$

$$
+ F_{22}\left(-\frac{4}{3h^2}\right)\left(\frac{\partial^2 \psi_y}{\partial y^2} + \frac{\partial^3 w}{\partial y^3}\right) - \left[A_{44}\left(\psi_y + \frac{\partial w}{\partial y}\right) + D_{44}\left(-\frac{4}{h^2}\right)\left(\psi_y + \frac{\partial w}{\partial y}\right) \right]
$$

$$
- \frac{4}{3h^2}\left[F_{66}\left(\frac{\partial^2 \psi_y}{\partial x^2} + \frac{\partial^2 \psi_x}{\partial y \, \partial x}\right) + H_{66}\left(\frac{\partial^2 \psi_x}{\partial x \, \partial y} + \frac{\partial^2 \psi_y}{\partial x^2} + 2\frac{\partial^3 w}{\partial x^2 \, \partial y}\right)\left(-\frac{4}{3h^2}\right) \right.
$$

$$
+ F_{12}\frac{\partial^2 \psi_x}{\partial x \, \partial y} + H_{12}\left(-\frac{4}{3h^2}\right)\left(\frac{\partial^2 \psi_x}{\partial x \, \partial y} + \frac{\partial^3 w}{\partial x^2 \, \partial y}\right)
$$

$$
\left. + F_{22}\frac{\partial^2 \psi_y}{\partial y^2} + H_{22}\left(-\frac{4}{3h^2}\right)\left(\frac{\partial^2 \psi_y}{\partial y^2} + \frac{\partial^3 w}{\partial y^3}\right) \right]
$$

$$
+ \frac{4}{h^2}\left[D_{44}\left(\frac{\partial w}{\partial y} + \psi_y\right) + F_{44}\left(-\frac{4}{h^2}\right)\left(\frac{\partial w}{\partial y} + \psi_y\right) \right]
$$

$$
= \bar{I}_3 \frac{\partial^2 \psi_y}{\partial t^2} - \frac{4}{3h^2}\bar{I}_5 \frac{\partial^3 w}{\partial y \, \partial t^2}
$$

(4.2.51c)

Following the Navier solution procedure, we assume the following solution form, which satisfies the boundary conditions in Eq. (4.2.49)

$$w = \sum_{m,n=1}^{\infty} W_{mn} \sin \alpha x \sin \beta y \ e^{-i\omega t}$$

$$\psi_x = \sum_{m,n=1}^{\infty} X_{mn} \cos \alpha x \sin \beta y \ e^{-i\omega t}$$

$$\psi_y = \sum_{m,n=1}^{\infty} Y_{mn} \sin \alpha x \cos \beta y \ e^{-i\omega t} \qquad (4.2.52)$$

where $\alpha = m\pi/a$ and $\beta = n\pi/b$, and ω is the frequency of the natural vibration. Further, we assume that the applied transverse load, q, can be expanded in the double-Fourier series as

$$q = \sum_{m,n=1}^{\infty} Q_{mn} \sin \alpha x \sin \beta y \qquad (4.2.53)$$

Substituting Eqs. (4.2.52) and (4.2.53) into Eq. (4.2.51), and collecting the coefficients, we obtain

$$\begin{bmatrix} M_{11} & M_{12} & M_{13} \\ M_{12} & M_{22} & M_{23} \\ M_{13} & M_{23} & M_{33} \end{bmatrix} \begin{Bmatrix} \ddot{W}_{mn} \\ \ddot{X}_{mn} \\ \ddot{Y}_{mn} \end{Bmatrix} + \begin{bmatrix} c_{11} & c_{12} & c_{13} \\ c_{12} & c_{22} & c_{23} \\ c_{13} & c_{23} & c_{33} \end{bmatrix} \begin{Bmatrix} W_{mn} \\ X_{mn} \\ Y_{mn} \end{Bmatrix} = \begin{Bmatrix} Q_{mn} \\ 0 \\ 0 \end{Bmatrix}$$

$$(4.2.54)$$

for any fixed values of m and n. The elements $c_{ij} = c_{ji}$ and $M_{ij} = M_{ji}$ of the coefficient matrices $[c]$ and $[M]$ are given by

$$c_{11} = \alpha^2 A_{55} + \beta^2 A_{44} - \frac{8}{h^2}(\alpha^2 D_{55} + \beta^2 D_{44})$$

$$+ \left(\frac{4}{h^2}\right)^2 (\alpha^2 F_{55} + \beta^2 F_{44})$$

$$+ \left(\frac{4}{3h^2}\right)^2 [\alpha^4 H_{11} + 2(H_{12} + 2H_{66})\alpha^2 \beta^2 + \beta^4 H_{22}]$$

$$c_{12} = \alpha A_{55} - \frac{8}{h^2}\alpha D_{55} + \left(\frac{4}{h^2}\right)^2 \alpha F_{55} - \frac{4}{3h^2}[\alpha^3 F_{11} + \alpha\beta^2(F_{12} + 2F_{66})]$$

$$+ \left(\frac{4}{3h^2}\right)^2 [\alpha^3 H_{11} + \alpha\beta^2(H_{12} + 2H_{66})]$$

$$c_{13} = \beta A_{44} - \frac{8}{h^2} \beta D_{44} + \left(\frac{4}{h^2}\right)^2 \beta F_{44} - \frac{4}{3h^2} \left[\alpha^2\beta(F_{12} + 2F_{66}) + \beta^3 F_{22}\right]$$

$$+ \left(\frac{4}{3h^2}\right)^2 \left[\alpha^2\beta(H_{12} + 2H_{66}) + \beta^3 H_{22}\right]$$

$$c_{22} = A_{55} + \alpha^2 D_{11} + \beta^2 D_{66} - \frac{8}{h^2} D_{55} + \left(\frac{4}{h^2}\right)^2 F_{55}$$

$$- \frac{8}{3h^2}(\alpha^2 F_{11} + \beta^2 F_{66}) + \left(\frac{4}{3h^2}\right)^2 (\alpha^2 H_{11} + \beta^2 H_{66})$$

$$c_{23} = \alpha\beta \left[D_{12} + D_{66} - \frac{8}{3h^2}(F_{12} + F_{66}) + \left(\frac{4}{3h^2}\right)^2 (H_{12} + H_{66}) \right]$$

$$c_{33} = A_{44} + \alpha^2 D_{66} + \beta^2 D_{22} - \frac{8}{h^2} D_{44} + \left(\frac{4}{h^2}\right)^2 F_{44}$$

$$- \frac{8}{3h^2}(\alpha^2 F_{66} + \beta^2 F_{22}) + \left(\frac{4}{3h^2}\right)^2 (\beta^2 H_{22} + \alpha^2 H_{66})$$

$$M_{11} = I_1 + I_7\left(\frac{4}{3h^2}\right)(\alpha^2 + \beta^2)$$

$$M_{12} = -\frac{4}{3h^2}\bar{I}_5\alpha, \qquad M_{13} = -\frac{4}{3h^2}\bar{I}_5\beta$$

$$M_{22} = \bar{I}_3, \qquad M_{33} = \bar{I}_3, \qquad M_{23} = 0 \qquad\qquad (4.2.55)$$

For static bending, Eq. (4.2.54) takes the form

$$[C]\begin{Bmatrix} W \\ X \\ Y \end{Bmatrix} = \begin{Bmatrix} Q \\ 0 \\ 0 \end{Bmatrix} \qquad\qquad (4.2.56)$$

and for free vibration we have

$$[C]\begin{Bmatrix} W \\ X \\ Y \end{Bmatrix} = \omega^2 [M]\begin{Bmatrix} W \\ X \\ Y \end{Bmatrix} \qquad\qquad (4.2.57)$$

where [simplified form of c_{ij} in Eq. (4.2.55)]

$$c_{11} = \frac{8}{15}h\left(\alpha^2 G_{13} + \beta^2 G_{23}\right) + \frac{h^3}{12}\left[\alpha^4 Q_{11} + 2(2G_{12} + Q_{12})\alpha^2\beta^2 + \beta^4 Q_{22}\right]$$

$$c_{12} = \frac{8}{15}h\alpha G_{12} - \frac{4h^3}{315}\left[\alpha^3 Q_{11} + \alpha\beta^2(2G_{12} + Q_{12})\right]$$

$$c_{13} = \frac{8}{15}h\beta G_{23} - \frac{4h^3}{315}\left[\alpha^2\beta(2G_{12} + Q_{12}) + \beta^3 Q_{22}\right]$$

$$c_{22} = \frac{8}{15}hG_{13} + \frac{17h^3}{315}\left(\alpha^2 Q_{11} + \beta^2 G_{12}\right)$$

$$c_{23} = \frac{17\alpha\beta h^3}{315}\left(G_{12} + Q_{12}\right)$$

$$c_{33} = \frac{8}{15}hG_{23} + \frac{17h^3}{315}\left(\alpha^2 G_{12} + \beta^2 Q_{22}\right) \tag{4.2.58}$$

and Q_{ij} are given by Eq. (4.2.38).

Example 4.11

Here we present numerical results for isotropic and orthotropic plates under uniformly distributed transverse load. We also solve Eq. (4.2.57) for the natural frequencies. The following orthotropic material properties [the elastic properties of aragonite crystals in Eq. (1.5.16) are converted to the engineering constants in million psi]:

$$E_1 = 20.83, \qquad E_2 = 10.94, \qquad G_{12} = 6.10,$$

$$G_{13} = 3.71, \qquad G_{23} = 6.19 \qquad \nu_{12} = 0.44, \qquad \nu_{21} = 0.23 \quad (4.2.59)$$

Bending Results

Numerical results of bending are presented in Tables 4.2 and 4.3 for isotropic ($\nu = 0.3$) and orthotropic plates, respectively. The elastic constant c_{11} used in Table 4.3 has the value of 23.2×10^6 psi. The following nondimensionalized

Table 4.2 Comparison of Deflections and Stresses in Isotropic ($\nu = 0.3$) Plates Under a Uniformly Distributed Transverse Load ($m, n = 1, 3, \ldots, 19$)

							Constitutive Equation		Equilibrium Equation	
b/a	a/h	Source	$\bar{w}$	$\pm\bar{\sigma}_1$	$\pm\bar{\sigma}_2$	$+\bar{\sigma}_6$	$-\bar{\sigma}_4$	$-\bar{\sigma}_5$	$-\bar{\sigma}_4$	$-\bar{\sigma}_5$
1	5	HSDPT[a]	0.0535	0.2944	0.2944	0.2112	0.4840	0.4840	0.3703	0.3703
				(0.2949)[b]	(0.2949)	(0.2124)	(0.4871)	(0.4871)	(0.3324)	(0.3324)
		FSDPT	0.0536	0.2873	0.2873	0.1946	0.3928	0.3928	0.4909	0.4909
	10	HSDPT	0.0467	0.2890	0.2890	0.1990	0.4890	0.4890	0.4543	0.4543
				(0.2893)	(0.2893)	(0.1996)	(0.4937)	(0.4937)	(0.4417)	(0.4417)
		FSDPT	0.0467	0.2873	0.2873	0.1946	0.3928	0.3928	0.4909	0.4909
	100	HSDPT	0.0444	0.2873	0.2873	0.1947	0.4909	0.4909	0.4905	0.4905
				(0.2874)	(0.2874)	(0.1948)	(0.4965)	(0.4965)	(0.4959)	(0.4959)
		FSDPT	0.0444	0.2873	0.2873	0.1946	0.3928	0.3928	0.4909	0.4909
							(0.3972)	(0.3972)	(0.4965)	(0.4965)
	CPT		0.0444	0.2873	0.2873	0.1946	0.0	0.0	0.4909	0.4909
									(0.4965)	(0.4965)
2	5	HSDPT	0.1248	0.6202	0.2818	0.2927	0.6745	0.5201	0.5615	0.4569
		FSDPT	0.1248	0.6100	0.2779	0.2769	0.5451	0.4192	0.6813	0.6813
	10	HSDPT	0.1142	0.6125	0.2789	0.2809	0.6794	0.5230	0.6448	0.5051
		FDSPT	0.1142	0.6100	0.2779	0.2769	0.5451	0.4192	0.6813	0.5240
	100	HSDPT	0.1106	0.6100	0.2779	0.2769	0.6813	0.5240	0.6809	0.5238
		FSDPT	0.1106	0.6100	0.2779	0.2769	0.5451	0.4192	0.6813	0.5240
	CPT		0.1106	0.6100	0.2779	0.2769	0.0	0.0	0.6813	0.5240

[a]HSDPT = Higher-order shear deformation plate theory; FSDPT = first-order shear deformation plate theory.
[b]Numbers in parentheses were obtained by $m, n = 1, 3, \ldots, 29$ in the series of Eq. (4.2.52).

378

Table 4.3 Comparison of Deflections and Stresses in Orthotropic Square Plates Under a Uniform Transverse Load ($m, n = 1, 3, \ldots, 19$).

b/a	h/a	$c_{11}w/hq_0$				σ_1/q_0				σ_5/q_0 [a]			
		Exact[b]	HSDPT	FSDPT	CPT	Exact	HSDPT	FSDPT	CPT	Exact	HSDPT	FSDPT	CPT
2	0.05	21,542	21,542	21,542	21,210	262.67	262.6	262.0	262.2	14.048	13.98 (13.57)	11.20 (14.00)	0.0 (14.00)
	0.10	1,408.5	1,408.5	1,408.5	1,326	65.975	65.95	65.38	65.55	6.927	6.958 (6.229)	5.599 (6.998)	0.0 (6.998)
	0.14	387.23	387.5	387.6	345.1	33.862	33.84	33.27	33.44	4.878	4.944 (4.027)	3.998 (4.997)	0.0 (4.999)
1	0.05	10,443	10,450.	10,450.	10,250.	144.31	144.3	143.9	144.4	10.873	10.85 (10.45)	8.701 (10.88)	0.0 (10.88)
	0.10	688.57	689.5	689.6	640.7	36.021	36.01	35.62	36.09	5.341	5.382 (4.657)	4.338 (5.442)	0.0 (5.442)
	0.14	191.07	191.6	191.6	166.8	18.346	18.34	17.94	18.41	3.731	3.805 (2.884)	3.806 (3.857)	0.0 (3.887)
0.5	0.05	2,048.7	2,051.0	2,051.0	1,989.0	40.657	40.67	40.50	40.84	6.243	6.163 (5.765)	4.948 (6.184)	0.0 (6.214)
	0.10	139.08	139.8	139.8	124.3	10.025	10.05	9.888	10.21	2,957	2.885 (2.893)	2.436 (3.044)	0.0 (3.107)
	0.14	39.790	40.21	40.23	32.36	5.036	5.068	4.903	5.209	1.999	2.080 (1.186)	1.705 (2.131)	0.0 (2.219)

[a] Numbers in parentheses denote the shear stress values obtained from the stress equilibrium equations.
[b] From Srinivas and Rao (1970).

379

deflections and stresses are tabulated in Table 4.2:

$$\bar{w} = w\left(\frac{a}{2}, \frac{b}{2}\right) \frac{h^3 E_2}{q_0 a^4}$$

$$\bar{\sigma}_i = \sigma_i\left(\frac{a}{2}, \frac{b}{2}, \pm\frac{h}{2}\right) \frac{h^2}{q_0 a^2}, \qquad i = 1, 2$$

$$\bar{\sigma}_6 = \sigma_6\left(0, 0, \pm\frac{h}{2}\right) \frac{h^2}{q_0 a^2}$$

$$\bar{\sigma}_4 = \sigma_4\left(\frac{a}{2}, 0, 0\right) \frac{h}{q_0 a}$$

$$\bar{\sigma}_5 = \sigma_5\left(0, \frac{b}{2}, 0\right) \frac{h}{q_0 a} \tag{4.2.60}$$

It should be noted that the first-order shear deformation theory presented in Section 4.2.2 can be obtained as a special case from the present theory by setting the plate stiffnesses, F_{ij}, H_{ij}, D_{44}, D_{55}, F_{44}, and F_{55}, and inertias, I_5 and I_7, to zero, and multiplying the stiffnesses, A_{44} and A_{55}, with shear correction coefficients.

Two pairs of transverse shear stresses, one obtained from the constitutive equations and the other from the equilibrium equations, are presented in the tables. In the first-order shear deformation theory, the shear correction factors are assumed to be $K_1^2 = K_2^2 = \frac{5}{6}$. The following conclusions can be drawn from the results of Tables 4.2 and 4.3.

1. Even for the isotropic plates, the effect of transverse shear deformation is significant. The classical plate theory (CPT) underpredicts (for $a/b = 1$) the deflections by 4.9% at $a/h = 10$ and 17% at $a/h = 5$ and the stress σ_1 by 0.7% at $a/h = 10$ and 2.6% at $a/h = 5$, when compared to the higher-order shear deformation plate theory (HSDPT) (see Table 4.2).
2. The first-order shear deformation plate theory (FSDPT) is quite accurate when the transverse deflections are concerned. But the stresses are no better than those predicted by CPT (see Tables 4.2 and 4.3).
3. The transverse shear stresses predicted by the constitutive equations of the higher-order theory are the most accurate of the three plate theories, when compared to the exact solution of Srinivas and Rao (1970) (see Table 4.2 and Fig. 4.11). It is interesting to note that FSDPT and CPT give more accurate transverse shear stresses than HSDPT when equilibrium equations are used.
4. The infinite series for deflections converges faster than that for stresses; the convergence is slower for thick plates than for thin plates (see Table 4.2).

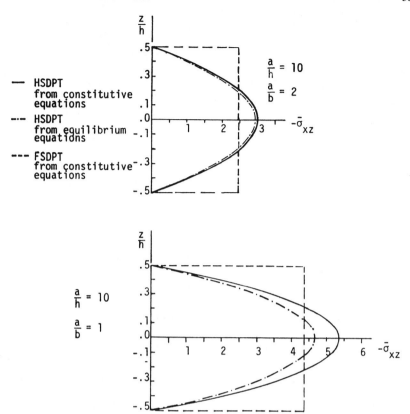

Figure 4.11 Distribution of the transverse shear stress across the thickness of simply supported, rectangular plates under uniformly distributed transverse load (orthotropic case).

Natural Vibration

The numerical results of the natural vibration of isotropic ($\nu = 0.3$) and orthotropic [see Eq. (4.2.59) for material properties] square plates are presented in Tables 4.4 and 4.5, respectively. The results are compared with the exact solutions of the three-dimensional elasticity theory. In Table 4.5, the first three eigenvalues obtained by the present theory are compared with the exact values and the values obtained by FSDPT and CPT. From the results presented in Tables 4.4 and 4.5, the following observations can be made:

1. The classical plate theory overestimates the frequencies. The errors increase with increasing mode numbers.

2. The frequencies predicted by FSDPT are fairly accurate; the error increases with increasing mode number.

3. The frequencies predicted by HSDPT are the most accurate of all.

Table 4.4 Comparison of Natural Frequencies, $\bar{\omega} = \omega h(\sqrt{\rho/G})$, of Isotropic ($\nu = 0.3$) Plates ($a/h = 10$)

m	n	Exact	HSDPT	FSDPT	CPT
			$b/a = 1$		
1	1	0.0932^a	0.0931	0.0930	0.0955
					$(0.0963)^b$
2	1	0.226	0.2222	0.2219	0.2360
					(0.2408)
2	2	0.3421	0.3411	0.3406	0.3732
					(0.3853)
1	3	0.4171	0.4158	0.4149	0.4629
					(0.4816)
2	3	0.5239	0.5221	0.5206	0.5951
					(0.6261)
1	4	—	0.6545	0.6520	0.7668
					(0.8187)
3	3	0.6889	0.6862	0.6834	0.8090
					(0.8669)
2	4	0.7511	0.7481	0.7446	0.8926
					(0.9632)
3	4	—	0.8949	0.8896	1.0965
					(1.2040)
1	5	0.9268	0.9230	0.9174	1.1365
					(1.2521)
2	5	—	1.0053	0.9984	1.2549
					(1.3966)
4	4	1.0889	1.0847	1.0764	1.3716
					(1.5411)
3	5	—	1.1361	1.1268	1.4475
					(1.6374)
			$b/a = \sqrt{2}$		
1	1	0.704^c	0.7038	0.7036	0.7180
					(0.7224)
1	2	1.376	1.3738	1.3729	1.4273
					(1.4448)
2	1	2.018	2.0141	2.0123	2.1281
					(2.1671)
1	3	2.431	2.4263	2.4235	2.5908
					(2.6487)
2	2	2.634	2.6283	2.6250	2.8207
					(2.8895)
2	3	3.612	3.6013	3.5948	3.9575
					(4.0935)
1	4	3.800	3.7891	3.7818	4.1822
					(4.3343)
3	1	3.987	3.9748	3.9666	4.4062
					(4.5751)

Table 4.4 (*Continued*)

m	n	Exact	HSDPT	FSDPT	CPT
3	2	4.535	4.5198	4.5089	5.0729
					(5.2974)
2	4	4.890	4.8737	4.8608	5.5133
					(5.7790)
3	3	5.411	5.3915	5.3754	6.1680
					(6.5014)
1	5	5.411	5.3915	5.3754	6.1680
					(6.5014)
2	5	6.409	6.3846	6.3609	7.4563
					(7.9462)

[a] Exact values are from Srinivas, Joga Rao, and Rao (1970).
[b] Numbers in parentheses denote natural frequencies obtained by omitting the rotatory inertia.
[c] Exact values are from Reismann and Lee (1969).

Table 4.5 Comparison of Natural Frequencies, $\bar{\omega} = \omega h(\sqrt{\rho/c_{11}})$, of an Orthotropic Square Plate ($a/h = 10$)

m	n	Exact[a] I	II[b]	III	HSDPT I	II[b]	III	FSDPT I	II[b]	III	CPT
1	1	0.0474	1.3077	1.6530	0.0474	1.3086	1.6550	0.0474	1.3159	1.6646	0.0493
											(0.0497)[c]
1	2	0.1033	1.3331	1.7160	0.1033	1.3339	1.7209	0.1032	1.3410	1.7305	0.1095
											(0.1120)
2	1	0.1188	1.4205	1.6805	0.1189	1.4216	1.6827	0.1187	1.4285	1.6921	0.1327
											(0.1354)
2	2	0.1694	1.4316	1.7509	0.1695	1.4323	1.7562	0.1692	1.4393	1.7655	0.1924
											(0.1987)
1	3	0.1888	1.3765	1.8115	0.1888	1.3772	1.8210	0.1884	1.3841	1.8305	0.2070
											(0.2154)
3	1	0.2180	1.5777	1.7334	0.2184	1.5789	1.7361	0.2178	1.5857	1.7450	0.2671
											(0.2779)
2	3	0.2475	1.4596	1.8523	0.2477	1.4603	1.8622	0.2469	1.4670	1.8714	0.2879
											(0.3029)
3	2	0.2624	1.5651	1.8195	0.2629	1.5658	1.8255	0.2619	1.5725	1.8341	0.3248
											(0.3418)
1	4	0.2969	1.4372	1.9306	0.2969	1.4379	1.9466	0.2959	1.4445	1.9560	0.3371
											(0.3599)
4	1	0.3319	1.7179	1.8548	0.3330	1.7186	1.8588	0.3311	1.7265	1.8657	0.4471
											(0.4773)
3	3	0.3320	1.5737	1.9289	0.3326	1.5744	1.9395	0.3310	1.5812	1.9479	0.4172
											(0.4470)
2	4	0.3476	1.5068	1.9749	0.3479	1.5076	1.9912	0.3463	1.5141	2.0002	0.4152
											(0.4480)
4	2	0.3070	1.6940	1.9447	0.3720	1.6947	1.9514	0.3696	1.7022	1.9586	0.5018
											(0.5415)

[a] From Srinivas and Rao (1970).
[b] Pure thick-twist modes.
[c] Numbers in parentheses indicate frequencies obtained by ommiting the rotary inertia.

4. The effect of transverse shear deformation increases with increasing mode number.

This completes the example.

Exercises 4.2

1. The displacement field in the shear deformation theory of axisymmetric bending of circular plates is given by

$$u = u_0 + z\psi, \qquad v = 0, \qquad w = w_0$$

Compute the strains and derive the governing equilibrium equations using the total potential energy principle.

2. The stationary variational principle for the axisymmetric bending of circular plates involves the determination of $\delta R = 0$, where

$$R = \pi \int_0^a \left[A_{11} \left(\frac{du_0}{dr} \right)^2 + 2 A_{12} \frac{u_0}{r} \frac{du_0}{dr} + A_{22} \left(\frac{u_0}{r} \right)^2 \right.$$

$$+ A_{55} \left(\psi + \frac{dw_0}{dr} \right)^2 + 2 M_r \frac{d\psi}{dr} + 2 M_\theta \frac{\psi}{r}$$

$$\left. + \bar{D}_{22} M_r^2 - 2 \bar{D}_{12} M_r M_\theta + \bar{D}_{11} M_\theta^2 - 2qw \right] r \, dr$$

where $\bar{D}_{ij} = D_{ij}/D_0$, $D_0 = D_{12}^2 - D_{11} D_{22}$. Determine the Euler equations of the functional.

3. Find the exact solution for the deflection of the axisymmetric bending of a clamped isotropic circular plate under uniform load. Use the results of Exercise 1.

4. Determine the fundamental frequency of a simply supported orthotropic square plate using the first-order shear deformation theory.

5. Consider the displacement field

$$u = -z \frac{\partial w}{\partial x} + F(z) \Phi_x$$

$$v = -z \frac{\partial w}{\partial y} + F(z) \Phi_y$$

where

$$F(z) = \int \left[1 - 4\left(\frac{z}{h}\right)^2 \right] dz$$

Show that the bending moments M_1, M_2, and M_6 are related to w, Φ_x, and Φ_y by

$$M_1 = -D_{11}\frac{\partial^2 w}{\partial x^2} - D_{12}\frac{\partial^2 w}{\partial y^2} + F_{11}\frac{\partial \Phi_x}{\partial x} + F_{12}\frac{\partial \Phi_y}{\partial y}$$

$$M_2 = -D_{12}\frac{\partial^2 w}{\partial x^2} - D_{22}\frac{\partial^2 w}{\partial y^2} + F_{12}\frac{\partial \Phi_x}{\partial x} + F_{22}\frac{\partial \Phi_y}{\partial y}$$

$$M_6 = -2D_{66}\frac{\partial^2 w}{\partial x \partial y} + F_{66}\left(\frac{\partial \Phi_x}{\partial y} + \frac{\partial \Phi_y}{\partial x} \right)$$

where

$$F_{ij} = \int_{-h/2}^{h/2} zF(z)c_{ij}\,dz \qquad (i, j = 1, 2, 6)$$

6. For a simply supported orthotropic plate under sinusoidal load

$$q = q_0 \sin\frac{m\pi x}{a} \sin\frac{n\pi y}{b}$$

assume a solution of the form in Eq. (4.2.23) and determine the amplitudes W, X, and Y in terms of the plate stiffness D_{ij} and parameters α and β [i.e., solve Eq. (4.2.24) using Cramer's rule and express the answer in terms of the said quantities].

7. Show that the strains in the refined higher-order theory of simply supported, rectangular plates can be expressed in terms of the amplitudes [see Eq. (4.2.52)] W, X, and Y (assume that the inplane displacements are zero) as

$$e_1 = \left[-\alpha Xz + k_1 z^3(\alpha^2 W + \alpha X) \right] \sin \alpha x \sin \beta y, \qquad k_1 = \frac{4}{3h^2}$$

$$e_2 = \left[-\beta Yz + k_1 z^3(\beta^2 W + \beta Y) \right] \sin \alpha x \sin \beta y$$

$$e_4 = \left[X + \alpha W - k_2 z^2(X + \alpha W) \right] \cos \alpha x \sin \beta y, \qquad k_2 = \frac{4}{h^2}$$

$$e_5 = \left[Y + \beta W - k_2 z^2(Y + \beta W) \right] \sin \alpha x \cos \beta y$$

$$e_6 = \left[z(\beta X + \alpha Y) - k_1 z^3(2\alpha\beta W + \alpha Y + \beta X) \right] \cos \alpha x \cos \beta y$$

8. Show that the transverse normal and shear stresses obtained from the stress equilibrium equations for the classical plate theory (for simply-supported, orthotropic rectangular plates) are given by

$$\sigma_z = -\frac{h^3}{48}\left\{\left[1+\left(\frac{2z}{h}\right)^3\right]+6\left[1+\left(\frac{2z}{h}\right)\right]\right\}$$

$$\times\left[\alpha^4 Q_{11}+2\alpha^2\nu^2(2G_{12}+Q_{12})+\beta^4 Q_{22}\right]W\sin\alpha x\sin\beta y$$

$$\sigma_{xz} = \frac{h^2}{8}\left[1-\left(\frac{2z}{h}\right)^2\right]\left[\alpha^3 Q_{11}+(2G_{12}+Q_{12})\alpha\beta^2\right]W\cos\alpha x\sin\beta y$$

$$\sigma_{yz} = \frac{h^2}{8}\left[1-\left(\frac{2z}{h}\right)^2\right]\left[\alpha^2\beta(2G_{12}+Q_{12})+\beta^3 Q_{22}\right]W\sin\alpha x\cos\beta y$$

where Q_{ij} are given by Eq. (4.2.38).

9. Show that the transverse normal and shear stresses obtained from the stress equilibrium equations for the first-order shear deformation theory (for simply supported, orthotropic rectangular plates) are given by

$$\sigma_z = \frac{h^3}{48}\left\{\left[1+\left(\frac{2z}{h}\right)^3\right]+6\left[1+\left(\frac{2z}{h}\right)\right]\right\}$$

$$\left\{\left[\alpha^3 Q_{11}+\alpha\beta^2(2G_{12}+Q_{12})\right]X\right.$$

$$\left.+\left[(2G_{12}+Q_{12})\alpha^2\beta+\beta^3 Q_{22}\right]Y\right\}\sin\alpha x\sin\beta y$$

$$\sigma_{xz} = -\frac{h^2}{8}\left[1-\left(\frac{2z}{h}\right)^2\right]\left[(\alpha^2 Q_{11}+\beta^2 G_{12})X\right.$$

$$\left.+(G_{12}+Q_{12})\alpha\beta Y\right]\cos\alpha x\sin\beta y$$

$$\sigma_{yz} = -\frac{h^2}{8}\left[1-\left(\frac{2z}{h}\right)^2\right]\left[(\alpha^2 G_{12}+\beta^2 Q_{22})Y\right.$$

$$\left.+(G_{12}+Q_{12})\alpha\beta X\right]\sin\alpha x\cos\beta y$$

where Q_{ij} are given by Eq. (4.2.38).

10. Show that the kinetic energy of the classical plate theory can be expressed as

$$K = \frac{1}{2}\int_R\left\{I_1\left[(\dot u_0)^2+(\dot v_0)^2+(\dot w)^2\right]+I_3\left[\left(\frac{\partial\dot w}{\partial x}\right)^2+\left(\frac{\partial\dot w}{\partial y}\right)^2\right]\right\}dx\,dy$$

where $\dot u_0 = \partial u_0/\partial t$, u_0, v_0, and w are the displacements of the midplane [see Eqs. (4.1.2) and (4.1.5)], and I_1 and I_3 are the principal and rotary inertias, respectively,

$$I_1 = \rho h, \qquad I_3 = \left(\frac{\rho h^3}{12}\right)$$

where ρ is the density.

11. *Plate buckling (or stability)*.   Plates used in many engineering structures are often subjected to normal and shearing forces acting in the plane of the plate (see Fig. E11). When these inplane forces are sufficiently small, the equilibrium is stable and the deformation involves compression and shear (in the absence of transverse loads). For certain intensities of the inplane loads and zero transverse load, simultaneously with inplane displacements, lateral displacements are introduced. In this situation, the originally *stable* equilibrium becomes *unstable* and the plate is said to have buckled. The least magnitude of the inplane load that causes the plate to buckle is called the *critical load*. The purpose of this exercise is to derive the equations of plate buckling problem by the variational method. Assuming the following strains

$$e_{11} = \frac{1}{2}\left(\frac{\partial w}{\partial x}\right)^2 - z\frac{\partial^2 w}{\partial x^2}, \qquad e_{22} = \frac{1}{2}\left(\frac{\partial w}{\partial y}\right)^2 - z\frac{\partial^2 w}{\partial y^2},$$

$$2e_{12} = \frac{\partial w}{\partial x}\frac{\partial w}{\partial y} - z\frac{\partial^2 w}{\partial x\,\partial y}$$

in the principle of virtual work, derive the following differential equation of plate buckling (CPT)

$$D\nabla^4 w + \frac{\partial}{\partial x}\left(n_1\frac{\partial w}{\partial x} + n_6\frac{\partial w}{\partial y}\right) + \frac{\partial}{\partial y}\left(n_6\frac{\partial w}{\partial x} + n_2\frac{\partial w}{\partial y}\right) = 0$$

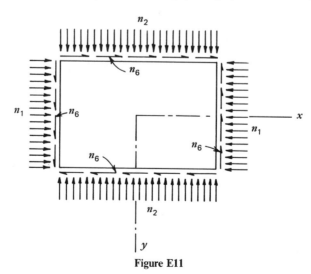

Figure E11

where n_1, n_2, and n_6 are the inplane forces (see Fig. E11). The underlined terms in the strains are contributions caused by the large lateral displacement introduced during buckling.

12. Show that the transverse stresses derived from the stress equilibrium equations for the refined higher-order theory (for simply supported, orthotropic rectangular plates) are given by

$$\bar{\sigma}_z = \sigma_z - \frac{h^3}{480}\left\{\left[1 + \left(\frac{2z}{h}\right)^5\right] + 10\left[1 + \left(\frac{2z}{h}\right)\right]\right\}$$

$$\cdot\left\{\left[\alpha^4 Q_{11} + 2(2G_{12} + Q_{12})\alpha^2\beta^2 + \beta^4 Q_{22}\right]W\right.$$

$$+ \left[\alpha^3 Q_{11} + \alpha\beta^2(2G_{12} + Q_{12})\right]X$$

$$\left. + \left[\alpha^2\beta(2G_{12} + Q_{12}) + \beta^3 Q_{22}\right]Y\right\}\sin\alpha x \sin\beta y$$

$$\bar{\sigma}_{xz} = \sigma_{xz} + \frac{h^2}{48}\left[1 - \left(\frac{2z}{h}\right)^4\right]$$

$$\cdot\left\{\left[\alpha^3 Q_{11} + (2G_{12} + Q_{12})\alpha\beta^2\right]W\right.$$

$$\left. + \left(\alpha^2 Q_{11} + \beta^2 G_{12}\right)X + \alpha\beta(G_{12} + Q_{12})Y\right\}\cos\alpha x \sin\beta y$$

$$\bar{\sigma}_{yz} = \sigma_{yz} + \frac{h^2}{48}\left[1 - \left(\frac{2z}{h}\right)^4\right]$$

$$\cdot\left\{\left[(2G_{12} + Q_{12})\alpha^2\beta + \beta^3 Q_{22}\right]W\right.$$

$$\left. + \alpha\beta(G_{12} + Q_{12})X + \left(\alpha^2 G_{12} + \beta^2 Q_{22}\right)Y\right\}\sin\alpha x \cos\beta y$$

where σ_z, σ_{xz}, and σ_{yz} are the expressions given in Exercise 9.

13. Show that the critical buckling load for a simply supported, rectangular plate subjected to edge compression on sides $x = 0$ and $x = a$ can be computed from (for the classical plate theory)

$$\frac{N_1}{n_1} = \left[D_{11}\alpha^2 + 2(D_{12} + 2D_{66})\beta^2 + D_{22}\left(\frac{\beta^4}{\alpha^2}\right)\right]$$

where $\alpha = (m\pi/a)$, $\beta = (n\pi/b)$, and n_1 is the compressive load.

14. Show that the natural frequencies in the classical plate theory of a simply supported rectangular plate can be computed from the expression

$$\rho\omega^2 = \left[D_{11}\alpha^4 + 2(D_{12} + 2D_{66})\alpha^2\beta^2 + D_{22}\beta^4\right]$$

where ρ is the density of the material of the plate, and α and β are the same parameters defined in Exercise 13.

4.3 LAMINATED COMPOSITE PLATES

4.3.1 Introduction

The phrase "composite material" refers to material that is obtained by combining two or more materials on a macroscopic scale. Composite materials, unlike metals, are man-made and, therefore, the constituents of composite materials can be selected and combined so as to produce a useful material that has desired properties, such as high strength, high stiffness, greater corrosion resistance, greater fatigue life, low weight, and so on. Some examples of laminated composites are: *bimetals*, which are laminates of two different metals; laminated glass (used for automobile windows), which consists of polyvinyl butyral sandwiched between two layers of glass; and *laminated fiber-reinforced composites*, which consist of layers of fibrous composites (composite materials that are made by combining long fibers in a matrix material) (see Fig. 4.12).

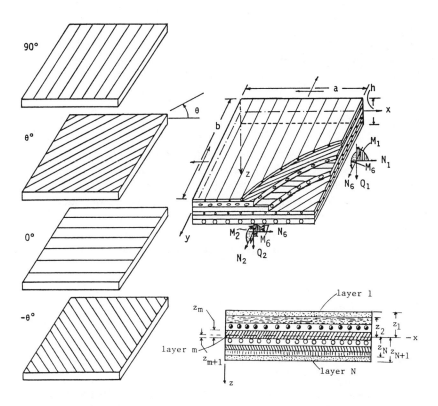

(a) Laminate construction

(b) Laminated plate geometry and the coordinate system

Figure 4.12 Laminated plates—construction, geometry, and the coordinate system.

The uses of composite materials in civil and aerospace structures in the last two decades have increased not only in number, but also in the breadth of applications, which range from medical prosthetic devices to sports equipment, electronics, appliances, automotive parts, and high performance aircraft structures. The use of fiber-reinforced composite laminates in structural applications results from the outstanding strength, stiffness, low specific gravity of fibers, and added degree of flexibility (the material properties can be tailored through the variation of the stacking sequence and fiber orientation). Coupled with these advantages is the disadvantage, from the analytical point of view, of bending–stretching coupling. In other words, for laminated composite plates, one should solve, in general, for both inplane displacements and bending deflections.

The first complete laminated anisotropic plate theory is attributed to Reissner and Stavsky (1961). The laminated version of the Levy–Reissner–Mindlin shear deformation plate theories is due to Yang, Norris, and Stavsky (1966) and Whitney and Pagano (1970). Higher-order laminated plate theories are reviewed by Librescu (1975) and Lo, Christensen, and Wu (1977). Also, the works of Green and Naghdi (1982) and Reddy (1985) should be mentioned.

The present study of laminated plates deals with the first-order shear deformation theory, and with a brief account of the refined theory, both of which account for stretching–bending coupling. Of course, one can obtain the theories in Sections 4.2.2 and 4.2.3 as special cases from the present theories.

4.3.2   A First-Order Theory

Consider a laminate consisting of N thin orthotropic layers, each having constant thickness. The mth layer ($m = 1, 2, \ldots, N$) is oriented at an arbitrary angle, θ_m, with respect to the plate coordinate axes (see Fig. 4.12). The xy plane coincides with the midplane of the plate. Here, we use the same notation as that introduced in Section 4.2.3 for displacements, strains, stresses, and resultants.

We begin with the following displacement field [see Eq. (4.2.4)],

$$u_1(x, y, z, t) = u(x, y, t) + z\psi_x(x, y, t)$$

$$u_2(x, y, z, t) = v(x, y, t) + z\psi_y(x, y, t)$$

$$u_3(x, y, z, t) = w(x, y, t) \qquad (4.3.1)$$

Here t is the time; u_1, u_2, and u_3 are the displacements in the x, y, and z directions, respectively; u, v, and w are the associated midplane displacements; and ψ_x and ψ_y are the rotations in the xz and yz planes owing to bending only. Assuming that the displacements are small, we derive the strain–displacement relations for the plate. Using Eq. (1.3.11), we obtain

$$e_1 \equiv e_{11} = \frac{\partial u}{\partial x} + z\frac{\partial \psi_x}{\partial x} = e_1^0 + z\kappa_1^0$$

$$e_2 \equiv e_{22} = \frac{\partial v}{\partial y} + z\frac{\partial \psi_y}{\partial y} = e_2^0 + z\kappa_2^0$$

$$e_3 \equiv e_{33} = 0$$

$$e_4 \equiv 2e_{23} = \psi_y + \frac{\partial w}{\partial y}$$

$$e_5 \equiv 2e_{13} = \psi_x + \frac{\partial w}{\partial x}$$

$$e_6 \equiv 2e_{12} = \frac{\partial u}{\partial y} + \frac{\partial v}{\partial x} + z\left(\frac{\partial \psi_x}{\partial y} + \frac{\partial \psi_y}{\partial x}\right) = e_6^0 + z\kappa_6^0 \qquad (4.3.2)$$

where e_i^0 and κ_i^0 are defined by Eq. (4.2.36).

To derive the equations of motion, we use Hamilton's principle:

$$\int_{t_0}^{t_1} \delta L \, dt = 0 \qquad (4.3.3)$$

where δL is the first variation of the Lagrangian,

$$\delta L = \delta\Pi - \delta K$$

$$= \int_R \left(\delta e_1^0 N_1 + \delta\kappa_1^0 M_1 + \delta e_2^0 N_2 + \delta\kappa_2^0 M_2\right.$$

$$\left. + \delta e_4 Q_2 + \delta e_5 Q_1 + \delta e_6^0 N_6 + \delta\kappa_6^0 M_6 - \delta w q\right) dx \, dy$$

$$- \int_{C_1} \hat{N}_n \delta u_n \, dS - \int_{C_2} \hat{N}_s \delta u_s \, dS - \int_{C_3} \hat{M}_n \delta\psi_n \, dS$$

$$- \int_{C_4} \hat{M}_s \delta\psi_s \, dS - \int_{C_5} \hat{Q}_n \delta w \, dS$$

$$- \int_V \rho\left[\dot{u}_1 \delta\dot{u}_1 + \dot{u}_2 \delta\dot{u}_2 + \dot{u}_3 \delta\dot{u}_3 + z^2\left(\dot{\psi}_x \delta\dot{\psi}_x + \dot{\psi}_y \delta\dot{\psi}_y\right)\right] dV \qquad (4.3.4)$$

and Π and K are the total potential and kinetic energies. Using Eq. (4.3.2) in Eq. (4.3.4), substituting the result into Eq. (4.3.3), and integrating by parts both with respect to time t and spatial coordinates (x, y, z), we obtain

$$\delta u: \quad N_{1,x} + N_{6,y} = I_1 \ddot{u} + I_2 \ddot{\psi}_x$$

$$\delta v: \quad N_{6,x} + N_{2,y} = I_1 \ddot{v} + I_2 \ddot{\psi}_y$$

$$\delta w: \quad Q_{1,x} + Q_{2,y} = q + I_1 \ddot{w} \qquad \text{in } R \quad (4.3.5)$$

$$\delta \psi_x: \quad M_{1,x} + M_{6,y} - Q_1 = I_3 \ddot{\psi}_x + I_2 \ddot{u}$$

$$\delta \psi_y: \quad M_{6,x} + M_{2,y} - Q_2 = I_3 \ddot{\psi}_y + I_2 \ddot{v}$$

and the boundary conditions in Eq. (4.2.9). Here I_1, I_2, and I_3 are the normal, coupled normal-rotary, and rotary inertia coefficients,

$$(I_1, I_2, I_3) = \int_{-h/2}^{h/2} (1, z, z^2) \rho \, dz = \sum_{m=1}^{N} \int_{z_m}^{z_{m+1}} (1, z, z^2) \rho^{(m)} \, dz \quad (4.3.6)$$

and $\rho^{(m)}$ is the material density of the mth layer, located between the $z = z_m$ and $z = z_{m+1}$ planes (see Fig. 4.12), q is the transverse distributed force, and N_i, Q_i, and M_i are the stress and moment resultants (see Fig. 4.13) given by Eqs. (4.2.40):

$$(N_i, M_i) = \int_{-h/2}^{h/2} (1, z) \sigma_i \, dz, \qquad (Q_1, Q_2) = \int_{-h/2}^{h/2} (\sigma_5, \sigma_4) \, dz \quad (4.3.7)$$

Here σ_i $(i = 1, 2, \dots)$ denote the in-plane stress components ($\sigma_1 = \sigma_x$, $\sigma_2 = \sigma_y$, $\sigma_4 = \sigma_{yz}$, $\sigma_5 = \sigma_{xz}$, and $\sigma_6 = \sigma_{xy}$).

The equations derived thus far are identical to those of ordinary (i.e., not laminated) plates, except for the definition of the inertia coefficients in Eq. (4.3.6). Indeed, the theory of layered composite plates differs from that of ordinary plates in the description of the constitutive behavior. The constitutive equations of an individual lamina m are of the form [see Eq. (1.5.3b)]

$$\sigma_i' = c_{ij}'^{(m)} e_j' \qquad (4.3.8)$$

where σ_i' and e_j' are referred to lamina material coordinates $x_1' = x'$ and $x_2' = y'$, and $c_{ij}'^{(m)}$ are the elastic constants of the mth lamina. To derive the plate constitutive equations, we transform Eq. (4.3.8) to stresses in plate coordinates (x, y, z). This is accomplished by using Eq. (1.5.29),

$$Q_{ij}^{(m)} = Q_{ij}^{(m)} \left(c_{ii}'^{(m)}, \theta_m \right) \qquad (4.3.9)$$

where $Q_{ij}^{(m)}$, equal to the c_{ij}' of Eq. (1.5.29), are the material coefficients of the mth lamina referred to the plate coordinates. Thus, we have

$$\sigma_i = Q_{ij}^{(m)} e_j \qquad (4.3.10)$$

where σ_i and e_j are the stresses and strains in the mth lamina referred to the

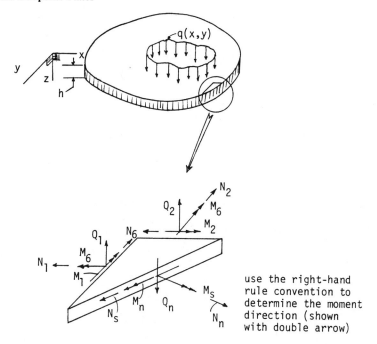

use the right-hand
rule convention to
determine the moment
direction (shown
with double arrow)

Boundary conditions for plates

clamped	SS-1 simply-supported-1	SS-2 simply-supported-2	free
$w = 0$	$w = 0$	$w = 0$	$Q_n = 0$
$\psi_s = 0$	$M_s = 0$	$\psi_s = 0$	$M_s = 0$
$\psi_n = 0$	$M_n = 0$	$M_n = 0$	$M_n = 0$

Figure 4.13 Geometry, moment, and shear force resultants and various boundary conditions for a plate element.

plate axes. Substitution of Eq. (4.3.10) into Eq. (4.3.7) gives, for example,

$$N_i = \int_{-h/2}^{h/2} \sigma_i \, dz = \sum_{m=1}^{N} \int_{z_m}^{z_{m+1}} \sigma_i \, dz$$

$$= \sum_{m=1}^{N} \int_{z_m}^{z_{m+1}} Q_{ij}^{(m)} e_j \, dz = \sum_{m=1}^{N} \int_{z_m}^{z_{m+1}} Q_{ij}^{(m)} \left(e_j^0 + z \kappa_j^0 \right) dz$$

$$= A_{ij} e_j^0 + B_{ij} \kappa_j^0 \tag{4.3.11a}$$

where N is the number of layers in the laminate (i.e., plate). Similarly, we have

$$M_i = \sum_{m=1}^{N} \int_{z_m}^{z_{m+1}} Q_{ij}^{(m)} \left(z e_j^0 + z^2 \kappa_j^0 \right) dz$$

$$= B_{ij} e_j^0 + D_{ij} \kappa_j^0 \qquad (4.3.11b)$$

where A_{ij}, B_{ij}, and D_{ij} $(i, j = 1, 2, 6)$ are the respective inplane, bending–stretching, the bending stiffnesses of the plate

$$\left(A_{ij}, B_{ij}, D_{ij} \right) = \sum_{m=1}^{N} \int_{z_m}^{z_{m+1}} Q_{ij}^{(m)} (1, z, z^2) \, dz \qquad (i, j = 1, 2, 6) \quad (4.3.12)$$

The laminate constitutive equations can be expressed in the form

$$\left\{ \begin{array}{c} \{N\} \\ \{M\} \end{array} \right\} = \begin{bmatrix} [A] & [B] \\ [B]^T & [D] \end{bmatrix} \left\{ \begin{array}{c} \{e\} \\ \{\kappa\} \end{array} \right\}$$

$$\left\{ \begin{array}{c} Q_1 \\ Q_2 \end{array} \right\} = \begin{bmatrix} A_{55} & A_{45} \\ A_{45} & A_{44} \end{bmatrix} \left\{ \begin{array}{c} e_5 \\ e_4 \end{array} \right\} \qquad (4.3.13)$$

where A_{ij} $(i, j = 4, 5)$ denote the stiffness coefficients

$$A_{ij} = \sum_{m=1}^{N} K_i K_j \int_{z_m}^{z_{m+1}} Q_{ij}^{(m)} \, dz, \qquad (i, j = 4, 5) \qquad (4.3.14)$$

and K_i are the shear correction coefficients.

Equations (4.3.13) and (4.3.14) can be used to rewrite Eq. (4.3.5) completely in terms of the generalized displacements u, v, w, ψ_x, and ψ_y. The resulting equations can be expressed in matrix form as

$$[L]\{\Delta\} = \{f\} \qquad (4.3.15a)$$

where $\{\Delta\} = \{u, v, w, \psi_x, \psi_y\}^T$, $[L]$ is the (symmetric) matrix of differential operators, and $\{f\}$ is the column of forces:

$$L_{11} = A_{11} d_{11} + 2 A_{16} d_{12} + A_{66} d_{22} - I_1 d_{tt}$$

$$L_{12} = (A_{12} + A_{66}) d_{12} + A_{16} d_{11} + A_{26} d_{22}$$

$$L_{13} = 0$$

$$L_{14} = B_{11} d_{11} + 2 B_{16} d_{12} + B_{66} d_{22} - I_2 d_{tt}$$

$$L_{15} = (B_{12} + B_{66}) d_{12} + B_{16} d_{11} + B_{26} d_{22} = L_{24}$$

$$L_{22} = 2 A_{26} d_{12} + A_{22} d_{22} + A_{66} d_{11} - I_1 d_{tt}$$

$$L_{23} = 0$$

$$L_{25} = 2 B_{26} d_{12} + B_{22} d_{22} + B_{66} d_{11} - I_2 d_{tt}$$

$$L_{33} = -A_{55} d_{11} - 2 A_{45} d_{12} - A_{44} d_{22} - I_1 d_{tt}$$

$$L_{34} = A_{55} d_1 - A_{45} d_2$$

$$L_{35} = -A_{45}d_1 - A_{44}d_2,$$

$$L_{44} = D_{11}d_{11} + 2D_{16}d_{12} + D_{66}d_{22} - A_{55} - I_3 d_{tt}$$

$$L_{45} = (D_{12} + D_{66})d_{12} + D_{16}d_{11} + D_{26}d_{22} - A_{45}$$

$$L_{55} = 2D_{26}d_{12} + D_{22}d_{22} + D_{66}d_{11} - A_{44} - I_3 d_{tt}$$

$$f_1 = 0, \qquad f_2 = 0, \qquad f_3 = q, \qquad f_4 = 0, \qquad f_5 = 0 \quad (4.3.15b)$$

where $d_{ij} = (\partial^2/\partial x_i \, \partial x_j)$ $(x_1 = x$ and $x_2 = y)$, and $d_{tt} = (\partial^2/\partial t^2)$. This completes the derivation of the equations of motion governing a laminated composite plate.

For classical plate theory (CPT), Eq. (4.3.15) has the following form:

$$\begin{bmatrix} L_{11} & L_{12} & L_{13} \\ L_{12} & L_{22} & L_{23} \\ L_{13} & L_{23} & L_{33} \end{bmatrix} \begin{Bmatrix} u \\ v \\ w \end{Bmatrix} = \begin{Bmatrix} 0 \\ 0 \\ q \end{Bmatrix} \qquad (4.3.16a)$$

where the coefficients L_{ij} are given by $[d_{ijkl} = d_{ij}d_{kl}, d_{ijk} = d_{ij}(\partial/\partial x_k)]$

$$L_{11} = A_{11}d_{11} + 2A_{16}d_{12} + A_{66}d_{22}$$

$$L_{12} = A_{16}d_{11} + (A_{12} + A_{66})d_{12} + A_{26}d_{22}$$

$$L_{13} = -B_{11}d_{111} - 3B_{16}d_{112} - (B_{12} + 2B_{66})d_{122} - B_{26}d_{222}$$

$$L_{22} = A_{66}d_{11} + 2A_{26}d_{12} + A_{22}d_{22}$$

$$L_{23} = -B_{16}d_{111} - (B_{12} + 2B_{66})d_{112} - 3B_{26}d_{122} - B_{22}d_{222}$$

$$L_{33} = -[D_{11}d_{1111} + 4D_{16}d_{111}d_2 + 2(D_{12} + 2D_{66})d_{1122}$$

$$+ 4D_{26}d_{1222} + D_{22}d_{2222}] \qquad (4.3.16b)$$

Exact Solution of the Equilibrium Equations

The boundary-value problem associated with the equilibrium of layered composite plates involves solving the operator equation (4.3.15) subjected to a given set of boundary conditions. It is not possible to construct exact solutions of Eq. (4.3.15) when the plate is of arbitrary geometry, constructed of arbitrarily-oriented layers, and subjected to an arbitrary loading or boundary conditions. However, series solutions of Eq. (4.3.15) can be obtained for rectangular plates made of two classes of lamination schemes:

1. *Cross-ply* laminated plates, which have N unidirectionally reinforced (orthotropic) layers with material directions oriented at $0°$ or $90°$ to the

plate coordinate axes such that the following plate stiffnesses are zero:

$$A_{16} = A_{26} = B_{16} = B_{26} = D_{16} = D_{26} = A_{45} = 0 \quad (4.3.17)$$

2. *Antisymmetric angle-ply* laminated plates, which have an even number N of orthotropic layers with material directions alternatingly oriented at θ and $-\theta$ to the plate coordinate axes. For antisymmetric angle-ply plates, the following plate stiffnesses are zero:

$$A_{16} = A_{26} = A_{45} = B_{11} = B_{12} = B_{22} = B_{66} = D_{16} = D_{26} = 0 \quad (4.3.18)$$

The solutions are obtained in the same way as in Navier's method, where appropriate forms of the solutions (with arbitrary coefficients) that satisfy the boundary conditions are assumed and the coefficients in the assumed solution are computed by substituting the assumed displacement functions into the equations of equilibrium. It turns out that for composite plates, this procedure yields solutions for only two types of plates, namely, rectangular plates for which Eqs. (4.3.17) or (4.3.18) hold. Of course, the solutions exist for specific boundary conditions for each of the two classes listed above. We discuss these solutions below. Both bending and free vibration of these two classes of plates are discussed.

Cross-Ply Plates in Bending and Vibration

1. *Boundary Conditions.* The plate is assumed to be simply supported in the following sense (see Fig. 4.6 for the geometry and coordinates):

$$u(x,0) = u(x,b) = 0, \qquad N_2(x,0) = N_2(x,b) = 0$$

$$v(0,y) = v(a,y) = 0, \qquad N_1(0,y) = N_1(a,y) = 0$$

$$w(x,0) = w(x,b) = w(0,y) = w(a,y) = 0$$

$$\psi_x(x,0) = \psi_x(x,b) = 0, \qquad M_2(x,0) = M_2(x,b) = 0$$

$$\psi_y(0,y) = \psi_y(a,y) = 0, \qquad M_1(0,y) = M_1(a,y) = 0 \quad (4.3.19)$$

2. *Loading.* The loading is assumed to be expandable in terms of the double-Fourier series,

$$q = \sum_{m,n=1}^{\infty} Q_{mn} \sin \alpha x \sin \beta y \quad (4.3.20)$$

where $\alpha = m\pi/a$ and $\beta = n\pi/b$.

3. *Solution.* The solution is given by

$$u = \sum_{m,n=1}^{\infty} U_{mn}\cos\alpha x \sin\beta y$$

$$v = \sum_{m,n=1}^{\infty} V_{mn}\sin\alpha x \cos\beta y$$

$$w = \sum_{m,n=1}^{\infty} W_{mn}\sin\alpha x \sin\beta y$$

$$\psi_x = \sum_{m,n=1}^{\infty} X_{mn}\cos\alpha x \sin\beta y$$

$$\psi_y = \sum_{m,n=1}^{\infty} Y_{mn}\sin\alpha x \cos\beta y \qquad (4.3.21)$$

where U_{mn}, V_{mn}, and so on, are parameters to be determined subjected to the condition that the solution in Eq. (4.3.21) satisfies the operator equation (4.3.15). Substituting Eq. (4.3.21) into Eq. (4.3.15) we obtain

$$[C]\begin{Bmatrix} U_{mn} \\ V_{mn} \\ W_{mn} \\ X_{mn} \\ Y_{mn} \end{Bmatrix} = \{F\} \qquad (4.3.22)$$

The elements of the coefficient matrix $[C]$ and column vector $\{F\}$ are given by

$$C_{11} = A_{11}\alpha^2 + A_{66}\beta^2, \qquad C_{12} = (A_{12} + A_{66})\alpha\beta$$

$$C_{13} = 0, \qquad C_{14} = \alpha^2 B_{11} + B_{66}\beta^2, \qquad C_{15} = (B_{12} + B_{66})\alpha\beta$$

$$C_{22} = A_{22}\beta^2 + A_{66}\alpha^2, \qquad C_{23} = 0, \qquad C_{24} = C_{15}$$

$$C_{25} = B_{22}\beta^2 + B_{66}\alpha^2, \qquad C_{33} = \alpha^2 A_{55} + \beta^2 A_{44}$$

$$C_{34} = \alpha A_{55}, \qquad C_{35} = \beta A_{44}, \qquad C_{44} = D_{11}\alpha^2 + D_{66}\beta^2 + A_{55}$$

$$C_{45} = (D_{12} + D_{66})\alpha\beta, \qquad C_{55} = D_{66}\alpha^2 + D_{22}\beta^2 + A_{44}$$

$$F_1 = 0, \qquad F_2 = 0, \qquad F_3 = Q_{mn}, \qquad F_4 = 0, \qquad F_5 = 0 \quad (4.3.23)$$

Thus, for a given $\alpha = m\pi/a$, $\beta = n\pi/b$, generalized forces F_i, and cross-ply construction, one needs to solve the 5×5 matrix equation (4.3.22) for the vector of amplitudes of the generalized displacements.

Natural vibration frequencies ω can also be obtained for the boundary conditions and cross-ply construction discussed above by assuming a solution of the form

$$u = U(x,y)\cos\omega t, \qquad v = V(x,y)\cos\omega t, \qquad w = W(x,y)\cos\omega t$$

$$\psi_x = X(x,y)\cos\omega t, \qquad \psi_y = Y(x,y)\cos\omega t \qquad (4.3.24)$$

where $U(x, y)$, $V(x, y)$, and so on, are the series expressions in Eq. (4.3.21). Substituting Eq. (4.3.24) into Eq. (4.3.15) (with $q = 0$), we obtain the eigenvalue equation

$$([C] - \omega^2[M]) \begin{Bmatrix} U_{mn} \\ V_{mn} \\ W_{mn} \\ X_{mn} \\ Y_{mn} \end{Bmatrix} = \{0\} \tag{4.3.25}$$

where the elements $M_{ij} = M_{ji}$ of the mass matrix $[M]$ are given by

$$M_{11} = M_{22} = M_{33} = I_1, \qquad M_{14} = M_{25} = I_2, \qquad M_{44} = M_{55} = I_3 \tag{4.3.26}$$

and the remaining elements are zero.

Angle-Ply Plates in Bending and Vibration

1. *Boundary Conditions.*

$$u(0, y) = u(a, y) = 0, \qquad N_6(0, y) = N_6(a, y) = 0$$
$$v(x, 0) = v(x, b) = 0, \qquad N_6(x, 0) = N_6(x, b) = 0$$
$$w(x, 0) = w(x, b) = w(0, y) = w(a, y) = 0$$
$$\psi_x(x, 0) = \psi_x(x, b) = 0, \qquad M_2(x, 0) = M_2(x, b) = 0$$
$$\psi_y(0, y) = \psi_y(a, y) = 0, \qquad M_1(0, y) = M_1(a, y) = 0 \tag{4.3.27}$$

2. *Loading.* The loading is assumed to be the same form as in Eq. (4.3.20).

3. *Solution.* The solution to the associated boundary-value problem is of the form (multiply the functions with $\cos \omega t$ for the eigenvalue problem)

$$u = \sum_{m,n=1}^{\infty} U_{mn} \sin \alpha x \cos \beta y$$

$$v = \sum_{m,n=1}^{\infty} V_{mn} \cos \alpha x \sin \beta y$$

$$w = \sum_{m,n=1}^{\infty} W_{mn} \sin \alpha x \sin \beta y$$

$$\psi_x = \sum_{m,n=1}^{\infty} X_{mn} \cos \alpha x \sin \beta y$$

$$\psi_y = \sum_{m,n=1}^{\infty} Y_{mn} \sin \alpha x \cos \beta y \tag{4.3.28}$$

Substituting Eq. (4.3.28) into Eq. (4.3.15), we obtain

$$[K]\begin{Bmatrix} U_{mn} \\ V_{mn} \\ W_{mn} \\ X_{mn} \\ Y_{mn} \end{Bmatrix} = \omega^2 [M] \begin{Bmatrix} U_{mn} \\ V_{mn} \\ W_{mn} \\ X_{mn} \\ Y_{mn} \end{Bmatrix} + \{F\} \qquad (4.3.29)$$

The coefficient matrices $[K]$ and $[M]$ and column vector $\{F\}$ are given by

$$K_{11} = A_{11}\alpha^2 + A_{66}\beta^2, \qquad K_{12} = \alpha\beta(A_{12} + A_{66})$$

$$K_{13} = 0, \qquad K_{14} = 2\alpha\beta B_{16}, \qquad K_{15} = \alpha^2 B_{16} + \beta^2 B_{26}$$

$$K_{22} = \alpha^2 A_{66} + \beta^2 A_{22}, \qquad K_{23} = 0, \qquad K_{24} = K_{15}$$

$$K_{25} = 2\alpha\beta B_{26}, \qquad K_{33} = \beta^2 A_{55} + \alpha^2 A_{44}$$

$$K_{34} = \alpha A_{55}, \qquad K_{35} = \beta A_{44}, \qquad K_{44} = \alpha^2 D_{11} + \beta^2 D_{66} + A_{55}$$

$$K_{45} = \alpha\beta(D_{12} + D_{66}), \qquad K_{55} = \alpha^2 D_{66} + \beta^2 D_{22} + A_{44}$$

$$M_{11} = M_{22} = M_{33} = I_1, \qquad M_{14} = M_{25} = I_2, \qquad M_{44} = M_{55} = I_3$$

$$F_1 = 0, \qquad F_2 = 0, \qquad F_3 = Q_{mn}, \qquad F_4 = 0, \qquad F_5 = 0 \qquad (4.3.30)$$

For plates in bending, set ω^2 to zero in Eq. (4.3.29) and solve the equation for the amplitudes U_{mn}, V_{mn}, W_{mn}, X_{mn}, and Y_{mn}, and then the solution is given by Eq. (4.3.28). For uniformly distributed and point loads, one should sum the series for $m, n = 1, 3, 5, \ldots$ to obtain the solution. For free vibration, set vector F to zero and solve the resulting eigenvalue problem for frequencies (and associated eigenvectors).

Now we consider some examples wherein we utilize equations (4.3.22), (4.3.25), and (4.3. 29) to obtain bending deflections and natural frequencies of simply supported, rectangular plates. Of course, the lamination schemes are restricted to those of cross-ply and antisymmetric angle-ply [see Eqs. (4.3.17) and (4.3.18)].

Example 4.12

Consider a laminated square plate constructed of an advanced composite material whose engineering constants are given by

$$\frac{E_1}{E_2} = 25, \qquad \frac{G_{23}}{E_2} = 0.2, \qquad \frac{G_{13}}{E_2} = 0.5, \qquad G_{13} = G_{12}, \qquad \nu_{12} = 0.25$$

$$(4.3.31)$$

The value of the shear correction factors $K_1^2 = K_2^2$ is taken to be $\frac{5}{6}$. Here we consider the bending of simply supported laminates under sinusoidal and uniformly distributed loads.

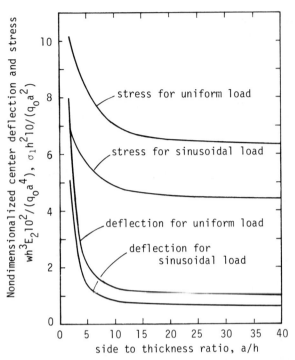

Figure 4.14 Nondimensionalized center deflection and stress versus side to thickness ratio for two-layer $(0°/90°)$ square plates.

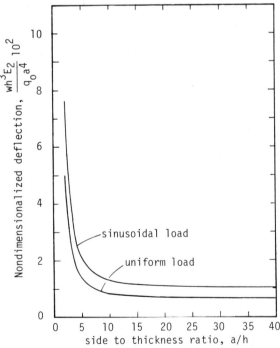

Figure 4.15 Nondimensionalized deflection versus side to thickness ratio of two-layer $(-45°/45°)$ square plates.

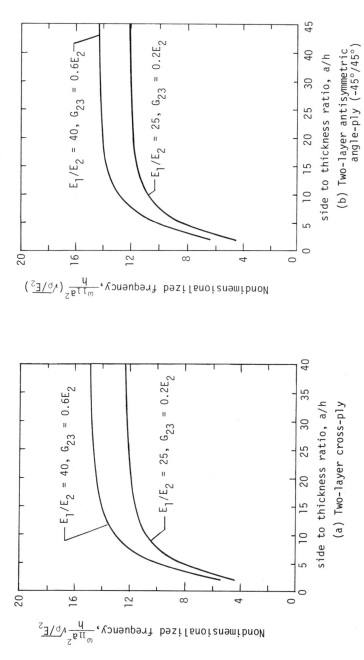

Figure 4.16 Nondimensionalized fundamental frequencies versus side to thickness ratio for two-layer square plates of high modulus composites.

[see Eqs. (4.3.8)–(4.3.14)] where A_{ij}, B_{ij}, and so on, are the plate stiffnesses, defined by

$$\left(A_{ij}, B_{ij}, D_{ij}, E_{ij}, F_{ij}, H_{ij} \right) = \sum_{m=1}^{N} \int_{z_m}^{z_{m+1}} Q_{ij}^{(m)} \left(1, z, z^2, z^3, z^4, z^6\right) dz$$

$$(i, j = 1, 2, 6)$$

$$\left(A_{ij}, D_{ij}, F_{ij} \right) = \sum_{m=1}^{N} \int_{z_m}^{z_{m+1}} Q_{ij}^{(m)} \left(1, z^2, z^4\right) dz \quad (i, j = 4, 5) \qquad (4.3.34)$$

and $Q_{ij}^{(m)}$ are the stiffnesses of the mth lamina in the plate coordinates.

Exact Solutions for Symmetric Cross-Ply Plates

Here we consider the exact solutions of Eqs. (4.2.51) for simply supported, symmetric cross-ply rectangular plates. For symmetric (about the midplane) cross-ply plates, the following plate stiffnesses are identically zero:

$$B_{ij} = E_{ij} = 0 \quad \text{for } i, j = 1, 2, 4, 5, 6$$

$$A_{16} = A_{26} = D_{16} = D_{26} = F_{16} = F_{26} = H_{16} = H_{26} = 0$$

$$A_{45} = D_{45} = F_{45} = 0 \qquad (4.3.35)$$

Thus, the stretching–bending coupling is zero. Therefore, Eqs. (4.2.51) are valid for the symmetric cross-ply plates; of course, the plate stiffnesses A_{ij}, D_{ij}, F_{ij}, and H_{ij} should be calculated using Eq. (4.3.34). For the boundary conditions in Eq. (4.2.49) and loading of the form in Eq. (4.2.53), the solution of Eqs. (4.2.51) is given by Eq. (4.2.52), wherein the amplitudes W_{mn}, X_{mn}, and Y_{mn} are obtained from Eq. (4.2.54).

Example 4.14

Here we present numerical results for two example problems and discuss the accuracy of the present solutions over those obtained by the first-order shear deformation theory by comparing both results with the three-dimensional elasticity solutions of Pagano (1970). The two problems are:

1. A three-ply rectangular laminate with layers of equal thickness, which is subjected to a sinusoidally distributed transverse load.
2. A three-ply square laminate with layers of equal thickness, which is subjected to a uniformly distributed transverse load.

In both problems, the lamina properties are assumed to be the same as those

Table 4.6 Nondimensionalized[a] Deflections and Stresses in a Rectangular Laminate Under Sinusoidal Load $(0°/90°/0°, h_i = h/3, b/a = 3)$

a/h	Source	$\bar{w}$	$\bar{\sigma}_1$	$\bar{\sigma}_2$	$\bar{\sigma}_4$	$\bar{\sigma}_5$	$-\bar{\sigma}_6$
4	Pagano[b]	2.82	1.10	0.119	0.0334	0.387	0.0281
	HSDPT	2.6411	1.0356	0.1028	0.0348	0.2724	0.0263
	FSDPT	2.3626	0.6130	0.0934	0.0308	0.1879	0.0205
10	Pagano	0.919	0.725	0.0435	0.0152	0.420	0.0123
	HSDPT	0.8622	0.6924	0.0398	0.0170	0.2859	0.0115
	FSDPT	0.803	0.6214	0.0375	0.0159	0.1894	0.0105
20	Pagano	0.610	0.650	0.0299	0.0119	0.434	0.0093
	HSDPT	0.5937	0.6407	0.0289	0.0139	0.2880	0.0091
	FSDPT	0.5784	0.6228	0.0283	0.0135	0.1896	0.0088
100	Pagano	0.508	0.624	0.0253	0.0108	0.439	0.0083
	HSDPT	0.507	0.624	0.0253	0.0129	0.2886	0.0083
	FSDPT	0.5064	0.6233	0.0253	0.0127	0.1897	0.0083
	CPT	0.503	0.623	0.0252	—	—	0.0083

[a] $\bar{w} = (wh^3 E_2/q_0 a^4)10^2$, $w = w(a/2, b/2)$; $\bar{\sigma}_1 = \sigma_1(a/2, b/2, h/2)(h^2/q_0 a^2)$; $\bar{\sigma}_2 = \sigma_2(a/2, b/2, h/6)(h^2/q_0 a^2)$; $\bar{\sigma}_4 = \sigma_4(a/2, 0, 0)(h/q_0 a)$; $\bar{\sigma}_5 = \sigma_5(0, b/2, 0)(h/q_0 a)$; $\bar{\sigma}_6 = \sigma_6(0, 0, h/2)(h^2/q_0 a^2)$.
[b] The values are taken from Pagano (1970).

given in Eq. (4.3.31). For the first-order shear deformation theory, the shear correction coefficients are taken to be $K_1^2 = K_2^2 = \frac{5}{6}$.

The results for the two problems are presented in Tables 4.6 and 4.7. From the results, one can conclude that the refined theory (HSDPT) gives more accurate solutions than the first-order shear deformation theory (FSDPT) when compared to the 3-D elasticity solution. The refined theory also gives a faster

Table 4.7 Nondimensionalized Deflections of Three-Layer Cross-Ply Square Plates Under Uniform Load $(0°/90°/0°, h_i = h/3)$

	Refined Theory			First-Order Theory		
a/h	$N = 9$[a]	$N = 29$	$N = 49$	$N = 9$	$N = 29$	$N = 49$
2	7.7681	7.7661	7.7661	7.7170	7.7066	7.7062
4	2.9103	2.9091	2.9091	2.6623	2.6597	2.6596
10	1.0903	1.0900	1.0900	1.0224	1.0220	1.0219
20	0.7761	0.7760	0.7760	0.7574	0.7573	0.7573
50	0.6839	0.6838	0.6838	0.6808	0.6807	0.6807
100	0.6705	0.6705	0.6705	0.6697	0.6697	0.6697

[a] Denotes the final value of the integers m and n in the series (4.2.52).

converging solution when compared to the FSDT. The results for uniform loading show that the series solutions converge faster for thin plates than for thick plates.

It should be noted that one can obtain series solutions to antisymmetric angle-ply plates using the present theory. The solution form is the same as that given for the first-order shear deformation theory. Stability and vibration of laminated plates using the refined theory, in view of the theory and analysis presented in this section, is straightforward (see Phan, 1984).

Exercises 4.3

1. Show that the plate stiffnesses of laminated plates constructed of homogeneous orthotropic lamina can be expressed as

$$A_{ij} = \sum_{k=1}^{N} Q_{ij}^{(k)}(z_k - z_{k-1})$$

$$B_{ij} = \frac{1}{2} \sum_{k=1}^{N} Q_{ij}^{(k)}(z_k^2 - z_{k-1}^2)$$

$$D_{ij} = \frac{1}{3} \sum_{k=1}^{N} Q_{ij}^{(k)}(z_k^3 - z_{k-1}^3)$$

2. For plane state of stress of an orthotropic lamina in principal material coordinates, the thermoelastic stress–strain relations are given by

$$\begin{Bmatrix} \sigma_1 \\ \sigma_2 \\ \sigma_6 \end{Bmatrix} = \begin{bmatrix} c_{11} & c_{12} & 0 \\ c_{12} & c_{22} & 0 \\ 0 & 0 & c_{66} \end{bmatrix} \begin{Bmatrix} e_1 - \alpha_1 T \\ e_2 - \alpha_2 T \\ 0 \end{Bmatrix}$$

where e_i is the total strain, $\alpha_i T$ is the free thermal strain, α_i $(i = 1, 2)$ are the coefficients of thermal expansion, and T is the temperature increase from a reference temperature. Derive the stress–strain relations in plate coordinates for a lamina that is oriented at θ degrees counter-clockwise from the plate axes.

3. Verify Eq. (4.3.16) for classical plate theory. The resulting operator equation is 3×3, relating u, v, and w.

4. Obtain an exact solution for cross-ply laminated, simply supported plates using classical lamination theory [i.e., obtain the solution of Eq. (4.3.16) for the classical theory of plates].

5. Repeat Exercise 4 for antisymmetric angle-ply plates.

6. Determine the exact solution of a simply supported, isotropic square plate under (a) uniform temperature change, T_0, and (b) sinusoidal

temperature change $T = T_0\sin(\pi x/a)\sin(\pi y/a)$. Use the classical plate theory.

7. Repeat Exercise 6 for an orthotropic plate.

8. The equation governing the buckling of an orthotropic plate is given by

$$D_{11}\frac{\partial^4 w}{\partial x^4} + 2(D_{12} + 2D_{66})\frac{\partial^4 w}{\partial x^2\,\partial y^2} + D_{22}\frac{\partial^4 w}{\partial y^4} + P\frac{\partial^2 w}{\partial x^2} = 0$$

where P is the compressive inplane load per unit length of the edges $x = 0$ and $x = a$. The critical value of P (i.e., buckling load) is determined by solving the differential equation. For simply supported plates with the boundary conditions

$$w = 0, \qquad M_1 = -D_{11}\frac{\partial^2 w}{\partial x^2} - D_{12}\frac{\partial^2 w}{\partial y^2} = 0 \quad \text{at } x = 0, a$$

$$w = 0, \qquad M_2 = -D_{12}\frac{\partial^2 w}{\partial x^2} - D_{22}\frac{\partial^2 w}{\partial y^2} = 0 \quad \text{at } y = 0, b$$

obtain an expression for the buckling load. *Hint:* Assume $w = \sum_{m,n=1}^{\infty} W_{mn}\sin(m\pi x/a)\sin(n\pi y/b)$, which satisfies the boundary conditions, and solve the resulting algebraic equation for P (which depends on m and n).

9. Show that the natural frequencies of a simply supported, cross-ply rectangular plate (in the classical plate theory) are given by

$$\omega_{mn}^2 = \frac{1}{\rho}\left[K_{33} + \frac{2K_{12}K_{23}K_{13} - K_{22}K_{13}^2 - K_{11}K_{23}^2}{K_{11}K_{22} - K_{12}^2}\right]$$

where

$$K_{11} = A_{11}\alpha^2 + A_{66}\beta^2$$

$$K_{12} = (A_{12} + A_{66})\alpha\beta$$

$$K_{13} = -B_{11}\alpha^3$$

$$K_{22} = A_{22}\beta^2 + A_{66}\alpha^2$$

$$K_{23} = -B_{22}\beta^3$$

$$K_{33} = D_{11}\alpha^4 + 2(D_{12} + 2D_{66})\alpha^2\beta^2 + D_{22}\beta^4$$

$$\alpha = \frac{m\pi}{a}, \quad \beta = \frac{n\pi}{b}$$

10. Repeat Exercise 9 for an antisymmetric angle-ply plate. The answer is the same as that given in Exercise 9 with the following changes:

$$K_{13} = -\left(3B_{16}\alpha^2 + B_{26}\beta^2\right)\beta$$

$$K_{23} = -\left(B_{16}\alpha^2 + 3B_{26}\beta^2\right)\alpha$$

11. Consider a rectangular orthotropic plate of dimensions a and b, simply supported at $y = 0$ and b, and clamped at $x = -(a/2)$ and $a/2$, and subjected to a uniformly distributed load of intensity q_0. Using Levy's method, derive the expression for the transverse deflection:

$$w(x, y) = \frac{q_0}{24D_{22}}\left(y^4 - 2by^3 + b^3 y\right)$$

$$+ \frac{4q_0 b^4}{\pi^5 D_{22}} \frac{1}{\alpha_1^2 - \alpha_2^2} \sum_{n=1,3,5,\ldots}^{\infty} W_n \sin\frac{n\pi y}{b}$$

where

$$W_n = \frac{1}{n^5}\left[\frac{\alpha_2\sinh\dfrac{n\pi\alpha_2 a}{2b}\cosh\dfrac{n\pi\alpha_1 x}{b} - \alpha_1\sinh\dfrac{n\pi\alpha_1 a}{2b}\cosh\dfrac{n\pi\alpha_2 x}{b}}{\alpha_1\sinh\dfrac{n\pi\alpha_1 a}{2b}\cosh\dfrac{n\pi\alpha_2 a}{2b} - \alpha_2\sinh\dfrac{n\pi\alpha_2 a}{2b}\cosh\dfrac{n\pi\alpha_1 a}{2b}}\right]$$

and α_1 and α_2 are the roots of the equation

$$D_{11}\alpha^4 - 2(D_{12} + 2D_{66})\alpha^2 + D_{22} = 0$$

12. Show that the result of the above problem can be applied to symmetric cross-ply plates.

4.4 THEORY OF SHELLS

4.4.1 Introduction

In Sections 4.1–4.3, we studied the theory and analysis of flat plates. We now extend the theory to curved plates and surfaces, better known as *shells*. Shells are common structural elements in many engineering structures, including pressure vessels, submarine hulls, ship hulls, wings and fuselages of airplanes, pipes, exteriors of rockets, missiles, automobile tires, concrete roofs, containers of liquids, and many other structures. The theory of laminated shells includes the theories of ordinary shells, flat plates, and curved beams as special cases.

Therefore, in the present study, we consider the theory of laminated composite shells.

A number of theories exist for layered anisotropic shells. Many of these theories were developed originally for thin shells, and are based on the Kirchhoff–Love kinematic hypothesis that straight lines normal to the undeformed midsurface remain straight and normal to the middle surface after deformation and undergo no thickness stretching. Surveys of various shell theories can be found in the works of Naghdi (1956) and Bert (1974b), and a detailed study of thin ordinary (i.e., not laminated) shells can be found in the monographs by Kraus (1967), Ambartsumyan (1964), and Vlasov (1964).

The first analysis that incorporated the bending–stretching coupling (owing to unsymmetric lamination in composites) is due to Ambartsumyan (1953, 1964). In his analyses, Ambartsumyan assumed that the individual orthotropic layers were oriented such that the principal axes of material symmetry coincided with the principal coordinates of the shell reference surface. Thus, Ambartsumyan's work dealt with what is now known as laminated orthotropic shells, rather than laminated anisotropic shells; in laminated anisotropic shells, the individual layers are, in general, anisotropic and the principal axes of material symmetry of the individual layers coincide with only one of the principal coordinates of the shell (the thickness normal coordinate).

Dong, Pister, and Taylor (1962) formulated a theory of thin shells laminated of anisotropic material that is an extension of the theory developed by Stavsky (1963) for laminated anisotropic plates to Donnell's shallow shell theory (see Donnell, 1933). Cheng and Ho (1963) presented an analysis of laminated anisotropic cylindrical shells using Flügge's shell theory (see Flügge, 1960). A first approximation theory for the unsymmetric deformation of nonhomogeneous, anisotropic, elastic cylindrical shells was derived by Widera and Chung (1970) by means of the asymptotic integration of the elasticity equations. For a homogeneous, isotropic material, the theory reduces to Donnell's equations.

All of the theories listed above are based on Kirchhoff–Love's hypotheses, in which the transverse shear deformation is neglected. These theories, known as the Love's first-approximation theories (see Love, 1888), are expected to yield sufficiently accurate results when (1) the radius-to-thickness ratio is large, (2) the dynamic excitations are within the low-frequency range, and (3) the material anisotropy is not severe. However, the application of such theories to layered anisotropic composite shells could lead to 30% or more errors in deflections, stresses, and frequencies.

The effects of transverse shear deformation and transverse isotropy, as well as thermal expansion through the thickness of cylindrical shells, were considered by Gulati and Essenberg (1967) and Zukas and Vinson (1971). Whitney and Sun (1974) developed a shear deformation theory for laminated cylindrical shells that includes both transverse shear deformation and transverse normal strain as well as expansional strains. Recently, Reddy (1982b, 1984b) presented a generalization of Sander's shell theory (1959) to laminated, doubly-curved anisotropic shells. The theory accounts for transverse shear strains and the von

Karman (nonlinear) strains. The present study is based on these papers and is limited to the small-displacement theory of cylindrical and spherical shells.

4.4.2 Geometric Relations

Figure 4.17a contains a differential element of a shell. Here (ξ_1, ξ_2, ζ) denote the orthogonal curvilinear coordinates (or shell coordinates) such that the ξ_1- and ξ_2-curves are lines of curvature on the midsurface $\zeta = 0$, and ζ-curves are straight lines perpendicular to the midsurface, Ω. For the doubly-curved shells discussed here, the lines of principal curvature coincide with the coordinate

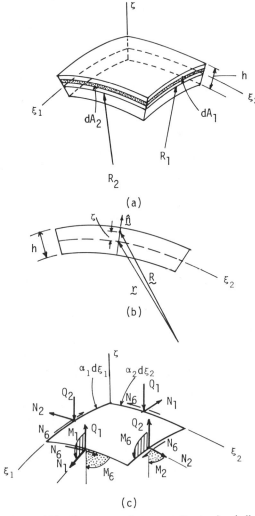

Figure 4.17   Geometry and stress resultants of a shell.

lines. The values of the principal radii of curvature of the middle surface are denoted by R_1 and R_2.

The position vector of a point on the middle surface is denoted by $\mathbf{r}$, and the position of a point at distance ζ from the middle surface is denoted by $\mathbf{R}$ (see Fig. 4.17b). The distance ds between points $(\xi_1, \xi_2, 0)$ and $(\xi_1 + d\xi_1, \xi_2 + d\xi_2, 0)$ is determined by

$$(ds)^2 = d\mathbf{r} \cdot d\mathbf{r}$$

$$= \alpha_1^2 (d\xi_1)^2 + \alpha_2^2 (d\xi_2)^2 \tag{4.4.1}$$

where $d\mathbf{r} = \mathbf{r}_1 \, d\xi_1 + \mathbf{r}_2 \, d\xi_2$, the vectors $\mathbf{r}_1$ and $\mathbf{r}_2$ ($\mathbf{r}_i = \partial \mathbf{r}/\partial \xi_i$) are tangent to the ξ_1 and ξ_2 coordinate lines, and α_1 and α_2 are the *surface metrics*

$$\alpha_1^2 = \mathbf{r}_1 \cdot \mathbf{r}_1, \qquad \alpha_2^2 = \mathbf{r}_2 \cdot \mathbf{r}_2 \tag{4.4.2}$$

An elemental area of the middle surface is determined by

$$dA = \alpha_1 \alpha_2 \, d\xi_1 \, d\xi_2 \tag{4.4.3}$$

and the unit vector normal to the surface is

$$\hat{\mathbf{n}} = \frac{\mathbf{r}_1 \times \mathbf{r}_2}{\alpha_1 \alpha_2} \tag{4.4.4}$$

Note that $\mathbf{r}_1 \cdot \mathbf{r}_2 = 0$ when the lines of principal curvature coincide with the coordinate lines. Further, we have

$$\frac{\partial \hat{\mathbf{n}}}{\partial \xi_1} = \frac{1}{R_1} \frac{\partial \mathbf{r}}{\partial \xi_1}, \qquad \frac{\partial \hat{\mathbf{n}}}{\partial \xi_2} = \frac{1}{R_2} \frac{\partial \mathbf{r}}{\partial \xi_2} \qquad \text{(a theorem of Rodrigues)}$$

$$\tag{4.4.5}$$

$$\frac{\partial}{\partial \xi_2} \left(\frac{\alpha_1}{\xi_1} \right) = \frac{1}{R_2} \frac{\partial \alpha_1}{\partial \xi_2}, \qquad \frac{\partial}{\partial \xi_1} \left(\frac{\alpha_2}{\xi_2} \right) = \frac{1}{R_1} \frac{\partial \alpha_2}{\partial \xi_1} \qquad \text{(Codazzi equations)}$$

$$\tag{4.4.6}$$

The position vector $\mathbf{R}$ of a point at distance ζ from the middle surface can be expressed in terms of $\mathbf{r}$ and $\hat{\mathbf{n}}$ by

$$\mathbf{R} = \mathbf{r} + \hat{\mathbf{n}}\zeta \tag{4.4.7}$$

By differentiation, we have

$$\frac{\partial \mathbf{R}}{\partial \xi_1} = \frac{\partial \mathbf{r}}{\partial \xi_1} + \frac{\partial \hat{\mathbf{n}}}{\partial \xi_1} \zeta, \qquad \frac{\partial \mathbf{R}}{\partial \xi_2} = \frac{\partial \mathbf{r}}{\partial \xi_2} + \frac{\partial \hat{\mathbf{n}}}{\partial \xi_2} \zeta \tag{4.4.8}$$

In view of Eq. (4.4.5), Eq. (4.4.8) takes the form

$$\frac{\partial \mathbf{R}}{\partial \xi_1} = \left(1 + \frac{\zeta}{R_1}\right)\frac{\partial \mathbf{r}}{\partial \xi_1}, \qquad \frac{\partial \mathbf{R}}{\partial \xi_2} = \left(1 + \frac{\zeta}{R_2}\right)\frac{\partial \mathbf{r}}{\partial \xi_2} \qquad (4.4.9)$$

Consequently, the distance between points (ξ_1, ξ_2, ζ) and $(\xi_1 + d\xi_1, \xi_2 + d\xi_2, \zeta + d\zeta)$ is given by

$$(dS)^2 = d\mathbf{R} \cdot d\mathbf{R}$$

$$= L_1^2(d\xi_1)^2 + L_2^2(d\xi_2)^2 + L_3^2(d\zeta)^2 \qquad (4.4.10)$$

where $d\mathbf{R} = (\partial \mathbf{R}/\partial \xi_1)\,d\xi_1 + (\partial \mathbf{R}/\partial \xi_2)\,d\xi_2 + (\partial \mathbf{R}/\partial \zeta)\,d\zeta$, and L_1, L_2, and L_3 are the *Lamé coefficients*

$$L_1 = \alpha_1\left(1 + \frac{\zeta}{R_1}\right), \qquad L_2 = \alpha_2\left(1 + \frac{\zeta}{R_2}\right), \qquad L_3 = 1 \qquad (4.4.11)$$

It should be noted that the vectors $\partial \mathbf{R}/\partial \xi_1$ and $\partial \mathbf{R}/\partial \xi_2$ are parallel to the vectors $\partial \mathbf{r}/\partial \xi_1 = \mathbf{r}_1$ and $\partial \mathbf{r}/\partial \xi_2 = \mathbf{r}_2$. This completes a review of basic geometric relations in the theory of shells.

4.4.3 Stress Resultants

From Fig. 4.17a, the elements of area of the cross sections are

$$dA_1 = L_1\,d\xi_1\,d\zeta = \alpha_1\left(1 + \frac{\zeta}{R_1}\right)d\xi_1\,d\zeta$$

$$dA_2 = L_2\,d\xi_2\,d\zeta = \alpha_2\left(1 + \frac{\zeta}{R_2}\right)d\xi_2\,d\zeta \qquad (4.4.12)$$

Let N_1 be the tensile force, measured per unit length along a ξ_2-coordinate line (see Fig. 4.17), on a cross section perpendicular to a ξ_1-coordinate line. Then the total tensile force on the differential element in the ξ_1-direction is $N_1\alpha_2\,d\xi_2$. This force is equal to the integral of $\sigma_1\,dA_2$ over the thickness,

$$N_1\alpha_2\,d\xi_2 = \int_{-h/2}^{h/2}\sigma_1\,dA_2 \qquad (4.4.13)$$

where h is the thickness of the shell ($\zeta = -h/2$ and $\zeta = h/2$ denote the bottom and top surfaces of the shell). Using Eq. (4.4.12), we can write

$$N_1 = \int_{-h/2}^{h/2}\sigma_1\left(1 + \frac{\zeta}{R_2}\right)d\zeta$$

Similarly, the remaining stress resultants per unit length can be derived. The complete set is given by (see Fig. 4.17c)

$$
\begin{Bmatrix}
N_1 \\
N_2 \\
N_{12} \\
N_{21} \\
Q_1 \\
Q_2 \\
M_1 \\
M_2 \\
M_{12} \\
M_{21}
\end{Bmatrix}
= \int_{-h/2}^{h/2}
\begin{Bmatrix}
\sigma_1\left(1 + \dfrac{\zeta}{R_2}\right) \\
\sigma_2\left(1 + \dfrac{\zeta}{R_1}\right) \\
\sigma_6\left(1 + \dfrac{\zeta}{R_2}\right) \\
\sigma_6\left(1 + \dfrac{\zeta}{R_1}\right) \\
\sigma_5\left(1 + \dfrac{\zeta}{R_2}\right) \\
\sigma_4\left(1 + \dfrac{\zeta}{R_1}\right) \\
\zeta\sigma_1\left(1 + \dfrac{\zeta}{R_2}\right) \\
\zeta\sigma_2\left(1 + \dfrac{\zeta}{R_1}\right) \\
\zeta\sigma_6\left(1 + \dfrac{\zeta}{R_2}\right) \\
\zeta\sigma_6\left(1 + \dfrac{\zeta}{R_1}\right)
\end{Bmatrix} d\zeta
\qquad (4.4.14)
$$

Note that, in contrast to the plate theory (which is obtained by setting $1/R_1 = 1/R_2 = 0$), the shear stress resultants N_{12} and N_{21}, and the twisting moments M_{12} and M_{21}, in general, are not equal. For *thin* shells (h/R_1, h/R_2 less than 1/20), however, one can neglect ζ/R_1 and ζ/R_2 in comparison with unity. Under this assumption, one has $N_{12} = N_{21} \equiv N_6$ and $M_{12} = M_{21} \equiv M_6$. It should be pointed out that no assumptions concerning the kinematics or deformation are made to this point. To derive shell equations that are tractable, we are required to make some simplifying assumptions.

4.4.4 Equations of Motion

The basic equations of three-dimensional elasticity theory can be simplified for thin flexible bodies. A set of simplifying assumptions that provides a reasonable description of the behavior of thin elastic shells is as follows:

1. The thickness of the shell is small compared to the principal radii of curvature (h/R_1, $h/R_2 \ll 1$).

2. The transverse normal stress is negligible.

3. Normals to the reference surface of the shell before deformation remain straight, but not necessarily normal, after deformation (a relaxed Kirchhoff–Love hypothesis).

The shell under consideration is composed of a finite number of orthotropic layers of uniform thickness, as shown in Fig. 4.18. In view of assumption 1, the stress resultants in Eq. (4.4.14) can be expressed as

$$(N_i, M_i) = \sum_{m=1}^{N} \int_{\zeta_m}^{\zeta_{m+1}} \sigma_i(1, \zeta)\, d\zeta, \qquad i = 1, 2, 6$$

$$(Q_1, Q_2) = \sum_{m=1}^{N} K_i^2 \int_{\zeta_m}^{\zeta_{m+1}} (\sigma_5, \sigma_4)\, d\zeta, \qquad\qquad (4.4.15)$$

where N is the number of layers in the shell, and ζ_m and ζ_{m+1} are the bottom and top ζ-coordinates of the mth lamina.

The strain–displacement equations of a shell are an approximation, within the assumptions made above, of the strain–displacement relations referred to orthogonal curvilinear coordinates. It is also assumed that the transverse displacement u_3 does not vary with ζ so that $u_3 = u_3(\xi_1, \xi_2)$. As in the shear deformation theory of flat plates, we begin with the displacement field

$$\bar{u}_1 = \frac{1}{\alpha_1}(L_1 u_1) + \zeta\phi_1, \qquad \bar{u}_2 = \frac{1}{\alpha_2}(L_2 u_2) + \zeta\phi_2, \qquad \bar{u}_3 = u_3$$

$$(4.4.16)$$

where $(\bar{u}_1, \bar{u}_2, \bar{u}_3)$ are the displacements of a point (ξ_1, ξ_2, ζ) along the (ξ_1, ξ_2, ζ) coordinates and (u_1, u_2, u_3) are the displacements of a point $(\xi_1, \xi_2, 0)$. Substituting Eq. (4.4.16) into the strain–displacement relations of an orthogonal

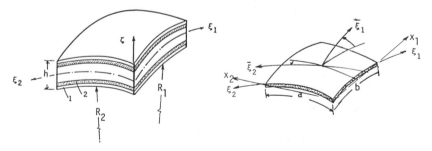

Figure 4.18   Laminated shell geometry and lamina details.

curvilinear coordinate system, one obtains (see Sanders, 1959)

$$\varepsilon_1 = \varepsilon_1^0 + \zeta\kappa_1^0$$

$$\varepsilon_2 = \varepsilon_2^0 + \zeta\kappa_2^0$$

$$\varepsilon_4 = \varepsilon_4^0$$

$$\varepsilon_5 = \varepsilon_5^0$$

$$\varepsilon_6 = \varepsilon_6^0 + \zeta\kappa_6^0 \tag{4.4.17}$$

where

$$\varepsilon_1^0 = \frac{1}{\alpha_1}\frac{\partial u_1}{\partial \xi_1} + \frac{u_3}{R_1}$$

$$\varepsilon_2^0 = \frac{1}{\alpha_2}\frac{\partial u_2}{\partial \xi_2} + \frac{u_3}{R_2}$$

$$\varepsilon_4^0 = \frac{1}{\alpha_2}\frac{\partial u_3}{\partial \xi_2} + \phi_2 - \frac{u_2}{R_2}$$

$$\varepsilon_5^0 = \frac{1}{\alpha_1}\frac{\partial u_3}{\partial \xi_1} + \phi_1 - \frac{u_1}{R_1}$$

$$\varepsilon_6^0 = \frac{1}{\alpha_1}\frac{\partial u_2}{\partial \xi_1} + \frac{1}{\alpha_2}\frac{\partial u_1}{\partial \xi_2}$$

$$\kappa_1^0 = \frac{1}{\alpha_1}\frac{\partial \phi_1}{\partial \xi_1}$$

$$\kappa_2^0 = \frac{1}{\alpha_2}\frac{\partial \phi_2}{\partial \xi_2}$$

$$\kappa_6^0 = \frac{1}{\alpha_1}\frac{\partial \phi_2}{\partial \xi_1} + \frac{1}{\alpha_2}\frac{\partial \phi_1}{\partial \xi_2} + \frac{1}{2}\left(\frac{1}{R_2} - \frac{1}{R_1}\right)\left(\frac{1}{\alpha_1}\frac{\partial u_2}{\partial \xi_1} - \frac{1}{\alpha_2}\frac{\partial u_1}{\partial \xi_2}\right) \tag{4.4.18}$$

where ϕ_1 and ϕ_2 are the rotations of the reference surface $\zeta = 0$ about the ξ_2- and ξ_1-coordinate axes, respectively.

The stress–strain relations for the mth orthotropic lamina in the material coordinate axes are given by Eq. (4.3.8). In the shell coordinates, the equation becomes

$$\{\sigma\} = [Q^{(m)}]\{\varepsilon\} \tag{4.4.19}$$

where $Q_{ij}^{(m)}$ are the material properties of the mth layer.

Hamilton's principle (or the principle of virtual work for bodies in motion) in the present case yields (at time t)

$$0 = \int_{-h/2}^{h/2}\left\{\int_{\Omega}\left[\sigma_1^{(m)}\,\delta\varepsilon_1 + \sigma_2^{(m)}\,\delta\varepsilon_2 + \sigma_6^{(m)}\,\delta\varepsilon_6 + \sigma_4^{(m)}\,\delta\varepsilon_4 + \sigma_5^{(m)}\,\delta\varepsilon_5\right.\right.$$

$$+\rho^{(m)}\left[\left(1 + \frac{\zeta}{R_1}\right)\ddot{u}_1 + \zeta\ddot{\phi}_1\right]\left[\left(1 + \frac{\zeta}{R_1}\right)\delta u_1 + \zeta\,\delta\phi_1\right]$$

$$+\rho^{(m)}\left[\left(1 + \frac{\zeta}{R_2}\right)\ddot{u}_2 + \zeta\ddot{\phi}_2\right]\left[\left(1 + \frac{\zeta}{R_2}\right)\delta u_2 + \zeta\,\delta\phi_2\right]$$

$$\left.\left.+\rho^{(m)}\ddot{u}_3\delta u_3 - q\,\delta u_3\right]\alpha_1\alpha_2\,d\xi_1\,d\xi_2\right\}\,d\zeta$$

$$= \int_{\Omega}\left[N_1\,\delta\varepsilon_1^0 + N_2\,\delta\varepsilon_2^0 + N_6\varepsilon_6^0 + M_1\,\delta\kappa_1^0 + M_2\,\delta\kappa_2^0 + M_6\,\delta\kappa_6^0 + Q_1\,\delta\varepsilon_5^0\right.$$

$$+Q_2\,\delta\varepsilon_4^0 + (P_{11}\ddot{u}_1 + P_1\ddot{\phi}_1)\,\delta u_1 + (P_{22}\ddot{u}_2 + P_2\ddot{\phi}_2)\,\delta u_2 + I_1\ddot{u}_3\,\delta u_3$$

$$\left.+(I_3\ddot{\phi}_1 + P_1\ddot{u}_1)\,\delta\phi_1 + (I_3\ddot{\phi}_2 + P_2\ddot{u}_2)\,\delta\phi_2 - q\,\delta u_3\right]\alpha_1\alpha_2\,d\xi_1\,d\xi_2$$

$$\tag{4.4.20}$$

where q is the distributed transverse load, I_i are the inertias defined in Eq. (4.2.41), and P_{11}, P_{22}, P_1, and P_2 are defined by

$$P_{ii} = \sum_{m=1}^{N}\int_{\zeta_m}^{\zeta_{m+1}}\rho^{(m)}\left(1 + \frac{\zeta}{R_i}\right)\left(1 + \frac{\zeta}{R_i}\right)d\zeta,$$

$$P_i = \sum_{m=1}^{N}\int_{\zeta_m}^{\zeta_{m+1}}\rho^{(m)}\left(1 + \frac{\zeta}{R_i}\right)\zeta\,d\zeta \tag{4.4.21}$$

We shall neglect terms involving the squares of curvatures in P's.

The governing equations of motion can be derived from Eq. (4.4.20) by integrating the displacement gradients in ε_i^0 by parts and setting the coefficients of δu_i $(i = 1, 2, 3)$ and $\delta\phi_i$ $(i = 1, 2)$ to zero separately. We obtain

$$\frac{\partial N_1}{\partial x_1} + \frac{\partial}{\partial x_2}(N_6 + c_0 M_6) + \frac{Q_1}{R_1} = \left(I_1 + \frac{2}{R_1}I_2\right)\frac{\partial^2 u_1}{\partial t^2} + \left(I_1 + \frac{I_3}{R_1}\right)\frac{\partial^2 \phi_1}{\partial t^2}$$

$$\frac{\partial}{\partial x_1}(N_6 - c_0 M_6) + \frac{\partial N_2}{\partial x_2} + \frac{Q_2}{R_2} = \left(I_1 + \frac{2}{R_2}I_2\right)\frac{\partial^2 u_2}{\partial t^2} + \left(I_1 + \frac{I_3}{R_2}\right)\frac{\partial^2 \phi_2}{\partial t^2}$$

$$\frac{\partial Q_1}{\partial x_1} + \frac{\partial Q_2}{\partial x_2} - \left(\frac{N_1}{R_1} + \frac{N_2}{R_2} - q\right) = I_1\frac{\partial^2 u_3}{\partial t^2}$$

$$\frac{\partial M_1}{\partial x_1} + \frac{\partial M_6}{\partial x_2} - Q_1 = I_3\frac{\partial^2 \phi_1}{\partial t^2} + \left(I_1 + \frac{I_3}{R_1}\right)\frac{\partial^2 u_1}{\partial t^2}$$

$$\frac{\partial M_6}{\partial x_1} + \frac{\partial M_2}{\partial x_2} - Q_2 = I_3\frac{\partial^2 \phi_2}{\partial t^2} + \left(I_1 + \frac{I_3}{R_2}\right)\frac{\partial^2 u_2}{\partial t^2} \qquad (4.4.22)$$

where we used the notation,

$$c_0 = \frac{1}{2}\left(\frac{1}{R_1} - \frac{1}{R_2}\right), \qquad dx_i = \alpha_i\, d\xi_i \qquad (4.4.23)$$

The boundary conditions involve the specification of expressions of the form in Eq. (4.2.9). The resultants (N_i, M_i, Q_i) are related to $(\varepsilon_i^0, \kappa_i)$ by Eq. (4.3.13).

The equations of Love's first-approximation shell theory can be obtained by setting $c_0 = 0$ in Eq. (4.4.22). Equations (4.4.22) can be specialized to flat plates, cylindrical shells, and spherical shells, respectively, by setting $1/R_1 = 1/R_2 = 0$, $1/R_1 = 0$ and $R_2 = R$ (the x_1-axis is taken along a generator of the cylinder), and $R_1 = R_2 = R$. The classical thin shell theory can be obtained by setting $\phi_1 = -\partial u_3/\partial x_1 + u_1/R_1$, and $\phi_2 = -\partial u_3/\partial x_2 + u_2/R_2$.

In the following sections, we consider exact solutions of some shell problems. These include both isotropic and laminated shell structures.

4.4.5 Exact Solutions

Equations (4.4.22) cannot be solved for arbitrary geometry, boundary conditions, or loading of a shell. Here we develop exact solutions for certain shells under specific geometry, boundary conditions, and loading. We begin with axisymmetrically loaded circular cylindrical shells, which arise in connection with pipes, tanks, boilers, and so on.

Axisymmetrically Loaded Cylindrical Shells

Consider a thin, isotropic, cylindrical shell of radius R and subjected to axisymmetric radial load q. Because of the symmetry (in geometry and loading), we have

$$u_2 = 0, \qquad N_6 = Q_2 = M_6 = 0, \qquad N_1 = N_1(x_1), \qquad Q_1 = Q_1(x_1),$$

$$N_2 = N_2(x_1), \qquad M_2 = M_2(x_1)$$

From Eq. (4.4.22) we obtain (for static equilibrium),

$$\frac{dN_1}{dx_1} = 0, \qquad \frac{dQ_1}{dx_1} - \frac{N_2}{R} + q = 0, \qquad \frac{dM_1}{dx_1} - Q_1 = 0 \qquad (4.4.24)$$

The resultants N_1, N_2, M_1, and M_2 are related to the displacements ($u_2 = 0$) u_1 and u_3 by

$$N_1 = \frac{Eh}{1 - \nu^2}\left(\frac{du_1}{dx_1} + \nu\frac{u_3}{R}\right), \qquad N_2 = \frac{Eh}{1 - \nu^2}\left(\frac{u_3}{R} + \nu\frac{du_1}{dx_1}\right)$$

$$M_1 = -D\frac{d^2 u_3}{dx_1^2}, \qquad\qquad M_2 = \nu M_1 \qquad\qquad (4.4.25)$$

where D is the flexural rigidity of the shell, $D = Eh^3/12(1 - \nu^2)$. Using M_1 from Eq. (4.4.25) in the third equation of (4.4.24), and then using the result in the second equation in (4.4.24), we get

$$\frac{d^2}{dx_1^2}\left(D\frac{d^2 u_3}{dx_1^2}\right) + \frac{Eh}{R^2}u_3 + \nu\frac{N_1}{R} - q = 0 \qquad (4.4.26)$$

For zero axial load, the first part of Eq. (4.4.24) gives $N_1 = 0$, and we have

$$\frac{d^4 u_3}{dx_1^4} + 4\beta^4 u_3 = \frac{q}{D}, \qquad \frac{du_1}{dx_1} = -\frac{\nu u_3}{R} \qquad (4.4.27)$$

where

$$\beta^4 = \frac{Eh}{4R^2 D} = \frac{3(1 - \nu^2)}{R^2 h^2} \qquad (4.4.28)$$

The general solution of Eq. (4.4.27) is given by

$$u_3 = e^{-\beta x_1}(c_1\cos\beta x_1 + c_2\sin\beta x_1) + e^{\beta x_1}(c_3\cos\beta x_1 + c_4\sin\beta x_1) + w_p$$

$$(4.4.29)$$

where w_p is the particular solution and c_1–c_4 are constants of integration to be determined for appropriate boundary conditions. We now consider some specific examples.

Example 4.15

Consider the bending of a (infinitely) long cylinder under circumferentially distributed line load Q_0 (see Fig. 4.19). We have $q = 0$, and hence $w_p = 0$ in Eq. (4.4.29). The boundary conditions of the problem are

$$Q_1|_{x_1=0} \equiv \left(-D\frac{d^3 u_3}{dx_1^3}\right)\bigg|_{x_1=0} = \frac{Q_0}{2}, \qquad \left(\frac{du_3}{dx_1}\right)\bigg|_{x_1=0} = 0 \quad (4.4.30)$$

The first boundary condition is the result of the requirement that each half of the cylinder carry one-half of the external load. The second boundary condition is an expression of the symmetry about $x_1 = 0$. The additional boundary conditions are provided by the requirement that the deflection u_3 and all derivatives of u_3 with respect to x_1 be zero for $x_1 = \infty$. These additional boundary conditions are satisfied only if $c_3 = c_4 = 0$. The boundary conditions in Eq. (4.4.30) give

$$c_1 = c_2 = -\frac{Q_0}{8\beta^3 D} \quad (4.4.31)$$

and for $x_1 \geq 0$ we have from Eqs. (4.4.29) and (4.4.25)

$$u_3 = -\frac{Q_0}{8\beta^3 D} e^{-\beta x_1}(\sin \beta x_1 + \cos \beta x_1) \quad (4.4.32)$$

$$N_2 = \frac{Eh}{R} u_3 = -\frac{EhQ_0}{8\beta^3 DR} e^{-\beta x_1}(\sin \beta x_1 + \cos \beta x_1) \quad (4.4.33)$$

$$M_2 = \nu M_1 = -\frac{\nu Q_0}{4\beta} e^{-\beta x_1}(\cos \beta x_1 - \sin \beta x_1) \quad (4.4.34)$$

$$Q_1 = \frac{Q_0}{2} e^{-\beta x_1}\cos \beta x_1 \quad (4.4.35)$$

The maximum deflection and bending moment occur at $x_1 = 0$. We have

$$(u_3)_{max} = -\frac{Q_0 R^2 \beta}{2 Eh}, \qquad (M_1)_{max} = -\frac{Q_0}{4\beta} \quad (4.4.36)$$

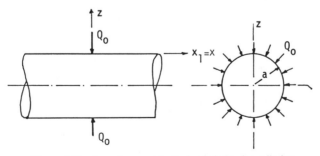

Figure 4.19 An axisymmetrically loaded circular cylinder.

Note that the quantities in Eqs. (4.4.32)–(4.4.35) decrease in value with increasing βx_1. Thus, the effect of concentrated loads can be neglected at locations $x_1 > (\pi/\beta)$.

Example 4.16

Consider a long cylindrical shell subjected to uniform internal pressure p_0. We wish to determine the stress and displacement fields for two sets of boundary conditions: (a) the ends are clamped, and (b) the ends are simply supported.

1. *Clamped Shell.* The solution can be obtained using the principle of superposition. The problem can be represented (see Fig. 4.20) as the sum of a shell subjected to internal pressure and a shell subjected to end moments M_1 and shear force Q_1 (both unknown). The radial expansion owing to the internal pressure (from elementary mechanics of materials) is

$$(u_3)_1 = \frac{p_0 R^2}{Eh} \tag{4.4.37}$$

The solution of the shell subjected to end moments and shear forces can be found from Eq. (4.4.29) by setting $w_p = 0$ (because $q = 0$) and $c_3 = c_4 = 0$ (because u_3 and its derivatives for $x_1 = \infty$ are zero). The constants c_1 and c_2 are determined by the boundary conditions

$$\left(-D\frac{d^2u_3}{dx_1^2}\right)\Bigg|_{x_1=0} = M_1, \qquad \left(-D\frac{d^3u_3}{dx_1^3}\right)\Bigg|_{x_1=0} = Q_1 \tag{4.4.38}$$

$$c_1 = -\frac{1}{2\beta^3 D}(Q_1 + \beta M_1), \qquad c_2 = \frac{M_1}{2\beta^2 D}$$

$$(u_3)_2 = -\frac{e^{-\beta x_1}}{2\beta^3 D}\left[\beta M_1(\sin\beta x_1 - \cos\beta x_1) - Q_1\cos\beta x_1\right] \tag{4.4.39}$$

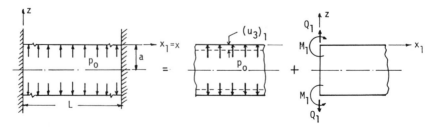

Figure 4.20   A clamped cylindrical shell subjected to internal pressure.

The total deflection is given by

$$u_3 = (u_3)_1 + (u_3)_2 = -\frac{e^{-\beta x_1}}{2\beta^3 D} \left[\beta M_1 (\sin \beta x_1 - \cos \beta x_1) - Q_1 \cos \beta x_1 \right]$$

$$+ \frac{p_0 R^2}{Eh} \tag{4.4.40}$$

The unknown reactions M_1 and Q_1 are determined by using the boundary conditions of the total problem:

$$u_3(0) = 0 \rightarrow -\frac{1}{2\beta^3 D}(\beta M_1 + Q_1) + \frac{p_0 R^2}{Eh} = 0$$

$$\frac{du_3}{dx_1}(0) = 0 \rightarrow \frac{1}{2\beta^2 D}(2\beta M_1 + Q_1) = 0$$

$$M_1 = -\frac{p_0}{2\beta^2}, \qquad Q_1 = \frac{p_0}{\beta} \tag{4.4.41}$$

The expression for u_3 is given by

$$u_3 = -\frac{p_0}{4\beta^4 D} e^{-\beta x_1}(\sin \beta x_1 + \cos \beta x_1) + \frac{p_0 R^2}{Eh} \tag{4.4.42}$$

The stress, $\sigma_1 = -6M_1(x_1)/h^2$, is given by

$$\sigma_1 = \frac{3p_0}{\beta^2 h^2} e^{-\beta x_1}(\cos \beta x_1 - \sin \beta x_1) \tag{4.4.43}$$

2. *Simply Supported Shell.* In this case, $M_1 = 0$, $Q_1 \neq 0$ in Eq. (4.4.38). From the first part of Eq. (4.4.41), we obtain $Q_1 = -(2\beta^3 Dp_0 R^2/Eh)$, and

$$u_3 = (1 - e^{-\beta x_1})\frac{p_0 R^2}{Eh}$$

$$\sigma_1 = -\frac{\beta^2 p_0 R^2}{Eh} e^{-\beta x_1} \tag{4.4.44}$$

This completes the example.

Simply Supported, Cylindrical and Spherical Panels

As in the case of laminated flat plates, exact solutions of Eq. (4.4.22) can be constructed for simply supported cross-ply laminated shells. More specifically, the exact solutions can be obtained under the following conditions:

(i) Symmetric or antisymmetric cross-ply laminates, that is, laminates with:

$$A_{16} = A_{26} = B_{16} = B_{26} = D_{16} = D_{26} = A_{45} = 0 \quad (4.4.45)$$

(ii) Freely supported boundary conditions (see Fig. 4.18):

$$N_1(0, x_2) = N_1(a, x_2) = M_1(0, x_2) = M_1(a, x_2) = 0$$

$$u_3(0, x_2) = u_3(a, x_2) = u_2(0, x_2) = u_2(a, x_2) = 0$$

$$N_2(x_1, 0) = N_2(x_1, b) = M_2(x_1, 0) = M_2(x_1, b) = 0$$

$$u_3(x_1, 0) = u_3(x_1, b) = u_1(x_1, 0) = u_1(x_1, b) = 0$$

$$\phi_2(0, x_2) = \phi_2(a, x_2) = \phi_1(x_1, 0) = \phi_1(x_1, b) = 0 \quad (4.4.46)$$

where a and b are the planeform dimensions of the shell.

(iii) Sinusoidal (spatial) distribution of the transverse load:

$$q = \sum_{m,n=1}^{\infty} Q_{mn}(t) \ \sin \alpha x_1 \ \sin \beta x_2,$$

$$\alpha = \frac{m\pi}{a}, \qquad \beta = \frac{n\pi}{b} \qquad (4.4.47)$$

where a and b are the dimensions of the shell middle surface along the x_1 and x_2 axes, respectively. The time variation of the load does not influence the spatial form of the solution.

Under these conditions, the exact form of the spatial variation of u_i ($i = 1, 2, 3$) and ϕ_i ($i = 1, 2$) is given by

$$u_1(x_1, x_2, t) = \sum_{m,n=1}^{\infty} U_{mn}(t) \cos \alpha x_1 \sin \beta x_2$$

$$u_2(x_1, x_2, t) = \sum_{m,n=1}^{\infty} V_{mn}(t) \sin \alpha x_1 \cos \beta x_2$$

$$u_3(x_1, x_2, t) = \sum_{m,n=1}^{\infty} W_{mn}(t) \sin \alpha x_1 \sin \beta x_2$$

$$\phi_1(x_1, x_2, t) = \sum_{m,n=1}^{\infty} X_{mn}(t) \cos \alpha x_1 \sin \beta x_2$$

$$\phi_2(x_1, x_2, t) = \sum_{m,n=1}^{\infty} Y_{mn}(t) \sin \alpha x_1 \cos \beta x_2 \qquad (4.4.48)$$

Clearly, the solution satisfies the boundary conditions in Eq. (4.4.46).

Substitution of Eqs. (4.4.47) and (4.4.48) into Eq. (4.4.22) yields a set of five linear algebraic equations in terms of the unknown amplitudes $U_{mn}(t)$, $V_{mn}(t)$, $W_{mn}(t)$, $X_{mn}(t)$, and $Y_{mn}(t)$. These equations can be expressed in matrix form as

$$-[c]\{\Delta\} + [M]\{\ddot{\Delta}\} = \{F\} \qquad (4.4.49)$$

where

$$\{\Delta\}^T = \{U_{mn}, V_{mn}, W_{mn}, X_{mn}, Y_{mn}\}, \qquad \{F\}^T = \{0, 0, Q_{mn}, 0, 0\}$$

$$(4.4.50)$$

and $c_{ij} = c_{ji}$ and $M_{ij} = M_{ji}$ are given by

$$c_{11} = -\alpha^2 A_{11} - \beta^2 A_{66} - c_0^2 \beta^2 D_{66} - \frac{A_{55}}{R_1^2} - 2c_0 \beta^2 B_{66}$$

$$c_{12} = -\alpha\beta\left(A_{12} + A_{66} - c_0^2 D_{66}\right)$$

$$c_{13} = \alpha\left(\frac{A_{11}}{R_1} + \frac{A_{12}}{R_2} + \frac{A_{55}}{R_1}\right)$$

$$c_{14} = -\alpha^2 B_{11} - \beta^2 B_{66} - c_0 \beta^2 D_{66} + \frac{A_{55}}{R_1}$$

$$c_{15} = -\alpha\beta\left(B_{12} + B_{66} + c_0 D_{66}\right)$$

$$c_{22} = -\alpha^2 A_{66} - \alpha^2 c_0^2 D_{66} - \beta^2 A_{22} - \frac{A_{44}}{R_2^2} + 2c_0 \alpha^2 B_{66}$$

$$c_{23} = \beta\left(\frac{A_{12}}{R_1} + \frac{A_{22}}{R_2} + \frac{A_{44}}{R_2}\right)$$

$$c_{24} = -\alpha\beta\left(B_{66} - c_0 D_{66} + B_{12}\right)$$

$$c_{25} = -\alpha^2\left(B_{66} - c_0 D_{66}\right) - \beta^2 B_{22} + \frac{A_{44}}{R_2}$$

$$c_{33} = -A_{55}\alpha^2 - A_{44}\beta^2 - \frac{1}{R_1}\left(\frac{A_{11}}{R_1} + \frac{A_{12}}{R_2}\right) - \frac{1}{R_2}\left(\frac{A_{12}}{R_1} + \frac{A_{22}}{R_2}\right)$$

$$c_{34} = \alpha\left(\frac{B_{11}}{R_1} + \frac{B_{12}}{R_2} - A_{55}\right)$$

$$c_{35} = \beta\left(\frac{B_{12}}{R_1} + \frac{B_{22}}{R_2} - A_{44}\right)$$

$$c_{44} = -\alpha^2 D_{11} - \beta^2 D_{66} - A_{55}$$

$$c_{45} = -\alpha\beta(D_{12} + D_{66})$$

$$c_{55} = -\alpha^2 D_{66} - \beta^2 D_{22} - A_{44}$$

$$M_{11} = M_{22} = M_{33} = I_1, \qquad M_{44} = M_{55} = I_3$$

$$M_{14} = I_1 + \frac{I_3}{R_1}, \qquad M_{25} = I_1 + \frac{I_3}{R_2} \quad \text{all other } M_{ij} = 0 \qquad (4.4.51)$$

As discussed in the analysis of laminated plates, one can obtain series solutions of cylindrical and spherical shells for bending, natural vibration, and buckling problems. Of course, the limitation is that the shell boundary conditions should be those listed in Eq. (4.4.46). In the next two examples, we discuss the results of bending and natural vibrations of shells.

Example 4.17

As a first example of bending of shells, we consider a spherical shell under point load at the center. The following geometric data is used:

$$R_1 = R_2 = 96.0 \text{ in.}, \qquad a = b = 32.0 \text{ in.}, \qquad h = 0.1 \text{ in.}$$

$$E_1 = E_2 = 10^7 \text{ psi}, \qquad \nu = 0.3, \qquad \text{intensity of load} = 100 \text{ lb} \qquad (4.4.52)$$

This problem was also solved using the finite-element method by Yang (1973).

Table 4.8 contains the center transverse deflection obtained using the shear deformation theory (SDT) and classical shell theory (CST) for various terms in the series. The numerical solution of Vlasov (1964) is taken from Yang's paper. It should be noted that both Vlasov and Yang did not consider transverse

Table 4.8   Center Transverse Deflection of a Spherical Shell Under Point Load at the Center[a]

Theory	$n = 9$[b]	$n = 49$	$n = 99$	$n = 149$	$n = 199$	$n = 249$
SDT	0.032594	0.039469	0.03972	0.039786	0.039814	0.039832
CST	—	—	0.03959	—	0.039647	0.039653
Vlasov (1964)	—	—	0.03956	—	—	—

[a] $R_1 = R_2 = 96$ in.; $a = b = 32$ in., load $= 100$ lb, $E = 10^7$ psi, $\nu = 0.3$.
[b] $n \times n$-Term series; finite-element solution of Yang (1973): 0.03867.

Table 4.9 Nondimensional[a] Center Deflection versus Radius to Thickness Ratio of Spherical Shells Under Uniformly Distributed Load (19-Term Solution)

R/a	$0°/90°$		$0°/90°/0°$		$0°/90°/90°/0°$	
	$a/h = 100$	$a/h = 10$	$a/h = 100$	$a/h = 10$	$a/h = 100$	$a/h = 10$
1	0.0718	6.0544	0.0718	4.8173	0.07151	4.8366
2	0.2855	12.668	0.2858	8.0210	0.2844	8.0517
3	0.6441	15.739	0.6224	9.1148	0.6246	9.1463
4	1.1412	17.184	1.0443	9.5686	1.0559	9.5999
5	1.7535	19.944	1.5118	9.7937	1.5358	9.8249
10	5.5428	19.065	3.6445	10.110	3.7208	10.141
10^{30}	16.980	19.469	6.6970	10.220	6.8331	10.251

[a] $\bar{w} = [wE_2h^3/(q_0a^4)]10^3$

shear strains. It is clear from the results that the series solution converges very slowly. The difference between the values predicted by SDT and CST theories is not significant, because the ratios $a/h = 320$ and $R/h = 960$ are very large (hence, the shell is essentially shallow and very thin).

Next, the results of bending of cross-ply laminated shells are discussed. The geometric parameters used in the isotropic shells are also used for composite shells. Individual layers are assumed to be orthotropic, with the following properties:

$$E_1 = 25E_2, \qquad G_{23} = 0.2E_2, \qquad G_{13} = G_{12} = 0.5E_2, \qquad \nu_{12} = 0.25$$

$$(4.4.53)$$

Table 4.9 contains nondimensionalized center transverse deflections of cross-ply spherical shells under a uniformly distributed transverse load. The results are tabulated for various ratios of radius to side and for two values of side to thickness. From these results, it follows that the center deflection varies more rapidly with the ratio R/a for deep shells (i.e., for large ratios of a/h) than for shallow shells (i.e., for small ratios of a/h).

Example 4.18

Here we present the results of the free vibration of cross-ply laminated spherical shells with the same material properties as in Eq. (4.4.53). The fundamental frequencies, in nondimensional form, are presented in Table 4.10.

Two sets of shear correction factors were used to investigate their influence on the fundamental frequencies. For a fixed ratio of R/a, the shear correction factors have little or no effect on the frequencies for $a/h = 100$. For $a/h = 10$, the effect of smaller values of shear correction factors is to lower the frequencies, the rate of decrease being the highest for the four-layer cross-ply.

Table 4.10 Nondimensionalized Fundamental Frequencies[a] ($\bar{\omega} = \omega a^2 \sqrt{\rho/E_2}/h$), versus Radius to Side Length Ratio of Spherical Shell ($a/b = 1$)

	0°/90°		0°/90°/0°		0°/90°/90°/0°	
R/a	$a/h = 100$	$a/h = 10$	$a/h = 100$	$a/h = 10$	$a/h = 100$	$a/h = 10$
1	125.93	14.481	125.99	16.115	126.33	16.172
	125.93	14.420	125.99	15.746	126.33	15.827
2	67.362	10.749	68.075	13.382	68.294	13.447
	67.361	9.5161	68.072	12.794	68.291	12.890
3	46.002	9.9608	47.265	12.731	47.415	12.795
	46.001	9.6473	47.260	12.077	47.410	12.178
4	35.228	9.4102	36.971	12.487	37.082	12.552
	35.227	9.2644	36.964	11.808	37.075	11.910
5	28.825	9.2309	30.993	12.372	31.079	12.437
	28.825	9.0791	30.986	11.680	31.071	11.783
10	16.706	8.9841	20.347	12.215	20.380	12.280
	16.704	8.8225	20.335	11.506	20.368	11.609
10^{30}	9.6873	8.8998	15.183	12.162	15.184	12.226
	9.6850	8.7317	15.167	11.447	15.168	11.551

[a]The first line of values corresponds to $K_1^2 = K_2^2 = \frac{5}{6}$, and the second line corresponds to $K_1^2 = 0.7$ and $K_2^2 = 0.6$.

This completes the study of the theory of shells. The closed-form solutions presented do not exhaust all that are available in the literature. For shells other than cylindrical and boundary conditions other than simply supported, analytical solutions are limited. In such cases, variational or numerical solutions can be developed. In the next section, we present the finite-element analysis of plates and shells. For detailed accounts of the theory and analysis of plates and shells (analytical as well as numerical), the reader is referred to the many books and papers listed in the Bibliography.

Exercises 4.4

1. A very long, isotropic cylinder of radius R and uniform thickness h is subjected to an axisymmetric uniform load q_0 per unit length over a length L_0 (the length of the cylinder is much larger than L_0). Derive an expression for the deflection u_3.

2. Consider a cylindrical tank of height b, radius a, and uniform thickness h, entirely filled with a liquid of specific weight γ. The tank bottom is assumed to be built in (i.e., clamped) and the top is open. Using Eq. (4.4.27) obtain the expression for the deflection u_3.

3. Consider a long, isotropic cylinder of radius R and uniform thickness h, which is clamped at the ends. Using the procedure presented in Example 4.16, derive the expressions for deflection u_3 and bending moment M_1 when the cylinder is subjected to uniform temperature change.

4. Specialize Eq. (4.4.51) for classical shell theory (i.e., without transverse shear effects).

5. Determine the transverse deflection of a simply supported, antisymmetric cross-ply, circular cylindrical shell of radius R, thickness h, and dimensions a and b, which is subjected to a sinusoidally distributed load. Use the results of Exercise 4.

6. Rework the problem in Example 4.16 for orthotropic cylinders.

4.5 FINITE-ELEMENT ANALYSIS OF PLATES AND SHELLS

4.5.1 Introduction

As pointed out in Sections 4.2–4.4, the partial differential equations governing plates and shells cannot be solved analytically for complex geometries, boundary conditions, and loadings. The use of numerical methods facilitates the solution of these equations for problems of practical importance. Among the numerical methods available for the solution of differential equations defined over arbitrary domains, the finite-element method is the most effective.

This section deals with the study of finite-element formulations of plates and shells. We begin with the discussion of finite-element models of classical plate theory. We then study the finite-element models of shear deformation theory of laminated plates and shells and present numerical results for a number of problems. As we shall see, the finite-element model based on the total potential energy principle for the classical plate theory requires cubic or higher-order interpolation functions (analogous to the classical beam models presented in Section 3.4) in the approximation of the transverse deflection. On the other hand, the finite-element model based on the total potential energy principle of the first-order shear deformation theory allows us to use linear or higher-order interpolation functions for all displacements. As a special case, for large values of the side to thickness ratio, the shear deformable plate element gives the same results as the thin plate theory. However, the shear deformable elements suffer from the so-called locking phenomenon, and one should use reduced integration techniques to circumvent the locking. Both of these models are called *displacement models*. In this section, we also study the finite-element models based on the Reissner functional of the classical plate theory. Such models are termed *mixed* models and they require linear or higher-order interpolation functions. With this general introduction, we now proceed to construct finite-element models of plates.

4.5.2 Displacement Models of The Classical Plate Theory

Consider the equations (4.1.26) of classical, small-deflection theory of elastic plates. Since the inplane displacements (u, v) are uncoupled from the transverse deflection w, one can solve Eq. (4.1.26a) independent of Eq. (4.1.26b). Equations (4.1.26a) are the same as those in Eq. (1.7.49) with $A_{ij} = c_{ij}h$. Consequently, the finite-element models presented in Section 3.4 can be used to determine the inplane displacements. Therefore, we focus our attention on Eq. (4.1.26b).

Toward constructing the finite-element model, we cast Eq. (4.1.26b) in a variational form. Since Eq. (4.1.26b) is derived from the principle of total potential energy, Eq. (4.1.17a), it is natural to derive the finite-element model from Eq. (4.1.17a). Retaining terms involving only w, we rewrite Eq. (4.1.17a) [and using the boundary condition from Eq. (4.1.24)] over an element R_e,

$$0 = -\int_{R_e} \left(M_x \frac{\partial^2 \delta w}{\partial x^2} + M_y \frac{\partial^2 \delta w}{\partial y^2} + 2 M_{xy} \frac{\partial^2 \delta w}{\partial x \, \partial y} + q \, \delta w \right) dx \, dy$$

$$+ \oint_{S_e} \left(M_n \frac{\partial \delta w}{\partial n} - V_n \delta w \right) dS \tag{4.5.1}$$

where R_e denotes the domain of a typical finite element and S_e its closed boundary. The geometry of the element is discussed later. Let [see Eqs. (4.1.6) and (4.1.15)]

$$\{e\} = z\{\kappa\}, \qquad \{\kappa\} = \left\{ \begin{array}{c} -\dfrac{\partial^2 w}{\partial x^2} \\[2mm] -\dfrac{\partial^2 w}{\partial y^2} \\[2mm] -2\dfrac{\partial^2 w}{\partial x \, \partial y} \end{array} \right\}, \qquad \left\{ \begin{array}{c} M_x \\ M_y \\ M_{xy} \end{array} \right\} = \int_{-h/2}^{h/2} z \left\{ \begin{array}{c} \sigma_x \\ \sigma_y \\ \sigma_{xy} \end{array} \right\} dz$$

$$\tag{4.5.2}$$

The stress–strain relations are [see Eq. (1.7.45)]

$$\{\sigma\} = [c]\{e\} \tag{4.5.3}$$

From Eqs. (4.5.2) and (4.5.3), we obtain

$$\{M\} = \int_{-h/2}^{h/2} z[c]\{e\} \, dz = [D]\{\kappa\} \tag{4.5.4}$$

where

$$[D] = \int_{-h/2}^{h/2} z^2 [c]\, dz \tag{4.5.5}$$

Next, we write Eq. (4.5.1) in terms of the column of curvatures $\{\kappa\}$ as

$$0 = \int_{R_e} \left\{ \begin{matrix} \delta\kappa_x \\ \delta\kappa_y \\ \delta\kappa_{xy} \end{matrix} \right\}^T \left\{ \begin{matrix} M_x \\ M_y \\ M_{xy} \end{matrix} \right\} dx\, dy - \int_{R_e} \delta w q\, dx\, dy + \oint_{S_e} \left(\frac{\partial \delta w}{\partial n} M_n - \delta w V_n \right) dS$$

$$= \int_{R_e} \{\delta\kappa\}^T [D]\{\kappa\}\, dx\, dy - \int_{R_e} \delta w q\, dx\, dy + \oint_{S_e} \left(\frac{\partial \delta w}{\partial n} M_n - \delta w V_n \right) dS$$

$$\tag{4.5.6}$$

Now let

$$w = \sum_{i=1}^{n} \Delta_i^e \psi_i^e = \{\Delta^e\}^T \left\{ \begin{matrix} \psi_1^e \\ \psi_2^e \\ \vdots \\ \psi_n^e \end{matrix} \right\} \tag{4.5.7}$$

where Δ_i denote nodal values of w and its derivatives, and ψ_i are appropriate interpolation functions. Substituting Eq. (4.5.7) into $\{\kappa\}$ of Eqs. (4.5.2), we obtain

$$\{\kappa\} = - \left\{ \begin{matrix} \sum_{i=1}^{n} \Delta_i^e \dfrac{\partial^2 \psi_i^e}{\partial x^2} \\[2mm] \sum_{i=1}^{n} \Delta_i^e \dfrac{\partial^2 \psi_i^e}{\partial y^2} \\[2mm] 2\sum_{i=1}^{n} \Delta_i^e \dfrac{\partial^2 \psi_i^e}{\partial x\, \partial y} \end{matrix} \right\} = -[B^e]\{\Delta^e\} \tag{4.5.8a}$$

where $[B^e]$ is the $3 \times n$ matrix

$$[B^e] = \begin{bmatrix} \dfrac{\partial^2 \psi_1}{\partial x^2} & \dfrac{\partial^2 \psi_2}{\partial x^2} & \cdots & \dfrac{\partial^2 \psi_n}{\partial x^2} \\[2mm] \dfrac{\partial^2 \psi_1}{\partial y^2} & \dfrac{\partial^2 \psi_2}{\partial y^2} & \cdots & \dfrac{\partial^2 \psi_n}{\partial y^2} \\[2mm] 2\dfrac{\partial^2 \psi_1}{\partial x\, \partial y} & 2\dfrac{\partial^2 \psi_2}{\partial x\, \partial y} & \cdots & 2\dfrac{\partial^2 \psi_n}{\partial x\, \partial y} \end{bmatrix} \tag{4.5.8b}$$

From Eqs. (4.5.8) and (4.5.6), we have

$$0 = \int_{R_e} \{\delta\Delta^e\}^T [B^e]^T [D][B^e]\{\Delta^e\}\, dx\, dy - \int_{R_e} \{\delta\Delta^e\}^T \begin{Bmatrix} \psi_1^e \\ \psi_2^e \\ \vdots \\ \psi_n^e \end{Bmatrix} q\, dx\, dy$$

$$+ \oint_{S_e} \{\delta\Delta^e\}^T \left(M_n \begin{Bmatrix} \dfrac{\partial\psi_1^e}{\partial n} \\ \dfrac{\partial\psi_2^e}{\partial n} \\ \vdots \end{Bmatrix} - V_n \begin{Bmatrix} \psi_1^e \\ \psi_2^e \\ \vdots \\ \psi_n^e \end{Bmatrix} \right) dS$$

$$0 = \{\delta\Delta^e\}^T ([K^e]\{\Delta^e\} - \{F^e\}) \tag{4.5.9}$$

and, since $\{\delta\Delta^e\}$ is arbitrary, we obtain

$$[K^e]\{\Delta^e\} = \{F^e\} \tag{4.5.10}$$

where

$$K_{ij}^e = \int_{R_e} [B^e]^T [D][B^e]\, dx\, dy$$

$$F_i^e = \int_{R_e} q \begin{Bmatrix} \psi_1^e \\ \psi_2^e \\ \vdots \\ \psi_n^e \end{Bmatrix} dx\, dy - \oint_{S_e} \left(M_n \begin{Bmatrix} \dfrac{\partial\psi_1^e}{\partial n} \\ \dfrac{\partial\psi_2^e}{\partial n} \\ \vdots \end{Bmatrix} - V_n \begin{Bmatrix} \psi_1^e \\ \psi_2^e \\ \vdots \\ \psi_n^e \end{Bmatrix} \right) dS \tag{4.5.11}$$

This completes the finite-element formulation. Next, we discuss various plate elements and the derivation of their interpolation functions.

Triangular Elements

Figure 4.21 shows a triangular finite element with three nodes at the vertices of the triangle. From the variational statement in Eq. (4.5.1), it is clear that the essential boundary conditions of the problem involve specification of w as well as $\partial w/\partial n$ (or, equivalently, $\partial w/\partial x$ and $\partial w/\partial y$) at the boundary points. Further, the shear force V_n can be approximated only if ψ_i are cubic functions of x and y. These considerations require us to select a cubic polynomial with

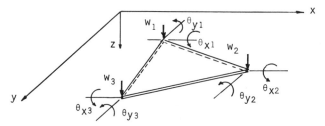

Figure 4.21 A triangular element for the displacement formulation of the classical theory of plates.

nine constants, because three conditions $(w, \partial w/\partial x = \theta_x, \partial w/\partial y = \theta_y)$ per node are available. Two such polynomials are

$$w(x, y) = a_1 + a_2 x + a_3 y + a_4 x^2 + a_5 y^2 + a_6 x^3 + a_7 x^2 y + a_8 xy^2 + a_9 y^3$$

$$= \{1 \quad x \quad y \quad x^2 \quad y^2 \quad x^3 \quad x^2 y \quad xy^2 \quad y^3\} \begin{Bmatrix} a_1 \\ a_2 \\ \vdots \\ a_9 \end{Bmatrix} \qquad (4.5.12)$$

$$w(x, y) = b_1 + b_2 x + b_3 y + b_4 xy + b_5 x^2 + b_6 y^2 + b_7 x^3$$
$$+ b_8 (x^2 y + xy^2) + b_9 y^3$$

$$= \{1 \quad x \quad y \quad xy \quad x^2 \quad y^2 \quad x^3 \quad x^2 y + xy^2 \quad y^3\} \begin{Bmatrix} b_1 \\ b_2 \\ \vdots \\ b_9 \end{Bmatrix} \qquad (4.5.13)$$

In each case, the interpolation functions can be derived using the procedure described in Section 3.4 [see Eqs. (3.4.64) and (3.4.65)]. The column of nodal variables $\{\Delta\}$ is given by

$$\{\Delta^{(e)}\}^T = \{w_1 \quad \theta_{x1} \quad \theta_{y1} \quad w_2 \quad \theta_{x2} \quad \theta_{y2} \quad w_3 \quad \theta_{x3} \quad \theta_{y3}\} \qquad (4.5.14)$$

and the element stiffness matrix in Eq. (4.5.11) is of the order 9×9 (see Szilard, 1974).

The shape functions thus obtained provide continuity of the displacements across the element boundaries. However, they violate, except at the nodes, the normal slope continuity across the element boundaries. Elements that violate the continuity of any primary unknowns across element boundaries are called *nonconforming* or *incompatible* elements. The generation of compatible elements for the classical theory of plates is algebraically complex and computationally expensive.

Rectangular Elements

Figure 4.22 shows a rectangular element with four nodes at the vertices. We have 12 conditions in this case. Therefore, we must select a 12-term polynomial,

$$w(x, y) = c_1 + c_2 x + c_3 y + c_4 xy + c_5 x^2 + c_6 y^2 + c_7 x^3 + c_8 x^2 y + c_9 xy^2$$

$$+ c_{10} y^3 + c_{11} x^3 y + c_{12} xy^3 \qquad (4.5.15)$$

The nodal displacement vector is given by

$$\{\Delta^e\}^T = \{ w_1 \quad \theta_{x1} \quad \theta_{y1} \quad w_2 \quad \theta_{x2} \quad \theta_{y2} \quad w_3 \quad \theta_{x3} \quad \theta_{y3} \quad w_4 \quad \theta_{x4} \quad \theta_{y4} \}$$

$$(4.5.16)$$

The resulting element stiffness matrix is of order 12×12 (see Szilard, 1974).

Although the resulting deflections are continuous across element interfaces, they violate the slope continuity at the edges opposite to those where the rotations have been introduced.

Another choice of interpolation functions for the rectangular element is provided by the (tensor) product of the cubic interpolation functions of a beam [see Eq. (3.4.48)]. For instance, the interpolation function corresponding to w_1 is given by

$$\psi_1 = \left[3\left(\frac{x}{a}\right)^2 - 2\left(\frac{x}{a}\right)^3 \right]\left[3\left(\frac{y}{b}\right)^2 - 2\left(\frac{y}{b}\right)^3 \right] \qquad (4.5.17a)$$

Similarly, the interpolation function associated with θ_{x1} is given by

$$\psi_2 = -\left[a\left(\frac{x}{a}\right)^2\left(1 - \frac{x}{a}\right) \right]\left[3\left(\frac{y}{b}\right)^2 - 2\left(\frac{y}{b}\right)^3 \right] \qquad (4.5.17b)$$

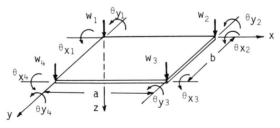

Figure 4.22 A rectangular element for the displacement formulation of the classical plate theory.

These interpolation functions do not contain the xy term, and therefore, the shearing strain cannot be represented. Although these interpolation functions are compatible, the resulting stiffness matrix gives solutions that converge to a number approximately 6% smaller than the exact solution. Therefore, the element is not often used.

In summary, the plate bending elements discussed in this section are nonconforming, and caution should be exercised when they are used in elastic stability analysis of plates and shells, bending of folded plates, and in the analysis of shells with a sudden change in curvature. In light of recent developments in the finite-element analysis of plates and shells, these elements can be considered obsolete.

4.5.3 Mixed Models of The Classical Plate Theory

As mentioned in the previous section, a compatible finite element based on the displacement formulation of classical plate theory is algebraically complex and computationally demanding. A quintic polynomial would satisfy the continuity requirements at element boundaries and result in 21 degrees of freedom per element. The continuity requirements at element boundaries are directly related to the order of the differential equation and dimension of the domain. The continuity requirements can be relaxed by reformulating the fourth-order equation (4.1.26b) as a set of second-order equations in terms of the deflection and bending moments. Such a formulation is called a *mixed formulation*. Here we discuss two such models (see Reddy and Tsay, 1977a, b).

Mixed Model I

Consider the equations [see Eqs. (4.1.21c) and (4.1.25b)]

$$-\left(\frac{\partial^2 M_x}{\partial x^2} + 2\frac{\partial^2 M_{xy}}{\partial x \, \partial y} + \frac{\partial^2 M_y}{\partial y^2} \right) = q \qquad (4.5.18)$$

$$
\left.
\begin{aligned}
M_x &= -\left(D_{11}\frac{\partial^2 w}{\partial x^2} + D_{12}\frac{\partial^2 w}{\partial y^2} \right) \\[2mm]
M_y &= -\left(D_{12}\frac{\partial^2 w}{\partial x^2} + D_{22}\frac{\partial^2 w}{\partial y^2} \right) \\[2mm]
M_{xy} &= -2D_{66}\frac{\partial^2 w}{\partial x \, \partial y}
\end{aligned}
\right\} \qquad (4.5.19)
$$

Substitution of Eqs. (4.5.19) into Eq. (4.5.18) leads to the fourth-order equation in Eq. (4.1.26b), whose finite-element models were discussed in Section 4.5.2. Here we discuss the finite-element model of Eqs. (4.5.18) and (4.5.19), which involve second-order equations relating w, M_x, M_y, and M_{xy}.

To facilitate the derivation of a variational form, Eqs. (4.5.19) are expressed in the alternative form,

$$\frac{\partial^2 w}{\partial x^2} = -\left(\overline{D}_{22} M_x - \overline{D}_{12} M_y\right)$$

$$\frac{\partial^2 w}{\partial y^2} = -\left(\overline{D}_{11} M_y - \overline{D}_{12} M_x\right)$$

$$2\frac{\partial^2 w}{\partial x \, \partial y} = -\left(D_{66}\right)^{-1} M_{xy} \qquad (4.5.20)$$

where

$$\overline{D}_{ij} = D_{ij}/D_0, \qquad D_0 = D_{11}D_{22} - D_{12}^2 \qquad (4.5.21)$$

The variational form of Eqs. (4.5.18) and (4.5.20) over a typical element R_e can be derived using the inverse procedure described in Section 3.4 (see Reddy and Rasmussen, 1982). We have

$$J_1^e\left(w, M_x, M_y, M_{xy}\right)$$

$$= \int_{R_e}\left[\frac{1}{2}\left(-\overline{D}_{22} M_x^2 + 2\overline{D}_{12} M_x M_y - \overline{D}_{11} M_y^2 - \frac{1}{D_{66}} M_{xy}^2\right)\right.$$

$$\left. + \frac{\partial w}{\partial x}\left(\frac{\partial M_x}{\partial x} + \frac{\partial M_{xy}}{\partial y}\right) + \frac{\partial w}{\partial y}\left(\frac{\partial M_{xy}}{\partial x} + \frac{\partial M_y}{\partial y}\right) - qw\right] dx \, dy$$

$$- \oint_{S_e}\left[M_x\left(\frac{\partial w}{\partial x} n_x\right) + M_y\left(\frac{\partial w}{\partial y} n_y\right) + M_{xy}\left(\frac{\partial w}{\partial x} n_y + \frac{\partial w}{\partial y} n_x\right) + Q_n w\right] dS$$

$$(4.5.22)$$

This is a stationary functional. The condition $\delta J_1^e = 0$ gives Eqs. (4.5.18) and (4.5.20) as the Euler equations. The boundary conditions of the formulation are given below:

Essential Boundary Conditions: Specify

$$w, \; M_x, \; M_y \; \text{and} \; M_{xy} \qquad (4.5.23a)$$

Natural Boundary Conditions: Specify

$$Q_n \equiv \left(\frac{\partial M_x}{\partial x} + \frac{\partial M_{xy}}{\partial y} \right) n_x + \left(\frac{\partial M_{xy}}{\partial x} + \frac{\partial M_y}{\partial y} \right) n_y,$$

$$\frac{\partial w}{\partial x} n_x, \quad \frac{\partial w}{\partial y} n_y, \quad \text{and} \quad \left(\frac{\partial w}{\partial x} n_y + \frac{\partial w}{\partial y} n_x \right) \tag{4.5.23b}$$

Note that the natural boundary conditions involving the bending moments in the displacement formulation become the essential boundary conditions in the mixed formulation.

Let w, M_x, M_y, and M_{xy} be interpolated by expressions of the form

$$w = \sum_{i=1}^{r} w_i \psi_i^1, \quad M_x = \sum_{i=1}^{s} M_{xi} \psi_i^2, \quad M_y = \sum_{i=1}^{p} M_{yi} \psi_i^3, \quad M_{xy} = \sum_{i=1}^{q} M_{xyi} \psi_i^4$$

$$\tag{4.5.24}$$

where ψ_i^α ($\alpha = 1, 2, 3, 4$) are appropriate interpolation functions. Substituting Eq. (4.5.24) into Eq. (4.5.22), and setting the partial variations of J_1^e with respect to w_j, M_{xj}, M_{yj}, and M_{xyj} to zero separately, we obtain

$$\begin{bmatrix} [K^{11}] & [K^{12}] & [K^{13}] & [K^{14}] \\ & [K^{22}] & [K^{23}] & [K^{24}] \\ \text{symmetric} & & [K^{33}] & [K^{34}] \\ & & & [K^{44}] \end{bmatrix} \begin{Bmatrix} \{w\} \\ \{M_x\} \\ \{M_y\} \\ \{M_{xy}\} \end{Bmatrix} = \begin{Bmatrix} \{F^1\} \\ \{F^2\} \\ \{F^3\} \\ \{F^4\} \end{Bmatrix} \tag{4.5.25}$$

where

$$K_{ij}^{11} = 0, \quad i, j = 1, 2, \ldots, r$$

$$K_{ij}^{12} = \int_{R_e} \frac{\partial \psi_i^1}{\partial x} \frac{\partial \psi_j^2}{\partial x} \, dx \, dy, \quad i = 1, 2, \ldots, r \quad \text{and} \quad j = 1, 2, \ldots, s$$

$$K_{ij}^{13} = \int_{R_e} \frac{\partial \psi_i^1}{\partial y} \frac{\partial \psi_j^3}{\partial y} \, dx \, dy, \quad i = 1, 2, \ldots, r \quad \text{and} \quad j = 1, 2, \ldots, p$$

$$K_{ij}^{14} = \int_{R_e} \left(\frac{\partial \psi_i^1}{\partial x} \frac{\partial \psi_j^4}{\partial y} + \frac{\partial \psi_i^1}{\partial y} \frac{\partial \psi_j^4}{\partial x} \right) dx \, dy,$$

$$i = 1, 2, \ldots, r \quad \text{and} \quad j = 1, 2, \ldots, q$$

$$K_{ij}^{22} = \int_{R_e} \left(-\overline{D}_{22} \right) \psi_i^2 \psi_j^2 \, dx \, dy, \quad i, j = 1, 2, \ldots, s$$

$$K_{ij}^{23} = \int_{R_e} \overline{D}_{12} \psi_i^2 \psi_j^3 \, dx \, dy, \quad i = 1, 2, \ldots, s \quad \text{and} \quad j = 1, 2, \ldots, p$$

$$K_{ij}^{24} = 0, \qquad i = 1, 2, \ldots, s \quad \text{and} \quad j = 1, 2, \ldots, q$$

$$K_{ij}^{33} = \int_{R_e} \left(-\overline{D}_{11} \right) \psi_i^3 \psi_j^3 \, dx \, dy, \qquad i, j = 1, 2, \ldots, p$$

$$K_{ij}^{34} = 0, \qquad i = 1, 2, \ldots, p \quad \text{and} \quad j = 1, 2, \ldots, q$$

$$K_{ij}^{44} = \int_{R_e} \left(D_{66} \right)^{-1} \psi_i^4 \psi_j^4 \, dx \, dy, \qquad i, j = 1, 2, \ldots, q$$

$$F_i^1 = \int_{R_e} q \psi_i^1 \, dx \, dy + \oint_{S_e} Q_n \psi_i^1 \, dS, \qquad i = 1, 2, \ldots, r$$

$$F_i^2 = \oint_{S_e} \left(\frac{\partial w}{\partial x} n_x \right) \psi_i^2 \, dS, \qquad i = 1, 2, \ldots, s$$

$$F_i^3 = \oint_{S_e} \left(\frac{\partial w}{\partial y} n_y \right) \psi_i^3 \, dS, \qquad i = 1, 2, \ldots, p$$

$$F_i^4 = \oint_{S_e} \left(\frac{\partial w}{\partial x} n_y + \frac{\partial w}{\partial y} n_x \right) \psi_i^4 \, dS, \qquad i = 1, 2, \ldots, q \qquad (4.5.26)$$

An examination of the variational form in Eq. (4.5.22) shows that the following minimum continuity conditions are required of ψ_i^α ($\alpha = 1, 2, 3, 4$):

$$\psi_i^1 = \text{linear in } x \text{ and linear in } y$$

$$\psi_i^2 = \text{linear in } x \text{ and constant in } y$$

$$\psi_i^3 = \text{linear in } y \text{ and constant in } x$$

$$\psi_i^4 = \text{linear in } x \text{ and linear in } y \qquad (4.5.27a)$$

For a rectangular element, the interpolation functions ψ_i^α that meet only the minimum requirements are

$$\psi_i^1 = \text{bilinear functions in Eq. } (3.4.66)$$

$$\psi_1^2 = 1 - \frac{\bar{x}}{a}, \qquad \psi_2^2 = \frac{\bar{x}}{a}$$

$$\psi_1^3 = 1 - \frac{\bar{y}}{b}, \qquad \psi_2^3 = \frac{\bar{y}}{b}$$

$$\psi_i^4 = \text{bilinear functions in Eq. } (3.4.66) \qquad (4.5.27b)$$

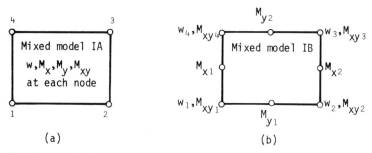

Figure 4.23 Rectangular elements for the mixed formulation of the classical plate theory (mixed models IA and IB).

The rectangular element associated with this choice of interpolation functions is shown in Fig. 4.23b. The coefficient matrices in Eq. (4.5.26) can be easily evaluated for this element. For instance, we have

$$
[K^{12}] = \frac{b}{2a}\begin{bmatrix} 1 & -1 \\ -1 & 1 \\ -1 & 1 \\ 1 & -1 \end{bmatrix}, \qquad [K^{13}] = \frac{a}{2b}\begin{bmatrix} 1 & -1 \\ 1 & -1 \\ -1 & 1 \\ -1 & 1 \end{bmatrix}
$$

$$
[K^{22}] = -\overline{D}_{22}\frac{a}{6}\begin{bmatrix} 2 & 1 \\ 1 & 2 \end{bmatrix}, \qquad [K^{33}] = -\overline{D}_{11}\frac{b}{6}\begin{bmatrix} 2 & 1 \\ 1 & 2 \end{bmatrix} \qquad (4.5.28)
$$

and so on. The element stiffness matrix is of the order 12×12.

Another choice of ψ_i^a is provided by (see Fig. 4.23a):

$$
\psi_i^1 = \psi_i^2 = \psi_i^3 = \psi_i^4 = \text{bilinear functions in Eq. (3.4.66)} \qquad (4.5.29)
$$

Computation of coefficient matrices is simple and straightforward. The resulting stiffness matrix is of the order 16×16.

Mixed Model II

A simplified mixed model can be derived by eliminating the twisting moment M_{xy} from equations (4.5.18) and (4.5.20). We have

$$
-\left(\frac{\partial^2 M_x}{\partial x^2} - 4D_{66}\frac{\partial^4 w}{\partial x^2 \partial y^2} + \frac{\partial^2 M_y}{\partial y^2} \right) = q \qquad (4.5.30a)
$$

$$
\left.\begin{aligned}
\frac{\partial^2 w}{\partial x^2} &= -\left(\overline{D}_{22} M_x - \overline{D}_{12} M_y \right) \\[2mm]
\frac{\partial^2 w}{\partial y^2} &= -\left(\overline{D}_{11} M_y - \overline{D}_{12} M_x \right)
\end{aligned}\right\} \qquad (4.5.30b)
$$

If we wish to solve the free vibration or buckling problem, we must modify the equilibrium equation (4.5.30a) by adding the contribution of the inertia term and inplane loads (see Exercise 4.2.11):

$$-\left(\frac{\partial^2 M_x}{\partial x^2} - 4D_{66}\frac{\partial^4 w}{\partial x^2 \partial y^2} + \frac{\partial^2 M_y}{\partial y^2} \right)$$

$$+ \lambda_v w - \lambda_b \left[n_{10}\frac{\partial^2 w}{\partial x^2} + 2n_{60}\frac{\partial^2 w}{\partial x \partial y} + n_{20}\frac{\partial^2 w}{\partial y^2} \right] = q \qquad (4.5.31)$$

where λ_v and λ_b are the vibration and buckling parameters, respectively,

$$\lambda_v = \frac{\rho h \omega^2}{D_{22}}, \qquad \lambda_b = -\frac{n_1}{n_{10}} = -\frac{n_2}{n_{20}} = -\frac{n_6}{n_{60}} \qquad (4.5.32)$$

and n_1, n_2, and n_6 are the inplane forces, and n_{10}, n_{20}, and n_{60} are the specified values of the normal (compressive) and shearing inplane edge forces (see Fig. 4.24). The specified edge forces should satisfy the equilibrium equations

$$\lambda_b\left(\frac{\partial n_{10}}{\partial x} + \frac{\partial n_{60}}{\partial y} \right) = 0, \qquad \lambda_b\left(\frac{\partial n_{60}}{\partial x} + \frac{\partial n_{20}}{\partial y} \right) = 0 \qquad (4.5.33)$$

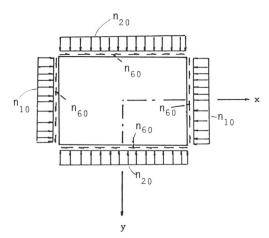

Figure 4.24   Inplane normal and shear forces for a plate.

The stationary functional associated with Eqs. (4.5.30b) and (4.5.31) is given by

$$J_2^e(w, M_x, M_y)$$

$$= \int_{R_e} \left[\frac{1}{2} \left(-\overline{D}_{22} M_x^2 + 2\overline{D}_{12} M_x M_y - \overline{D}_{11} M_y^2 \right) \right.$$

$$+ \frac{\partial w}{\partial x} \frac{\partial M_x}{\partial x} + \frac{\partial w}{\partial y} \frac{\partial M_y}{\partial y} + 2D_{66} \left(\frac{\partial^2 w}{\partial x \partial y} \right)^2 - qw \right] dx \, dy$$

$$+ \frac{1}{2} \int_{R_e} \left\{ \lambda_v w^2 + \lambda_b \left[n_{10} \left(\frac{\partial w}{\partial x} \right)^2 + 2n_{60} \frac{\partial w}{\partial x} \frac{\partial w}{\partial y} + n_{20} \left(\frac{\partial w}{\partial y} \right)^2 \right] \right\} dx \, dy$$

$$- \oint_{S_e} \left[M_x \frac{\partial w}{\partial x} n_x + M_y \frac{\partial w}{\partial y} n_y - 2D_{66} \frac{\partial^2 w}{\partial x \partial y} \left(\frac{\partial w}{\partial x} n_y + \frac{\partial w}{\partial y} n_x \right) + Q_n w \right] dS$$

$$(4.5.34)$$

An examination of the functional reveals that the minimum continuity conditions on the interpolation functions ψ_i^α ($\alpha = 1, 2, 3$) are the same as those listed in Eq. (4.5.27a). Therefore, interpolation functions of Eq. (4.5.27b) or Eq. (4.5.29) can be used. The corresponding rectangular elements are shown in Fig. 4.25.

The finite-element model of Eqs. (4.5.30b) and (4.5.31) is obtained by substituting Eq. (4.5.24) into Eq. (4.5.34). We obtain

$$\begin{bmatrix} [K^{11}] & [K^{12}] & [K^{13}] \\ & [K^{22}] & [K^{23}] \\ \text{symmetric} & & [K^{33}] \end{bmatrix} \begin{Bmatrix} \{w\} \\ \{M_x\} \\ \{M_y\} \end{Bmatrix}$$

$$= \begin{Bmatrix} \{F^1\} \\ \{F^2\} \\ \{F^3\} \end{Bmatrix} + \lambda_v \begin{Bmatrix} [S]\{w\} \\ \{0\} \\ \{0\} \end{Bmatrix} + \lambda_b \begin{Bmatrix} [G]\{w\} \\ \{0\} \\ \{0\} \end{Bmatrix}$$

$$(4.5.35)$$

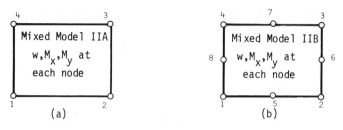

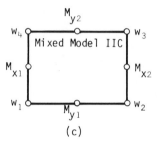

Figure 4.25 Rectangular elements for the mixed formulation of the classical plate theory (Mixed models IIA, IIB, and IIC).

where

$$K_{ij}^{11} = 4D_{66} \int_{R_e} \frac{\partial^2 \psi_i^1}{\partial x\,\partial y} \frac{\partial^2 \psi_j^1}{\partial x\,\partial y}\, dx\,dy, \qquad i,j = 1,2,\ldots,r$$

$$K_{ij}^{12} = \int_{R_e} \frac{\partial \psi_i^1}{\partial x} \frac{\partial \psi_j^2}{\partial x}\, dx\,dy, \qquad i = 1,2,\ldots,r \quad \text{and} \quad j = 1,2,\ldots,s$$

$$K_{ij}^{13} = \int_{R_e} \frac{\partial \psi_i^1}{\partial y} \frac{\partial \psi_j^3}{\partial y}\, dx\,dy, \qquad i = 1,2,\ldots,r \quad \text{and} \quad j = 1,2,\ldots,p$$

$$K_{ij}^{22} = \int_{R_e} \left(-\overline{D}_{22}\right) \psi_i^2 \psi_j^2\, dx\,dy, \qquad i,j = 1,2,\ldots,s$$

$$K_{ij}^{23} = \int_{R_e} \overline{D}_{12} \psi_i^2 \psi_j^3\, dx\,dy, \qquad i = 1,2,\ldots,s \quad \text{and} \quad j = 1,2,\ldots,p$$

$$K_{ij}^{33} = \int_{R_e} \left(-\overline{D}_{11}\right) \psi_i^3 \psi_j^3\, dx\,dy, \qquad i,j = 1,2,\ldots,p$$

$$F_i^1 = \int_{R_e} q\psi_i^1\, dx\,dy + \oint_{S_e} Q_n \psi_i^1\, dS, \qquad i = 1,2,\ldots,r$$

$$F_i^2 = \oint_{S_e} \left(\frac{\partial w}{\partial x} n_x \right) \psi_i^2 \, dS, \qquad i = 1, 2, \ldots, s$$

$$F_i^3 = \oint_{S_e} \left(\frac{\partial w}{\partial y} n_y \right) \psi_i^3 \, dS, \qquad i = 1, 2, \ldots, p$$

$$S_{ij} = \int_{R_e} \psi_i^1 \psi_j^1 \, dx \, dy, \qquad i, j = 1, 2, \ldots, r$$

$$G_{ij} = \int_{R_e} \left[n_{10} \frac{\partial \psi_i^1}{\partial x} \frac{\partial \psi_j^1}{\partial x} + n_{60} \left(\frac{\partial \psi_i^1}{\partial x} \frac{\partial \psi_j^1}{\partial y} + \frac{\partial \psi_i^1}{\partial y} \frac{\partial \psi_j^1}{\partial x} \right) + n_{20} \frac{\partial \psi_i^1}{\partial y} \frac{\partial \psi_j^1}{\partial y} \right] dx \, dy$$

$$(4.5.36)$$

Note that the term $2D_{66}(\partial^2 w/\partial x \partial y)(\partial w/\partial x \, n_y + \partial w/\partial y \, n_x)$ is omitted in the finite-element model. This amounts to the assumption that either the twisting moment or the tangential rotation is specified to be zero on the edges of a plate. This completes the development of the mixed finite-element models. We now present some applications of the mixed elements for bending, buckling, and vibration of rectangular plates (see Reddy and Tsay, 1977a and b and Tsay and Reddy, 1978).

Bending Analysis ($\lambda_v = \lambda_b = 0$)

A number of plate problems are analyzed using the mixed rectangular elements developed in this section. The present results are compared with known analytical and approximate solutions. The bending analysis is carried out using the following types of mixed elements.

Mixed Model I. Bilinear approximations are used for deflection w, and moments M_x, M_y, and M_{xy} (mixed model IA). Thus, the element has four nodes and four degrees of freedom at each node (see Fig. 4.23a), resulting in a 16×16 element stiffness matrix.

Mixed Model II. This model is subdivided into three models depending on the type of interpolation used. In mixed model IIA, bilinear interpolation of w, M_x, and M_y is used; in mixed model IIB, quadratic (serendipity) interpolation is used. These elements have stiffness matrices of orders 12×12 and 24×24 respectively (see Fig. 4.25). In mixed model IIC, bilinear approximations are used for deflection w, and partial linear approximations for M_x and M_y. The four corner nodes each have one deflection degree of freedom. Midnodes on the sides perpendicular to the y-axis have one degree of freedom M_y, and midnodes on the sides perpendicular to the x-axis have one degree of freedom M_x (see Fig. 4.25c). This element results in a 8×8 element stiffness matrix.

The four mixed plate elements discussed above are used to analyze the bending of rectangular plates with simply supported and clamped boundary

Table 4.11 Simply Supported, Square Plate Under Uniformly Distributed Load ($a/b = 1$, $\nu = 0.3$)

Mesh Size	IA	Mixed Models Linear, IIA	Mixed Models Quadratic, IIB	Mixed Models IIC	Herrmann (1966)	Hybrid Model Allman (1971a)	Compatible Cubic Displa. Model[c]
			Central Deflection, $(wD/q_0a^4)10^2$ (0.4062)[a]				
1 × 1	0.4613 (16)[b]	0.3906 (12)	0.3867 (24)	0.4943 (8)	0.9018 (6)	0.347 (12)	0.220 (12)
2 × 2	0.4237 (36)	0.4082 (27)	0.4353 (63)	0.4289 (21)	0.5127 (15)	0.392 (27)	0.371 (27)
4 × 4	0.4106 (100)	0.4069 (75)	0.4062 (195)	0.4117 (65)	0.4316 (45)	0.403 (75)	0.392 (75)
6 × 6	0.4082 (196)	0.4066 (147)	0.4062 (399)	0.4087 (133)	0.4174 (101)	—	—
			Bending Moment at the Center, $(M_x/q_0a^2)10$ (0.479)[a]				
1 × 1	0.7196	0.6094	0.3813	0.3482	0.328	0.604	—
2 × 2	0.5246	0.5049	0.4818	0.4498	0.446	0.515	—
4 × 4	0.4891	0.4849	0.4788	0.4721	0.471	0.487	—
6 × 6	0.4834	0.4815	0.4789	0.4759	0.476	—	—

[a]Exact solution from Table 4.2, CPT solution.
[b]Numbers in parentheses indicate the number of degrees of freedom in the mesh of a quarter plate.
[c]From Clough and Tocher (1966).

Table 4.12 Clamped Square Plate Under Uniformly Distributed Load ($a/b = 1$, $\nu = 0.3$)

| Mesh Size | Mixed Models | | | | | Hybrid Model | Compatible Cubic Displa. |
	IA	Linear, IIA	Quadratic, IIB	IIC	Herrmann (1966)	Allman (1971a)	Model[c]
	Central Deflection, $(wD/q_0a^4)10^2$ (0.1265)[a]						
1 × 1	0.1664 (16)[b]	0.1563 (12)	0.1466 (24)	0.2278 (8)	0.7440 (6)	0.087 (12)	0.026 (12)
2 × 2	0.1529 (36)	0.1480 (27)	0.1260 (63)	0.1627 (21)	0.2854 (15)	0.132 (27)	0.120 (27)
4 × 4	0.1339 (100)	0.1325 (75)	0.1264 (195)	0.1359 (65)	0.1696 (45)	0.129 (75)	0.121 (75)
6 × 6	0.1299 (196)	0.1292 (147)	0.1265 (399)	0.1307 (133)	0.1463 (101)	—	—
	Bending Moment at the Center, $(M_x/q_0a^2)10$ (0.231)[a]						
1 × 1	0.5193	0.4875	0.2056	0.2487	0.208	0.344	—
2 × 2	0.3166	0.2899	0.2248	0.2432	0.242	0.314	—
4 × 4	0.2478	0.2443	0.2287	0.2339	0.235	0.250	—
6 × 6	0.2374	0.2358	0.2290	0.2313	0.232	—	—

[a]Exact solution from Timoshenko and Woinowski-Krieger (1959); see also Example 4.8.
[b]Number of degrees of freedom in the mesh of a quarter plate.
[c]From Clough and Tocher (1966).

443

Table 4.13 Simply Supported, Orthotropic, Square Plate Under Uniformly Distributed Load

Mesh	Glass–Epoxy[b]		Graphite–Epoxy[c]	
	Linear	Quadratic	Linear	Quadratic
Central Deflection,[a] $(wH/q_0 a^4)\,10^3$				
1×1	2.9737	2.9412	0.9348	0.9207
2×2	3.1041	3.3081	0.9346	0.9224
4×4	3.0930	3.0872	0.9220	0.9198
6×6	3.0901	3.0876	0.9208	0.9199
8×8	3.0890	3.0876	0.9204	0.9200
Exact[d]	3.0876		0.9200	
Bending Moment at the Center $(M_x/q_0 a^2)\,10$				
1×1	0.9435	0.5718	1.6152	0.9491
2×2	0.8078	0.7687	1.3570	1.3094
4×4	0.7737	0.7627	1.2845	1.2661
6×6	0.7676	0.7628	1.2740	1.2658
8×8	0.7654	0.7628	1.2704	1.2658
Exact[d]	0.7628		1.2659	
Bending Moment at the Center, $(M_y/q_0 a^2)\,10^2$				
1×1	3.6290	2.2297	0.6720	0.3800
2×2	2.8937	2.7849	0.2860	0.3370
4×4	2.7851	2.7556	0.3143	0.3274
6×6	2.7685	2.7556	0.3252	0.3272
8×8	2.7628	2.7556	0.3240	0.3272
Exact[d]	2.7556		0.3271	

[a]$H = D_{12} + 2D_{66}$.
[b]Glass-epoxy: $E_1 = 7.8 \times 10^6$ psi, $E_2 = 2.6 \times 10^6$ psi, $\nu_{12} = 0.25$, $G_{12} = 1.3 \times 10^6$ psi.
[c]Graphite-epoxy: $E_1 = 31.8 \times 10^6$ psi, $E_2 = 1.02 \times 10^6$ psi, $\nu_{12} = 0.31$, $G_{12} = 0.96 \times 10^6$ psi.
[d]From Eq. (4.1.61).

conditions. The exact solutions for these two boundary conditions, based on the classical plate theory, are available from Section 4.1.2 for uniformly distributed loads. The mixed finite-element solutions are presented in Tables 4.11 and 4.12 for isotropic plates and in Table 4.13 for orthotropic plates. The present finite-element solutions for center deflection and bending moments are compared with various finite-element solutions available in the literature. From the results, it is clear that mixed models IIA and IIB give the best results. The numerical convergence of various finite-element models is illustrated in Fig. 4.26.

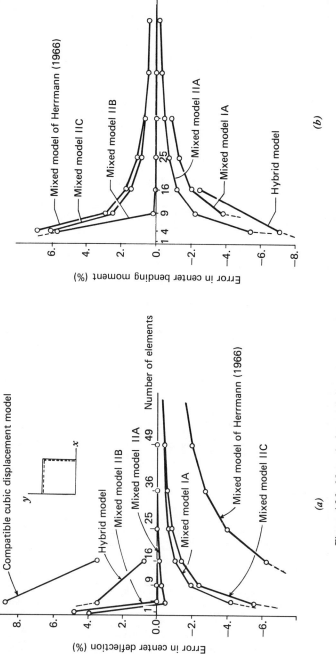

Figure 4.26 Numerical convergence character of various finite-element models: (*a*) convergence of center deflection; (*b*) convergence of center bending moment.

Table 4.14 Buckling Coefficients for Simply Supported, Square Plate ($\lambda_b = Na^2/\pi^2 D$)

Load Case (exact λ)[a]	Mesh	Triangular Elements			Rectangular Elements			Present Analysis	
		Allman (1971b)	Clough and Fellippa (1968)	Anderson et al. (1968)	Kapur and Hartz (1966)	Dawe (1979)	Carson and Newton (1969)	Linear, IIA	Quadratic, IIB
Uniform uniaxial compression (4.00)	2 × 2	—	—	3.22	—	—	4.01575	4.2095	4.0063
	3 × 3	—	—	—	3.645	—	4.00324	4.0922	4.0012
	4 × 4	4.031	4.126	3.72	3.770	3.978	4.0010	4.0517	4.0004
	8 × 8	4.006	4.031	3.94	3.933	3.993	4.0007	4.0129	4.0000
Uniform biaxial compression (2.00)	2 × 2	—	—	—	—	—	—	2.1048	2.0031
	3 × 3	—	—	—	—	—	—	2.0461	2.0006
	4 × 4	2.016	—	—	—	1.989	—	2.0258	2.0002
	8 × 8	2.003	—	—	—	1.997	—	2.0064	2.0000
Uniform shear[b] (9.34)	2 × 2	—	—	—	—	—	10.016	—	14.4124
	4 × 4	10.131	—	—	—	9.481	9.418	14.3220	9.6758
	6 × 6	—	—	—	—	—	—	11.1499	9.3816
	8 × 8	9.468	—	—	—	—	—	10.2898	9.3403

[a]From Timoshenko and Gere (1961); see also Exercise 4.2.13.
[b]Obtained using the full plate.

Table 4.15 Buckling Coefficients for Clamped Square Plate ($\lambda_b = Na^2/\pi^2 D$)

Load Case (Exact λ)[a]	Mesh	Triangular Elements					Rectangular Elements	Present Analysis	
		Allman (1971b)	Clough and Fellippa (1968)	Anderson et al. (1968)	Kapur and Hartz (1966)	Dawe (1969)	Carson and Newton (1969)	Linear, IIA	Quadratic, IIB
Uniform uniaxial compression (10.07)	2 × 2	—	—	—	—	—	—	11.530	10.2009
	3 × 3	—	—	9.30	—	—	—	10.688	10.1107
	4 × 4	10.990	—	—	9.284	10.147	—	10.412	10.0866
	8 × 8	10.252	—	—	9.782	10.065	—	10.156	10.0748
Uniform biaxial compression (5.30)	2 × 2	—	—	—	4.901	—	5.36223	6.0592	5.3746
	3 × 3	—	—	—	—	—	5.36599	5.6339	5.3233
	4 × 4	5.602	5.625	5.043	4.975	—	5.32713	5.4868	5.3102
	8 × 8	5.356	5.399	5.119	5.160	—	5.30544	5.3487	5.3040
Uniform shear[b] (14.71)	2 × 2	—	—	—	—	—	23.264	—	29.3967
	4 × 4	17.382	—	—	—	—	15.043	25.7773	15.4640
	6 × 6	—	—	—	—	—	—	18.5902	14.8361
	8 × 8	15.172	—	—	—	—	—	16.7104	14.7123

[a]From Timoshenko and Gere (1961).
[b]Obtained using the full plate.

Buckling (Stability) Analysis

The buckling parameters for simply supported and clamped square plates are presented in Tables 4.14 and 4.15, respectively. Mixed model IIB gives very accurate results. The only result that comes close to the present results are those obtained by Carson and Newton (1969), who used a *complete* cubic displacement finite-element model.

Natural Vibration

Fundamental frequencies of square plates for the following edge conditions are presented:

1. Simply supported square plate (Table 4.16).
2. Clamped square plate (Tables 4.17 and 4.19).
3. Cantilevered rectangular plate (Table 4.18).

The results obtained using the mixed model IIB are as accurate as those

Table 4.16   Natural Frequencies for Simply Supported, Square Plate $(\lambda_v = \omega^2 \rho h\, a^4/D)^a$

	Present Analysis		Szilard	Tabarrok
Mesh	Model IIA	Model IIB	(1974)	et al. (1974)
1×1	576.0	408.439	—	—
2×2	431.529	390.461	—	389.83
3×3	407.809	389.792	255.041	—
4×4	399.766	389.685	291.385	389.67

^{a}Exact value: $4\pi^4 = 389.636$, from Szilard (1974).

Table 4.17   Natural Frequencies for Clamped Square Plate $(\lambda_v = \omega^2 \rho h\, a^4/D)$

	Present Analysis		Szilard	Tabarrok	
Mesh	Model IIA	Model IIB	(1974)	et al. (1974)	Exact^a
1×1	1440.0	1355.077	—		
2×2	1361.283	1303.380	—	1296.8	1295.28
3×3	1325.837	1298.051	772.84		
4×4	1312.403	1296.114	906.1		

^{a}From Szilard (1974).

Table 4.18 Natural Frequencies for Cantilever Plate ($a/b = 2$, $\nu = 0.3$)[a]
($\lambda_v^{1/2} = \omega a^2 \sqrt{\rho h / D}$)

Mode	Mesh	Present Analysis Model IIA	Present Analysis Model IIB	Anderson et al. (1968)	Barton (1951)[b] Ritz method	Barton (1951)[b] Test	Plunkett (1963)
1		3.27	3.46	3.99	3.47	3.42	3.50
2		14.46	14.60	15.30	14.93	14.52	14.50
3	2×1	22.89	22.49	21.16	21.26	20.86	21.70
4		51.28	49.07	49.47	48.71	46.90	48.10
1		3.41	3.44	3.44	3.47	3.42	3.50
2		14.55	14.76	14.76	14.93	14.52	14.50
3	4×2	22.64	21.51	21.60	21.26	20.86	21.70
4		49.93	48.13	48.28	48.71	46.90	48.10

[a] The plate is clamped along edge $x = 0$ and free on edges $x = a$ and $y = 0, b$.
[b] Results are independent of mesh.

obtained by Tabarrok et al. (1974), who used a refined compatible triangular element (a *quintic* polynomial for the displacement is used). Table 4.18 contains natural frequencies of a cantilevered plate with length-to-width ratio equal to 2. In this case, the results are compared with those of Anderson et al. (1968), Barton (1951), and Plunkett (1963). Clearly, the results obtained using mixed model IIB are closer to the test results. Table 4.19 contains the vibration frequencies of a clamped orthotropic square plate with various ratios of α and β ($\alpha = D_{11}/H$, $\beta = D_{22}/H$, $H = D_{12} + 2D_{66}$). The results are in good agreement with those obtained by Elishakoff (1974), Lekhnitskii (1968), Hearmon (1959), and Kanazawa and Kawai (1952).

In summary, the mixed finite-element models of plates give accurate results for deflections, stresses, buckling loads, and natural frequencies. The mixed models give more accurate results when compared to certain conforming and nonconforming elements. Further, the mixed finite elements are algebraically less complex, and involve less "element forming" efforts.

4.5.4 Shear Deformable Plate Elements

The shear deformable theory of laminated composite plates (see Section 4.3) involves five dependent unknowns (u, v, w, ψ_x, ψ_y). Here, we develop the finite-element model of the theory using the variational statement in Eq. (4.3.3). Numerical examples of plate bending, vibration, and transient response will be discussed (see Reddy, 1980, Reddy and Chao, 1981, and Reddy, 1982c 1983b and 1983c).

Suppose that the midplane R of the plate is subdivided into a finite number of elements, R_e ($e = 1, 2, \ldots$). Over each element, R_e, the generalized dis-

Table 4.19 Natural Frequencies for Clamped Square Orthotropic Plate with Various α and β^a ($\lambda_v^{1/2} = \omega a^2\sqrt{\rho h/H}$)

β	Methods	$\alpha = \frac{1}{3}$	$\alpha = \frac{1}{2}$	$\alpha = 1$	$\alpha = 2$	$\alpha = 3$
$\frac{1}{3}$	Linear[b]	6.290	7.198	9.880	15.199	20.506
	Quadratic	6.327	7.210	9.814	14.977	20.127
	Elishakoff	6.105	7.935	9.449	14.540	19.659
	Lekhnitskii	6.404	7.266	9.853	15.027	20.201
	Hearmon	6.541	7.398	9.969	15.113	20.256
	Kanazawa and Kawai	6.434	7.341	10.016	15.355	20.638
$\frac{1}{2}$	Linear		8.108	10.790	16.110	21.417
	Quadratic		8.095	10.699	15.863	21.013
	Elishakoff		7.750	10.235	15.300	20.408
	Lekhnitskii		8.128	10.715	15.890	21.064
	Hearmon		8.255	10.827	15.970	21.113
	Kanazawa and Kawai		8.249	10.927	15.242	21.553
1	Linear			13.473	18.793	24.100
	Quadratic			13.306	18.469	23.619
	Elishakoff			12.658	17.661	22.740
	Lekhnitskii			13.302	18.477	23.651
	Hearmon			13.398	18.541	23.685
	Kanazawa and Kawai			13.608	18.928	24.232
2	Linear				24.112	29.419
	Quadratic				23.634	28.784
	Elishakoff				22.596	27.640
	Lekhnitskii				23.651	28.825
	Hearmon				23.685	28.828
	Kanazawa and Kawai				24.251	28.561
3	Linear					34.725
	Quadratic					33.936
	Elishakoff (1974)					32.665
	Lekhnitskii (1968)					33.999
	Hearmon (1959)					33.991
	Kanazawa and Kawai (1952)					34.871

[a] $\alpha = D_{11}/H$, $\beta = D_{22}/H$, $H = D_{11}\nu_{21} + 2D_{66}$, $\omega(\alpha,\beta) = \omega(\beta,\alpha)$.
[b] Mixed models IIA and IIB.

placements (u, v, w, ψ_x, ψ_y) are interpolated by expressions of the form

$$u = \sum_{i=1}^{n} u_i\phi_i, \qquad v = \sum_{i=1}^{n} v_i\phi_i, \qquad w = \sum_{i=1}^{n} w_i\phi_i,$$

$$\psi_x = \sum_{i=1}^{n} \psi_{xi}\phi_i, \qquad \psi_y = \sum_{i=1}^{n} \psi_{yi}\phi_i \qquad (4.5.37)$$

where ϕ_i are the interpolation functions and n is the number of nodes per element. Note that the same interpolation functions are used for all five displacements. The total Lagrangian in Eq. (4.3.4) can be expressed in terms of the displacements $(u, v, w, \psi_x, \psi_y)$ by substituting Eq. (4.3.2) into Eq. (4.3.13), and the result into Eq. (4.3.4). The total potential energy and the total Lagrangian (in Hamilton's principle) are given by

$$\Pi = U + V, \qquad L = \Pi - K \tag{4.5.38a}$$

where V is the potential energy owing to applied loads, and U and K are the strain and kinetic energies:

$$U = U_1 + U_2 \tag{4.5.38b}$$

$$K = \frac{1}{2}\int_\Omega \left\{ I_1\left[\left(\frac{\partial u}{\partial t}\right)^2 + \left(\frac{\partial v}{\partial t}\right)^2 + \left(\frac{\partial w}{\partial t}\right)^2\right] + I_3\left[\left(\frac{\partial \psi_x}{\partial t}\right)^2 + \left(\frac{\partial \psi_y}{\partial t}\right)^2\right] \right.$$

$$\left. + 2I_2\left[\frac{\partial \psi_x}{\partial t}\frac{\partial u}{\partial t} + \frac{\partial \psi_y}{\partial t}\frac{\partial v}{\partial t}\right] \right\} dx\, dy \tag{4.5.38c}$$

$$U_1 = \frac{1}{2}\int_\Omega \left\{ A_{11}\left(\frac{\partial u}{\partial x}\right)^2 + 2A_{16}\frac{\partial u}{\partial x}\frac{\partial u}{\partial y} + A_{66}\left(\frac{\partial u}{\partial y}\right)^2 \right.$$

$$+ \left(A_{12}\frac{\partial v}{\partial y} + A_{16}\frac{\partial v}{\partial x}\right)\frac{\partial u}{\partial x}$$

$$+ \frac{\partial v}{\partial y}\left(A_{12}\frac{\partial u}{\partial x} + A_{26}\frac{\partial u}{\partial y}\right) + \frac{\partial u}{\partial y}\left(A_{26}\frac{\partial v}{\partial y} + A_{66}\frac{\partial v}{\partial x}\right)$$

$$+ \frac{\partial v}{\partial x}\left(A_{16}\frac{\partial u}{\partial x} + A_{66}\frac{\partial u}{\partial y}\right) + A_{22}\left(\frac{\partial v}{\partial y}\right)^2 + 2A_{26}\frac{\partial v}{\partial y}\frac{\partial v}{\partial x}$$

$$+ A_{66}\left(\frac{\partial v}{\partial x}\right)^2 + \frac{\partial u}{\partial x}\left(B_{11}\frac{\partial \psi_x}{\partial x} + B_{16}\frac{\partial \psi_x}{\partial y} + B_{12}\frac{\partial \psi_y}{\partial y} + B_{16}\frac{\partial \psi_y}{\partial x}\right)$$

$$+ \left(\frac{\partial u}{\partial y} + \frac{\partial v}{\partial x}\right)\left(B_{16}\frac{\partial \psi_x}{\partial x} + B_{66}\frac{\partial \psi_x}{\partial y} + B_{26}\frac{\partial \psi_y}{\partial y} + B_{66}\frac{\partial \psi_y}{\partial x}\right)$$

$$+ \frac{\partial v}{\partial y}\left(B_{12}\frac{\partial \psi_x}{\partial x} + B_{26}\frac{\partial \psi_x}{\partial y} + B_{22}\frac{\partial \psi_y}{\partial y} + B_{26}\frac{\partial \psi_y}{\partial x}\right)$$

$$+ \frac{\partial \psi_x}{\partial x}\left(B_{11}\frac{\partial u}{\partial x} + B_{16}\frac{\partial u}{\partial y} + B_{12}\frac{\partial v}{\partial y} + B_{16}\frac{\partial v}{\partial x}\right)$$

$$+ \left(\frac{\partial \psi_x}{\partial y} + \frac{\partial \psi_y}{\partial x}\right)\left(B_{16}\frac{\partial u}{\partial x} + B_{66}\frac{\partial u}{\partial y} + B_{26}\frac{\partial v}{\partial y} + B_{66}\frac{\partial v}{\partial x}\right)$$

$$+ \frac{\partial \psi_y}{\partial y}\left(B_{12}\frac{\partial u}{\partial x} + B_{26}\frac{\partial u}{\partial y} + B_{22}\frac{\partial v}{\partial y} + B_{26}\frac{\partial v}{\partial x}\right)$$

$$+ D_{11}\left(\frac{\partial\psi_x}{\partial x}\right)^2 + 2D_{16}\frac{\partial\psi_x}{\partial x}\frac{\partial\psi_x}{\partial y} + D_{66}\left(\frac{\partial\psi_x}{\partial y}\right)^2$$

$$+ \frac{\partial\psi_x}{\partial x}\left(D_{12}\frac{\partial\psi_y}{\partial y} + 2D_{16}\frac{\partial\psi_y}{\partial x}\right) + \frac{\partial\psi_y}{\partial y}\left(D_{12}\frac{\partial\psi_x}{\partial x} + 2D_{26}\frac{\partial\psi_x}{\partial y}\right)$$

$$+ 2D_{66}\frac{\partial\psi_x}{\partial y}\frac{\partial\psi_y}{\partial x} + D_{22}\left(\frac{\partial\psi_y}{\partial y}\right)^2$$

$$\left. + 2D_{26}\frac{\partial\psi_y}{\partial x}\frac{\partial\psi_y}{\partial y} + D_{66}\left(\frac{\partial\psi_y}{\partial x}\right)^2\right\} dx\, dy \qquad (4.5.38d)$$

$$U_2 = \frac{1}{2}\int_\Omega \left\{\left[A_{55}\left(\frac{\partial w}{\partial x} + \psi_x\right) + A_{45}\left(\frac{\partial w}{\partial y} + \psi_y\right)\right]\left(\frac{\partial w}{\partial x} + \psi_x\right)\right.$$

$$\left. + \left[A_{45}\left(\frac{\partial w}{\partial x} + \psi_x\right) + A_{44}\left(\frac{\partial w}{\partial y} + \psi_y\right)\right]\left(\frac{\partial w}{\partial y} + \psi_y\right)\right\} dx\, dy$$

$$(4.5.38e)$$

Note that U_1 denotes energy caused by bending, and U_2 denotes energy resulting from transverse shear deformation.

Substituting Eq. (4.5.37) into Eq. (4.3.3), we obtain the element equations

$$[M^e]\{\ddot{\Delta}^e\} + [K^e]\{\Delta^e\} = \{F^e\} \qquad (4.5.39a)$$

where

$$\{\Delta^e\} = \begin{Bmatrix} \{u^e\} \\ \{v^e\} \\ \{w^e\} \\ \{\psi_x^e\} \\ \{\psi_y^e\} \end{Bmatrix}, \qquad [M^e] = \begin{bmatrix} I_1[S] & [0] & [0] & I_2[S] & [0] \\ & I_1[S] & [0] & [0] & I_2[S] \\ & & I_1[S] & [0] & [0] \\ & & & I_3[S] & [0] \\ \text{symmetric} & & & & I_3[S] \end{bmatrix}_e$$

$$(4.5.39b)$$

$$[K^e] = \begin{bmatrix} [K^{11}] & [K^{12}] & [K^{13}] & [K^{14}] & [K^{15}] \\ & [K^{22}] & [K^{23}] & [K^{24}] & [K^{25}] \\ & & [K^{33}] & [K^{34}] & [K^{35}] \\ & & & [K^{44}] & [K^{45}] \\ & \text{symmetric} & & & [K^{55}] \end{bmatrix}_e, \qquad \{F^e\} = \begin{Bmatrix} \{F^1\} \\ \{F^2\} \\ \{F^3\} \\ \{F^4\} \\ \{F^5\} \end{Bmatrix}_e$$

$$(4.5.39c)$$

The elements of stiffness matrix $[K^e]$, mass matrix $[M^e]$, and force vector $\{F^e\}$ are given below.

$$[K^{11}] = A_{11}[S^{xx}] + A_{16}\left([S^{xy}] + [S^{xy}]^T\right) + A_{66}[S^{yy}]$$

$$[K^{12}] = A_{12}[S^{xy}] + A_{16}[S^{xx}] + A_{26}[S^{yy}] + A_{66}[S^{xy}]^T$$

$$[K^{13}] = [0]$$

$$[K^{14}] = B_{11}[S^{xx}] + B_{16}\left([S^{xy}] + [S^{xy}]^T\right) + B_{66}[S^{yy}]$$

$$[K^{15}] = B_{12}[S^{xy}] + B_{16}[S^{xx}] + B_{26}[S^{yy}] + B_{66}[S^{xy}]^T$$

$$[K^{22}] = A_{22}[S^{yy}] + A_{26}\left([S^{xy}]^T + [S^{xy}]\right) + A_{66}[S^{xx}]$$

$$[K^{23}] = [0]$$

$$[K^{24}] = B_{12}[S^{xy}]^T + B_{26}[S^{yy}] + B_{16}[S^{xx}] + B_{66}[S^{xy}]$$

$$[K^{25}] = B_{22}[S^{yy}] + B_{26}\left([S^{xy}] + [S^{xy}]^T\right) + B_{66}[S^{xx}]$$

$$[K^{33}] = A_{55}[S^{xx}] + A_{45}\left([S^{xy}] + [S^{xy}]^T\right) + A_{44}[S^{yy}]$$

$$[K^{34}] = A_{55}[S^{x0}] + A_{45}[S^{y0}]$$

$$[K^{35}] = A_{45}[S^{x0}] + A_{44}[S^{y0}]$$

$$[K^{44}] = D_{11}[S^{xx}] + D_{16}\left([S^{xy}] + [S^{xy}]^T\right) + D_{66}[S^{yy}] + A_{55}[S]$$

$$[K^{45}] = D_{12}[S^{xy}] + D_{16}[S^{xx}] + D_{26}[S^{yy}] + D_{66}[S^{xy}]^T + A_{45}[S]$$

$$[K^{55}] = D_{26}\left([S^{xy}] + [S^{xy}]^T\right) + D_{66}[S^{xx}] + D_{22}[S^{yy}] + A_{44}[S]$$

$$S_{ij}^{\xi\eta} = \int_{R_e} \frac{\partial \phi_i}{\partial \xi} \frac{\partial \phi_j}{\partial \eta}\, dx\, dy, \qquad \xi, \eta = 0, x, y; \quad S_{ij} \equiv S_{ij}^{00}$$

$$F_i^\alpha = \int_{R_e} f_\alpha \phi_i\, dx\, dy + P_i, \qquad \alpha = 1, 2, \ldots, 5 \qquad (4.5.40)$$

Here P_i denotes the nodal contributions of the boundary forces of the element, and $f_1, f_2, \ldots, f_5$ are given by Eq. (4.3.15b).

Equation (4.5.39a) can be reduced to appropriate forms depending on the type of analysis. For static analysis, we set the inertia term, $\{\ddot{\Delta}\}$, to zero. For free vibration problems, we replace the inertia term by

$$\{\ddot{\Delta}\} = -\omega^2\{\Delta\}$$

Then Eq. (4.5.39a) takes the form of an eigenvalue problem,

$$([K^e] - \omega^2[M^e])\, \{\Delta\} = \{0\} \tag{4.5.41}$$

For transient analysis, Eq. (4.5.39a) should be further approximated. We discuss such an approximation in the next few paragraphs. In all three cases (i.e., bending, vibration, and transient analyses), the element equations are assembled and boundary and initial conditions are imposed, as described in Section 3.4.

To solve the time-dependent problem, one must approximate the time derivatives in Eq. (4.5.39a) to obtain algebraic equations relating $\{\Delta\}$ at time $t + \Delta t$ to $\{\Delta\}$ at time t, where Δt is the time step. Then the algebraic equations are assembled, boundary conditions are applied, and equations are solved for the global nodal values of the displacements at time $t + \Delta t$. There exist a number of time integration schemes for the solution of equations of the form in Eq. (4.5.39); see the bibliography. Among them the Newmark direct integration scheme (see Newmark, 1959) is commonly used in structural dynamics. In the Newmark integration scheme, the vectors $\{\Delta\}$ and $\{\dot{\Delta}\}$ at time $t = (n + 1)\,\Delta t$ are approximated by the expressions (see Reddy, 1984),

$$\{\dot{\Delta}\}_{n+1} = \{\dot{\Delta}\}_n + \left[(1 - \alpha)\{\ddot{\Delta}\}_n + \alpha\{\ddot{\Delta}\}_{n+1}\right]\Delta t$$

$$\{\Delta\}_{n+1} = \{\Delta\}_n + \{\dot{\Delta}\}_n\Delta t + \left[\left(\frac{1}{2} - \beta\right)\{\ddot{\Delta}\}_n + \beta\{\ddot{\Delta}\}_{n+1}\right](\Delta t)^2$$

$$\tag{4.5.42}$$

where α and β are parameters that control the accuracy and stability of the scheme, and the subscript n indicates that the vectors are evaluated at nth time step (i.e., at time, $t = n\,\Delta t$).

Rearranging Eqs. (4.5.39a) and (4.5.42), we obtain

$$[\hat{K}]\{\Delta\}_{n+1} = \{\hat{F}\} \tag{4.5.43}$$

where $\{\Delta\}_{n+1}$ denotes the value of $\{\Delta\}$ at time $t = (n + 1)\Delta t$, n being the time step number, and

$$[\hat{K}] = [K] + a_0[M],$$

$$\{\hat{F}\} = \{F\}_{n+1} + [M](a_0\{\Delta\}_n + a_1\{\dot{\Delta}\}_n + a_2\{\ddot{\Delta}\}_n)$$

$$a_0 = \frac{1}{\beta\Delta t^2}, \qquad a_1 = a_0\Delta t, \qquad a_2 = \frac{1}{2\beta} - 1 \qquad (4.5.44)$$

Once the solution $\{\Delta\}$ is known at time $t_{n+1} = (n + 1)\Delta t$, the first and second derivatives of $\{\Delta\}$ (velocity and accelerations) at t_{n+1} can be computed from

$$\{\ddot{\Delta}\}_{n+1} = a_0(\{\Delta\}_{n+1} - \{\Delta\}_n) - a_1\{\dot{\Delta}\}_n - a_2\{\ddot{\Delta}\}_n$$

$$\{\dot{\Delta}\}_{n+1} = \{\dot{\Delta}\}_n + a_3\{\ddot{\Delta}\}_n + a_4\{\ddot{\Delta}\}_{n+1} \qquad (4.5.45)$$

where $a_3 = (1 - \alpha)\Delta t$, and $a_4 = \alpha\Delta t$.

The parameters α and β in the Newmark scheme can be determined to obtain integration accuracy and stability. The following two choices are commonly used:

Linear Acceleration Method

$$\alpha = \tfrac{1}{2}, \qquad \beta = \tfrac{1}{6} \qquad (4.5.46a)$$

Constant Acceleration Method

$$\alpha = \tfrac{1}{2}, \qquad \beta = \tfrac{1}{4} \qquad (4.5.46b)$$

Newmark (1959) originally proposed the constant acceleration method as an unconditionally stable scheme (also called trapezoidal rule). Although the values of α and β in Eqs. (4.5.46a) and (4.5.46b) guarantee stability of the numerical scheme, the accuracy of the scheme is largely dictated by the choice of the time step. Small time steps generally give more accurate solutions than large time steps. An estimate of time step in a problem can be obtained from

$$\Delta t \le 0.25(\rho h D)^{1/2} d^2 \qquad (4.5.47)$$

where ρ is the density, h the thickness, D is the flexural rigidity (smaller of the two, D_{11} and D_{22} for orthotropic plates), and d is the minimum distance between any two global nodes of the finite-element mesh.

In the present study, rectangular elements with four, eight, and nine nodes are employed with the same interpolation for all of the variables. The resulting stiffness matrices are 20×20 for the four-node element, 40×40 for the eight-node element, and 45×45 for the nine-node element. Since the element is based on a shear deformation theory, reduced integration (see Zienkiewicz et al., 1971 and 1976) should be used to evaluate the stiffness coefficients associated with the energy caused by transverse shear (i.e., coefficients in $[K]$ containing A_{44}, A_{45}, and A_{55} terms). For instance, when the four-node rectangular element is used, the 1×1 Gauss rule should be used to evaluate K_{ij} corresponding to the transverse shear strains, and the 2×2 Gauss rule should be used to evaluate K_{ij} associated with the bending strains and inertia coefficients M_{ij}. For a quadratic element, the 2×2 Gauss rule should be used to integrate the shear energy terms, and the 3×3 Gauss rule for the bending and inertia terms.

This completes the derivation of the shear deformable element. In the following paragraphs, we discuss the application of the element to the bending, free vibration, and transient response of elastic plates. The results for deflections, stresses, and frequencies are presented (with few exceptions) in nondimensional form. In the case of orthotropic plates and layered composite plates, the following material properties are used:

$$\text{Material I:} \quad \frac{E_1}{E_2} = 25, \quad \frac{G_{12}}{E_2} = 0.5, \quad \frac{G_{23}}{E_2} = 0.2, \quad \nu_{12} = 0.25$$

$$\text{Material II:} \quad \frac{E_1}{E_2} = 40, \quad \frac{G_{12}}{E_2} = 0.6, \quad \frac{G_{23}}{E_2} = 0.5, \quad \nu_{12} = 0.25$$

$$(4.5.48)$$

Further, it is assumed that $G_{12} = G_{13}$ and $\nu_{12} = \nu_{13}$. A value of $5/6$ was used for the shear correction coefficients, $K_1^2 = K_2^2$. All of the numerical results presented here were obtained on an IBM 370 computer, using the double-precision arithmetics.

Bending Analysis

The first example problem is designed to investigate the effect of element type (i.e., linear L, eight-node quadratic Q8, and nine-node quadratic Q9), mesh (2L = 2×2 mesh of linear elements, 2Q8 = 2×2 mesh of eight-node quadratic elements), and reduced integration on the deflections and stresses.

The finite-element solutions of linear and quadratic elements are compared with the exact solutions for deflections and stresses (from Section 4.3) of a three-layer ($h_1 = h_3 = h/4$, $h_2 = h/2$) cross-ply ($0°/90°/0°$) square plate (material I) subjected to sinusoidal transverse loading (see Table 4.20). This problem is also equivalent to a four-layer (equal thickness) cross-ply ($0°/90°/90°/0°$) plate. The stresses are computed at the Gaussian points.

Table 4.20 Effect of Reduced Integration on the Maximum Deflection and Stresses of Three-Ply $(0°/90°/0°)$ Square Plate Subjected to Sinusoidal Loading $(h_1 = h_3 = h/4, h_2 = h/2,$ Material I)

a/h		Source	$\bar{w}$	$\bar{\sigma}_x$	$\bar{\sigma}_y$	$\bar{\sigma}_{xy}$	$\bar{\sigma}_{xz}$	$\bar{\sigma}_{yz}$
10		3-DES	7.434	0.599	0.403	0.0276	0.301	0.196
		Exact	6.627	0.499	0.361	0.0241	0.417	0.129
		Shell element[a]	6.299	0.532	0.307	0.0250	—	—
	FEM	4Q8-R	6.627	0.495	0.359	0.0240	0.414	0.128
		4L-R	6.5986	0.4668	0.3466	0.0227	0.395	0.2078
		4L-F	6.4267	0.4512	0.3280	0.0219	0.389	0.129
		2Q8-R	6.6152	0.4842	0.3509	0.0234	0.404	0.1255
		2Q8-F	6.6047	0.4831	0.3492	0.0234	0.4043	0.1258
		2Q9-R	6.5610	0.4800	0.3410	0.0231	0.400	0.126
		2Q9-F	6.5510	0.4790	0.3400	0.0231	0.399	0.126
		2Q9-FR	6.5580	0.4800	0.3410	0.0230	0.400	0.126
		2L-R	6.5079	0.3799	0.2838	0.0187	0.335	0.1067
		2L-F	5.9011	0.3339	0.2454	0.0163	0.125	0.316
20		3-DES	5.173	0.543	0.308	0.0230	0.328	0.156
		Exact	4.912	0.527	0.296	0.0221	0.437	0.109
		Shell element	4.847	0.557	0.307	0.0231	—	—
	FEM	4Q8-R	4.911	0.524	0.294	0.0219	0.434	0.108
		4L-R	4.8633	0.4940	0.2777	0.0207	0.415	0.1034
		4L-F	4.3461	0.4365	0.2451	0.0183	0.3949	0.1233
		2Q8-R	4.0914	0.5112	0.2870	0.0214	0.424	0.1057
		2Q8-F	4.8764	0.5082	0.2837	0.0214	0.4237	0.1063
		2Q9-R	4.8470	0.5050	0.2780	0.0213	0.418	0.106
		2Q9-F	4.8270	0.5020	0.2750	0.0211	0.418	0.107
		2Q9-FR	4.8470	0.5050	0.2780	0.0212	0.417	0.106
		2L-R	4.7117	0.4040	0.2290	0.0171	0.3529	0.0886
		2L-F	3.2355	0.2645	0.1491	0.0111	0.3028	0.1388
100		3-DES	4.385	0.539	0.271	0.0214	0.339	0.139
		Exact	4.337	0.538	0.271	0.0213	0.445	0.101
		Shell element	4.363	0.566	0.284	0.0223	—	—
	FEM	4Q8-R	4.336	0.535	0.269	0.0212	0.442	0.1002
		4L-R	4.2811	0.5045	0.2535	0.0200	0.422	0.0958
		4L-F	1.0339	0.1203	0.0604	0.0048	0.2975	0.2206
		2Q8-R	4.3192	0.5214	0.2621	0.0206	0.435	0.1020
		2Q8-F	4.1425	0.4900	0.2435	0.0199	0.430	0.1002
		2Q9-R	4.3006	0.4180	0.2550	0.0207	0.428	0.0989
		2Q9-F	4.1580	0.4900	0.2420	0.0201	0.427	0.0996
		2Q9-FR	4.2990	0.5180	0.2550	0.0206	0.426	0.0985
		2L-R	4.1072	0.4129	0.2076	0.0164	0.356	0.08163
		2L-F	0.3149	0.0299	0.0151	0.0012	0.230	0.211
		Classical thin-plate solution	4.350	0.539	0.269	0.213	0.339	0.138

[a]Panda and Natarajan (1979).

The nondimensionalized quantities are defined by

$$\bar{w} = \frac{w(a/2, a/2) E_2 h^3 \times 10^3}{q_0 a^4}, \qquad \bar{\sigma}_x = \left(\frac{h}{a}\right)^2 \frac{1}{q_0} \sigma_x\left(A, A, \frac{\pm h}{2}\right)$$

$$\bar{\sigma}_y = \left(\frac{h}{a}\right)^2 \frac{1}{q_0} \sigma_y\left(A, A, \frac{\pm h}{4}\right), \qquad \bar{\sigma}_{xy} = \left(\frac{h}{a}\right)^2 \frac{1}{q_0} \sigma_{xy}\left(B, B, \frac{\pm h}{2}\right)$$

$$\bar{\sigma}_{xz} = \frac{h}{aq_0} \sigma_{xz}(B, A) \text{ in layers 1 and 3}, \qquad \bar{\sigma}_{yz} = \frac{h}{aq_0} \sigma_{yz}(A, B) \text{ in layer 2}$$

$$(4.5.49)$$

where A and B are the Gauss-point coordinates (with respect to the center of the plate) given below:

	2 × 2 Linear	4 × 4 Linear	2 × 2 Quadratic	4 × 4 Quadratic
A	$0.125a$	$0.0625a$	$0.05283a$	$0.02642a$
B	$0.375a$	$0.4375a$	$0.4472a$	$0.4736a$

In Table 4.20, R denotes reduced integration for all terms, F is full integration for all terms, and FR is full integration for bending terms and reduced integration for the shear terms. The following observations can be made from the results of Table 4.20:

1.   The nine-node element gives virtually the same results for full (3 × 3 Gauss rule), reduced (2 × 2 Gauss rule), and mixed (3 × 3 and 2 × 2 Gauss rules) integrations. However, the results obtained by using the reduced integration are closest to the exact solution.

2.   The integration rule has a more significant effect on the accuracy in the eight-node element than in the nine-node element. Full integration gives less accurate results than the reduced integration, and the error increases with an increase in side-to-thickness ratio. This implies that the reduced integration is a must for thin plates.

3.   Full integration results in smaller errors for quadratic elements and refined meshes than for linear elements and/or coarse meshes. That is, the error between the solutions obtained by 2Q8-F and 2Q8-R is smaller than those obtained by 4L-F and 4L-R, and the error between the solutions obtained by 4L-F and 4L-R is smaller than those obtained by 2L-F and 2L-R.

4.   The shear deformable finite element gives accurate results when compared to the exact solutions (see Section 4.3).

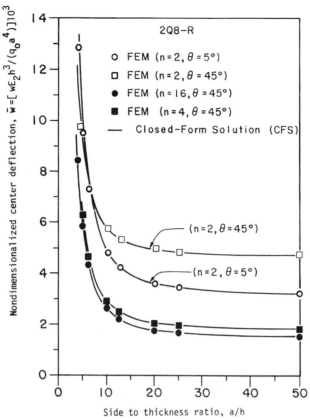

Figure 4.27 Comparison of the finite-element solution with the closed-form solution of angle-ply (antisymmetric) square plates under sinusoidal load (material II).

The nondimensionalized center deflection as a function of side-to-thickness ratio of antisymmetric angle-ply plates, with simply supported boundary conditions and sinusoidal distribution of transverse load, is presented in Fig. 4.27. The finite-element solutions are in excellent agreement with the exact solutions. It should be noted that for antisymmetric angle-ply plates, a full plate model should be used to account for the stretching–bending coupling [i.e., antisymmetric angle-ply plates do not have biaxial symmetry for rectangular plates with biaxial loading. This should be clear from the solution in Eq. (4.3.28); see Reddy, 1984c].

For additional numerical results, the reader is referred to the papers by Reddy (1980) and Reddy and Chao (1981).

Free Vibration Analysis

Figure 4.28a shows the effect of side-to-thickness ratio on the nondimensionalized fundamental frequency of a cross-ply rectangular plate (material II).

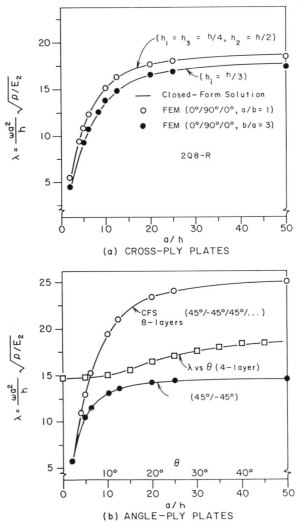

Figure 4.28 Nondimensionalized fundamental frequency versus side to thickness ratio of cross-ply and angle-ply plates (material II).

Similar results are presented for angle-ply plates in Fig. 4.28b, which also contains the plot of nondimensionalized frequency versus the lamination angle for four-layer, angle-ply plates ($a/b = 1$, $a/h = 10$, material II). The finite-element solutions are in close agreement with the exact solution.

Transient Analysis

As a first example, we consider a simply supported, isotropic square plate with suddenly applied (and kept indefinitely) uniform pressure loading. The follow-

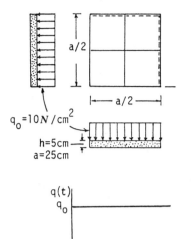

$q_0 = 10 N/cm^2$

$h = 5cm$
$a = 25cm$

$q(t)$
q_0

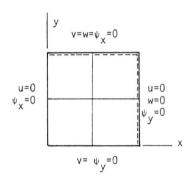

$v = w = \psi_x = 0$

$u = 0$
$\psi_x = 0$

$u = 0$
$w = 0$
$\psi_y = 0$

x

$v = \psi_y = 0$

Figure 4.29 Geometry, loading, finite-element mesh, and boundary conditions of a quarter plate.

ing material properties are used:

$$\rho = 8 \times 10^{-6} \text{ N-sec}^2/\text{cm}^4, \qquad \nu = 0.25, \qquad E = 2.1 \times 10^6 \text{ N/cm}^2$$

$$(4.5.50)$$

The plate geometry, finite-element mesh, boundary conditions for a quarter plate, and applied loading are shown in Fig. 4.29.

Figure 4.30 shows a comparison of the plot of the center deflection (for two different time steps) versus time with the mixed finite-element solution of Akay (1980). The small phase difference between the present solution and that of Akay can be attributed to the difference in the formulations. The classical plate theory (i.e., not accounting for transverse shear strains) solution is also given in the figure to show the influence of the shear deformation on the center

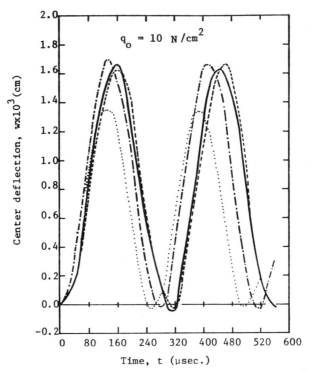

Figure 4.30 Center deflection versus time for a simply supported, square plate subjected to uniform pulse loading (mesh; 2×2). Present FEM: $-\Delta t = 10 \, \mu\text{sec}$; $---\Delta t = 5 \, \mu\text{sec}$. Akay (1980): $-\cdot-\Delta t = 5 \, \mu\text{sec}$. Classical solution: $....\Delta t = 5 \, \mu\text{sec}$.

deflection. Figure 4.31 shows the plot of center normal stress versus time for the same problem.

Next, we consider a simply supported, rectangular plate under suddenly applied patch loading (see Fig. 4.32). The following data are used:

$$\frac{a}{b} = \sqrt{2}, \qquad \Delta t = 0.1, \qquad q = q_0 H(t)$$

$$q_0 = \begin{cases} 1, & 0 < (x, y) \leq 0.2 \\ 0, & (x, y) > 0.2 \end{cases}$$

$$\nu = 0.3 \tag{4.5.51}$$

A 4×4 (nonuniform) mesh of nine-node elements is used in the quarter plate with three degrees of freedom (w, ψ_x, ψ_y) per node. The center deflection and bending moments of the present linear analysis are compared with the analytical thick-plate and thin-plate solutions of Reismann and Lee (1969) in Fig. 4.33. We note significant difference between the solutions of the two

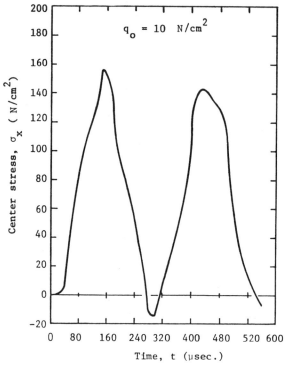

Figure 4.31 Center normal stress versus time for a simply supported, square plate subjected to uniform pulse loading (mesh: 2×2; $\Delta t = 5$ μsec).

theories. The deviation between the solutions of the shear deformation theory and the classical plate theory increases with plate thickness. The present finite element solutions for the center deflection and bending moment are in excellent agreement with the thick-plate solution of Reismann and Lee. Since the bending moment in the finite-element method is calculated at the Gauss points, it is not expected to match exactly with the analytical solution at the center of the plate.

4.5.5 Shear Deformable Shell Elements

A doubly-curved isoparametric rectangular shell element based on the dynamic version of Eq. (4.4.22) is presented here. Over the typical shell element, R_e, the displacements $(u_1, u_2, u_3, \phi_1, \phi_2)$ are interpolated by expressions of the form,

$$u_i = \sum_{j=1}^{n} u_i^j \psi_j(\xi_1, \xi_2), \qquad i = 1, 2, 3$$

$$\phi_i = \sum_{j=1}^{n} \phi_i^j \psi_j(\xi_1, \xi_2), \qquad i = 1, 2 \qquad (4.5.52)$$

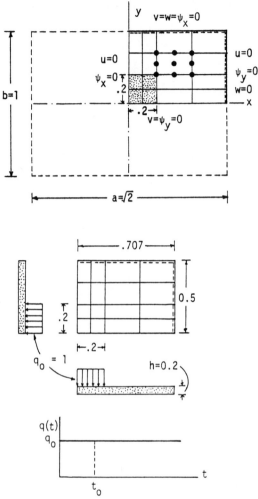

Figure 4.32 Geometry, loading, boundary conditions, and finite-element mesh of the quarter plate.

where ψ_j are the interpolation functions, and u_i^j and ϕ_i^j are the nodal values of u_i and ϕ_i, respectively. For a linear isoparametric element ($n = 4$), this interpolation results in a stiffness matrix of order 20×20.

Substitution of Eq. (4.5.52) into the virtual work principle, Eq. (4.4.20), yields an element equation of the form

$$[K]\{\Delta\} + [M]\{\ddot{\Delta}\} = \{F\} \qquad (4.5.53)$$

where $\{\Delta\} = \{\{u_1\}, \{u_2\}, (u_3), \{\phi_1\}, \{\phi_2\}\}^T$, $[K]$ and $[M]$ are element

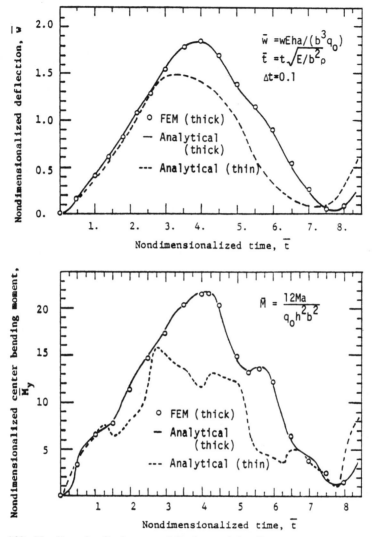

Figure 4.33 Nondimensionalized center deflection and bending moment versus time for a rectangular plate under suddenly applied uniform patch loading.

stiffness and mass matrices, respectively, and $\{F\}$ is the force vector:

$$[K] = \begin{bmatrix} [K^{11}] & [K^{12}] & [K^{13}] & [K^{14}] & [K^{15}] \\ [K^{21}] & [K^{22}] & [K^{23}] & [K^{24}] & [K^{25}] \\ [K^{31}] & [K^{32}] & [K^{33}] & [K^{34}] & [K^{35}] \\ [K^{41}] & [K^{42}] & [K^{43}] & [K^{44}] & [K^{45}] \\ [K^{51}] & [K^{52}] & [K^{53}] & [K^{54}] & [K^{55}] \end{bmatrix} \qquad (4.5.54a)$$

$$[M] = \begin{bmatrix} [M^{11}] & [0] & [0] & [M^{14}] & [0] \\ [0] & [M^{22}] & [0] & [0] & [M^{25}] \\ [0] & [0] & [M^{33}] & [0] & [0] \\ [M^{41}] & [0] & [0] & [M^{44}] & [0] \\ [0] & [M^{52}] & [0] & [0] & [M^{55}] \end{bmatrix} \quad (4.5.54b)$$

The matrix coefficients are given by

$$[K^{\beta\alpha}] = [K^{\alpha\beta}]^T, \qquad [M^{\beta\alpha}] = [M^{\alpha\beta}]^T, \qquad \alpha, \beta = 1, 2, 3, 4, 5$$

$$[K^{11}] = A_{11}[S^{11}] + (A_{16} + c_0 B_{16})([S^{12}] + [S^{12}]^T)$$

$$+ (A_{66} + 2c_0 B_{66} + c_0^2 D_{66})[S^{22}] + \frac{1}{R_1^2} A_{55}[S^{00}]$$

$$[K^{12}] = (A_{16} - c_0 B_{16})[S^{11}] + (A_{66} - c_0^2 D_{66})[S^{12}]^T + A_{12}[S^{12}]$$

$$+ (A_{26} + c_0 B_{26})[S^{22}] + \frac{1}{R_1 R_2} A_{45}[S^{00}]$$

$$[K^{13}] = \left(\frac{A_{11}}{R_1} + \frac{A_{12}}{R_2}\right)[S^{10}] + \left[\frac{A_{16}}{R_1} + \frac{A_{26}}{R_2} + c_0\left(\frac{B_{16}}{R_1} + \frac{B_{26}}{R_2}\right)\right][S^{20}]$$

$$- \frac{1}{R_1}(A_{45}[S^{02}] + A_{55}[S^{01}])$$

$$[K^{14}] = B_{11}[S^{11}] + B_{16}[S^{12}] + (B_{16} + c_0 D_{16})[S^{12}]^T$$

$$+ (B_{66} + c_0 D_{66})[S^{22}] - \frac{A_{55}}{R_1}[S^{00}]$$

$$[K^{15}] = B_{12}[S^{12}] + B_{16}[S^{11}] + (B_{26} + c_0 D_{26})[S^{22}]$$

$$+ (B_{66} + c_0 D_{66})[S^{12}]^T - \frac{1}{R_1} A_{45}[S^{00}]$$

$$[K^{22}] = A_{22}[S^{22}] + (A_{26} - c_0 B_{26})([S^{12}] + [S^{12}]^T)$$

$$+ (A_{66} - 2c_0 B_{66} + c_0^2 D_{66})[S^{11}] + \frac{A_{44}}{R_2^2}[S^{00}]$$

$$[K^{23}] = \left[\frac{A_{16}}{R_1} + \frac{A_{26}}{R_2} - c_0\left(\frac{B_{16}}{R_1} + \frac{B_{26}}{R_2} \right) \right][S^{10}]$$

$$+ \left(\frac{A_{12}}{R_1} + \frac{A_{22}}{R_2} \right)[S^{20}] - \frac{1}{R_2}\left(A_{44}[S^{02}] + A_{45}[S^{01}] \right)$$

$$[K^{24}] = (B_{16} - c_0 D_{16})[S^{11}] + (B_{66} - c_0 D_{66})[S^{12}] + B_{12}[S^{12}]^T$$

$$+ B_{26}[S^{22}] - \frac{1}{R_2}A_{45}[S^{00}]$$

$$[K^{25}] = B_{22}[S^{22}] + B_{26}[S^{12}]^T + (B_{26} - c_0 D_{26})[S^{12}]$$

$$+ (B_{66} - c_0 D_{66})[S^{11}] - \frac{A_{44}}{R_2}[S^{00}]$$

$$[K^{33}] = A_{45}[S^{12}] + A_{55}[S^{11}] + A_{44}[S^{22}] + A_{45}[S^{12}]^T$$

$$+ \left[\left(\frac{A_{11}}{R_1} + \frac{A_{12}}{R_2} \right)\frac{1}{R_1} + \left(\frac{A_{12}}{R_1} + \frac{A_{22}}{R_2} \right)\frac{1}{R_2} \right][S^{00}]$$

$$[K^{34}] = A_{55}[S^{10}] + A_{45}[S^{20}] + \left(\frac{B_{11}}{R_1} + \frac{B_{12}}{R_2} \right)[S^{01}]$$

$$+ \left(\frac{B_{16}}{R_1} + \frac{B_{26}}{R_2} \right)[S^{02}]$$

$$[K^{35}] = A_{45}[S^{10}] + A_{44}[S^{20}] + \left(\frac{B_{12}}{R_1} + \frac{B_{22}}{R_2} \right)[S^{02}]$$

$$+ \left(\frac{B_{16}}{R_1} + \frac{B_{26}}{R_2} \right)[S^{01}]$$

$$[K^{44}] = D_{11}[S^{11}] + D_{16}\left([S^{12}] + [S^{12}]^T \right) + D_{66}[S^{22}] + A_{55}[S^{00}]$$

$$[K^{45}] = D_{12}[S^{12}] + D_{16}[S^{11}] + D_{26}[S^{22}] + D_{66}[S^{12}]^T + A_{45}[S^{00}]$$

$$[K^{55}] = D_{26}[S^{12}] + D_{66}[S^{11}] + D_{22}[S^{22}] + D_{26}[S^{12}]^T + A_{44}[S^{00}]$$

$$[M^{11}] = \left(I_1 + \frac{2I_2}{R_1}\right)[S^{00}]$$

$$[M^{22}] = \left(I_1 + \frac{2I_2}{R_2}\right)[S^{00}]$$

$$[M^{33}] = I_1[S^{00}]$$

$$[M^{44}] = [M^{55}] = I_3[S^{00}]$$

$$[M^{14}] = \left(I_1 + \frac{I_3}{R_1}\right)[S^{00}]$$

$$[M^{25}] = \left(I_1 + \frac{I_3}{R_2}\right)[S^{00}]$$

$$S_{ij}^{\alpha\beta} = \int_{R_e} D_\alpha \psi_i D_\beta \psi_j \, dx_1 \, dx_2, \quad (\alpha,\beta = 0,1,2), \quad D_\alpha \equiv \frac{\partial}{\partial x_\alpha}, \quad D_0 = 1$$

$$(4.5.55)$$

The element gives as a special case the shear deformable plate element discussed in Section 4.5.4. The comments concerning the numerical behavior of the plate element also apply to the shell element. Application of the shell element to some typical shell problems are presented next. Here we limit the application to bending of shells.

A comparison of the center deflection of an orthotropic clamped cylindrical shell ($R_1 = 10^{30}$ in., $R_2 = 20$ in., $a = 20$ in., $h = 1$ in.; see Fig. 4.34) subjected to internal pressure, $p_0 = (6.41/\pi)$ psi, is presented in Table 4.21.

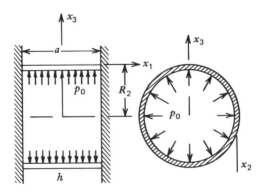

Figure 4.34 A clamped cylindrical shell subjected to internal pressure ($R_1 = 10^{30}$ in., $R_2 = 20.0$ in., $a = 20.0$ in., $p_0 = 6.4/\pi^2$ psi).

Table 4.21 Comparison of the Center Deflection of a Pressurized, Clamped Cylindrical Shell

Lamination Scheme	Present Solution		K. P. Rao (1978)	Analytical[a]
	4L	2Q9		
0°	0.0003754	0.0003727	0.0003666	0.000367
0°/90°	0.0001870	0.0001803	—	—

[a] Timoshenko and Woinowsky–Krieger (1959); also see Exercise 4.4.6.

The material properties used are

$$E_1 = 7.5 \times 10^6 \text{ psi}, \qquad E_2 = 2 \times 10^6 \text{ psi}, \qquad G_{12} = 1.25 \times 10^6 \text{ psi}$$

$$G_{12} = G_{13} = G_{23}, \qquad \nu_{12} = \nu_{13} = 0.25$$

It should be noted that both K. P. Rao and Timoshenko and Woinowsky–Krieger did not include the transverse shear strains in their analyses.

The center deflections under a point load (100 lb) at the center of a spherical shell ($R_1 = R_2 = 96$ in., $a = b = 32$ in., $h = 0.1$ in.), obtained by various investigators, are compared in Table 4.22. For an isotropic shell, the material properties used are: $E = 10 \times 10^6$ psi and $\nu = 0.3$. For orthotropic and layered anisotropic shells, the lamina material properties are the same as those of material I [see Eq. (4.5.48)]. The finite-element solution converges with refinement of the mesh to the series solution of Vlasov (1964). See also Table 4.8 for the series solution.

Exercises 4.5

1. Compute the coefficient matrices $[K^{14}]$ and $[K^{44}]$ of Eq. (4.5.26) for the choice of ψ_i in Eq. (4.5.27b).
2. Verify Eq. (4.5.28).
3. The element equation (4.5.35) can be assembled in the usual manner to obtain the global matrices. Before we proceed to solve the eigenvalue problem, we must impose the kinematic (or essential) boundary conditions on the global system of equations; the procedure is the same as in the case of equilibrium problems, except that the constrained nodal degrees of freedom are eliminated (by deleting the corresponding rows and columns). After imposing the boundary conditions, the global

Table 4.22 Comparison of the Center Deflection of a Spherical Shell Under Point Load at the Center

Lamination Scheme	4 × 4 Linear (Uniform)	2 × 2 Quadratic (Uniform)	4 × 4 Quadratic (Uniform)	4 × 4 Quadratic (Nonuniform)	K. P. Rao (1978)	Yang (1973)	Series Solution[a]
Isotropic[b]	−0.03506	−0.03726	−0.03904	−0.03935	−0.03866	−0.03867	−0.03956
Orthotropic	−0.09373	−0.10349	—	−0.12644	—	—	—
Cross-ply (0°/90°)	—	−0.10217	—	−0.12376	—	—	—
Angle-ply (45°/−45°)	—	−0.05504	—	—	—	—	—

[a]From Vlasov (1964).
[b]$\nu = 0.3$.

470

equation corresponding to Eq. (4.5.35) has the form

$$
\begin{bmatrix} [\hat{K}^{11}] & [\hat{K}^{12}] \\ [\hat{K}^{12}]^T & -[\hat{K}^{22}] \end{bmatrix} \begin{Bmatrix} \{\hat{W}\} \\ \{\hat{M}\} \end{Bmatrix} = \lambda_v \begin{Bmatrix} [\hat{S}]\{\hat{W}\} \\ \{0\} \end{Bmatrix}
$$

$$
+ \lambda_b \begin{Bmatrix} [\hat{G}]\{\hat{W}\} \\ \{0\} \end{Bmatrix} \qquad \text{(a)}
$$

where a caret over the quantities indicates that they are global quantities, after imposing the boundary conditions. Condense out $\{M\}$ and rewrite Eq. (a) in the form

$$
[\bar{K}]\{\hat{W}\} = \lambda_v [\hat{S}]\{\hat{W}\} + \lambda_b [\hat{G}]\{\hat{W}\} \qquad \text{(b)}
$$

4. Derive the finite-element model of Eq. (4.2.14).

5. Derive the displacement finite-element model of the equations governing an annular plate of inner radius b and outer radius a. The variational principle for the problem (for geometrically nonlinear case) is given by

$$
\delta\Pi = 2\int_b^a \left\{ \left[A_{11}\left(\frac{du_0}{dr} + \frac{1}{2}\left(\frac{dw}{dr}\right)^2\right) + A_{12}\frac{u_0}{r} \right] \right.
$$

$$
\times \left(\frac{d\delta u_0}{dr} + \frac{dw}{dr}\frac{d\delta w}{dr} \right)
$$

$$
+ \frac{1}{r}\left[A_{12}\left(\frac{du_0}{dr} + \frac{1}{2}\left(\frac{dw}{dr}\right)^2\right) + A_{22}\frac{u_0}{r} \right]\delta u_0
$$

$$
+ \left(D_{11}\frac{d\psi}{dr} + D_{12}\frac{\psi}{r} \right)\frac{d\delta\psi}{dr} + \frac{1}{r}\left(D_{12}\frac{d\psi}{dr} + D_{22}\frac{\psi}{r} \right)\delta\psi
$$

$$
\left. + A_{33}\left(\psi + \frac{dw}{dr} \right)\left(\delta\psi + \frac{d\delta w}{dr} \right) \right\} r\, dr
$$

$$
- 2\pi\int_b^a qr\,\delta w\, dr \qquad \text{(a)}
$$

Show that the finite-element model is of the form (i.e., define the matrix coefficients of the following equation)

$$
\begin{bmatrix} 2[K^{11}] & [K^{12}] & [0] \\ & [K^{22}] & [K^{23}] \\ \text{symmetric} & & [K^{33}] \end{bmatrix} \begin{Bmatrix} \{u\} \\ \{w\} \\ \{\psi\} \end{Bmatrix} = \begin{Bmatrix} \{0\} \\ \{F\} \\ \{0\} \end{Bmatrix} \qquad \text{(b)}
$$

6. Derive the mixed finite-element model for the vibration of annular plates using the following variational principle,

$$\delta R = 2\pi \int_b^a \left\{ \left[A_{11}\left(\frac{du_0}{dr} + \frac{1}{2}\left(\frac{dw}{dr}\right)^2 \right) + A_{12}\frac{u_0}{r} \right]\left(\frac{d\delta u_0}{dr} + \frac{dw}{dr}\frac{d\delta w}{dr} \right) \right.$$

$$+ \frac{1}{2}\left[A_{12}\left(\frac{du_0}{dr} + \frac{1}{2}\left(\frac{dw}{dr}\right)^2 \right) + A_{22}\frac{u_0}{r} \right]\delta u_0$$

$$\left. + A_{33}\left(\psi + \frac{dw}{dr} \right)\frac{d\delta w}{dr} \right\} r\,dr$$

$$+ 2\pi \int_b^a \left\{ rM_r \frac{d\delta\psi}{dr} + (M_\theta + Qr)\delta\psi \right.$$

$$+ \left[\frac{d\psi}{dr} + \bar{D}_{22}M_r - \bar{D}_{12}M_\theta \right] r\,\delta M_r$$

$$\left. + \left[\frac{\psi}{r} + (\bar{D}_{11}M_\theta - \bar{D}_{12}M_r) \right] r\,\delta M_\theta - qr\,\delta w \right\} dr$$

where $Q = A_{33}[\psi + (dw/dr)]$ and $\bar{D}_{ij} = D_{ij}/D_0$, $D_0 = D_{12}D_{12} - D_{11}D_{22}$. Show that the finite-element model is of the form

$$\begin{bmatrix} 2[K^{11}] & [K^{12}] & [0] & [0] & [0] \\ & [K^{22}] & [K^{23}] & [0] & [0] \\ & & [\bar{K}^{33}] & [K^{34}] & [K^{35}] \\ & \text{symmetric} & & [K^{44}] & [K^{45}] \\ & & & & [K^{55}] \end{bmatrix} \begin{Bmatrix} \{u\} \\ \{w\} \\ \{\psi\} \\ \{M_r\} \\ \{M_\theta\} \end{Bmatrix} = \begin{Bmatrix} \{0\} \\ \{F\} \\ \{0\} \\ \{0\} \\ \{0\} \end{Bmatrix}$$

7. Verify that Eqs. (4.5.18) and (4.5.20) are the Euler equations of the functional J_1^e in Eq. (4.5.22).

8. Verify Eqs. (4.5.25) and (4.5.26) by substituting Eq. (4.5.24) into the first variation of the functional J_1^e.

9. Verify Eqs. (4.5.35) and (4.5.36) by substituting Eq. (4.5.24) into the first variation of the functional J_2^e.

10. Verify Eq. (4.5.43) by algebraic manipulation of Eqs. (4.5.39a) and (4.5.42).

11. Show that the cubic displacement models discussed in Section 4.5.2 are [see Eqs. (4.5.12) and (4.5.15)] not compatible. More specifically, show that the slope dw/dx on the interface of two elements is not continuous.

12. Are the mixed elements and shear deformable elements discussed in Sections 4.5.3 and 4.5.4 compatible? Why?

13. Comment on the finite-element interpolation functions for the refined shear deformation theory presented in Section 4.2.3.

14. Derive the finite-element model of the refined theory using the variational form in Eq. (4.2.39). You are required to give the element equation and the algebraic form of the coefficient matrix.

BIBLIOGRAPHY

A list of books and technical papers on topics covered in the present study is included in this bibliography. Obviously, the list is incomplete. Reference to additional books and papers can be found in the works given here. For the convenience of the reader, the books and papers are catalogued by topic. Therefore, some of the references are entered under more than one topic.

CONTINUUM MECHANICS

Eringen, A. C., *Mechanics of Continua*, Wiley, New York, 1967.

Frederick, D. and Chang, T. S., *Continuum Mechanics*, Scientific Publishers, Boston, 1965.

Fung, Y. C., *Foundations of Solid Mechanics*, Prentice-Hall, Englewood Cliffs, N.J., 1965.

Fung, Y. C., *A First Course in Continuum Mechanics*, 2nd ed., Prentice-Hall, Englewood Cliffs, N.J., 1977.

Jaumzemis, W., *Continuum Mechanics*, Macmillan, New York, 1967.

Malvern, L. E., *Introduction to the Mechanics of a Continuous Medium*, Prentice-Hall, Englewood Cliffs, N.J., 1969.

Prager, W., *Introduction to Mechanics of Continua*, Ginn, Boston, 1961.

ELASTICITY, SOLID AND STRUCTURAL MECHANICS

Boresi, A. P., and Lynn, P. P., *Elasticity in Engineering Mechanics*, Prentice-Hall, Englewood Cliffs, N.J., 1974.

Budynas, R. G., *Advanced Strength and Applied Stress Analysis*, McGraw-Hill, New York, 1977.

Den Hartog, J. P., *Strength of Materials*, McGraw-Hill, New York, 1949.

Dym, C. L. and Shames, I. H., *Solid Mechanics: A Variational Approach*, McGraw-Hill, New York, 1973.

Lekhnitskii, S. G., *Theory of Elasticity of an Anisotropic Elastic Body*, Holden-Day, San Francisco, 1963.

Oden, J. T. and Ripperger, E. A., *Mechanics of Elastic Structures*, 2nd ed., Hemisphere/McGraw-Hill, New York, 1981.

Meirovitch, L., *Computational Methods in Structural Dynamics*, Sijthoff & Noordhoff, The Netherlands, 1980.

Novozhilov, V. V., *Theory of Elasticity*, Pergamon, New York, 1961.

Sechler, E. E., *Elasticity in Engineering*, Wiley, New York, 1952.

Sokolnikoff, I. S., *Mathematical Theory of Elasticity*, 2nd ed., McGraw-Hill, New York, 1956.

Southwell, R. V., *Theory of Elasticity*, Clarendon, Oxford, 1936, 1941.

Timoshenko, S. P. and Gere, J. M., *Theory of Elastic Stability*, 2nd ed., McGraw-Hill, New York, 1961.

Timoshenko, S. P. and Goodier, J. N., *Theory of Elasticity*, McGraw-Hill, New York, 1970.

Todhunter, I. and Pearson, K., *A History of the Theory of Elasticity*, Vols. 1 and 2, Dover, New York, 1960.

Ugural, A. C. and Fenster, S. K., *Advanced Strength and Applied Elasticity*, American Elsevier, New York, 1975.

Volterra, E. and Gaines, J. H., *Advanced Strength of Materials*, Prentice-Hall, Englewood Cliffs, N.J., 1971.

Zener, C., *Elasticity and Anelasticity of Metals*, University of Chicago Press, Chicago, 1948.

VECTORS AND TENSORS (WITH APPLICATIONS TO MECHANICS)

Aris, R., *Vectors, Tensors, and the Basic Equations in Fluid Mechanics*, Prentice-Hall, Englewood Cliffs, N.J., 1962.

Bowen, R. M. and Wang, C. C., *Introduction to Vectors and Tensors*, Plenum, New York, 1976.

Brillouin, L., *Tensors in Mechanics and Elasticity*, Academic Press, New York, 1969.

Jeffreys, H., *Cartesian Tensors*, Cambridge University Press, New York, 1952.

Karamcheti, K., *Vector Analysis and Cartesian Tensors*, Holden Day, San Francisco, 1967.

Lichnerowicz, A., *Elements of Tensor Analysis*, Mathuen, London and Wiley, New York, 1962

Reddy, J. N. and Rasmussen, M. L., *Advanced Engineering Analysis*, Wiley, New York, 1982.

ENERGY AND VARIATIONAL PRINCIPLES IN MECHANICS

Argyris, J. H. and Dunne, P. C., *Structural Principles and Data*, Pitman, New York, 1952.

Argyris, J. H. and Kelsey, S., *Energy Theorems and Structural Analysis*, Butterworth, London, 1960.

Bliss, G. A., *Calculus of Variations*, American Mathematical Society, 1944.

Courant, R., *Calculus of Variations and Supplementary Notes and Exercises*, revised and amended by J. Moser, New York University Press, New York, 1956.

Davies, G. A. O., *Virtual Work in Structural Analysis*, Wiley, New York, 1982.

Dym, C. L. and Shames, I. H., *Solid Mechanics: A Variational Approach*, McGraw-Hill, New York, 1973.

Forray, M. J., *Variational Calculus in Science and Engineering*, McGraw-Hill, New York, 1968.

Gelfand, I. M. and Fomin, S. V., *Calculus of Variations*, Prentice-Hall, Englewood Cliffs, N.J., 1963.

Lanczos, C., *The Variational Principles of Mechanics*, The University of Toronto Press, Toronto, 1964.

Langhaar, H. L., *Energy Methods in Applied Mechanics*, Wiley, New York, 1962.

Oden, J. T. and Reddy, J. N., *Variational Methods in Theoretical Mechanics*, 2nd ed., Springer-Verlag, Berlin, 1982.

Oden, J. T. and Ripperger, E. A., *Mechanics of Elastic Structures*, 2nd ed., Hemisphere/McGraw-Hill, New York, 1981.

Reddy, J. N. and Rasmussen, M. L., *Advanced Engineering Analysis*, Wiley, New York, 1982.

Rhodes, J., *Virtual Work and Energy Concepts*, Chatto & Windus, London, 1975.

Richards, T. H., *Energy Methods in Stress Analysis*, Ellis Horwood/Wiley, New York, 1977.

Volterra, E. and Gaines, J. H., *Advanced Strength of Materials*, Prentice-Hall, Englewood Cliffs, N.J., 1971.

Washizu, K., *Variational Methods in Elasticity and Plasticity*, 3rd ed., Pergamon Press, New York, 1982.

ENERGY AND VARIATIONAL PRINCIPLES

Betti, E., "Teoria della' Elasticita," *Nuovo Cimento*, Serie 2, Tom VII and VIII (1872).

Castigliano, A. C., "Intorno ai Sistemi Elastici," Dissertazione di Laurea presentata alla Commissione Esaminatrice della Reale Scuola degli Ingegneri di Torino, Vincenzo Bona, Turin, Italy (1873).

Clapeyron, E. B. P., "Memoire sur le Travail des Forces Elastiques dans un Corps Solide Elastique Deforme par l'action des Forces Exterieures," *Comp Rend. Paris*, Vol. XLVI (1858).

Engesser, F., "Über statisch unbestimmte Träger bei beliebigen Formänderungs gesetze und über den Satz von der kleinsten Erganzungsarbeit," *Z. Archit. Ing. Ver. zu Hanover*, Vol. 35 (1889).

Gurtin, M. D., "A Note on the Principle of Minimum Potential Energy for Linear Anisotropic Elastic Solids," *Q. Appl. Math.*, Vol. 20, pp. 379–382 (1963).

Hellinger, E., "Die allegemeinen Ansätze der Mechanik der Kontinua," Art 30, in *Encyklopädie der Mathematischen Wissenschaften*, Vol. IV/4, pp. 654–655, F. Klein and C. Muller (eds.), Leipzig, Teubner, 1914.

Hu, Hai-Chang, "On Some Variational Principles in the Theory of Elasticity and the Theory of Plasticity," *Sci. Sin.*, Vol. 4, pp. 33–54 (1955).

Lamb, E. H., "The Principle of Virtual Velocities and its Application to the Theory of Elastic Structures," *Inst. Civ. Eng.*, "Selected Engineering Papers," No. 10 (1923).

Maxwell, J. C., "On the Calculation of the Equilibrium and Stiffness of Frames," *Phil. Mag.*, Vol. 27 (1864).

Oden, J. T. and Reddy, J. N., "On Dual Complementary Variational Principles in Mathematical Physics," *Int. J. Eng. Sci.*, Vol. 12, pp. 1–29 (1974).

Reddy, J. N., "Modified Gurtin's Variational Principles in the Linear Dynamic Theory of Viscoelasticity," *Int. J. Solids & Struct.*, Vol. 12, pp. 227–235 (1976a).

Reddy, J. N., "On Complementary Variational Principles for the Linear Theory of Plates," *J. Struct. Mech.*, Vol. 4, No. 4, pp. 417–436 (1976b).

Reissner, E., "On a Variational Theorem in Elasticity," *J. Math. Phys.*, Vol. 29, pp. 90–95 (1950).

Reissner, E., "On Variational Principles of Elasticity," *Proc. Symp. Appl. Math.*, Vol. 8, pp. 1–6 (1958).

Tonti, E., "On the Mathematical Structure of a Large Class of Physical Theories," *Acad. Naz. dei Lincei*, Serie III, Vol. LII, pp. 48–56 (1972).

Washizu, K., "On the Variational Principles of Elasticity and Plasticity," Aeroelastic Research Laboratory, MIT, Tech. Report 25-18 (1955).

Washizu, K., "Variational Principles in Continuum Mechanics," Department of Aeronautical Engineering, Report, 62-2, University of Washington (1962).

Williams, D., "The Relations Between the Energy Theorems Applicable in Structural Theory," *Phil. Mag. J. Sci.*, 7th Ser., Vol. 26, pp. 617–635 (1938).

TRADITIONAL VARIATIONAL METHODS

Collatz, L., *The Numerical Treatment of Differential Equations*, 3rd ed., Springer, Berlin, 1959.

Courant, R., "Variational Methods for the Solution of Problems of Equilibrium and Vibration," *Bull. Am. Math. Soc.*, Vol. 49, pp. 1–43 (1943).

Crandall, S. H., *Engineering Analysis*, McGraw-Hill, New York, 1956.

Duncan, W. J., "The Principles of the Galerkin Method," *Brit. Aero. Res. Comm. R and M* (1848, 1938).

Finlayson, B. A., *The Method of Weighted Residuals and Variational Principles*, Academic Press, New York, 1972.

Galerkin, B. G., "Series–Solutions of Some Cases of Equilibrium of Elastic Beams and Plates" (in Russian), *Vestn. Inshenernov.*, Vol.1, pp. 897–903 (1915).

Galerkin, B. G., "Berechung der frei gelagerten elliptischen Platte auf Biegung," *Z. Angew. Math. Mech.* (1923).

Goodier, J. N., "Applications of a Reciprocal Theorem of Linear Thermoelasticity," Division of Applied Mechanics, Stanford University, TR-128 (AD260476) (1976).

Hildebrand, F. B., *Methods of Applied Mathematics*, 2nd ed., Prentice-Hall, Englewood Cliffs, N.J. 1965.

Kantorovich, L. V., "Some Remarks on Ritz's Method" (in Russian), *Tr. Vyssh. Voen. Morsk. Inzh. Stroit Uchil.*, No. 3 (1941).

Kantorovich, L. V. and Krylov, V. I., *Approximate Methods of Higher Analysis*, 4th ed., translated from Russian by C. D. Benster, Interscience Publishers, New York, 1958.

Krylov, N. M., "Sur la solution approchee des problemes de la physique mathematique et de la sciende d'ingenier," *Bull. Acad. Sci. USSR*, Vol. 7, pp. 1089–1114 (1930).

Lamb, H., "On Reciprocal Theorem in Dynamics," *Proc. Lond. Math. Soc.*, Vol. 19, pp. 144–151 (1888).

Lanczos, C., *The Variational Principles of Mechanics*, University of Toronto Press, Toronto, 1964.

Langhaar, H. L., *Energy Methods in Applied Mechanics*, Wiley, New York, 1962.

Mikhlin, S. G., *Variational Methods in Mathematical Physics*, Pergamon Press (distributed by Macmillan), New York, 1964.

Mikhlin, S. G., *The Numerical Performance of Variational Methods*, Wolters-Noordhoff, Groningen, the Netherlands, 1971.

Oden, J. T. and Reddy, J. N., *Variational Methods in Theoretical Mechanics*, 2nd ed., Springer-Verlag, New York, 1982.

Rayleigh, J. W. S., "Some General Theorems Relating to Vibrations," *Proc. Lond. Math. Soc.*, Vol. 4, pp. 357–368 (1873).

Rayleigh, J. W. S., *Theory of Sound*, 1st ed. (1877); 2nd rev. ed., Dover, New York, 1945.

Reddy, J. N. and Rasmussen, M. L., *Advanced Engineering Analysis*, Wiley, New York, 1982.

Rektorys, K., *Variational Methods in Mathematics, Science and Engineering*, Reidel, Boston, 1977.

Ritz, W., "Ueber eine neue Methode zur Losung gewisser Variationsprobleme der mathematischen Physik," *J. Reine Angew. Math.*, Vol. 135, pp. 1–61 (1908).

Ritz, W., "Theorie der Transversalschwingungen einer quadratischen Platte mit freien Randern," *Ann. Phys.*, Vol. 28, pp. 737–786 (1909).

Schechter, R. S., *The Variational Methods in Engineering*, McGraw-Hill, New York, 1967.

Southwell, R. V., "Some Extensions of Rayleigh Principle," *Q. J. Mech. Appl. Math.*, Vol. 6(3), pp. 257–272 (1953).

Trefftz, E., "Zur Theorie der Stabilitat des elastischen Gleichqewichts," *Z. Angew. Math. Mech.*, Vol. 13, pp. 160–165 (1923).

Trefftz, E., "Ein Gegenstuck zum Ritzschen Verfahren," *Proc. 2nd Int. Congr. Appl. Mech.*, E. Meissner (ed.), Zurich, pp. 131–137 (1926).

Washizu, K., *Variational Methods in Elasticity and Plasticity*, 3rd ed., Pergamon Press, New York, 1982.

THE FINITE ELEMENT METHOD

Ashwell, D. G. and Gallagher, R. H. (eds.), *Finite Elements for Thin Shells and Curved Members*, Wiley, London, 1976.

Atluri, S. N., Gallagher, R. H., and Zienkiewicz, O. C. (eds.), *Hybrid and Mixed Finite Element Methods*, Wiley, New York, 1982.

Bathe, K. J., *Finite Element Procedures in Engineering Analysis*, Prentice-Hall, Englewood Cliffs, N.J., 1982.

Becker, E. B., Carey, G. F., and Oden, J. T., *Finite Elements: An Introduction*, Prentice-Hall, Englewood Cliffs, N.J. 1981.

Bernadou, M. and Boisserie, J. M., *The Finite Element Method in Thin Shell Theory: Application to Arch Dam Simulations*, Birkhauser, Boston, 1982.

Brebbia, C. A. and Connor, J. J., *Fundamentals of Finite Element Techniques for Structural Engineers*, Butterworths, London, 1975.

Cheung, Y. K. and Yeo, M. F., *A Practical Introduction to Finite Element Analysis*, Pitman, London, 1979.

Cook, R. D., *Concepts and Applications of Finite Element Analysis*, Wiley, New York, 1974; 2nd ed. 1981.

Desai, C. S., *Elementary Finite Element Method*, Prentice-Hall, Englewood Cliffs, N.J., 1979.

Desai, C. S. and Abel, J. F., *Introduction to the Finite Element Method*, Van Nostrand Reinhold, New York, 1972.

Gallagher, R. H., *Finite Element Analysis Fundamentals*, Prentice-Hall, Englewood Cliffs, N.J., 1975.

Glowinski, R., Rodin, E. Y., and Zienkiewicz, O. C. (eds)., *Energy Methods in Finite Element Analysis*, Wiley, New York, 1979.

Hinton, E. and Owen, D. R. J., *Finite Element Programming*, Academic Press, London, 1977.

Huebner, K. H. and Thornton, E. A., *The Finite Element Method for Engineers*, 2nd ed., Wiley, New York, 1982.

Rao, S. S., *The Finite Element Method in Engineering*, Pergamon Press, Oxford, 1982.

Reddy, J. N., *An Introduction to the Finite Element Method*, McGraw-Hill, New York, 1984.

Segerlind, L. J., *Applied Finite Element Analysis*, Wiley, New York, 1976.

Zienkiewicz, O. C., *The Finite Element Method*, 3rd ed., McGraw-Hill, London, 1977.

Zienkiewicz, O. C. and Morgan, K., *Finite Elements and Approximation*, Wiley, New York, 1983.

SURVEYS OF FINITE-ELEMENT SOFTWARE

Belytschko, T., "A Survey of Numerical Methods and Computer Programs for Dynamic Structural Analysis," *Nucl. Eng. Des.*, Vol. 37, pp. 23–24 (1976).

Fredriksson, B. and Mackerle, J., "Structural Mechanics Finite Element Computer Programs, Surveys and Availability," LiTH-IKP-R-054, Linkoping Institute of Technology, Department of Mechanical Engineering, Division of Solid Mechanics (1975, revised in 1976).

Marcal, P. V. (ed.), In *General Purpose Finite Element Computer Programs*, ASME Special Publication, ASME, New York, 1970.

Marcal, P. V., "Survey of General Purpose Programs for Finite Element Analysis," in *Advances in Computational Methods in Structural Mechanics and Design*, Oden, J. T., Clough, R. W., and Yamamoto, Y. (eds.), UAH Press, Huntsville, Al, pp. 517–528, 1972.

Pilkey, W., Saczalski, K. and Sehaeffer, H. (eds.), *Structural Mechanics Computer Programs, Surveys, Assessments and Availability*, The University Press of Virginia, Charlottesville, 1974.

THEORY OF PLATES

Books

Agarwal, B. D. and Broutman, L. G., *Analysis and Performance of Fiber Composites*, Wiley, New York, 1980.

Ambartsumyan, S. A., *Theory of Anisotropic Plates* (English translation), Technomic, Stamford, Conn., 1970.

Ashton, J. E. and Whitney, J. M., *Theory of Laminated Plates*, Progress in Material Science Series, Vol. 4, Technomic, Stamford, Conn., 1970.

Bares, R., *Tables for the Analysis of Plates*, *Slabs and Diaphrams Based on the Elastic Theory*: *Berechnungstafeln fur Platten und Wandscheiben* (German–English edition), Bauverlag GmbH, Wiesbaden, 1969.

Boley, B. A. and Weiner, J. H., *Theory of Thermal Stresses*, Wiley, New York, 1960.

Bresse, J. A., *Cours de Mecanique Appliquee*, Mallet-Bachelier, Paris, 1859; 2nd ed., Gauthier-Villaris, Paris, 1866.

Brush, D. O. and Almroth, B. O., *Buckling of Bars*, *Plates, and Shells*, McGraw-Hill, New York, 1975.

Bulson, P. S., *The Stability of Flat Plates*, American Elsevier, New York, 1969.

Calcote, L. R., *The Analysis of Laminated Composite Structures*, Van Nostrand Reinhold, New York, 1969.

Christensen, R. M., *Mechanics of Composite Materials*, Wiley-Interscience, New York, 1979.

Cox, H. L., *The Buckling of Plates and Shells*, Pergamon, New York, 1963.

Donnell, L. H., *Beams, Plates and Shells*, McGraw-Hill, New York, 1976.

Galerkin, B. G., *Thin Elastic Plates* (in Russian), Gostrojisdat, Leningrad, 1933.

Gatewood, B. E., *Thermal Stresses*, McGraw-Hill, New York, 1957.

Hodge, P. G., *Limit Analysis of Rotationally Symmetric Plates and Shells*, Prentice-Hall, Englewood-Cliffs, N.J., 1963.

Jaeger, L. G., *Elementary Theory of Elastic Plates*, Macmillan, New York, 1964.

Johns, D. J., *Thermal Stress Analysis*, Pergamon Press, N.Y., 1965.

Jones, R. M., *Mechanics of Composite Materials*, McGraw-Hill, New York, 1975.

Langhaar, H. L., *Energy Methods in Applied Mechanics*, Wiley, New York, 1962.

Leissa, A. W., *Vibration of Plates*, NASA SP-160, NASA, Washington, D.C., 1969.

Lekhnitskii, S. G., *Anisotropic Plates*, Gordon and Breach, New York, 1968.

Mansfield, E. H., *The Bending and Stretching of Plates*, Macmillan, New York, 1964.

Marguerre, K. and Woernle, H. T., *Elastic Plates*, Ginn/Blaisdell, Massachusetts, 1969.

McFarland, D., Smith, B. L., and Bernhart, W. D., *Analysis of Plates*, Spartan Books, New York, 1972.

Meirovitch, L., *Analytical Methods in Vibrations*, Macmillian, New York, 1967.

Morley, L. S. D., *Skew Plates and Structures*, Macmillian, New York, 1963.

Oden, J. T. and Ripperger, E. A., *Mechanics of Elastic Structures*, 2nd ed., Hemisphere/McGraw-Hill, New York, 1981.

Panc, V., *Theories of Elastic Plates*, Noordhoff, Leyden, The Netherlands, 1975.

Szilard, R., *Theory and Analysis of Plates*, Prentice-Hall, Englewood Cliffs, N.J., 1974.

Timoshenko, S. and Gere, J. M., *Theory of Elastic Stability*, 2nd ed., McGraw-Hill, New York, 1961.

Timoshenko, S. and Woinowsky-Krieger, S., *Theory of Plates and Shells*, 2nd ed., McGraw-Hill, New York, 1959.

Troitsky, M. S., *Stiffened Plates—Bending, Stability, and Vibrations*, Elsevier, New York, 1976.

Ugural, A. C., *Stresses in Plates and Shells*, McGraw-Hill, New York, 1981.

Vlasov, V. Z. and Leontev, N. N., *Beams, Plates and Shells on Elastic Foundation* (Translation of Balki, plity, i oblochki na uprugom osnavanii) NASA TT F-357, National Aeronautics and Space Administration, Washington, D.C., 1966.

Volterra, E. and Gaines, J. H., *Advanced Strength of Materials*, Prentice-Hall, Englewood Cliffs, N.J., 1971.

Way, S., "Plates," in *Handbook of Engineering Mechanics*, McGraw-Hill, New York, 1962.

Papers

Bert, C. W., "Analysis of Plates," Chapter 4 in *Structural Design and Analysis*, Part I, C. C. Chamis (ed.), Volume 7 in *Composite Materials*, L. J. Broutman and R. H. Krock (eds.), Academic Press, New York, pp. 149–206 (1974a).

Bernoulli, J., "Essai theorique sur les vibrations de plaques elastiques rectangulaires et libres," *Nova Acta Acad.-Petropolit.*, Vol. 5, pp. 197–219 (1789).

Cauchy, A. L., "Sur L'equilibre Le Mouvement D'une Plaque Solide," *Exercises Math.*, Vol. 3, p. 328 (1828).

Euler, L., "De motu vibratorio tympanorum," *Novi Comment. Acad. Petropolit.*, Vol. 10, pp. 243–260 (1766).

Galerkin, B. G., "State of Stress in Bending of Rectangular Plate, According to the Theory of Thick Plates and the Theory of Thin Plates" (in Russian), *Tr. Lenigradskogo Inst. Sooruzh.*, No. 2 pp. 3–21 (1935).

Goodier, J. N., "On the Problem of the Beam and the Plate in the Theory of Elasticity," *Trans. R. Soc. Can.*, Vol. 32, p. 65 (1938).

Green, A.E., "On Reissner's Theory of Bending of Elastic Plates," *Q. Appl. Math.*, Vol. 7, pp. 223–228 (1949).

Green, A. E. and Naghdi, P. M., "A Theory of Laminated Composite Plates," *IMA J. Appl. Math.*, Vol. 29, p. 1 (1982).

Hencky, H., "Über die Berucksichtigung der Schubverzenung in ebenen Platten," Ing.-Arch., Vol. 16, pp. 72–76 (1947).

Kirchhoff, G., "Vorlesungen über Mathematische Physik," *Mechanik*, B. G. Teubner, Leipzig (1877).

Kirchhoff, A., "Über das Gleichgewicht und die Bewegung einer elastischen Scheibe," *J. Angew. Math.*, Vol. 40, p. 51 (1850).

Kromm, A., "Verallgemeinerte Theorie der Plattenstatik," *Ing. Arch.*, Vol. 21 (1953).

Kromm, A., "Über die Randquerkrafte bei gestutzten Platten," *Z. Angew. Math. Mech.*, Vol. 35, pp. 231–242 (1955).

Levinson, M., "An Accurate Simple Theory of the Statistics and Dynamics of Elastic Plates," *Mech. Res. Commun.*, Vol. 7, p. 343 (1980).

Levy, M., "Memoire Sur La Theorie Des Plaques Elastiques Planes," *J. Math. Pures Appl.*, Vol. 3, p. 219 (1877).

Lo, K. H., Christensen, R. M., and Wu, E. M., "A Higher-Order Theory of Plate Deformation," *J. Appl. Mech.*, Vol. 44, pp. 662–676 (1977).

Mindlin, R. D., "Influence of Rotary Inertia and Shear on Flexural Motions of Isotropic, Elastic Plates," *J. Appl. Mech.*, Vol. 18, pp. 31–38 (1951).

Nelson, R. B. and Lorch, D. R., "A Refined Theory for Laminated Orthotropic Plates," *J. Appl. Mech.*, Vol. 41, p. 171 (1974).

Nowacki, W., "The State of Stress in Thin Plate due to the Action of Sources of Heat," *Publ. Int. Assoc. Bridge Struct. Eng.*, Vol. 16, pp. 373–398 (1956).

Poisson, S. D., "Memoire Sur L'equilibre Et Le Mouvement Des Corps Elastique," *Mem. Acad. Sci.*, Vol. 8, p. 357 (1829).

Reddy, J. N., "Simple Finite Elements with Relaxed Continuity for Nonlinear Analysis of Plates," in *Finite Element Methods in Engineering*, A. P. Kabaila and V. A. Pulmano (eds.), University of New South Wales, Sydney, Australia, pp. 265–281 (1979a).

Reddy, J. N., "A Refined Nonlinear Theory of Plates with Transverse Shear Deformation," *J. Solids Struct.* (in press) (1984a).

Reddy, J. N., "A Simple Higher Order Theory for Laminated Composite Plates," *J. Appl. Mech.*, to appear (1985).

Reissner, E., "On the Theory of Bending of Elastic Plates," *J. Math. Phys.*, Vol. 23, p. 184 (1944).

Reissner, E., "The Effect of Transverse Shear Deformation on the Bending of Elastic Plates," *J. Appl. Mech. Trans. ASME*, Vol. 12, No.55, A69–A77 (1945).

Reissner, E., "On Bending of Elastic Plates," *Q. Appl. Math.*, Vol. 5, pp. 55–68 (1947).

Reissner, E., "On Transverse Bending of Plates, Including the Effect of Transverse Shear Deformation," *Int. J. Solids Struct.*, Vol. 11, pp. 569–573 (1975).

Reissner, E. and Stavsky, Y., "Bending and Stretching of Certain Types of Heterogeneous Aeolotropic Elastic Plates," *J. Appl. Mech.*, Vol. 28, pp. 402–408 (1961).

Reissner, E. and Stein, M., "Torsion and Transverse Bending of Cantilever Plates," National Advisory Commission on Aeronautics, NACA Technical Note No. 2639 (June 1951).

Stavsky, Y., "Thermoelasticity of Heterogeneous Aeolotropic Plates," *J. Eng. Mech. Div.*, Proc. ASCE, EM2, pp. 89–105 (1963).

Sun, C. T. and Whitney, J. M., "Theories for Dynamic Response of Laminated Plates," *AIAA J.*, Vol. 11(2), pp. 178–184 (1973).

Tiffen, R. and Lowe, P. G., "An Exact Theory of Generally Loaded Elastic Plates in Terms of Moments of the Fundamental Equations," *Proc. Lond. Math. Soc.*, Vol. 13, p. 653 (1963).

Timoshenko, S. P., "On the Correction for Shear of the Differential Equation for Transverse Vibrations of Prismatic Bars," *Phil. Mag.*, Vol. 41, p. 744 (1921).

von Karman, T., "Festigkeitsprobleme in Maschinenbau," *Encycl. Math. Wiss.*, Vol. 4, pp. 348–351 (1910).

von Karman, T., Sechler, E. E., and Donnel, L. H., "The Strength of Thin Plates in Compression," *Trans. ASME*, Vol. 54, pp. 53–57 (1932).

Whitney, J. M. and Pagano, N. J., "Shear Deformation in Heterogeneous Anisotropic Plates," *J. Appl. Mech.*, Vol. 37, pp. 1031–1036 (1970).

Yang, P. C., Norris, C. H., and Stavsky, Y., "Elastic Wave Propagation in Heterogeneous Plates," *Int. J. Solids Struct.*, Vol. 2, pp. 665–684 (1966).

THEORY OF SHELLS

Books (See also the books listed under "Theory of Plates")

Ambartsumyan, S. A., *Theory of Anisotropic Shells*, Moscow, 1961; English translation, NASA-TT-F-118, 1964.

Chamis, C. C. (ed.), *Structural Design and Analysis*, Part I, Academic Press, New York, 1974.

Dikeman, M., *Theory of Thin Elastic Shells*, Pitman, Boston, 1982.

Dym, C. L., *Introduction to the Theory of Shells*, Pergamon, New York, 1974.

Flügge, W., *Stresses in Shells*, Springer, Berlin, 1960.

Gould, P. L., *Static Analysis of Shells*, Lexington Books, Massachusetts, 1977.

Heyman, J., *Equilibrium of Shell Structures*, Oxford University Press, United Kingdom, 1977.

Kraus, H., *Thin Elastic Shells*, Wiley, New York, 1967.

Kuhn, P., *Stresses in Aircraft and Shell Structures*, McGraw-Hill, New York, 1956.

Librescu, L., *Elastostatics and Kinetics of Anisotropic and Heterogeneous Shell-Type Structures*, Noordhoff, Leyden, The Netherlands, 1975.

Novozhilov, V. V., *The Theory of Thin Shells*, Noordhoff, Groningen, The Netherlands, 1959.

Vlasov, V. Z., *General Theory of Shells and Its Applications in Engineering* (Translation of *Obshcaya teoriya oboloscheck i yeye prilozheniya v tekhnike*), NASA TT F-99, National Aeronautics and Space Administration, Washington, D. C., 1964.

Papers (See also "Analysis of Shells")

Ambartsumyan, S. A., "Calculation of Laminated Anisotropic Shells," *Izv. Akad. Nauk. Armenskoi SSR, Ser. Fiz. Mat. Est. Tekh. Nauk.*, Vol. 6, No. 3, p. 15 (1953).

Bert, C. W., "Analysis of Shells," Chapter 5 in *Structural Design and Analysis*, Part I, C. C. Chamis (ed.), Vol. 7 of *Composite Materials*, L. J. Broutman and R. H. Krock (eds.), Academic Press, New York, pp. 207–258 (1974b).

Dong, S. B., Pister, K. S., and Taylor, R. L., "On the Theory of Laminated Anisotropic Shells and Plates," *J. Aerosp. Sci.*, Vol. 29, pp. 969–975 (1962).

Dong, S. B. and Tso, K. W., "On a Laminated Orthotropic Shell Theory Including Transverse Shear Deformation," *J. Appl. Mech.*, Vol. 39, pp. 1091–1096 (1972).

Donnell, L. H., "Stability of Thin Walled Tubes in Torsion," *NACA Rep.* 479 (1933).

Green, A. E., "On the Linear Theory of Thin Elastic Shells," *Proc. Roy. Soc.*, Series A, Vol. 266 (1962).

Hsu, T. M. and Wang, J. T. S., "A Theory of Laminated Cylindrical Shells Consisting of Layers of Orthotropic Laminae," *AIAA J.*, Vol. 8(12), p. 2141 (1970).

Koiter, W. T., "Foundations and Basic Equations of Shell Theory. A Survey of Recent Progress," *Theory of Shells*, F. I. Niordson (ed.), IUTAM Symposium, Copenhagen, pp. 93–105 (1967).

Librescu, L., "On the Theory of Anisotropic Shells and Plates," *Int. J. Solids Struct.*, Vol. 3, pp. 53–68 (1967).

Logan, D. L. and Widera, G. E. O., "Refined Theories for Nonhomogeneous Anisotropic Cylindrical Shells: Part II-Application," *Eng. Mech. Div.*, Vol. 106, No. EM6, pp. 1075–1090 (1980).

Loo, T. T., "An Extension of Donnell's Equation for a Circular Cylindrical Shell," *J. Aeronaut. Sci.*, Vol. 24, pp. 390–391 (1957).

Love, A. E. H., "On the Small Free Vibrations and Deformations of the Elastic Shells," *Philos. Trans. Roy. Soc. (London)*, Ser. A, Vol. 17, pp. 491–546 (1888).

Morley, L. S. D., "An Improvement of Donnell's Approximation of Thin-Walled Circular Cylinders," *Q. J. Meth. Appl. Math.*, Vol. 8, pp. 169–176 (1959).

Naghdi, P. M., "A Survey of Recent Progress in the Theory of Elastic Shells," *Appl. Mech. Rev.*, Vol. 9, No. 9, pp. 365–368 (1956).

Naghdi, P. M., "Foundations of Elastic Shell Theory," *Prog. Solid Mech.*, Vol. 4, I. N. Sneddon and R. Hill (eds.), North-Holland, Amsterdam, p. 1 (1963).

Nash, W. A., "Bibliography on Shell and Shell-Like Structures," David Taylor Model Basin Report 863 (1954); Department of Engineering Mechanics, University of Florida (1957).

Reissner, E., "On the Derivation of the Theory of Elastic Shells," *J. Math. Phys.*, Vol. 42 (4), pp. 263–277 (1963).

Reissner, E., "On the Foundations of the Theory of Elastic Shells," *Proc. 11th Int. Congr. Appl. Mech.*, H. Gortler (ed.), Springer, West Germany (1966).

Sanders Jr., J. L., "An Improved First Approximation Theory for Thin Shells," *NASA TR-24* (1959).

Vlasov, V. Z., "Some New Problems on Shells and Thin Structures," National Advisory Commission on Aeronautics, NACA Technical Memo No. 1204 (March 1949).

Whitney, J. M. and Sun, C. T., "A Higher Order Theory for Extensional Motion of Laminated Anisotropic Shells and Plates," *J. Sound Vib.*, Vol. 30, p. 85 (1973).

Whitney, J. M. and Sun, C. T., "A Refined Theory for Laminated Anisotropic, Cylindrical Shells," *J. App. Mech.*, Vol. 41, pp. 471–476 (1974).

Widera, G. E. O. and Chung, S. W., "A Theory for Non-Homogeneous Anisotropic Cylindrical Shells," *Z. Angew. Math. Phys.*, Vol. 21, pp. 378–399 (1970).

Widera, G. E. O. and Logan, D. L., "Refined Theories for Nonhomogeneous Anisotropic Cylindrical Shells: Part I-Derivation," *J. Eng. Mech. Div.*, Vol. 106, No. EM6, pp. 1053–1074 (1980).

EXACT AND FINITE-ELEMENT ANALYSIS OF PLATES

Allman, D. J., "Triangular Finite Elements for Plate Bending with Constant and Linearly Varying Bending Moments," *High Speed Computing of Elastic Structures I*, Vol. 61, B. Fraeijs de Veubeke (ed.), Universite de Liege, pp. 106–136 (1971).

Allman, D. J., "Finite Element Analysis of Plate Buckling Using a Mixed Variational Principle," *Proc. 3rd Conf. Matrix Meth. Struct. Mech.*, Wright-Patterson AFB, Ohio, October 19–21, 1971.

Anderson, B. W., "Vibration of Triangular Cantilever Plates," *J. Appl. Mech. Trans. Am. Soc. Mech. Engr.*, Vol. 18, No. 4, pp. 129–134 (1954).

Anderson, R. G., The Application of Non-conforming Triangular Plate Bending Element to Plate Vibration Problems, M. S. Thesis, University of Wales, Swansea (1966).

Anderson, R. G., Irons, B. M., and Zienkiewicz, O. C., "Vibration and Stability of Plates Using Finite Elements," *Int. J. Solids Struct.*, Vol. 4, No. 10, pp. 1031–1055 (1968).

Bapu Rao, M. N. and Kumaran, K. S. S., "Finite Element Analysis of Mindlin (circular) Plates," *J. Mech. Des.*, Vol. 101, pp. 619–623 (1979).

Barker, R. M., Lin, F. T., and Dara, J. R., "Three-Dimensonal Finite-Element Analysis of Laminated Composites," *National Symposium on Computerized Structural Analysis and Design*, George Washington University (1972).

Barton, M. V., "Vibration of Rectangular and Skew Cantilever Plates," *J. Appl. Mech.*, Vol. 18, pp. 129–134 (1951).

Bazeley, G. P., Cheung, Y. K., Irons, B. M., and Zienkiewicz, O. C., "Triangular Elements in Plate Bending-Conforming and Non-Conforming Solutions," *Proc. 1st Conf. Matrix Methods Struct. Mech.*, Wright-Patterson AFB, Dayton, Ohio, AFFDL-TR-66-80, pp. 547–576 (1966).

Bell, K., "A Refined Triangular Plate Bending Finite Element," *Int. J. Numer. Method Eng.*, Vol. 1, No. 1, pp. 101–122 (1969).

Bell, R. and Kirkhope, J., "Vibration Analysis of Polar Orthotropic Discs Using the Transfer Matrix Method," *J. Sound Vib.*, Vol. 71, No. 3, pp. 421–428 (1980).

Bert, C. W., *Nonhomogeneous Polar–Orthotropic Circular Disks of Varying Thickness*, Engineering Experiment Station Bulletin No. 190, The Ohio State University, Columbus (1962).

Bert, C. W., "Recent Research in Composite and Sandwich Plate Dynamics," *Shock Vib. Dig.*, Vol. 11, No. 10, pp. 13–23 (1979).

Bert, C. W., "Vibration of Composite Structures. Recent Advances in Structural Dynamics," M. Petyt (ed.), Vol. 2, Proceeding International Conference, University of Southhampton, Southhampton, England, pp. 693–712, July 7–11, 1980.

Bert, C. W. and Chen, T. L. C., "Effect of Shear Deformation on Vibration of Antisymmetric Angle-Ply Laminated Rectangular Plates," *Int. J. Solids Struct.*, Vol. 14, pp. 465–473 (1978).

Bert, C. W. and Mayberry, B. L., "Free Vibrations of Unsymmetrically Laminated Anisotropic Plate with Clamped Edges," *J. Compos. Mater.*, Vol. 3, pp. 282–293 (1969).

Best, G. C., "Vibration Analysis of Cantilevered Square Plate by the Stiffness Matrix Method," *Proc. 1st Conf. Matrix Methods Struct. Mech.*, Wright-Patterson AFB, Dayton, Ohio, AFFDL-TR-66-80, pp. 849–886 (1966).

Bron, J. and Dhatt, G., "Mixed Quadrilateral Elements for Bending," *AIAA J.*, Vol. 19, pp. 1359–1361 (1972).

Butlin, G. and Ford, R., "A Compatible Triangular Plate Bending Finite Element," Department of Engineering Report 68-15, University of Leicester, Leicester (1969).

Butlin, G. A. and Leckie, F. A., "A Study of Finite Elements Applied to Plate Flexure," Proceedings Symposium on Numerical Methods for Vibrational Problems, Institute of Sound and Vibration Research, University of Southhampton, England, pp. 26–37 (1967).

Carrier, G. F., "Stress Distributions in Cylindrically Aeolotropic Plates," *J. Appl. Mech.*, Vol. 10, pp. A117–A122 (1943).

Carrier, G. F., "The Bending of the Cylindrically Aeolotropic Plate," *J. Appl. Mech.*, Vol. 11, No. 3, pp. A129–A133 (1944).

Carson, W. G. and Newton, R. E., "Plate Buckling Analysis Using a Fully Compatible Finite Element," *AIAA J.*, Vol. 7, No. 3, pp. 527–529 (1969).

Chang, C. C. and Conway, H. D., "Marcus Method Applied to Solution of Uniformly Loaded and Clamped Rectangular Plates Subjected to Forces in Its Plane," *J. Appl. Mech.*, Vol. 19, pp. 179–184 (1952).

Chatterjee, A. and Setlur, A. V., "A Mixed Finite Element Formulation for Plate Problems," *Int. J. Numer. Methods Eng.*, Vol. 4, pp. 67–84 (1972).

Clough, R. W. and Fellippa, C. A., "A Refined Quadrilateral Element for Analysis of Plate Bending," *Proc. 2nd Conf. Matrix Methods Struct. Mech.*, Wright-Patterson AFB, Dayton, Ohio (1968).

Clough R. W. and Tocher, J. L., "Finite Element Stiffness Matrices for Analysis of Plate Bending," *Proc. 1st Conf. Matrix Methods Struct. Mech.*, Wright-Patterson AFB, Dayton, Ohio, AFFDL-TR-66-80, pp. 515–546 (1966).

Cook, R. D., "Eigenvalue Problems with a Mixed Plate Element," *AIAA J.*, Vol. 7, No. 5, pp. 982–983 (1969).

Cook, R. D., "Some Elements for Analysis of Plate Bending," *ASCE J. Eng. Mech. Div.*, Vol. 93, No. EM6, pp. 1452–1470 (1972).

Conway, H. D., "The Bending of Symmetrically Loaded Circular Plates with Variable Thickness," *J. Appl. Mech.*, Vol. 15, pp. 1–6 (1948).

Conway, H. D., "Note on the Bending of Circular Plates of Variable Thickness," *J. Appl. Mech.*, Vol. 16, pp. 209–210 (1949).

Conway, H. D., "Bending of Rectangular Plates Subjected to a Uniformly Distributed Load and to Tensile or Compressive Forces in the Plane of the Plate," *J. Appl. Mech.*, Vol. 17, pp. 99–100 (1950).

Conway, H. D., "Axially Symmetrical Plates with Linearly Varying Thickness," *J. Appl. Mech.*, Vol. 18, No. 2, pp. 140–142 (1951).

Conway, H. D., "Closed Form Solutions for Plates of Variable Thickness," *J. Appl. Mech.*, Vol. 20, pp. 564–565 (1953).

Conway, H. D., "Non-Axial Bending of Ring Plates of Varying Thickness," *J. Appl. Mech.*, Vol. 25, No. 3, pp. 386–388 (1958).

Conway, H. D., "A Levy-type Solution for a Rectangular Plate of Variable Thickness," *J. Appl. Mech.*, Vol. 25, pp. 297–298 (1958).

Cowper, G. R., Kosko, E., Lindberg, G., and Olson, M., "Static and Dynamic Applications of High-Precision Triangular Plate Bending Element," *J. AIAA*, Vol. 7, No. 10, pp. 1957–1965 (1969).

Das, Y. C. and Rath, B. K., "Thermal Bending of Moderately Thick Rectangular Plate," *AIAA J.*, Vol. 10, No. 10, pp. 1349–1351 (1972).

Dawe, D. J., "A Finite Element Approach to Plate Vibration Problems," *J. Mech. Eng. Sci.*, Vol. 7, No. 1, pp. 28–32 (1965).

Dawe, D. J., "Vibration of Rectangular Plates of Variable Thickness," *J. Mech. Eng. Sci.*, Vol. 8, No. 1, pp. 43–51 (1966).

Dawe, D. J., "Application of the Discrete Element Method to the Buckling Analysis of Rectangular Plates Under Arbitrary Membrane Loading," *Aeronaut. Q.*, Vol. 20, pp. 114–128 (1969).

Deresiewicz, H. and Mindlin, R. D., "Axially Symmetric Flexural Vibrations of a Circular Disk," *J. Appl. Mech.*, Vol. 22, pp. 86–88 (1955).

Dickinson, S. M., "The Flexural Vibration of Rectangular Orthotropic Plates Subject to In-Plane Forces," *J. Appl. Mech.*, Vol. 36, No. 1, pp. 101–102 (1967).

Dickinson, S. M. and Henshell, R. D., "Clough-Tocher Triangular Plate-Bending Element in Vibration," *AIAA J.*, Vol. 7, pp. 560–561 (1969).

Dokainish, M. A., "A New Approach for Plate Vibrations: Combination of Transfer Matrix and Finite-Element Technique," *J. Eng. Ind. Trans. ASME*, Vol. 94, No. 2, pp. 526–530 (1972).

Dokainish, M. A. and Rawtani, S., "Vibration Analysis of Rotating Cantilever Plates," *Int. J. Numer. Methods Eng.*, Vol. 3, pp. 233–248 (1971).

Dungar, R. and Severn, R. T., "Triangular Finite Elements of Variable Thickness and Their Application to Plate and Shell Problems," *J. Strain Anal.*, Vol. 4, No. 1, pp. 10–21 (1969).

Dungar, R., Severn, R. T., and Taylor, P., "Vibration of Plate and Shell Structures Using Triangular Finite Elements," *J. Strain Anal.*, Vol. 2, No. 1, pp. 78–83 (1977).

Elishakoff, I. B., "Vibration Analysis of Clamped Square Plate," *AIAA J.*, Vol. 12, No. 7, pp. 921–924 (1974).

Felgar, R. P., "Formulas for Integrals Containing Characteristic Functions of a Vibrating Beam," Circular 14, Bureau of Engineering Research, University of Texas, Austin, 1950.

Forsyth, E. M. and Warburton, G. B., "Transient Vibration of Rectangular Plates," *J. Mech. Eng. Sci.*, Vol. 2, No. 4, pp. 325–330 (1960).

Fortier, R. C. and Rossettos, J. N., "On the Vibration of Shear Deformable Curved Anisotropic Composite Plates," *J. Appl. Mech.*, Vol. 40, pp. 299–301 (1973).

Fraeijs de Veubeke, B., "A Conforming Finite Element for Plate Bending," *Int. J. Solids Struct.*, Vol. 4, pp. 95–108 (1968).

Fried, I., "Shear in C^0 and C^1 Plate Bending Elements," *Int. J. Solids Struct.*, Vol. 9, No. 4, pp. 449–460 (1963).

Fung, Y. C., "Bending of Thin Elastic Plates of Variable Thickness," *J. Aeronaut. Sci.*, Vol. 20, pp. 445–468 (1953).

Gallego Juarez, J. A., "Axisymmetric Vibrations of Circular Plates with Stepped Thickness," *J. Sound Vib.*, Vol. 26, pp. 411–416 (1973).

Gerard, G., "Buckling of Composite Elements," Part II of *Handbook of Structural Stability*, NACA TN 3785, National Advisory Committee for Aeronautics, Washington, D.C. (1957).

Ghali, A. and Bathe, K. J., "Analysis of Plates Subjected to In-Plane Forces Using Large Finite Elements," *Publ. IABSE*, Vol. 30-I, pp. 61–72 (1970).

Ghali, A. and Bathe, K. J., "Analysis of Plate Bending Problem Using Large Finite Elements," *Publ. IABSE*, Vol. 30-II pp. 29–40 (1970).

Gopalacharyulu, S., "A Higher Order Conforming Rectangular Element," *Int. J. Numer. Meth. Eng.*, Vol. 6, No. 2, pp. 305–308 (1973).

Greimann, L. F. and Lynn, P. P., "Finite Element Analysis of Plate Bending with Transverse Shear Deformation," *Nucl. Eng. Des.*, Vol. 14, pp. 223–230 (1970).

Habata, U., "On the Lateral Vibration of a Rectangular Plate Clamped at Four Edges," *Trans. Japan Soc. Mech. Eng.*, Vol. 23, No. 131 (1957).

Hartz, B. J., "Matrix Formulation of Structural Stability Problems," *Proc. ASCE*, Vol. 91, *J. Struct. Div.*, ST. Vol. 6, pp. 141–157 (1965).

Hearmon, R. F. S., "The Frequency of Flexural Vibration of Rectangular Orthotropic Plates with Clamped or Supported Edges," *J. Appl. Mech.*, Vol. 26, No. 4, pp. 537–540 (1959).

Hellan, K., "Analysis of Elastic Plates in Flexure by a Simplified Finite Element Method," *Acta Polytech. Scand.*, Civil Engineering Series, No. 46, Trondheim (1967).

Heller, C. O., "Behavior of Stiffened Plates," Engineering Department Report 67-1, Vol. 1, "Analysis," U.S. Naval Academy, Annapolis (1967).

Henshell, R. D., Walters, D., and Warburton, G. B., "A New Family of Curvilinear Plate Bending Elements for Vibration and Stability," *J. Sound Vib.*, Vol. 20, No. 3, pp. 381–397 (1972).

Herrmann, L. R., "A Bending Analysis of Plates," *Proc. 1st Conf. Matrix Methods Struct. Mech.*, Wright-Patterson AFB, AFFDL-TR-66-80, Dayton, Ohio, pp. 577–604 (1966).

Herrmann, L. R., "Finite Element Bending Analysis for Plates," *Proc. ASCE*, Vol. 93, *J. Em. Div.*, No. 8, pp. 13–26 (1967).

Hinton, E., "Flexure of Composite Laminates Using the Thick Finite Strip Method," *Comput. Struct.*, Vol. 7, pp. 217–220 (1977).

Holand, I. and Moan, T., "The Finite Element Method in Plate Buckling," in *Finite Element in Stress Analysis* (Holland, I. and Bell, K., eds.), Technical University of Norway Press, Trondheim, Norway, pp. 475–500 (1969).

Hughes. T. J. R., Cohen, M. and Haroun, M., "Reduced and Selective Integration Techniques in the Finite Element Analysis of Plates," *Nucl. Eng. Des.*, Vol. 46, No. 1, pp. 203–222 (1978).

Hughes, T. J. R., Taylor, R. L., and Kanoknukulchai, W., "A Simple and Efficient Finite Element for Plate Bending," *Int. J. Numer. Methods Eng.*, Vol. 11, pp. 1529–1543 (1977).

Hussainy, S. A. and Srinivas, S., "Flexure of Rectangular Composite Plates," *Fibre Sci. Technol.*, Vol. 8, pp. 59–76 (1975).

Irie, T. and Yamada, G., "Free Vibration of a Mindlin Annular Plate of Varying Thickness," *J. Sound Vib.*, Vol. 66, No. 2, pp. 187–197 (1979).

Irons, B. M., "A Conforming Quadratic Triangular Element for Plate Bending," *Int. J. Numer. Methods Eng.*, Vol. 1, pp. 29–46 (1969).

Jones, R. E., "A Generalization of the Direct-Stiffness Method of Structural Analysis," *AIAA J.*, Vol. 2, No. 5, pp. 821–826 (1964).

Kanazawa, T. and Kawai, T., "On the Vibration of Anisotropic Rectangular Plates (Studied by Integral Equations), *Proceedings of 2nd Japan National Congress for Applied Mechanics*, pp. 333–338 (1952).

Kapur, K. K. and Hartz, B. J., "Stability of Plates Using the Finite Element Method," *J. Eng. Mech. Div., Proc. ASCE*, Vol. 92, No. EM2, pp. 175–195 (1966).

Kaul, R. K. and Cadambe, V., "The Natural Frequencies of Thin Skew Plates," *Aeronaut. Q.*, Vol. 7, pp. 337–352 (1956).

Kennedy, J. B. and Huggins, M. V., "Series Solution of Skewed Stiffened Plates," *Proc. ASCE*, Vol. 90, *J. Eng. Mech. Div., No. EM1*, pp. 1–22 (1964).

Kikuchi, F. and Ando, Y., "Rectangular Finite Element for Plate Bending Analysis Based on Hellinger-Reissner's Variational Principle," *J. Nucl. Sci. Technol.*, Vol. 9, pp. 28–35 (1972).

Kirkhope, J. and Wilson, G. J., "Vibration of Circular and Annular Plates Using Finite Elements," *Int. J. Numer. Methods Eng.*, Vol. 4, No. 2, pp. 181–193 (1972).

Kirkhope, J. and Wilson, G. J., "Vibration and Stress Analysis of Thin Rotating Discs Using Annular Elements," *J. Sound Vib.*, Vol. 44, No. 4, pp. 461–474 (1976).

Krishna Murty, A. V., "Finite Element Modeling of Natural Vibration Problems," *Shock Vib. Dig.*, Vol. 9, No. 2, pp. 19–37 (1977).

Leckie, F. A., "The Application of Transfer Matrices to Plate Vibrations," *Ing. Arch.*, Vol. 32, pp. 100–111 (1963).

Lehnhoff, T. F. and Miller, R. E., "The Influence of Transverse Shear on the Small Displacement Theory of Circular Plates," *AIAA J.*, Vol. 7, pp. 1499–1505 (1969).

Leissa, A. W., *Vibration of Plates*, NASA SP-160, NASA, Washington, D.C. (1969).

Leissa, A. W., "Plate Vibration Research, 1976–80: Complicating Effects," *Shock Vib. Dig.*, Vol. 13, No. 10, pp. 19–36 (1981).

Leissa, A. W., "Advances in Vibration, Buckling and Postbuckling Studies on Composite Plates," *Composite Structures*, Proceedings, 1st International Conference, Paisley College of Technology, Paisley, Scotland, Sept. 16–18, 1981, I. H. Marhall (eds.), Applied Science Publishers, London, pp. 312–334 (1981).

Leissa, A. W. et al., "A Comparison of Approximate Methods for the Solution of Plate Bending Problems," *AIAA J.*, Vol. 7, No. 5, pp. 920–927 (1969).

Levy, M., "Sur l'equilibre elastique d'une plaque rectangulaire," *C. R. Acad. Sci.*, Vol. 129, pp. 535–539 (1899).

Lindberg, G. M., Olson, M. D., and Tulloch, H. A., "Closed Form, Finite Element Solutions for Plate Vibrations," National Aeronautical Establishment, Ottawa, Canada, NRC-LR-518 (1969).

Lynn, P. P. and Dhillon, B. S., "Convergence of Eigenvalue Solutions in Conforming Plate Bending Finite Elements," *Int. J. Numer. Methods Eng.*, Vol. 4, No. 2, pp. 217–234 (1972).

Mansfield, E. H., "On the Analysis of Elastic Plates of Variable Thickness," *Q. J. Appl. Mech. Math.*, Vol. 15, pp. 167–192 (1962).

Mason, V., "Rectangular Finite Elements for Analysis of Plate Vibrations," *J. Sound Vib.*, Vol. 7, pp. 436–448 (1968).

Mau, S. T., Tong, P., and Pian, T. H. H., "Finite Element Solutions for Laminated Thick Plates," *J. Compos. Mater.*, Vol. 6, pp. 304–311 (1972).

Mau, S. T. and Witmer, E. A., "Static Vibration, and Thermal Stress Analyses of Laminated Plates and Shells by the Hybrid-Stress Finite Element Method with transverse Shear Deformation Effects Included," Aeroelastic and Structures Research Laboratory, Report ASRL TR 169-2, Department of Aeronautics and Astronautics, MIT, Cambridge, MA (1972).

Mawenya, A. S. and Davies, J. D., "Finite Element Bending Analysis of Multilayer Plates," *Int. J. Numer. Methods Eng.*, Vol. 8, pp. 215–225 (1974).

Mendelson, A. and Hirschberg, M., "Analysis of Elastic Thermal Stress in Thin Plates with Spanwise and Chordwise Variation of Temperature and Thickness," Technical Note No. 3778, National Advisory Committee for Aeronautics, Washington, D.C. (1956).

Mindlin, R. D. and Deresiewicz, H., "Thickness Shear and Flexural Vibrations of a Circular Disk," *J. Appl. Phys.*, Vol. 25, pp. 1329–1332 (1954).

Morley, L. S. D., "The Approximate Solution of Plate Problems," *Proc. 9th Int. Congr. Appl. Mech.*, Brussels, Vol. VI, pp. 22–29 (1956).

Morley, L. S. D., "A Triangular Equilibrium Element with Linear Varying Bending Moments for Plate Bending Problems," *J. R. Aeronaut. Soc.*, Vol. 71, pp. 715–719 (1967).

Morley, L. S. D., "A Triangular Equilibrium Element in the Solution of Plate Bending Problems," *Aeronaut. Q.*, Vol. 49, pp. 149–169 (1968).

Morley, L. S. D., "On the Constant Moment Plate Bending Element," *J. Strain Anal.*, Vol. 6, pp. 20–24 (1971).

Neal, B. K., Henshell, R. D., and Edwards, G., "Hybrid Plate Bending Elements," *J. Sound Vib.*, Vol. 23, No. 1, pp. 101–112 (1972).

Noor, A. K. and Mathers, M. D., "Anisotropy and Shear Deformation in Laminated Composite Plates," *AIAA J.*, Vol. 14, pp. 282–285 (1976).

Noor, A. K. and Mathers, M. D., "Finite Element Analysis of Anisotropic Plates," *Int. J. Numer. Methods Eng.*, Vol. 11, pp. 289–307 (1977).

Pagano, N. J., "Exact Solutions for Composite Laminates in Cylindrical Bending," *J. Comp. Mater.*, Vol. 3 (3), pp. 398–411 (1969).

Pagano, N. J., "Exact Solutions for Rectangular Bidirectional Composites and Sandwich Plates," *J. Comp. Mater.*, Vol. 4, pp. 20–24 (1970).

Pagano, N. J. and Hatfield, S. J., "Elastic Behavior of Multilayer Bidirectional Composites," *AIAA J.*, Vol. 10, pp. 931–933 (1972).

Panda, S. C. and Natarajan, R., "Finite Element Analysis of Laminated Composite Plates," *Int. J. Numer. Methods Eng.*, Vol. 14, pp. 69–79 (1979).

Pardoen, G. C., "Static, Vibration, and Buckling Analysis of Axisymmetric Circular Plates Using Finite Elements," *Comput. Struct.*, Vol. 3, No. 2, pp. 355–375 (1973).

Pell, W. H., "Thermal Deflections of Anisotropic Thin Plates," *Q. Appl. Math.*, Vol. 4, pp. 27–44 (1946).

Perakatte, G. J. and Lehnhoff, T. F., "Flexure of Symmetrically Loaded Circular Plates Including the Effect of Transverse Shear," *J. Appl. Mech.*, Vol. 38, No. 4, pp. 1036–1041 (1971).

Pettersson, O., "Plates Subjected to Compression or Tension and to Simultaneous Transverse Load," *Ing. Vetensk. Akad. Tidskr. Teknol. Forsk.*, Vol. 25, pp. 78–82 (1954).

Phan, N. D., "Linear Analysis of Laminated Composite Plates Using a Higher-Order Shear Deformation Theory," M.S. Thesis, Department of Engineering Science and Mechanics, Virginia Polytechnic Institute and State University, Blacksburg, VA, 1984.

Pian, T. H. H. and Tong, P., "Basis of Finite Element Methods for Solid Continua," *Int. J. Numer. Methods Eng.*, Vol. 1, pp. 3–28 (1969).

Pickett, G., "Solution of Rectangular Clamped Plate with Lateral Load by Generalized Energy Method," *J. Appl. Mech.*, pp. 168–170 (1939).

Pilkey, W. D., "Stepped Circular Kirchhoff Plate," *AIAA J.*, Vol. 3, No. 9, pp. 1768–1770 (1965).

Pister, K.S. and Dong, S. B., "Elastic Bending of Layered Plates," Proc. ASCE, Vol. 84, *J. Eng. Mech. Div.*, pp. 1–10 (1959).

Plunkett, R., "Natural Frequencies of Uniform and Non-Uniform Rectangular Cantilever Plates," *J. Mech. Eng. Sci.*, Vol. 5, pp. 146–156 (1963).

Poppewell, N. and McDonald, D., "Conforming Rectangular and Triangular Plate-Bending Elements," *J. Sound Vib.*, Vol. 19, pp. 333–347 (1971).

Pryor, Jr., C. W. and Barker, R. M., "A Finite Element Analysis Including Transverse Shear Effects for Applications to Laminated Plates," *AIAA J.*, Vol. 9, pp. 912–917 (1971).

Pryor, Jr., C. W., Barker, R. M., and Frederick, D., "Finite Element Bending Analysis of Reissner Plates," *Proc. ASCE J. Eng. Mech. Div.*, Vol. 96, No. EM6, pp. 967–983 (1970).

Przemieniecki, J. S., "Equivalent Mass Matrices for Rectangular Plates in Bending," *J. AIAA*, Vol. 4, No. 5, pp. 949–950 (1966).

Ramaiah, G. K. and Vijayakumar, K., "Natural Frequencies of Polar Orthotropic Annular Plates," *J. Sound Vib.*, Vol. 26, pp. 517–531 (1973).

Rao, G. V., Prakash Rao, B., and Raju, I. S., "Vibrations of Inhomogeneous Thin Plates Using a High Precision Triangular Element," *J.Sound Vib.*, Vol. 34, No. 3, pp. 444–445 (1974).

Rao, S. S. and Prasad, A. S., "Vibration of Annular Plates Including the Effects of Rotary Inertia and Transverse Shear Deformation," *J. Sound Vib.*, Vol. 42, pp. 305–324 (1975).

Rao, G. V. and Raju, I. S., "A Comparative Study of Variable and Constant Thickness High Precision Triangular Plate Bending Elements in the Analysis of Variable Thickness Plates," *Nucl. Eng. Des.*, Vol. 26, No. 2, pp. 299–304 (1974).

Rao, G. V., Raju, K. K., and Raju, I. S., "Finite Element Formulations of Beams and Orthotropic Circular Plates," *Comput. Struct.*, Vol. 6, p. 169 (1976).

Rao, G. V., Venkataramana, J. and Rao, B. P., "Vibrations of Thick Plates Using a High Precision Triangular Element," *Nucl. Eng. Des.*, Vol. 31, No. 1, pp. 102–105 (1974).

Ratzersdofer, J., "Rectangular Plates with Stiffeners," *Aircr. Eng.*, Vol. 14, pp. 260–263 (1942).

Reddy, J. N., "On Complementary Variational Principles for the Linear Theory of Plates," *J. Struct. Mech.*, Vol. 4, pp. 417–436 (1976c).

Reddy, J. N., "Simple Finite Elements with Relaxed Continuity for Nonlinear Analysis of Plates," in *Finite Element Methods in Engineering*, Kabaila, A. P. and Pulmano, V. A. (eds.), The University of New South Wales, Sydney, Australia, pp. 265–281 (1979a).

Reddy, J. N., "Free Vibration of Antisymmetric, Angle-Ply Laminated Plates, Including Transverse Shear Deformation by the Finite Element Method," *J. Sound Vib.*, Vol. 66 (4), pp. 565–576 (1979b).

Reddy, J. N., "Finite-Element Modeling of Structural Vibrations: A Review of Recent Advances," *Shock Vib. Dig.*, Vol. 11, No. 1, pp. 25–39 (1979c).

Reddy, J. N., "A Penalty Plate-Bending Element for the Analysis of Laminated Anisotropic Composite Plates," *Int. Numer. Methods Eng.*, Vol. 15, pp. 1187–1206 (1980).

Reddy, J. N., "Finite-Element Modeling of Layered Anisotropic Composite Plates and Shells: A Review of Recent Research," *Shock Vib. Dig.*, Vol. 13, No. 12, pp. 3–12 (1981).

Reddy, J. N., "Analysis of Layered Composite Plates Accounting for Large Deflections and Transverse Shear Strains," in *Recent Advances in Non-Linear Computational Mechanics*, Hinton, E., Owen, D. R. J., and Taylor, C. (eds.), Pineridge Press, Swansea, U. K., pp. 155–202 (1982a).

Reddy, J. N., "A Simple Higher-Order Theory for Laminated Composite Plates," Report VPI-E-83.28, ESM Department, Virginia Polytechnic Institute, Blacksburg, VA (1983a).

Reddy, J. N. and Chao, W. C., "A Comparison of Closed-Form and Finite Element Solutions of Thick, Laminated, Anisotropic Rectangular Plates," *Nucl. Eng. Des.*, Vol. 64, pp. 153–167 (1981).

Reddy, J. N. and Hsu, Y. S., "Effects of Shear Deformation and Anisotropy on the Thermal Bending of Layered Composite Plates," *J. Therm. Stresses*, Vol. 3, pp. 475–493 (1980).

Reddy, J. N. and Huang, C. L., "Nonlinear Axisymmetric Bending of Annular Plates with Varying Thickness," *Int. J. Solids Struct.*, Vol. 17, pp. 811–825 (1981a).

Reddy, J. N. and Huang, , C. L., "Large Amplitude Free Vibrations of Annular Plates of Varying Thickness," *J. Sound. Vib.*, Vol. 79, No. 3, pp. 387–396 (1981b).

Reddy, J. N., Huang, C. L., and Singh, I. R., "Large Deflection and Large Amplitude Vibrations of Axisymmetric Circular Plates," *Int. J. Numer. Methods Eng.*, Vol. 17, pp. 527–541 (1981).

Reddy, J. N., and Tsay, C. S., "Mixed Rectangular Finite Elements for Plate Bending," *Proc. Oklahoma Acad. Sci.*, Vol. 47, pp. 144–148 (1977a).

Reddy, J. N., and Tsay, C. S., "Stability and Vibration of Thin Rectangular Plates by Simplified Mixed Finite Elements," *J. Sound Vib.*, Vol. 55, No. 2, pp. 289–302 (1977b).

Reismann, H., "Bending of Clamped Wedge Plates," *J. Appl. Mech.*, Vol. 20, p. 141 (1953).

Reismann, H. and Lee, Yu-Chung, "Forced Motion of Rectangular Plates," *Developments in Theoretical and Applied Mechanics*, Vol. 4, Frederick, D. (ed.), Pergamon Press, New York, p. 3, 1969.

Rock, T. and Hinton, E., "Free Vibration and Transient Response of Thick and Thin Plates Using the Finite Element Method," *Int. J. Earthquake Eng. Struct. Dyn.*, Vol. 3, No. 1, pp. 51–63 (1974).

Rock, T. A. and Hinton, E., "A Finite Element Method for the Free Vibration of Plates Allowing for Transverse Shear Deformation," *Comput. Struct.*, Vol. 7, pp. 37–44 (1976).

Salzman, A. P. and Patel, A. S., "Natural Frequencies of Orthotropic Circular Plates of Variable Thickness," *AIAA J.*, Vol. 6, pp. 1625–1626 (1968).

Schmit, L. A. and Monforton, G. R., "Finite Deflection Discrete Element Analysis of Sandwich Plates and Cylindrical Shells with Laminated Faces," *AIAA J.*, Vol. 8, No. 8, pp. 1454–1461 (1970).

Seide, P. and Stein, M., "Compressive Buckling of Simply Supported Plate with Longitudinal Stiffeners," NACA TN 1825, National Advisory Committee for Aeronautics, Washington, D.C. (1949).

Sherbourne, A. N. and Murthy, D. N. S., "Elastic Bending of Anisotropic Plates of Variable Thickness," *Int. J. Mech. Sci.*, Vol. 12, pp. 1023–1035 (1970).

Sherbourne, A. N. and Murthy, D. N. S., "Bending of Circular Plates with Variable Profile," *Comput. Struct.*, Vol. 11, pp. 355–361 (1980).

Sinha, P. K. and Rath, A. K., "Vibration and Buckling of Cross-Ply Laminated Circular Cylindrical Panels," *Aeronaut. Q.*, Vol. 26, pp. 211–218 (1975).

Smith, I. M., "A Finite Element Analysis for 'Moderately Thick' Rectangular Plates in Bending," *Int. J. Mech. Sci.*, Vol. 10, pp. 563–570 (1968).

Smith, I. M. and Duncan, W., "The Effectiveness of Nodal Continuities in Finite Element Analysis of Thin Rectangular and Skew Plates in Bending," *Int. J. Numer. Methods Eng.*, Vol. 2, pp. 253–258 (1970).

Spilker, R. L., "Higher Order Three-Dimensional Hybrid-Stress Elements for Thick-Plate Analyses," *Int. J. Numer. Methods Eng.*, Vol. 17, pp. 53–69 (1981).

Spilker, R. L., Chou, S. C., and Orringer, O., "Alternate Hybrid Stress Elements for Analysis of Multilayer Composite Plates," *J. Compos. Mater.*, Vol. 11, pp. 51–70 (1977).

Srinivas, S., Joga Rao, C. V., and Rao, A. K., "An Exact Analysis for Vibration of Simply Supported Homogeneous and Laminated Thick Rectangular Plates," *J. Sound Vib.*, Vol. 12, pp. 187–199 (1970).

Srinivas, S. and Rao, A. K., "Bending, Vibration, and Buckling of Simply Supported Thick Orthotropic Rectangular Plates and Laminates," *Int. J. Solids Struct.*, Vol. 6, pp. 1463–1481 (1970).

Svec, O. J. and Gladwell, J., "A Triangular Plate Bending Element for Contact Problems," *Int. J. Solids Struct.*, Vol. 9, pp. 435–446 (1973).

Tabarrok, B. Fenton, R. G., and Elsale, A. M., "Application of a Refined Plate Bending Element to Buckling Problems," *Comput. Struct.*, Vol. 4, pp. 1313–1321 (1974).

Tsay, C. S. and Reddy, J. N., "Bending, Stability, and Free Vibration of Thin Orthotropic Plates by Simplified Mixed Finite Elements," *J. Sound Vib.*, Vol. 59, No. 2, pp. 307–311 (1978).

Visser, W., "A Refined Mixed Type Plate Bending Element," *AIAA J.*, Vol. 7, No. 9, pp. 1801–1802 (1969).

Wang, T. K., "Buckling of Transverse Stiffened Plates in Shear," *J. Appl. Mech.*, Vol. 14, pp. A269–A274 (1947).

Warburton, G. B., "The Vibration of Rectangular Plates," *Proc. Inst. Mech. Eng.*, *London*, Vol. 168, pp. 371–381 (1954).

Warburton, G. B., "Recent Advances in Structural Dynamics," *Symp. Dyn. Anal. Struct.*, N.E.L., East Kilbridge, Scotland (1975).

Weinstein, A. and Chien, W. Z., "On the Vibrations of a Clamped Plate Under Tension," *Q. Appl. Math.*, Vol. 1, pp. 61–68 (1943).

Westbrook, D. R., Chakrabarti, S., and Cheung, Y. K., "A Three-Dimensional Finite Element Method for Plate Bending," *Int. J. Mech. Sci.*, Vol. 16, pp. 479–487 (1974).

Westbrook, D. R., Chakrabarti, S., and Cheung, Y. K., "A Three-Dimensional or Penalty Finite Element Method for Plate Bending," *Int. J. Mech. Sci.*, Vol. 18, pp. 347–350 (1976).

Whitney, J. M., "The Effect of Transverse Shear Deformation on the Bending of Laminated Plates," *J. Compos. Mater.,* Vol. 3, pp. 534–547 (1969).

Whitney, J. M., "Stress Analysis of Thick Laminated Composite and Sandwich Plates," *J. Compos. Mater.*, Vol. 6, pp. 426–440 (1972).

Whitney, J. M. and Leissa, A. W., "Analysis of Heterogeneous Anisotropic Plates," *J. Appl. Mech.*, Vol. 36, pp. 261–266 (1969).

Whitney, J. M. and Pagano, N. J., "Shear Deformation in Heterogeneous Anisotropic Plates," *J. Appl. Mech.*, Vol. 37, pp. 1031–1036 (1970).

Wilson, G. J. and Kirkhope, J., "Finite Element Bending Analysis of Nonuniform Circular Plates," *Comput. Struct.*, Vol. 6, pp. 459–466 (1976).

Wilson, G. J. and Kirkhope, J., "Vibration Analysis of Axial Flow Turbine Disks Using Finite Elements," *J. Eng. Ind.*, Vol. 98, pp. 1008–1013 (1976).

Woo, H. K., Kirmser, P. G., and Huang, C. L., "Vibration of Orthotropic Circular Plates with a Concentric Isotropic Core," *AIAA J.*, Vol. 11, pp. 1421–1422 (1973).

Yoshida, D. M. and Weaver, W., "Finite Element Analysis of Beams and Plates with Moving Loads," *Publ. IASBE*, Vol. 31-1, pp. 179–195 (1971).

Young, D., "Vibration of Rectangular Plates by the Ritz Method," *J. Appl. Mech.*, *Trans. Am. Soc. Mech. Eng.*, Vol. 17, pp. 448–453 (1950).

Young, D. and Felgar, R. P., "Tables of Characteristic Functions Representing Normal Modes of Vibration of a Beam," Publication No. 4913, University of Texas, Austin, July 1949.

Zienkiewicz, O. C., Taylor, R. L., and Too, J. M., "Reduced Integration Technique in General Analysis of Plates and Shells," *Int. J. Numer. Methods Eng.*, Vol. 3, pp. 575–586 (1971).

Zienkiewicz, O. C. and Hinton, E., "Reduced Integration, Function Smoothing and Non-Conformity in Finite Element Analysis," *J. Franklin Inst.*, Vol. 302, pp. 443–461 (1976).

ANALYSIS OF SHELLS (SEE ALSO "THEORY OF SHELLS")

Ahmad, S., Anderson, R. G., and Zienkiewicz, O. C., "Vibration of Thick, Curved, Shells with Particular Reference to Turbine Blades," *J. Strain Anal.*, Vol. 5, pp. 200–206 (1970).

Ahmad, S., Irons, B. M., and Zienkiewicz, O. C., "Analysis of Thick and Thin Shell Structures by Curved Finite Elements," *Int. J. Numer. Methods Eng.*, Vol. 2, pp. 419–451 (1970).

Argyris, J. H., "Matrix Displacement Analysis of Anisotropic Shells by Triangular Elements," *Aeronaut. J.*, Vol. 69, pp. 801–805 (1965).

Ashwell, D. G. and Gallagher, R. H., *Finite Elements for Thin Shells and Curved Members*, Wiley, New York, 1976.

Bert, C. W., "Analysis of Shells," Chapter 5 in *Structural Design and Analysis, Part I*, C. C. Chamis (ed.), Vol. 7, of *Composite Materials*, L. J. Broutman and R. H. Krock (eds.), Academic Press, New York, pp. 207–258, 1974b.

Bert, C. W., "Dynamics of Composite and Sandwich Panels—Parts I and II," (corrected title), *Shock Vib. Dig.*, Vol. 8, No. 10, pp. 37–48, 1976; Vol. 8, No. 11, pp. 15–24 (1976).

Bert, C. W. and Francis, P. H., "Composite Material Mechanics: Structural Mechanics," *AIAA J.*, Vol. 12, No. 9, pp. 1173–1186 (1974).

Bonnes, G., et al., "Curved Triangular Elements for the Analysis of Thin Shells," *Proc. 2nd Conf. Matrix Methods Struct. Mech.*, Wright-Patterson AFB, Dayton, Ohio, AFFDL-TR-68-150, pp. 617–639 (1969).

Cheng, S. and Ho, B. P. C., "Stability of Heterogeneous Aeolotropic Cylindrical Shells under Combined Loading," *J. Am. Inst. Aeronaut. Astronaut.*, Vol. 1 (4), pp. 892–898 (1963).

Clough, R. W. and Johnson, C. P., "A Finite Element Approximation for the Analysis of Thin Shells," *Int. J. Solids Struct.*, Vol. 4, pp. 42–60 (1968).

Dong, S. B., "Analysis of Laminated Shells of Revolution," *J. Eng. Mech. Div. ASCE*, Vol. 92, EM6, p. 135 (1966).

Dong, S. B. and Selna, L. G., "Natural Vibrations of Laminated Orthotropic Shells of Revolution," *J. Compos. Mat.*, Vol. 4 (1), pp. 2–19 (1970).

Gladwell, G. M. L. and Vijay, D. K., "Errors in Shell Finite Element Models for the Vibration of Circular Cylinders," *J. Sound Vib.*, Vol. 43 (4), pp. 511–528 (1970).

Gulati, S. T. and Essenberg, F., "Effects of Anisotropy in Axisymmetric Cylindrical Shells," *J. Appl. Mech.*, Vol. 34, pp. 659–666 (1967).

Hensell, R. D., Neale, B. K., and Warburton, G. B., "A New Hybrid Cylindrical Shell Finite Element," *J. Sound Vib.*, Vol. 16, No. 4, pp. 519–531 (1971).

Irons, B. M., "The Semiloof Shell Element," *Finite Elements for Thin Shells and Curved Members*, P. G. Ashwell and R. H. Gallagher (eds.), Wiley, 1976.

Jones, R. E. and Strome, D. R., "A Survey of Analysis of Shells by the Displacement Method," *Proc. 1st Conf. Matrix Methods Struct. Mech.*, Wright-Patterson AFB, Dayton, Ohio, AFFDL-TR-66-80, pp. 205–299 (1966).

Martins, R. A. F. and Owen, D. R. J., "Structural Instability and Natural Vibration Analysis of Thin Arbitrary Shells by Use of the Semiloof Element," *Intl. J. Numer. Methods Eng.*, Vol. 11 (3), pp. 481–498 (1977).

Noor, A. K. and Anderson, C. M., "Mixed Isoparametric Finite Element Models of Laminated Composite Shells," *Comput. Methods Appl. Mech. Eng.*, Vol. 11 (3), pp. 255–280 (1977).

Noor, A. K. and Camin, R. A., "Symmetry Considerations for Anisotropic Shells," *Comput. Methods Appl. Mech. Eng.*, Vol. 9, pp. 317–335 (1976).

Noor, A. K. and Mathers, M. D., "Shear-Flexible Finite-Element Models of Laminated Composite Plates and Shells," NASA TAN D-8044, Langley Research Center, Hampton, VA.

Olson, M. D. and Lindberg, G. M., "A Finite Cylindrical Shell Element and the Vibrations of a Curved Fan Blade," NAE-LR-497, National Aeronautical Establishment, Ottawa, Canada, p. 46 (1968).

Padovan, J., "Quasi-Analytical Finite Element Procedures for Axisymmetric Anisotropic Shells and Solids," *Comput. Struct.*, Vol. 4, pp. 467–483 (1974).

Padovan, J., "Numerical Analysis of Asymmetric Frequency and Buckling Eigenvalues of Pre-stressed Rotating Anisotropic Shells of Revolution," *Comput. Struct.*, Vol. 5, pp. 145–154 (1975).

Panda, S. C. and Natarajan, R., "Finite Element Analysis of Laminated Shells of Revolution," *Comput. Struct.*, Vol. 6, pp. 61–64 (1976).

Rao, G. V., Sundaramaiah, V. and Raju, I. S., "Finite Element Analysis of Vibrations of Initially Stressed Thin Shells of Revolution," *J. Sound Vib.*, Vol. 37, pp. 57–64 (1974).

Rao, K. P., "A Rectangular Laminated Anisotropic Shallow Thin Shell Finite Element," *Comput. Methods Appl. Mech. Eng.*, Vol. 15, pp. 13–33 (1978).

Reddy, J. N., "Finite Element Modeling of Structural Vibrations: A Review of Recent Advances," *Shock Vib. Dig.*, Vol. 11, No. 1, pp. 25–39 (1979d).

Reddy, J. N., "Finite-Element Modeling of Layered, Anisotropic Composite Plates and Shells: A Review of Recent Research," *Shock Vib. Dig.*, Vol. 13, No. 12, pp. 3–12 (1981).

Reddy, J. N., "Bending of Laminated Anisotropic Shells by a Shear Deformable Finite Element," *Fibre Sci. Tech.*, Vol. 17, pp. 9–24 (1982b).

Reddy, J. N., "On the Exact Solutions of Bending and Vibration of Thick Laminated Shells," *J. Eng. Mech.*, ASCE, to appear (1984b).

Ross, C. T. F., "Finite Elements for the Vibration of Cones and Cylinders," *Int. J. Numer. Methods Eng.*, Vol. 9 (4), pp. 833–845 (1975).

Schmit, L. A. and Monforton, G. R., "Finite-Element Analysis of Sandwich Plate and Laminate Shells with Laminated Faces," *Am. Inst. Aeronaut. Astronaut. J.*, Vol. 8, pp. 1454–1461 (1970).

Sen, S. K. and Gould, P. L, "Free Vibration of Shells of Revolution using FEM," *ASCE J. Eng. Mech. Div.*, Vol. 100 (EM2), pp. 283–303 (1974).

Shivakumar, K. N. and Krishna Murty, A. V., "A High Precision Ring Element for Vibrations of Laminated Shells," *J. Sound Vib.*, Vol. 58, No. 3, pp. 311–318 (1978).

Siede, P. and Chang, P. H. H., "Finite Element Analysis of Laminated Plates and Shells," NASA CR-157106 (1978).

Sinha, P. K. and Rath, A. K., "Vibration and Buckling of Cross-Ply Laminated Circular Cylindrical Panels," *Aeronaut. Q.*, Vol. 26, pp. 211–218 (1975).

Tabarrok, B. and Gass, N., "Vibration of Cylindrical Shells by Hybrid Finite Element Method," *AIAA J.*, Vol. 10 (12), pp. 1553–1554 (1972).

Tabarrok, B. and Gass, N., "A New Variational Principle with Application to the Vibration of Shallow Shells by the Method of Finite Elements," *Math. Mech.*, Vol. 53 (12), pp. 773–781 (1973).

Venkatesh, A. and Rao, K. P., "A Doubly-Curved Quadrilateral Finite Element for the Analysis of Laminated Anisotropic Thin Shell of Revolution," *Comput. Struct.*, Vol. 12, pp. 825–832 (1980).

Wempner, G., Oden, J. T., and Kross, D. A., "Finite Element Analysis of Thin Shells," *Proc. ASCE, J. Eng. Mech. Div.*, Vol. 95, pp. 1273–1294 (1968).

Yang, T. Y., "High Order Rectangular Shallow Shell Finite Element," *J. Eng. Mech. Div., ASCE*, Vol. 99, EM1, pp. 157–181 (1973).

Yang, T. Y. and Kim, H. W., "Asymmetrical Bending and Vibration of Conical Shell Finite Element," *AIAA J.*, Vol. 12 (3), pp. 257–258 (1974).

Zukas, J. A. and Vinson, J. R., "Laminated Transversely Isotropic Cylindrical Shells," *J. Appl. Mech. Trans.*, ASME, pp. 400–407 (1971).

TIME APPROXIMATIONS IN STRUCTURAL DYNAMICS

Bathe, K. J., *Finite Element Procedures in Engineering Analysis*, Prentice-Hall, Englewood Cliffs, N.J., 1982.

Belytschko, T. and Mullen, R., "Mesh Partitions of Explicit–Implicit Time Integrations," in *Formulations and Computational Algorithms in Finite Element Analysis*, K. J. Bathe, J. T. Oden, and W. Wunderlich (eds.), M.I.T. Press, Massachusetts, 1977.

Belytschko, T., Yen, H. J., and Mullen, R., "Mixed Methods for Time Integration," *Comput. Meth. Appl. Mech. Eng.*, Vol. 17/18, pp. 259–275 (1979).

Chan, S. P., Cox, H. L., and Benefield, W. A., "Transient Analysis of Forced Vibrations of Complex Structural Mechanical Systems," *J. R. Aeronaut. Soc.*, Vol. 66, pp. 457–460 (1962).

Clough, R. W. and Penzien, J., *Dynamics of Structures*, McGraw-Hill, New York, 1975.

Goudreau, G. L. and Taylor, R. L., "Evaluation of Numerical Integration Methods in Elastodynamics," *J. Comput. Methods Appl. Mech. Eng.*, Vol. 2, No. 1, pp. 69–97 (1973).

Houbolt, J. C., "A Recurrence Matrix Solution for the Dynamic Response of Elastic Aircraft," *J. Aeronaut. Sci.*, Vol. 17, pp. 540–550 (1950).

Hughes, T. J. R. and Liu, W. K., "Implicit–Explicit Finite Elements in Transient Analysis: Stability Theory, and Implementation and Numerical Examples," Vol. 45, No. 2, pp. 371–378 (1978).

Johnson, D. E., "A Proof of the Stability of the Houbolt Method," *AIAA J.*, Vol. 4, pp. 1450–1451 (1966).

Krieg, R. D., "Unconditional Stability in Numerical Time Integration Methods," *J. Appl. Mech.*, pp. 417–421 (1973).

Lax, P. D. and Richtmyer, R. D., "Survey of the Stability of Finite Difference Equations," *Communications in Pure and Applied Mathematics*, Vol. 9, pp. 267–293 (1956).

Leech, J. W., "Stability of Finite-Difference Equations for the Transient Response of a Flat Plate," *AIAA J.*, Vol. 3, No. 9, pp. 1772–1773 (1965).

Levy, S. and Kroll, W. D., "Errors Introduced by Finite Space and Time Increments in Dynamic Response Computation," *J. Res. Nat. Bur. Stand.*, Vol. 51, pp. 57–68 (1953).

Newmark, N. M., "A Method of Computation for Structural Dynamics," *ASCE Eng. Mech. Div.*, Vol. 85, pp. 67–94 (1959).

Nickell, R. E., "On the Stability of Approximation Operators in Problems of Structural Dynamics," *Int. J. Solids and Struct.*, Vol. 7, pp. 301–319 (1971).

Nickell, R. E., "Direct Integration Methods in Structural Dynamics," *J. Eng. Mech. Div. ASCE*, Vol. 99, No. EM2 (1972).

TRANSIENT ANALYSIS OF PLATES

Akay, H. U., "Dynamic Large Deflection Analysis of Plates Using Mixed Finite Elements," *Comput. Struct.*, Vol. 11, pp. 1–11 (1980).

Chow, T. S., "On the Propagation of Flexural Waves in an Orthotropic Laminated Plate and Its Response to an Impulsive Load," *J. Compos. Mater.*, Vol. 5, pp. 306–319 (1971).

Lee, Y. and Reismann, H., "Dynamics of Rectangular Plates," *Int. J. Eng. Sci.*, Vol. 7, pp. 93–113 (1969).

Moon, F. C., "Wave Surfaces Due to Impact on Anisotropic Plates," *J. Compos. Mater.*, Vol. 6, pp. 62–79 (1972).

Moon, F. C., "One-Dimensional Transient Waves in Anisotropic Plates," *J. Appl. Mech.*, Vol. 40, pp. 485–490 (1973).

Reddy, J. N., "On the Solutions to Forced Motions of Rectangular Composite Plates," *J. Appl. Mech.*, Vol. 49, pp. 403–408 (1982c).

Reddy, J. N., "Dynamic (Transient) Analysis of Layered Anisotropic Composite-Material Plates," *Int. J. Numer. Methods Eng.*, Vol. 19, pp. 237–255 (1983b).

Reddy, J. N., "Geometrically Nonlinear Transient Analysis of Laminated Composite Plates," *AIAA J.*, Vol. 21, No. 4, pp. 621–629 (1983c).

Reddy, J. N., "A Note on Symmetry Considerations in the Transient Response of Unsymmetrically Laminated Anisotropic Plates," *Int. J. Numer. Meth. Eng.*, Vol. 20, pp. 175–181 (1984c).

Reismann, H., "Forced Motion of Elastic Plates," *J. Appl. Mech.*, Vol. 35, pp. 510–515 (1968).

Reismann, H. and Lee, Y., "Forced Motions of Rectangular Plates," in *Developments in Theoretical and Applied Mechanics*, Vol. 4, D. Frederick (ed.), Pergamon Press, New York, pp. 3–18 (1969).

Rock, T. and Hinton, E., "Free Vibration and Transient Response of Thick and Thin Plates Using the Finite Element Method," *Earthquake Eng. Struct. Dyn.*, Vol. 3, pp. 51–63 (1974).

Sun, C. T., "Propagation of Shock Waves in Anisotropic Composite Plates," *J. Compos. Mater.*, Vol. 7, pp. 366–382 (1973).

Sun, C. T. and Whitney, J. M., "Forced Vibrations of Laminated Composite Plates in Cylindrical Bending," *J. Acoust. Soc. Am.*, Vol. 55 (5), pp. 1003–1008 (1974).

Sun, C. T. and Whitney, J. M., "Dynamic Response of Laminated Composite Plates," *AIAA J.*, Vol. 13 (10), pp. 1250–1260 (1975).

Tsui, T. Y. and Tong, P., "Stability of Transient Solution of Moderately Thick Plate by Finite Difference Method," *AIAA J.*, Vol. 9, pp. 2062–2063 (1971).

Volterra, E. and Zachmanoglou, E. C., *Dynamics of Vibrations*, Merrill, Columbus, Ohio, 1965.

ANSWERS TO SELECTED EXERCISES

CHAPTER 1

Exercises 1.1

1. Properties (i) and (ii) follow easily from the definition of $d(\cdot, \cdot)$. Property (iii) follows from the triangle inequality

$$|y_1(x,t) - y_3(x,t)| = |y_1(x,t) - y_2(x,t) + y_2(x,t) - y_3(x,t)|$$

$$\leq |y_1(x,t) - y_2(x,t)| + |y_2(x,t) - y_3(x,t)|$$

3. $d(w,0) = \frac{3}{2}L(c_1^2 + c_2^2) + 4c_1 c_2 L$

4. (a) $m = 4$ (2 per ball), $p = 0$, $n = 4$.
   (b) $m = 2$, $p = 1$ ($l =$ constant), $n = 1$.
   (c) $m = 2$, $p = 1$ ($x_1 + x_2 =$ constant), $n = 1$.
   (d) $m = 3$, $p = 2$ ($x_1 + x_2 =$ constant, $\dot{x}_1 - \dot{x}_3 = v_0$), $n = 1$.
   (e) $m = 4$, $p = 2$ ($l_1, l_2 =$ constant), $n = 2$.
   (f) $m = 4$, $p = 2$, $n = 2$.

5. $\phi_1 = x(x - L)$, $\phi_2 = x^2(x - L), \ldots, \phi_n = x^n(x - L)$.

6. (c) $A_i A_j \varepsilon_{ijk} = \frac{1}{2}(A_i A_j + A_i A_j)\varepsilon_{ijk}$

$= \frac{1}{2}(A_i A_j + A_j A_i)\varepsilon_{ijk}$ (interchanged A_i with A_j in the second term)

$= \frac{1}{2}(A_i A_j \varepsilon_{ijk} + A_j A_i \varepsilon_{ijk})$

$= \frac{1}{2}(A_i A_j \varepsilon_{ijk} + A_i A_j \varepsilon_{jik})$ (renamed i to be j and j to be i in the second term)

$= \frac{1}{2}A_i A_j (\varepsilon_{ijk} + \varepsilon_{jik})$

$= \frac{1}{2}A_i A_j (\varepsilon_{ijk} - \varepsilon_{ijk}) = 0$

(e) See Reddy and Rasmussen (1982), p. 33.

7. (a) See Reddy and Rasmussen (1982), p. 20

(c) $\text{grad}(r^n) = \hat{\mathbf{e}}_i \frac{\partial}{\partial x_i}(x_j x_j)^{n/2}$

$= \hat{\mathbf{e}}_i \frac{n}{2}(x_j x_j)^{(n/2)-1} \cdot \frac{\partial}{\partial x_i}(x_j x_j)$

$= \hat{\mathbf{e}}_i \frac{n}{2} r^{n-2} \cdot (\delta_{ij} x_j + x_j \delta_{ij})$

$= n\hat{\mathbf{e}}_i r^{n-2} x_i$

$= nr^{n-2}\mathbf{r}$

8. (a) $\hat{\mathbf{e}}_1' = \frac{\hat{\mathbf{e}}_1 - \hat{\mathbf{e}}_2 + \hat{\mathbf{e}}_3}{\sqrt{3}}$

$\hat{\mathbf{e}}_2' = \frac{\text{grad}(2x_1 + 3x_2 + x_3 - 5)}{|\text{grad}(2x_1 + 3x_2 + x_3 - 5)|}$

$= \frac{2\hat{\mathbf{e}}_1 + 3\hat{\mathbf{e}}_2 + \hat{\mathbf{e}}_3}{\sqrt{14}}$

$\hat{\mathbf{e}}_3' = \hat{\mathbf{e}}_1' \times \hat{\mathbf{e}}_2'$

$= \frac{-4\hat{\mathbf{e}}_1 + \hat{\mathbf{e}}_2 + 5\hat{\mathbf{e}}_3}{\sqrt{42}}$

(b) $\hat{\mathbf{e}}_1' = \frac{\hat{\mathbf{e}}_1 - \hat{\mathbf{e}}_2 + \hat{\mathbf{e}}_3}{\sqrt{3}}, \qquad \hat{\mathbf{e}}_2' = \hat{\mathbf{e}}_3' \times \hat{\mathbf{e}}_1'$

(c) $[A] = \begin{bmatrix} \frac{\sqrt{3}}{2} & \frac{1}{2} & 0 \\ -\frac{1}{2} & \frac{\sqrt{3}}{2} & 0 \\ 0 & 0 & 1 \end{bmatrix}$

9. Set $\phi = 1$ in Eq. (1.1.15) to get

$$\oint_S \hat{n}\, dS = 0$$

Set $A = r$ in Eq. (1.1.16) to get

$$\oint_S \hat{n}\cdot r\, dS = \int_V \text{div}\, r\, dV$$

$$= \delta_{ii}V$$

10. (a) $\displaystyle\int_V v\,\text{div}(\text{grad}\, u)\, dx = \int_V \{\text{div}(v\,\text{grad}\, u) - \text{grad}\, v\cdot\text{grad}\, u\}\, dx$

$$= \oint_S \hat{n}\cdot(v\,\text{grad}\, u)\, dS - \int_V \text{grad}\, v\cdot\text{grad}\, u\, dx$$

$$= \oint_S v\hat{n}\cdot\nabla u\, dS - \int_V \text{grad}\, v\cdot\text{grad}\, u\, dx$$

11. (a) $\displaystyle\int_0^L v\frac{d^2u}{dx^2}\, dx = -\int_0^L \frac{dv}{dx}\frac{du}{dx}\, dx + \left(v\frac{du}{dx}\right)\Big|_0^L.$

Exercises 1.2

1. $t^{(n)}\, dl - t^{(1)}\, dl\cos\alpha - t^{(2)}\, dl\sin\alpha = 0$

$t^{(n)} = t^{(1)}\cos\alpha + t^{(2)}\sin\alpha$

where dl is the length of the slant edge.

3. The normal and tangential forces on sides $x_2 = 0$ and $x_1 = 0$ are:

$$\text{side } x_2 = 0: \quad \sigma_{22}\, dl\sin\alpha,\ \sigma_{21}\, dl\sin\alpha$$

$$\text{side } x_1 = 0: \quad \sigma_{11}\, dl\cos\alpha,\ \sigma_{12}\, dl\cos\alpha$$

By resolving all forces on the wedge along σ_N and σ_S, obtain the result.

4. For maximum or minimum of $\sigma_N = \sigma_N(\alpha)$, set the first derivative of σ_N (in Exercise 3), with respect to α, to zero and compute $\cos\alpha$ and $\sin\alpha$ from the result. Substitute for $\cos\alpha$ and $\sin\alpha$ into the expression for σ_N in Exercise 3 to obtain the required result.

5. (a) Not possible (unless $c_1 = 0$).
   (b) Not possible.
   (c) Possible.

6. $f_1 = 0$, $f_2 = 0$, $f_3 = -4$.

7. $\sigma_{12} = -\dfrac{P}{2I_3}(h^2 - x_2^2)$, $\sigma_{22} = 0$.

8. $\sigma_{12} = \dfrac{-f_0 x_1 (h^2 - x_2^2)}{2I_3}$, $\sigma_{22} = \dfrac{f_0}{2I_3}\left(-\dfrac{x_2^3}{3} + h^2 x_2 - \dfrac{2}{3}h^3\right)$.

9. $t_1^{(n)} = (-2x_1^2 - 6 + 4x_1 x_2 + x_3 + x_1 - 3x_2)/\sqrt{3}$.

    $t_2^{(n)} = (-7 + 4x_1 x_2 + 3x_1^2 - 2x_2^2 - 4x_3)/\sqrt{3}$.

    $t_3^{(n)} = (2x_1 - 4 + 3x_3)/\sqrt{3}$.

    $t_N = -3.33$, $t_S = 8.578$.

10. $[\sigma] = \begin{bmatrix} -2 & 2 & -4 \\ 2 & -10 & 0 \\ -4 & 0 & 5 \end{bmatrix}$

    $\lambda_1 = -10.533$, $\hat{\mathbf{n}}^{(1)} = (-0.257, 0.964, -0.066)$.

    $\lambda_2 = 3.323$, $\hat{\mathbf{n}}^{(2)} = (0.870, 0.261, 0.418)$.

    $\lambda_3 = 6.856$, $\hat{\mathbf{n}}^{(3)} = (-0.420, -0.05, 0.906)$.

12. $(\sigma)_{max} = 11.824$ psi, $\mathbf{n} = (-0.73001, -0.33721, -0.59446)$.

13. See Reddy and Rasmussen (1982), p. 142.

Exercises 1.3

1. $e_{11} = 3x_1^2 x_2 + c_1(2c_2^3 + 3c_2^2 x_2 - x_2^3)$.

    $e_{12} = \frac{1}{2}\left[x_1^3 + (3c_1 c_2^2 - 6c_1 x_2^2)x_1\right]$.

    $e_{22} = x_2^3 - 3x_2(c_2^2 + c_1 x_1^2) - 2c_2^3$.

2. $e_{11} = 2x_2(1 - x_1)$, $e_{12} = \frac{1}{2}\left[x_2^2(3c_2 + c_3) - 2c_1\right]$,

    $e_{22} = -2c_3 x_2(1 - x_1)$.

3. $u_1 = \dfrac{e_0}{b}x_2$, $u_2 = 0$.

    $\varepsilon_{11} = 0$, $\varepsilon_{12} = \frac{1}{2}\left(\dfrac{e_0}{b}\right)$, $\varepsilon_{22} = \frac{1}{2}\left(\dfrac{e_0}{b}\right)^2$.

4. $u_1 = e_0\left(\dfrac{x_2}{b}\right)^2$, $u_2 = 0$.

    $\varepsilon_{11} = 0$, $\varepsilon_{12} = \dfrac{e_0 x_2}{b^2}$, $\varepsilon_{22} = 2\left(\dfrac{e_0 x_2}{b^2}\right)^2$

5. See Eq. (4.2.5).

6. (a) Yes.
   (b) No.

7. $e_n = \dfrac{e_0}{2b}$, $e_s = 0$.

8. $e_1 = 802.65 \times 10^{-6}$ in./in., $e_2 = 197.35 \times 10^{-6}$ in./in.
   $\alpha_1 = -26.57°$, $\alpha_2 = 63.43°$.

Exercises 1.5

1. Compatibility met.

2. $e_{11} = -\dfrac{f_0}{2EI_3}\left[x_1^2 x_2 - \nu\left(-\dfrac{x_2^3}{3} + h^2 x_2 - \dfrac{2}{3}h^3\right)\right]$.

   $e_{12} = -\dfrac{f_0 x_1}{4GI_3}(h^2 - x_2^2)$.

   $e_{22} = -\dfrac{f_0}{2EI_3}\left(-\dfrac{x_2^3}{3} + h^2 x_2 - \dfrac{2}{3}h^3 - \nu x_1^2 x_2\right)$.

   Compatibility not satisfied.

3. $\sigma_{11} = \dfrac{P}{A}$, all other $\sigma_{ij} = 0$.

   $u_1 = \dfrac{\sigma_{11}}{E}x_1$, $u_2 = -\dfrac{\nu\sigma_{11}}{E}x_2$.

4. $\sigma_{11} = -\dfrac{M_0 x_2}{I_3}$, all other $\sigma_{ij} = 0$.

   $u_1 = -\dfrac{M_0 x_2 x_1}{EI_3} + c_1 x_2 + c_2 x_3 + c_3$.

   $u_2 = \dfrac{M_0}{2EI_3}\left[\nu\left(x_2^2 - x_3^2\right) + x_1^2\right] + c_4 x_3 + c_5 x_1 + c_6$.

5. $\nabla^2\sigma_{ij} + \dfrac{1}{1+\nu}\Theta_{,ij} + \dfrac{\nu}{1-\nu}\delta_{ij}f_{k,k} + (f_{i,j} + f_{j,i}) = 0$ where $\Theta = \sigma_{kk}$.

6. $u_1 = -\dfrac{e_0 x_1}{a}$, $u_2 = 0$, $u_3 = \dfrac{\nu}{1-\nu}\dfrac{e_0 x_3}{a}$.

7. $u_1 = \dfrac{-f_0}{6EI_3}x_1\left[x_1^2 x_2 + \nu\left(2h^3 + 3h^2 x_2 - x_2^3\right)\right] + f(x_2)$.

   $u_2 = \dfrac{f_0}{6EI_3}x_2\left(2h^3 + \dfrac{3}{2}h^2 x_2 - \dfrac{1}{4}x_2^3 + \dfrac{3}{2}\nu x_1^2 x_2\right) + g(x_1)$.

8. The Navier equations are the equilibrium equations for the displacements of an elastic body. In the absence of temperature changes, the equations are obtained, for an isotropic body, by substituting Eq. (1.3.11) into Eq. (1.5.22), and the result into Eq. (1.2.34):

$$(\lambda + \mu)u_{i,ij} + \mu u_{j,ii} + f_j = \rho\frac{\partial^2 u_j}{\partial t^2}$$

which are the same as those in Eq. (1.6.14). When temperature changes are considered, we obtain

$$(\lambda + \mu)u_{i,ij} + \mu u_{j,ii} - (3\lambda + 2\mu)\alpha T_{,j} + f_j = \rho\frac{\partial^2 u_j}{\partial t^2}$$

where μ and λ are the Lamé constants, and α is the coefficient of thermal expansion.

9. Assume that $\sigma_{11} = \sigma_{11}(x_2)$ and $\sigma_{22} = \sigma_{12} = 0$, and obtain the compatibility equation

$$\frac{d^2}{dx_2^2}(\sigma_{11} + E\alpha T) = 0$$

Integrate the equation to obtain

$$\sigma_{11} = e\alpha\left[-T + \frac{t}{A}\int_{-h}^{h}T\,dx_2 + \frac{x_2 t}{I_3}\int_{-h}^{h}Tx_2\,dx_2\right]$$

$$e_{11} = \frac{\sigma_{11}}{E} + \alpha T, \qquad e_{22} = -\frac{\nu\sigma_{11}}{E} + \alpha T, \qquad e_{12} = 0$$

The constants of integration were evaluated by requiring that the resultant force and moment at $x = \pm(L/2)$ are zero.

10. (a) Temperature increase required to close the gap: $\Delta T = e_{11}/\alpha L = 55.59°F$. Temperature at which the gap closes: $35.59°F$.
    (b) $\sigma_{11} = E\alpha(95 - 35.59) = 7129.2$ psi.

11. (a) $e_{11} = e_{22} = e_{33} = \alpha T$.
    (b) $e_{11} = e_{22} = e_{33} = \alpha T(x_1)$.
       $u_1 = \alpha\int T(x_1)\,dx_1 + c_1$.
       $u_2 = \alpha T(x_1)x_2 + c_2$.
       $u_3 = \alpha T(x_1)x_3 + c_3$.

18. See Eq. (4.2.59).

Exercises 1.6

1. Mixed (third kind).

2. Stress (second kind).

3. Stress (second kind).

4. Mixed (third kind).

5. Mixed (third kind).

6. (a) $c_1 = 2c_3$, c_2 and c_3 are arbitrary constants.
   (b) $t_i = \sigma_{ij} n_j$; on $x_2 = \pm h : t_1 = t_2 = 0$

   at $x_1 = L : t_1 = 0 \rightarrow c_1 L + c_2 = 0$

   $$\int_{-h}^{h} \sigma_{11} x_2 \, dx_2 = P(L - x_1) \rightarrow c_1 = -\frac{P}{I_3}, \qquad c_2 = \frac{PL}{I_3}$$

   where $I_3 = 2bh^3/3$.

7. (a) $c_1 = -2c_4$, $c_2 = c_5$.
   (b) Edge $x_2 = 0$: $t_1 = \int_{-a}^{a} \sigma_{12} \, dx_1 = \frac{c_1 a^5}{5}$

   $$t_2 = \int_{-a}^{a} \sigma_{22} \, dx_1 = 0$$

   Edge $x_2 = b$: $t_1 = c_1 a^3 \left(b^2 - \frac{a^2}{5} \right) - 2c_2 ab^2$

   $$t_2 = 0$$

9. $U_0 = \frac{1}{2}\sigma_{ij} e_{ij}$; the orthotropic constitutive equations are given by Eq. (1.5.3b) with c_{ij} given by Eq. (1.5.15), where $c_{16} = c_{26} = c_{36} = c_{45} = 0$.

Exercises 1.7

1. $u = (\rho g/2E)(2xL - x^2 - \nu y^2)$, $v = -(\nu \rho g y/E)(L - x)$, $w = 0$; yes.

2. $e_s = 3.2 \times 10^{-6} P$; $e_a = 6 \times 10^{-6} P$; total elongation $= 9.2 \times 10^{-6} P$.

3. $e_s = -e_a$; $\sigma_s = \frac{15}{38} P$ (tensile); $\sigma_a = -\frac{4}{38} P$ (compression).

4. $\sigma_y = \gamma y/3$; $e_y = \gamma y/3E$; $v = \int_0^h e_y \, dy = \gamma h^2/6E$.

5. $V(x) = \begin{cases} 4600 - 1000x, & 0 \le x \le 10 \text{ ft} \\ 2000, & 10' \le x \le 12 \text{ ft} \end{cases}$

       $M(x) = \begin{cases} 4600x - 500x^2, & 0 \le x \le 10 \text{ ft} \\ 2000x - 24{,}000, & 10' \le x \le 12 \text{ ft} \end{cases}$

       $\sigma_x = -\dfrac{My}{I}, \qquad \sigma_{xy} = \dfrac{VQ}{Ib}, \qquad I = 290.67 \text{ in.}^4 \ (\bar{y} = 6.5 \text{ in.})$

       $Q(y) = \begin{cases} 2\displaystyle\int_{y_0}^{1.5} y \, dy + 8 \int_{1.5}^{3.5} y \, dy, & 0 \le y_0 \le 1.5 \text{ in.} \\[2mm] 8\displaystyle\int_{y_0}^{3.5} y \, dy, & y_0 \ge 1.5 \text{ in.} \end{cases}$

       $(\sigma_x)_{\max}^{\text{tensile}} = 2839.1 \text{ psi}; \qquad (\sigma_x)_{\max}^{\text{compression}} = -1528.75 \text{ psi}$

       $|\sigma_{xy}| = 392.46 \ psi$

8.   The elasticity solution from Eq. (1.5.27a), is

$$u_2(0,0) = \frac{PL^3}{3EI} + \frac{Ph^2L}{2GI}$$

       From the classical beam theory, we have

$$u_2 = \frac{PL^3}{3EI}$$

       The ratio of the shear deflection, $Ph^2L/2GI$, to the bending deflection, $PL^3/3EI$, is $(3/4)(1 + v)(2h/L)^2$.

10. $\Phi(x, y) = -\dfrac{a^2b^2G\theta}{(a^2 + b^2)} \left(\dfrac{x^2}{a^2} + \dfrac{y^2}{b^2} - 1 \right)$

       $\sigma_{zx} = -\dfrac{2My}{\pi ab^3}, \qquad \sigma_{zy} = \dfrac{2Mx}{\pi a^3b}, \qquad M = \dfrac{G\theta\pi a^3b^3}{a^2 + b^2}$

       $w(x, y) = \dfrac{M}{G} \dfrac{(b^2 - a^2)xy}{\pi a^3b^3}$

11. $M = \dfrac{Ga^4\theta}{15\sqrt{3}}, \ \sigma_{zx} = -G\theta y \left(1 + \dfrac{3x}{a}\right), \ \sigma_{zy} = G\theta \left[x \left(1 - \dfrac{3x}{2a}\right) + \dfrac{3y^2}{2a} \right]$

       $(\sigma_{yz})_{\max}$ occurs at $x = -a/3$ and $y = 0$: $\sigma_{yz}[-(a/3), 0] = -(G\theta a/2)$.

12. $\Phi(x, y) = c\left\{ y - \left[1 - \left(\dfrac{2x}{a} \right)^2 \right] \right\}^2$, c is a constant.

13. $\Phi(x, y) = (a^2 - x^2)(b^2 - y^2)(c_1 + c_2x^2 + c_3y^2 + \cdots)$

       where $c_1, c_2, \ldots$, are constants. Exact and approximate solutions for this problem are presented in Chapter 3.

16. $\nabla^4\phi = 0.$

17. $\dfrac{d^2u}{dr^2} + \dfrac{c}{r}\dfrac{du}{dr} - \dfrac{u}{r^2} = 0,\ c = \dfrac{1-2\nu}{1-\nu}.$

   $u = Ar^{k_1} + Br^{k_2},\qquad k_{1,2} = -\dfrac{c \pm \sqrt{c^2+8}}{2}.$

18. $e_x = z\dfrac{d\psi}{dx} - \dfrac{4}{3}\left(\dfrac{z}{h}\right)^2\left(\dfrac{d\psi}{dx} + \dfrac{d^2w}{dx^2}\right).$

   $e_y = 0,\qquad e_z = 0,\qquad e_{xy} = 0,\qquad e_{yz} = 0.$

   $2e_{xz} = \left[1 - 4\left(\dfrac{z}{h}\right)^2\right]\left(\psi + \dfrac{dw}{dx}\right).$

CHAPTER 2

Exercises 2.1

1. Minimize the total time, $T(y) = \displaystyle\int_0^a \sqrt{\dfrac{1+(dy/dx)^2}{2gy}}\,dx$ (see Section 1.1).

2. Minimize $L(y) = \displaystyle\int_a^b \sqrt{1 + \left(\dfrac{dy}{dx}\right)^2}\,dx.$

3. Maximize $A(y) = \displaystyle\int_a^b y(x)\,dx$, subject to the constraint

$$(b-a) - \int_a^b \sqrt{1 + \left(\dfrac{dy}{dx}\right)^2}\,dx = 0.$$

4. $U^* = \displaystyle\int_0^L \left(\dfrac{N^2}{2EA} + f_s\dfrac{V^2}{2GA} + \dfrac{T^2}{2GJ}\right)dx$

   where $f_s = (A/I^2)\displaystyle\int_A (Q^2/b^2)\,dA$, and Q is the first moment of area.

5. (a) $U^* = \displaystyle\int_0^L \dfrac{P^2}{2EA}\,dx,\qquad A = b\left[h_1 - \left(\dfrac{h_1-h_2}{L}\right)x\right].$

   (b) $U^* = \displaystyle\int_0^L \dfrac{T^2}{2GJ}\,dx,\qquad J = \dfrac{\pi}{32}\left[d_1^4 - \left(\dfrac{d_1^4-d_0^4}{L^4}\right)x^4\right].$

6. $\delta W = F_x[l_1\cos\theta_1\,\delta\theta_1 + l_2\cos\theta_2(\delta\theta_1 + \delta\theta_2)]$
   $- F_y[l_1\sin\theta_1\,\delta\theta_1 + l_2\sin\theta_2(\delta\theta_1 + \delta\theta_2)].$

7. $\delta W_E = -\int_0^L f_0\,\delta w\,dx_1 - P_0\,\delta w(0) - M_0\delta\left(\frac{dw}{dx_1}\right)\bigg|_{x_1=0}.$

$\delta W_I = \int_0^L \frac{M\delta M}{EI}\,dx_1, \qquad M = M_0 + \left(P_0 + \frac{f_0 x_1}{2}\right)x_1$

Exercises 2.2

1. $\delta I = \int_0^L \left(EA\frac{du}{dx}\frac{d\delta u}{dx} - f\sin u\,\delta u\right)dx - P\,\delta u(L).$

2. $\delta I = \int_0^L 2\left(\frac{du}{dx}\frac{d\delta u}{dx} + \frac{d\delta u}{dx}\frac{dv}{dx} + \frac{du}{dx}\frac{d\delta v}{dx} + \frac{dv}{dx}\frac{d\delta v}{dx} + f\delta u + g\delta v\right)dx.$

3. $\delta I = \int_V (\mathrm{grad}\,u\cdot\mathrm{grad}\,\delta u - f\delta u)\,dV - \oint_S \delta u\,\hat{\imath}\,dS.$

4. $\delta I = \int_0^a\int_0^b \left\{D\left[\frac{\partial^2 w}{\partial x^2}\frac{\partial^2\delta w}{\partial x^2} + \frac{\partial^2 w}{\partial y^2}\frac{\partial^2\delta w}{\partial y^2} + \nu\left(\frac{\partial^2 w}{\partial x^2}\frac{\partial^2\delta w}{\partial y^2} + \frac{\partial^2\delta w}{\partial x^2}\frac{\partial^2 w}{\partial y^2}\right)\right.\right.$

$\left.\left. +2(1-\nu)\frac{\partial^2 w}{\partial x\,\partial y}\frac{\partial^2\delta w}{\partial x\,\partial y}\right] + kw\delta w - q_0\delta w\right\}dx\,dy.$

5. $\delta I = \int_R \left\{\frac{\mu}{2}\left[(\delta u_{i,j} + \delta u_{j,i})u_{i,j} + (u_{i,j} + u_{j,i})\delta u_{i,j}\right]\right.$

$\left. - \delta P u_{i,i} - P\delta u_{i,i} - f_i\delta u_i\right\}dx_1\,dx_2 + \oint_S \hat{\imath}_i\delta u_i\,dS.$

6. $-EA\frac{d^2u}{dx^2} - f\sin u = 0, \qquad 0 < x < L.$

$EA\frac{du}{dx}(0) = 0, \qquad EA\frac{du}{dx}(L) - P = 0.$

7. $\delta u: -\frac{d}{dx}\left(\frac{du}{dx} + \frac{dv}{dx}\right) + f = 0, \qquad 0 < x < L.$

$\delta v: -\frac{d}{dx}\left(\frac{du}{dx} + \frac{dv}{dx}\right) + g = 0, \qquad 0 < x < L.$

EBC: specify u, v at $x = 0$, L.

NBC: $\frac{du}{dx} + \frac{dv}{dx} = 0$ at $x = 0$, L.

8. $-\mathrm{div}(\mathrm{grad}\,u) - f = 0$ in V, $\frac{\partial u}{\partial n} - \hat{\imath} = 0$ on S.

9. $D \nabla^4 w + kw - q_0 = 0$ in R (rectangular domain)

EBC	NBC
$[\delta w]_0^a = 0$ for any y	$-D \left[\dfrac{\partial^3 w}{\partial x^3} + \dfrac{\partial^3 w}{\partial x \partial y^2} + (1-\nu) \dfrac{\partial^3 w}{\partial x \partial y^2} \right]_0^a = 0$
$[\delta w]_0^b = 0$ for any x	$-D \left[\dfrac{\partial^3 w}{\partial y^3} + \dfrac{\partial^3 w}{\partial x^2 \partial y} + (1-\nu) \dfrac{\partial^3 w}{\partial y \partial x^2} \right]_0^b = 0$
$\left[\dfrac{\partial \delta w}{\partial x} \right]_0^a = 0$ for any y	$D \left[\dfrac{\partial^2 w}{\partial x^2} + \nu \dfrac{\partial^2 w}{\partial y^2} \right]_0^a = 0$
$\left[\dfrac{\partial \delta w}{\partial y} \right]_0^b = 0$ for any x	$D \left[\dfrac{\partial^2 w}{\partial y^2} + \nu \dfrac{\partial^2 w}{\partial x^2} \right]_0^b = 0$

10. $\begin{aligned} &\delta u_i: \quad -\mu\left(u_{i,j} + u_{j,i}\right)_{,j} + P_{,i} - f_i = 0 \\ &\delta P: u_{i,i} = 0 \end{aligned} \Bigg\}$ in R

$$\left[\mu\left(u_{i,j} + u_{j,i}\right) - P\delta_{ij} \right] n_j + \hat{t}_i = 0 \quad \text{on} \quad S$$

11. $I_L(w, \psi_x, \psi_y, \lambda_1, \lambda_2) = I(w, \psi_x, \psi_y)$

$$+ \int_0^a \int_0^b \left[\lambda_1 \left(\frac{\partial w}{\partial x} + \psi_x \right) + \lambda_2 \left(\frac{\partial w}{\partial y} + \psi_y \right) \right] dx\, dy$$

Setting $\delta I_L = 0$ we get

$$\delta \psi_x: \; -D \left[\frac{\partial}{\partial x} \left(\frac{\partial \psi_x}{\partial x} + \nu \frac{\partial \psi_y}{\partial y} \right) + (1-\nu) \frac{\partial}{\partial y} \left(\frac{\partial \psi_x}{\partial y} + \frac{\partial \psi_y}{\partial x} \right) \right] + \lambda_1 = 0$$

$$\delta \psi_y: \; -D \left[(1-\nu) \frac{\partial}{\partial x} \left(\frac{\partial \psi_x}{\partial y} + \frac{\partial \psi_y}{\partial x} \right) + \frac{\partial}{\partial y} \left(\frac{\partial \psi_y}{\partial y} + \nu \frac{\partial \psi_x}{\partial x} \right) \right] + \lambda_2 = 0$$

$$\delta w: \; -\left(\frac{\partial \lambda_1}{\partial x} + \frac{\partial \lambda_2}{\partial y} \right) - q = 0$$

$$\delta \lambda_1: \; \frac{\partial w}{\partial x} + \psi_x = 0, \quad \delta \lambda_2: \; \frac{\partial w}{\partial y} + \psi_y = 0$$

One can show that λ_1 and λ_2 are the shear forces, Q_x and Q_y.

EBC	NBC
Specify $[\delta \psi_x]_0^a = 0$	$D \left[\left(\dfrac{\partial \psi_x}{\partial x} + \nu \dfrac{\partial \psi_y}{\partial y} \right) \right]_0^a = 0$
$[\delta \psi_x]_0^b = 0$	$D(1-\nu) \left[\dfrac{\partial \psi_x}{\partial y} + \dfrac{\partial \psi_y}{\partial x} \right]_0^b = 0$

and so on.

12. $I_p(w, \psi_x, \psi_y)$

$$= I(w, \psi_x, \psi_y) + \frac{1}{2}\int_0^a\int_0^b\left[\gamma_1\left(\frac{\partial w}{\partial x} + \psi_x\right)^2 + \gamma_2\left(\frac{\partial w}{\partial y} + \psi_y\right)^2\right]dx\,dy$$

setting $\delta I_p = 0$, we get the first three equations of Exercise 11 with λ_1 and λ_2 replaced by

$$\lambda_1 = \gamma_1\left(\frac{\partial w}{\partial x} + \psi_x\right), \qquad \lambda_2 = \gamma_2\left(\frac{\partial w}{\partial y} + \psi_y\right)$$

The last two relations are the constitutive relations between the shear forces, $\lambda_1 = Q_x$ and $\lambda_2 = Q_y$, and the shear strains $\gamma_{xz} = \gamma_1[(\partial w/\partial x) + \psi_x]$ and $\gamma_{yz} = \gamma_2[(\partial w/\partial y) + \psi_y]$. The penalty parameters are given by $\gamma_1 = G_{13}K_1^2$, $\gamma_2 = G_{23}K_2^2$.

13. See Reddy and Rasmussen (1982), pp. 329–330.

Exercises 2.3

1. $\delta W = \int_0^L\left(EI\frac{d^2w}{dx^2}\frac{d^2\delta w}{dx^2} + f\delta w\right)dx - F_1\delta w(0) - F_2\delta w(L)$
$$- M_1\frac{d\delta w}{dx}(0) - M_2\frac{d\delta w}{dx}(L).$$

2. See Reddy and Rasmussen (1982), pp. 321–323. Note that the specified essential boundary conditions of a cantilever beam are:
$$w(0) = \frac{dw}{dx}(0) = 0.$$

3. $\delta\Pi = \int_0^L EI\frac{d^2w}{dx^2}\frac{d^2\delta w}{dx^2}dx + \int_0^b f_0\delta w\,dx.$
$$\frac{d^2}{dx^2}\left(EI\frac{d^2w}{dx^2}\right) + f_0H(b-x) = 0.$$
$$\frac{d}{dx}\left(EI\frac{d^2w}{dx^2}\right)\Big|_{x=0} = 0, \qquad \left(EI\frac{d^2w}{dx^2}\right)\Big|_{x=0} = 0.$$

5. $\delta\Pi = \int_0^L\left(EI\frac{d^2w}{dx^2}\frac{d^2\delta w}{dx^2} + f_0\delta w\right)dx + kw(L)\delta w(L).$
$$\frac{d^2}{dx^2}\left(EI\frac{d^2w}{dx^2}\right) + f_0 = 0, \qquad 0 < x < L.$$
$$-\frac{d}{dx}\left(EI\frac{d^2w}{dx^2}\right)\Big|_{x=L} + kw(L) = 0.$$
$$\left(EI\frac{d^2w}{dx^2}\right)\Big|_{x=L} = 0.$$

6. $\dfrac{1}{r}\dfrac{d}{dr}\left[-D_{11}\dfrac{d}{dr}\left(r\dfrac{d^2w}{dr^2}\right)+\dfrac{D_{22}}{r}\dfrac{dw}{dr}\right]+q=0,\ a<r<b.$

7. $e_0=\dfrac{1}{EA}\displaystyle\sum_{i=1}^{5}N_in_iL_i.$

$$N_1=N_2=\dfrac{P}{2b}\sqrt{a^2+b^2},\qquad N_4=N_5=-\dfrac{a}{\sqrt{a^2+b^2}}N_1,\qquad N_3=P.$$

$$n_1=n_2=\dfrac{a^2+b^2}{2b},\qquad n_3=1,\qquad n_4=n_5=-\dfrac{a}{2b}.$$

$$L_1=L_2=\sqrt{a^2+b^2},\qquad L_3=b,\qquad L_4=L_5=a.$$

8. $v=\dfrac{P_0h}{EA}(1+2\sqrt{2}),\ u=\dfrac{P_0h}{AE}.$

10. $\dfrac{w_0}{2}-w_c=-\left(\dfrac{25P_0+f_0L}{768EI}\right)L^3$ (the dummy unit load is taken at the center).

11. $w_0=-\dfrac{5P_0a^3}{6EI}.$

12. $\theta=\dfrac{7\times10^5}{6EI}$ (EI in lb-ft^2).

Exercises 2.4

1. $\left.\begin{array}{l}-\operatorname{div}\mathbf{v}-f=0\\[4pt]\operatorname{grad}u-\dfrac{1}{k}\mathbf{v}=0\end{array}\right\}$ in V

$\mathbf{n}\cdot\mathbf{v}-g=0$ on S_2, $u-\hat{u}=0$ on S_1

2. $\left.\begin{array}{r}\dfrac{dw}{dx}+\psi=0\\[6pt]-\dfrac{d}{dx}\left(EI\dfrac{d\psi}{dx}\right)+\lambda=0\\[6pt]-\dfrac{d\lambda}{dx}+f=0\end{array}\right\}$ $0<x<L$

$EI\dfrac{d\psi}{dx}=0$ or $\delta\psi=0$ at $x=0,L$

$\lambda(L)-V_0=0,\qquad \lambda(0)=0$ or $\delta w(0)=0$

3. See Exercise 2.2.5.

4. See Exercise 2.2.11 (with $\psi_1 = \psi_x$ and $\psi_2 = \psi_y$).

6. The stationary functional is the same as that in Exercise 2.4.3, with $\mu = 1$
    and $P = -\lambda$.

8. $\delta \sigma_{ij}$: $\frac{1}{2}(u_{i,j} + u_{j,i} + u_{m,i} u_{m,j}) - e_{ij} = 0$

    δe_{ij}: $-\sigma_{ij} + \rho \dfrac{\partial \Psi}{\partial e_{ij}} = 0$

    δu_i: $[\sigma_{mj}(\delta_{mi} + u_{i,m})]_{,j} + f_i = 0$

    The boundary conditions are the same as those in Eq. (2.4.13).

Exercises 2.5

1. $L = \frac{1}{2}k_1 x_1^2 + \frac{1}{2}(k_2 + k_3)(x_1 - x_2)^2 + \frac{1}{2}k_4(x_3 - x_2)^2.$

2. $\delta L^* = \dfrac{\dot{F}}{k}\,\delta F + \dfrac{F}{\eta}\,\delta F.$

3. $\delta L = kx\,\delta x + \eta \dot{x}\,\delta x$

4. $L = \frac{1}{2}m_1(l_1\dot{\theta}_1)^2 + \frac{1}{2}m_2\left[(l_1\dot{\theta}_1)^2 + (l_2\dot{\theta}_2)^2 + 2l_1 l_2 \dot{\theta}_1 \dot{\theta}_2 \cos(\theta_1 - \theta_2)\right]$
    $\qquad + \left[m_1 g l_1 \cos\theta_1 + m_2 g(l_1\cos\theta_1 + l_2\cos\theta_2)\right]$
    $\qquad (m_1 + m_2)l_1\ddot{\theta}_1 + m_2 l_2\ddot{\theta}_2 + (m_1 + m_2)g\theta_1 = 0$
    $\qquad l_1\ddot{\theta}_1 + l_2\ddot{\theta}_2 + g\theta_2 = 0$

5. $L = \frac{1}{2}m_1[l^2\dot{\theta}^2 + \dot{x}^2 + 2\dot{x}l\dot{\theta}\sin\theta] + \frac{1}{2}m_2\dot{x}^2$
    $\qquad + m_1 g(x + l\cos\theta) + m_2 gx + \frac{1}{2}k(x + h)^2$

    where h is the elongation in the spring due to the masses,

    $$h = (m_1 + m_2)g/k.$$

6. $L = K - V$
    $K = \frac{1}{2}m_1(\dot{f}^2 - 2l_1\dot{\theta}_1\dot{f}\cos\theta_1 + l_1^2\dot{\theta}_1^2)$
    $\qquad + \frac{1}{2}m_2[\dot{f}^2 - 2\dot{f}(l_1\dot{\theta}_1\cos\theta_1 + l_2\dot{\theta}_2\cos\theta_2) + l_1^2\dot{\theta}_1^2 + l_2^2\dot{\theta}_2^2$
    $\qquad + 2l_1 l_2 \dot{\theta}_1 \dot{\theta}_2 \cos(\theta_1 - \theta_2)]$
    $V = -m_1 g l_1 \cos\theta_1 - m_2 g(l_1\cos\theta_1 + l_2\cos\theta_2)$

7. $L = K - V,\ K = \frac{1}{2}m_1\dot{x}_1^2 + \frac{1}{2}m_2\dot{x}_2^2,\ V = -m_1 gx_1 - m_2 gx_2$

    $$x_1 + x_2 = \text{constant},\ l$$

or

$$L = \tfrac{1}{2}(m_1 + m_2)\dot{x}_1^2 + (m_1 - m_2)gx_1 + m_2gl$$

$$\ddot{x}_1 + \left(\frac{m_2 - m_1}{m_1 + m_2}\right)g = 0$$

8. $K = \tfrac{1}{2}(m_1\dot{x}_1^2 + m_2\dot{x}_2^2 + m_3\dot{x}_3^2).$

$V = -m_1gx_1 - m_2gx_2 - m_3gx_3.$

$x_1 + x_2 = l, \qquad \dot{x}_1 - \dot{x}_3 = v_0.$

$$\ddot{x}_1 = -\ddot{x}_2 = \ddot{x}_3 = \left(\frac{m_1 - m_2 + m_3}{m_1 + m_2 + m_3}\right)g.$$

9. $\ddot{x}_1 = g/23.$

10. $$\frac{\partial^2}{\partial x^2}\left(EI\frac{\partial^2 w}{\partial x^2}\right) + f_0\sin\frac{\pi x}{L} = \rho\frac{\partial^2 w}{\partial t^2}.$$

11. $T_0\nabla^2 u = \rho\dfrac{\partial^2 u}{\partial t^2} + f.$

14. $\sigma_{ij,j} + f_i = \rho\dfrac{\partial^2 u_i}{\partial t^2}$

$e_{ij} = \tfrac{1}{2}(u_{i,j} + u_{j,i})$

$\sigma_{ij} = \lambda\Theta\delta_{ij} + 2\mu e_{ij} - (3\lambda + 2\mu)\alpha(T - T_0)\delta_{ij}$

$k\nabla^2 T = \rho c\dfrac{\partial T}{\partial t} + (3\lambda + 2\mu)\alpha T_0\dfrac{\partial D}{\partial t}$

where $\Theta = e_{ii} = u_{i,i}$, T_0 = reference temperature.

15. $L = K - H$, where H is given by Eq. (2.4.7) and K is given by

$$K = \tfrac{1}{2}\int_V \rho\dot{u}_i\dot{u}_i\, dV.$$

Exercises 2.6

1. $P = \dfrac{AEv_0}{6}(9 + 4\sqrt{3})L_3$ when $v_0^2 \approx 0.$

2. $P = \dfrac{AE}{6}\sqrt{\dfrac{v_0}{L_3}}\left(\sqrt{\dfrac{3\sqrt{3}}{2}} + \sqrt{\dfrac{2}{\sqrt{3}}}\right)$ when $v_0^2 \approx 0.$

3. $e_V = \dfrac{1}{EI}\left(P_H\dfrac{ba^2}{2} + P_V\dfrac{a^3}{3}\right) + \dfrac{P_V b}{AE}$.

$e_H = \dfrac{1}{EI}\left[P_V\dfrac{ba^2}{2} + \dfrac{P_H b^2}{3}(3a + b)\right] + \dfrac{P_H a}{AE}$

4. Set $P_H = \frac{4}{5}P_0$, $P_V = -\frac{3}{5}P_0$, $a = 8'$, and $b = 3'$ in Exercise 2.6.3.

5. $w = \dfrac{Pb^2}{3EI}(a + b)$.

6. $w_c = \dfrac{L^3}{8}\sqrt{\dfrac{P}{EA}}$, $A = $ area of cross section.

7. $M = P_0(2h - x)$, $\dfrac{\partial M}{\partial P_0} = 2h - x$

$v = \dfrac{47}{24}\dfrac{P_0 h^3}{EI_1}$

8. $w(0) = \dfrac{-f_0 L^4}{8EI\left(1 + kL^3/3EI\right)}$

10. Let A and B denote the left end and midspan, respectively. The deflection at the center of the beam resulting from the moment is

$$w_B = \frac{M_0 L^2}{16EI}$$

The rotation of the end A caused by the point load is

$$\theta_A = \frac{PL^2}{16EI}$$

The work done, W_{BA}, by force P in going through displacement w_B due to the moment M_0 is

$$W_{BA} = Pw_B = \frac{PM_0 L^2}{16EI}$$

The work done, W_{AB}, by moment M_0 in rotating through slope θ_A owing to the load P is

$$W_{BA} = M_0\theta_A = \frac{M_0 PL^2}{16EI}$$

11. $M(x) = \begin{cases} (Pb - M_0)\dfrac{x}{a} + M_0, & 0 < x < a \\ (Pb - M_0)\dfrac{x}{a} + M_0 + [M_0 - P(a + b)]\dfrac{x}{a}, \\ \qquad\qquad\qquad\qquad a < x < b + a \end{cases}$

$\dfrac{\partial U^*}{\partial M_0}\bigg|_{M_0=0} = \theta_0$

12. $R = wL + S$, $M = -SL + \dfrac{wL^2}{2}$, $S = kw(0)$.

13. $w_A = 0.656$ in.

14. $w\left(\dfrac{L}{2}\right) = -\left(\dfrac{5PL^3}{48EI} + \dfrac{17f_0 L^4}{384EI}\right)$.

15. $u = \dfrac{P^2 h}{A^2 K^2}$, $v = \dfrac{5P^2 h}{A^2 K^2}$.

CHAPTER 3

Exercises 3.1

1. Yes.

2. Yes.

3. Yes.

4. No.

5. No.

6. Both.

7. $3L^3\sqrt{L/5}$.

8. $\sqrt{(9L^7 + 24L^5)/5}$.

10. $\|u - v\|_0 = \sqrt{123/105}$ (for $0 < x < 1$).

11. Yes.

12. $a = 121/160$, $b = -558/160$.

13. Linear operator.

14. Not a linear operator (because of an additive constant).

15. Linear functional.

16. Bilinear functional.

17. Not bilinear; linear only in v.

19. Linearly independent.

20. Linearly dependent.

21. Linearly dependent.

Exercises 3.2

1. $\phi_0 = h\left(\dfrac{x}{L}\right)$, $\phi_1 = x(x - L)$, $\phi_2 = x^2(x - L)$.

2. $\phi_0 = 0$, $\phi_1 = x^2$, $\phi_2 = x^3$; $\phi_1 = 1 - \cos\dfrac{\pi x}{L}$, $\phi_2 = 1 - \cos\dfrac{3\pi x}{L}$.

3. $\phi_0 = 0$, $\phi_1 = x(x - L)$, $\phi_2 = x\left(\tfrac{3}{4}L^2 - x^2\right)$; $\phi_1 = \sin\dfrac{\pi x}{L}$, $\phi_2 = \sin\dfrac{3\pi x}{L}$.

4. $\phi_0 = 0$, $\phi_1 = x^2(x - L)$, $\phi_2 = x^3(x - L)$; $\phi_1 = 1 - \cos\dfrac{2\pi x}{L}$,

   $\phi_2 = 1 - \cos\dfrac{4\pi x}{L}$.

5. $\phi_0 = 0$, $\phi_1 = (L - x)^2$, $\phi_2 = (L - x)^3$; $\phi_1 = 1 + \cos\dfrac{\pi x}{L}$,

   $\phi_2 = 1 + \cos\dfrac{3\pi x}{L}$.

6. $\phi_0 = 0$, $\phi_1 = x(a - x)y(a - y)$, $\phi_2 = x^2 y^2(a - x)(a - y)$;

   $\phi_1 = \sin\dfrac{\pi x}{a}\sin\dfrac{\pi y}{a}$, $\phi_2 = \sin\dfrac{3\pi x}{a}\sin\dfrac{3\pi y}{a}$,

   where a is the side of the square membrane; the origin of the coordinate system is taken at the lower left corner of the square region.

7. $\phi_0 = 0$, $\phi_1 = (a^2 - 4x^2)(a^2 - 4y^2)$, $\phi_2 = (a^3 - 8x^3)(a^3 - 8y^3)$;

   $\phi_1 = \cos\dfrac{\pi x}{2a}\cos\dfrac{\pi y}{2a}$, $\phi_2 = \cos\dfrac{3\pi x}{2a}\cos\dfrac{3\pi y}{2a}$

   (the quadrant bounded by $0 \le x \le \dfrac{a}{2}$ and $0 \le y \le \dfrac{a}{2}$ is used).

8. $c_1 = -(f_0/2T)$, $c_2 = 0$, where ϕ_0, ϕ_1, and ϕ_2 are as given in Exercise 1, and T is the tension in the cable $\{(-d/dx)[T(du/dx)] = f_0\}$. The Ritz solution coincides with the exact solution.

9. $c_1 = -5f_0 L^2/24EI$, $c_2 = f_0 L/12EI$ ($\phi_1 = x^2$, $\phi_2 = x^3$). The exact solution is

$$w = -\frac{f_0}{24EI}(6x^2 L^2 - 4L^3 + x^4)$$

see Reddy (1984), pp. 41–43 and Reddy and Rasmussen (1982), pp. 414–419.

10. $w = c_1 x(x - L) + c_2 x(\frac{3}{4}L^2 - x^2)$.

$c_1 = f_0 L^2/96EI$, $c_2 = -f_0 L/24EI$.

11. $w = c_1 x^2(x - L) + c_2 x^3(x - L)$.

$c_1 = -\dfrac{7}{64}\dfrac{P}{EI}$, $c_2 = \dfrac{5}{64}\dfrac{P}{EIL}$.

12. $w = c_1(x - L)^2 + c_2(x - L)^3$.

$c_1 = \dfrac{(30EI + kL^3)f_0 L^2}{48(3EI + kL^3)}$, $c_2 = \dfrac{f_0 L}{12EI}$.

13. $c_1 = \dfrac{5}{4}\dfrac{f_0}{a^2}$, for a one-parameter algebraic polynomial.

$c_1 = \dfrac{4a^2}{\pi^2}f_0$, $c_2 = \dfrac{8a^2 f_0}{81\pi^4}$, for two-parameter trigonometric functions.

14. For the choice of trigonometric functions given in Exercise 7, the matrix equations for the Ritz parameters are given by Eq. (3.365) of Reddy and Rasmussen, 1982.

15. Use the functional

$$R(w, M) = \int_0^L \left(-\frac{dw}{dx}\frac{dM}{dx} - \frac{M^2}{2EI} + f_0 w\right) dx$$

with the Ritz approximation, $w = c_1 x(x - L)$ and $M = c_2 x(x - L)$;

$$c_1 = f_0 L^2/20EI \quad \text{and} \quad c_2 = f_0/2EI.$$

16. The eigenvalue problem is given by ($w = c_1 x^2 + c_2 x^3$)

$$
\begin{bmatrix} \dfrac{4}{b}\ln\!\left(\dfrac{a+bL}{a}\right) & 12\left[\dfrac{L}{b} - \dfrac{a}{b^2}\ln\!\left(\dfrac{a+bL}{a}\right)\right] \\[2mm] \text{symmetric} & \dfrac{36}{b^3}\left[bL(bL-a) + a^2\ln\!\left(\dfrac{a+bL}{a}\right)\right] \end{bmatrix}\begin{Bmatrix} c_1 \\ c_2 \end{Bmatrix}
$$

$$
= \omega^2 \begin{bmatrix} \dfrac{L^5}{5} & \dfrac{L^6}{6} \\[2mm] \dfrac{L^6}{6} & \dfrac{L^7}{7} \end{bmatrix}\begin{Bmatrix} c_1 \\ c_2 \end{Bmatrix}
$$

17. (b) $c_1 = -\dfrac{4f_0 L^4}{\pi(EI\pi^4 + kL^4)}$, $c_2 = -\dfrac{4f_0 L^4}{3\pi(81EI\pi^4 + kL^4)}$.

18. (a) $\sqrt{\dfrac{L^9}{362880}\left(\dfrac{f_0}{EI}\right)}$ (b) $\sqrt{\dfrac{L^7}{210}\left(\dfrac{f_0}{EI}\right)}$.

19. Use the variational statement

$$
0 = \int_0^L \left(EI\dfrac{d^2 w}{dx^2}\dfrac{d^2 \delta w}{dx^2} - P\dfrac{dw}{dx}\dfrac{d\delta w}{dx}\right) dx,
$$

and $\phi_1 = \sin(\pi x/L)$ and $\phi_2 = \sin(3\pi x/L)$ to obtain the first two critical buckling loads,

$$
P_1 = EI\left(\dfrac{\pi}{L}\right)^2, \qquad P_2 = EI\left(\dfrac{3\pi}{L}\right)^2.
$$

20. $[B]\{c\} = \{f\}$.

$$
b_{ij} = EI\dfrac{ij(i+1)(j+1)}{(i+j-1)}(L)^{i+j-1}.
$$

$$
f_i = -\dfrac{f_0(L)^{i+2}}{i+2}.
$$

21. Variational form:

$$
0 = \int_0^1 \left(u\dfrac{d\delta u}{dx}\dfrac{du}{dx} + \delta u\right) dx
$$

For a one-parameter Ritz solution, take $\phi_0 = \sqrt{2}$ and $\phi_1 = 1 - x$.

22. Variational form:

$$0 = \int_0^L \left(EI \frac{d^2 \delta w}{dx^2} \frac{d^2 w}{dx^2} - P \frac{d \delta w}{dx} \frac{dw}{dx} \right) dx$$

For a one-parameter Ritz solution, take $w = c_1 \sin(\pi x / L)$. See Exercise 19 above.

23. Variational form:

$$0 = \int_0^1 \int_0^1 \left(\frac{\partial \delta u}{\partial x} \frac{\partial u}{\partial x} + \frac{\partial \delta u}{\partial y} \frac{\partial u}{\partial y} - \delta u \right) dx\, dy$$

For a one-parameter Ritz solution, assume [see Exercises 7 and 14 above]

$$u = c_1 \cos \frac{\pi x}{2} \cos \frac{\pi y}{2} \quad \text{or} \quad u = c_1 (1 - x^2)(1 - y^2)$$

Exercises 3.3

1. $u = c_1 x(2 - x) + c_2 x(3 - x^2);$ $\psi_1 = 1,$ $\psi_2 = x.$
   $c_1 = 3 - \ln 2,$ $c_2 = -2 + \ln 2.$
   $u_{\text{exact}} = x(1 + \ln 2) - (1 + x)\ln(1 + x).$

3. $w = c_1(a^2 - r^2)^2 + c_2 r^2(a^2 - r^2)^2,$ $\psi_1 = 1,$ $\psi_2 = r.$
   $c_1 = -\dfrac{f_0}{64D},$ $c_2 = 0.$
   The Petrov–Galerkin solution coincides with the exact solution.

4. $w = c_1 x^2,$ $M = M_0 + c_2(L - x)^2;$ $\psi_1 = \psi_2 = 1.$
   $c_1 = \dfrac{1}{2EI} \left(M_0 - \dfrac{f_0 L^2}{6} \right),$ $c_2 = -\dfrac{f_0}{2}.$

   For $\psi_1 = 1$ and $\psi_2 = x,$ we get $c_1 = (1/24EI)(12 M_0 - f_0 L^2),$
   $c_2 = -(f_0/2).$ The exact solution is

   $$w_{\text{exact}} = \left(\frac{2M_0 - f_0 L^2}{4EI} \right) x^2 + \frac{f_0 L}{6EI} x^3 - \frac{f_0}{24EI} x^4$$

   $$M_{\text{exact}} = \frac{2M_0 - f_0 L^2}{2} + f_0 L x - \frac{f_0}{2} x^2$$

5. $\phi_0 = (1 - y)\sin \pi x$, $\phi_1 = \sin \pi x \sin \pi y$, $\phi_2 = \sin 2\pi x \sin 2\pi y$.

$\psi_1 = 1$, $\psi_2 = xy$.

$c_1 = \dfrac{\pi}{8}$, $c_2 = -\dfrac{5\pi}{24}$.

$u_{exact} = \dfrac{\sin \pi x \sinh[\pi(1 - y)]}{\sinh \pi}$.

6. $\phi_0 = 0$, $\phi_1 = x$, $\phi_2 = x^2$; $\psi_1 = x$, $\psi_2 = x^2$.

8. See Exercise 3.

9. $w = c_1 x + c_2 x^2$; $M = M_0 + b_1(x - L) + b_2(x - L)^2$. Weight functions for the first equation: $\psi_1 = 1$, $\psi_2 = x$. Weight functions for the second equation: $\psi_1 = 1$, $\psi_2 = x$.

10. See Exercise 5.

11. (i) $u = \sqrt{2} + c_1(1 - x^2)$ and $\psi_1 = 1$ gives $c_1 = -\dfrac{1}{2\sqrt{2}} = -0.35355$.

   (ii) $u = \sqrt{2} + c_1(1 - x^2)$ and $\psi_1 = x$ gives $(c_1)_1 = \sqrt{2} - \sqrt{3}$ and $(c_1)_2 = \sqrt{2} + \sqrt{3}$.

   (iii) $u = \sqrt{2} + c_1(1 - x^2)$, $\psi_1 = \cos(\pi x/2)$.

   $(c_1)_1 = -0.40317$, $(c_1)_2 = -2.8727$.

   In (ii) and (iii) constant $(c_1)_1$ minimizes the residual.

12. (i) $u = \sqrt{2} + c_1(1 - x)$ and $\psi_1 = 1$ gives $(c_1)^2 = 1$.

   (ii) $u = \sqrt{2} + c_1(1 - x)$ and $\psi_1 = x$ gives $(c_1)^2 = 1$.

   Constant $(c_1)_1$ minimizes the residual.

13. $u = c_1(3x - 2x^2) + c_2(x + 2x^2 - 2x^3)$ gives the characteristic polynomial $5\lambda^2 - 148\lambda + 525 = 0$ or $\lambda_1 = 4.121$ and $\lambda_2 = 25.48$. The exact solution is given by the roots of the transcendental equation $\lambda + \tan \lambda = 0$. The first two roots of this equation are

$$(\lambda_1)_{exact} = 4.116, \qquad (\lambda_2)_{exact} = 24.14.$$

14. $c_{ij} = \begin{cases} \dfrac{16 f_0}{\pi^4} \left\{ ij \left[(i)^2 + (j)^2 \right] \right\}^{-1}, & i, j \ \ \text{odd} \\ 0, & i, j \ \ \text{even} \end{cases}$

15. $u = \sqrt{2} x^2 + c_1(1 - x^2)$ gives the quadratic equation $4c_1^2 + 2\sqrt{2} c_1 - 7 = 0$. The root $c_1 = \sqrt{2}(-1 + \sqrt{15})/2$ minimizes the residual.

16. $w_1 = c_1 x^2 (x - L)^2$ gives $c_1 = -f_0/24EI$. Select $w_2 = c_1 x^2 (x - L)^2 + c_2 x^3 (x - L)^2$ and obtain the solution.

18. $u_2 = c_1(3x - 2x^2) + c_2(x + 2x^2 - 2x^3)$

$$E = Au_2 - \lambda u_2, \qquad A = -\frac{d^2}{dx^2}$$

$$\int_0^1 EA(\phi_i)\, dx = 0, \qquad i = 1, 2.$$

$\lambda_1 = 4.212$ and $\lambda_2 = 34.188$.

19. $u = \sqrt{2}\, x^2 + c_1(1 - x^2)$ gives the cubic polynomial in c_1: $16c_1^3 - 48\sqrt{2}\, c_1^2 + 116c_1 - 47\sqrt{2} = 0$. The root $c_1 = 1.0885$ minimizes the residual.

21. $u_3 = c_1 x^2 + c_2 x^3 + c_3 x^4$ with collocation points at $x = L/4$, $L/2$ and $3L/4$ gives

$$c_1 = f_0 L^2/4a_0,$$

$$c_2 = -(f_0 L/36a_0), \text{ and } c_3 = 0.$$

22. For collocation points at $x = 1/3$ and $2/3$, the eigenvalue problem is given by

$$\begin{bmatrix} 4 & 0 \\ 4 & 4 \end{bmatrix} \begin{Bmatrix} c_1 \\ c_2 \end{Bmatrix} - \frac{\lambda}{27} \begin{bmatrix} 21 & 13 \\ 30 & 26 \end{bmatrix} \begin{Bmatrix} c_1 \\ c_2 \end{Bmatrix} = \begin{Bmatrix} 0 \\ 0 \end{Bmatrix}$$

The first two eigenvalues are $\lambda_1 = 3.785$ and $\lambda_2 = 19.7533$.

23. $u = \sqrt{2} + c_1(1 - x^2)$ with collocation at $x = \frac{1}{2}$ gives $(c_1)_1 = -2\sqrt{2} + \sqrt{6}$ and $(c_1)_2 = -2\sqrt{2} - \sqrt{6}$. The value of $(c_1)_1$ minimizes the residual.

24. $u = c_1(2x - x^2) + c_2(3x - x^3)$; subintervals: $(0, \frac{1}{2})$ and $(\frac{1}{2}, 1)$; $c_1 = -0.4711$, $c_2 = 0.2195$.

25. $c_{11} = f_0/2\pi^2$.

26. $u(x, y) = \dfrac{16a^2 f_0}{\pi^3} \left\{ \left[1 - \dfrac{\cosh(\pi y/2a)}{\cosh(\pi/2a)} \right] \cos\dfrac{\pi x}{2a} \right.$

$\left. + \dfrac{1}{27} \left[1 - \dfrac{\cosh(3\pi y/2a)}{\cosh(3\pi/2a)} \right] \cos\dfrac{3\pi x}{2a} \right\}$

27. $\dfrac{d^2c_1}{dt^2} + \dfrac{(2\pi)^4}{3}c_1 = 0$, $c_1(0) = 0.1107$, $\dfrac{dc_1}{dt}(0) = 0$

The solution is given by $w(x,t) = 0.1107\,(1 - \cos 2\pi x)\cos \alpha t$, $\alpha = [(2\pi)^2/\sqrt{3}]$.

28. $\dfrac{d^2c_1}{dy^2} - 10c_1 = 0$, $c_1(0) = 1$, $c_1(\infty) = 0$

The solution is, $c_1(y) = e^{-\sqrt{10}\,y}$.

29. $c_1 = (53/180)fa^2$ and $c_2 = -(7/144)\dfrac{f}{a^2}$ [see Sokolnikoff (1950), p. 430]

Exercises 3.4

1. $K_{ij}^e = \displaystyle\int_{x_e}^{x_{e+1}}(1+x)\dfrac{d\psi_i}{dx}\dfrac{d\psi_j}{dx}\,dx$, $F_i^e = \displaystyle\int_{x_e}^{x_{e+1}}f\psi_i\,dx + P_i$

2. $[K^e] = \dfrac{1}{h_e}\left(\begin{bmatrix} 1 & -1 \\ -1 & 1 \end{bmatrix} + \dfrac{(x_e + x_{e+1})}{2}\begin{bmatrix} 1 & -1 \\ -1 & 1 \end{bmatrix}\right)$

Assembled equations:

$$\begin{bmatrix} 2.5 & -2.5 & 0 \\ -2.5 & 6.0 & -3.5 \\ 0 & -3.5 & 3.5 \end{bmatrix}\begin{Bmatrix} U_1 \\ U_2 \\ U_3 \end{Bmatrix} = \begin{Bmatrix} P_1^{(1)} \\ P_2^{(1)} + P_1^{(2)} \\ P_2^{(2)} \end{Bmatrix}, \qquad P_2^{(1)} + P_1^{(2)} = 0$$

Boundary conditions:

$$U_1 = 1, \qquad P_2^{(2)} + U_3 = 0$$

Condensed equations:

$$\begin{bmatrix} 6.0 & -3.5 \\ -3.5 & 3.5 \end{bmatrix}\begin{Bmatrix} U_2 \\ U_3 \end{Bmatrix} = \begin{Bmatrix} 2.5U_1 \\ -U_3 \end{Bmatrix}$$

Solution:

$$U_2 = \dfrac{2.5 \times 4.5}{6 \times 4.5 - 3.5 \times 3.5} = 0.7627$$

$$U_3 = \dfrac{2.5 \times 3.5}{6 \times 4.5 - 3.5 \times 3.5} = 0.5932$$

$$u_{\text{exact}} = 1 - \dfrac{\ln(1+x)}{1 + \ln 2}$$

3. $\psi_1 = -\left(\dfrac{3x}{h} - 1\right)\left(\dfrac{3x}{2h} - 1\right)\left(\dfrac{x}{h} - 1\right)$

$\psi_2 = 9\dfrac{x}{h}\left(\dfrac{3x}{2h} - 1\right)\left(\dfrac{x}{h} - 1\right)$

$\psi_3 = -\dfrac{9}{2}\dfrac{x}{h}\left(\dfrac{3x}{h} - 1\right)\left(\dfrac{x}{h} - 1\right)$

$\psi_4 = \dfrac{x}{h}\left(\dfrac{x}{h} - 1\right)\left(\dfrac{3x}{2h} - 1\right)$

4. The governing equation for the one-dimensional model is given by Eq. (3.4.1) with $c = f = 0$ and $a = EA$, where

$$A = 1.2 + \frac{x}{15}$$

The coefficient matrix can be computed as in Exercise 1. The assembled equations for the two (equal) element case (with the boundary conditions imposed) are

$$\frac{E}{10}\begin{bmatrix} \tfrac{7}{3} & -\tfrac{7}{3} & 0 \\ -\tfrac{7}{3} & -\tfrac{7}{3} + 3 & -3 \\ 0 & -3 & 3 \end{bmatrix}\begin{Bmatrix} U_1 \\ U_2 \\ U_3 \end{Bmatrix} = \begin{Bmatrix} P_1^{(1)} \\ 0 \\ 32{,}000 \end{Bmatrix}, \qquad U_1 = 0$$

The displacements are:

$$U_2 = 0.6625 \times 10^{-2} \text{ cm}, \qquad U_3 = 1.1778 \times 10^{-2} \text{ cm}.$$

The stresses are:

$$\sigma^{(1)} = 22{,}856.94 \text{ N/cm}^2, \qquad \sigma^{(2)} = 17{,}777.78 \text{ N/cm}^2.$$

The solution for the four-element case is:

$$U_2 = 0.003567 \text{ cm}, \qquad U_3 = 0.006659 \text{ cm},$$

$$U_4 = 0.009387 \text{ cm}, \qquad U_5 = 0.011828 \text{ cm}$$

$$\sigma^{(1)} = 24{,}614.37 \text{ N/cm}^2, \ \sigma^{(2)} = 21{,}333.42 \text{ N/cm}^2,$$

$$\sigma^{(3)} = 18{,}823.55 \text{ N/cm}^2, \ \sigma^{(4)} = 16{,}842.14 \text{ N/cm}^2.$$

5. Two-elements: $A^{(1)} = 1.4 \text{ cm}^2$, $A^{(2)} = 1.8 \text{ cm}^2$.

The answers are the same as in Exercise 4 (because the taper is small).

Four-elements: $A^{(1)} = 1.3 \text{ cm}^2$, $A^{(2)} = 1.5 \text{ cm}^2$, $A^{(3)} = 1.7 \text{ cm}^2$, $A^{(4)} = 1.9$ `
cm^2.

The answers are the same as in Exercise 4.

6. $U_2 = -1.333(10^{-4})$ in., $U_3 = -6.6667(10^{-4})$ in., $U_4 = -12(10^{-4})$ in.
    $\sigma_s = 250$ psi(compression), $\sigma_a = 333.33$ psi(compression),
    $\sigma_b = 500$ psi(compression).

7. $U_2 = 0.00222$ in., $U_3 = 0.0111$ in.
    $\sigma_s = 4166.67$ psi, $\sigma_a = 5{,}555.56$ psi, $\sigma_b = 8{,}333.33$ psi.

8. $U_2 = 0.0204742$ in., $\sigma^{(1)} = 10{,}237$ psi, $\sigma^{(2)} = -38{,}389$ psi.

9. Element equations: $w = \displaystyle\sum_{i=1}^{n} w_i \psi_i,$ $M = \displaystyle\sum_{i=1}^{m} M_i \phi_i$

$$[A^e]\{w\} + [B^e]\{M\} = \{P^e\}$$

$$[A^e]^T\{M\} = \{F^e + G^e\}$$

where

$$A_{ij}^e = \int_{x_e}^{x_{e+1}} \frac{d\phi_i}{dx}\frac{d\psi_j}{dx}\,dx, \qquad B_{ij}^e = \int_{x_e}^{x_{e+1}} \frac{1}{EI}\phi_i\phi_j\,dx$$

$$F_i^e = \int_{x_e}^{x_{e+1}} f_0\psi_i\,dx, \qquad P_1^e = -\left(\frac{dw}{dx}\right)_{x_e}, \qquad P_2^e = \left(\frac{dw}{dx}\right)_{x_{e+1}}$$

$$G_1^e = -\left(\frac{dM}{dx}\right)_{x_e}, \qquad G_2^e = \left(\frac{dM}{dx}\right)_{x_{e+1}}$$

Also, see Reddy (1984), pp. 418–421.

10. $[A^e] = \dfrac{1}{h_e}\begin{bmatrix} 1 & -1 \\ -1 & 1 \end{bmatrix},$ $[B^e] = \dfrac{h_e}{6EI}\begin{bmatrix} 2 & 1 \\ 1 & 2 \end{bmatrix}$

$\{F^e\} = \dfrac{f_0 h}{2}\begin{Bmatrix} 1 \\ 1 \end{Bmatrix}$

11. $[A^e]\{w\} - [B^e]\{M\} = \{0\},$ $[C^e]\{M\} + \{F^e\} = \{0\}$

$$A_{ij}^e = \int_{x_e}^{x_{e+1}} \frac{d^2\psi_i}{dx^2}\frac{d^2\psi_j}{dx^2}\,dx, \qquad B_{ij}^e = \int_{x_e}^{x_{e+1}} \frac{1}{EI}\frac{d^2\psi_i}{dx^2}\phi_j\,dx$$

$$C_{ij}^e = \int_{x_e}^{x_{e+1}} \frac{d^2\phi_i}{dx^2}\frac{d^2\phi_j}{dx^2}\,dx, \qquad F_i^e = \int_{x_e}^{x_{e+1}} f_0\frac{d^2\phi_i}{dx^2}\,dx$$

Note that ψ_i and ϕ_i cannot be the linear interpolation functions.

12. Full beam model (3 elements):

$$U_2 = 0.00312, \ U_3 = -0.01077 \text{ in.}, \ U_4 = 0.001834$$

$$U_5 = U_3, \ U_6 = U_4, \ U_7 = U_2$$

Half beam model (2 elements):

$$U_2 = 0.00312, \ U_3 = -0.01077 \text{ in.},$$

$$U_4 = 0.001834, \ U_5 = -0.01673 \text{ in.}$$

13. $U_3 = 0.3030$ cm, $U_4 = -0.03795$, $U_6 = 0.15235$.

14. $U_2 = 0.002075$, $U_4 = -0.0021$, $U_5 = 0.1028$ in., $U_6 = -0.00218$.

15. $U_3 = -0.0008$ in., $U_4 = -0.00204$.

16. $U_2 = 0.0045$, $U_3 = -0.00675$ m, $U_4 = -0.001125$, $U_6 \approx 0.0$
    $U_7 = -0.00675$ m, $U_8 = 0.001125$, $U_{10} = -0.0045$ (4 elements are used).

17. $([K^e] - P[G^e])\{w\} = \{F^e\}$

$$K_{ij}^e = \int_{x_e}^{x_{e+1}} b \frac{d^2\phi_i}{dx^2} \frac{d^2\phi_j}{dx^2} \, dx, \qquad G_{ij}^e = \int_{x_e}^{x_{e+1}} \frac{d\phi_i}{dx} \frac{d\phi_j}{dx} \, dx$$

$$F_i^e = \int_{x_e}^{x_{e+1}} f\phi_i \, dx + P_i^e$$

18. $[G^e] = \dfrac{1}{30h} \begin{vmatrix} 36 & -3h & -36 & -3h \\ -3h & 4h^2 & 3h & -h^2 \\ -36 & 3h & 36 & 3h \\ -3h & -h^2 & 3h & 4h^2 \end{vmatrix}$

19. After imposing the boundary conditions (of a half beam), the eigenvalue problem becomes

$$\left(\begin{bmatrix} 2h^2 & 3h \\ 3h & 6 \end{bmatrix} - \lambda \begin{bmatrix} 4h^2 & 3h \\ 3h & 36 \end{bmatrix} \right) \begin{Bmatrix} \theta_1 \\ w_2 \end{Bmatrix} = \begin{Bmatrix} 0 \\ 0 \end{Bmatrix}$$

where $\lambda = Ph^2/60EI$. The minimum critical buckling load is

$$P_{cr} = 9.944 \left(\frac{EI}{L^2} \right)$$

which compares very well with the exact value $(P_{cr})_{\text{exact}} = \pi^2(EI/L^2)$.

20. $K_{ij}^e = \int_{r_e}^{r_{e+1}} \left[D_{11} \dfrac{d^2\phi_i}{dr^2} \dfrac{d^2\phi_j}{dr^2} + D_{22} \dfrac{1}{r^2} \dfrac{d\phi_i}{dr} \dfrac{d\phi_j}{dr} \right] r \, dr$

$F_i^e = \int_{r_e}^{r_{e+1}} rf\phi_i \, dr$

21. Program FEM2D from Reddy (1984) is used to solve the problem. Typical results are listed below (for $2G\theta = 2$):

$$\Phi(0,0) = 0.0724, \qquad \Phi(0.0, -0.125) = 0.06415$$

$$\Phi(0.0, -0.255) = 0.039961, \qquad \Phi(0.225, -0.125) = 0.0539$$

22. The results from FEM2D (see Reddy, 1984) are listed below (for $2G\theta = 2$):

   (a) $\Phi(3,0) = 4.9756$, $\Phi(3,1) = 4.9026$, $\Phi(3,2) = 4.2448$, $\Phi(3,3) = 2.8112$.

      $\partial\Phi/\partial x$ in the element with nodes at $(3,0)$ and $(4,0)$ is -4.9391.

   (b) $\Phi(3,0) = 4.9111$, $\Phi(3,1) = 4.8222$, $\Phi(3,2) = 4.3778$, $\Phi(3,3) = 2.6889$.

      $\partial\Phi/\partial x$ in the element with nodes at $(3,0)$, $(3,1)$, and $(4,1)$ is -4.9111.

23. *Variational formulation*:

$$0 = \int_{R^e} \left(K_1 \dfrac{\partial v}{\partial x} \dfrac{\partial u}{\partial x} + K_2 \dfrac{\partial v}{\partial y} \dfrac{\partial u}{\partial y} \right) dx\,dy + \oint_{S^e} v \left[h(u - u_0) + q \right] dS$$

$$+ \int_{R^e} vf\,dx\,dy$$

Finite element model:

$$[K^e]\{u\} = \{F^e\}$$

$$K_{ij}^e = \int_{R^e} \left(K_1 \dfrac{\partial\psi_i}{\partial x} \dfrac{\partial\psi_j}{\partial x} + K_2 \dfrac{\partial\psi_i}{\partial y} \dfrac{\partial\psi_j}{\partial y} \right) dx\,dy + \oint_{S^e} h\psi_i\psi_j \, dS$$

$$F_i^e = \int_{R^e} \psi_i f\,dx\,dy + \oint_{S^e} \psi_i (hu_0 - q)\,dS$$

24. See Reddy (1984), pp. 232–235.

26. Triangular elements (geometry is exactly represented):

$$u(12,0) = 0.011585 \text{ cm}, \qquad u(12,1) = 0.011627 \text{ cm},$$

$$v(12,1) = -0.00029 \text{ cm}.$$

27. Rectangular elements (with constant width of 1.6 cm): $u(12,0) = 0.009275$ cm, $u(12,0.8) = 0.009275$ cm, $v(12,0.8) = -0.0001855$ cm.

28. $u_1 = 0.86387 \times 10^{-5}$ in., $u_2 = 0.47413 \times 10^{-5}$ in.,

$u_3 = 0.35437 \times 10^{-5}$ in., $u_4 = 0.30489 \times 10^{-5}$ in.,

$\sigma_x^{(1)} = -36.434$ psi, $\sigma_y^{(1)} = 52.648$ psi.

29. $u = \sum\limits_{i=1}^{3} u_i \psi_i; \ \psi_i = \dfrac{1}{2A}(\alpha_i + \beta_i x + \gamma_i y)$

$A = 24, \qquad \alpha_1 = 80, \qquad \alpha_2 = -16, \qquad \alpha_3 = -16$

$\beta_1 = -5, \qquad \beta_2 = 7, \qquad \beta_3 = -2,$

$\gamma_1 = -4, \qquad \gamma_2 = -4, \qquad \gamma_3 = 4$

$2A\dfrac{\partial u}{\partial x} = \sum\limits_{i=1}^{3} u_i\beta_i = 3.5, \qquad 2A\dfrac{\partial u}{\partial y} = \sum\limits_{i=1}^{3} u_i\gamma_i = -4$

30. The results of the mesh refinement are presented in the table below:

Mesh	Tip Deflection (in.) $u_2(0,0)$	Bending Stress (psi.) (Location)
5 × 2	0.00363	1425.0 (9, 0.5)
10 × 2	0.00461	1907.50 (9.5, 0.5)
20 × 2	0.00495	2095.58 (9.75, 0.5)
Exact[a]	0.00500	2193.75 (9.75, 0.5)

[a]See Example 1.6; the exact deflection is $u_2(0,0) = PL^3/3EI$ and the exact stress is given by ($P = 300$ lb) $\sigma_{11} = -(Px_1x_2/I)$.

CHAPTER 4

Exercises 4.1

1. $\Sigma F_z = 0$:

$$-Q_x\, dy - Q_y\, dx + \left(Q_y + \frac{\partial Q_y}{\partial y}\, dy\right) dx + \left(Q_x + \frac{\partial Q_x}{\partial x}\, dx\right) dy + q\, dx\, dy = 0$$

or

$$\frac{\partial Q_x}{\partial x} + \frac{\partial Q_y}{\partial y} + q = 0$$

2.
$$\Pi(u_0, v_0, w) = \int_R \frac{1}{2}\left[A_{11}\left(\frac{\partial u_0}{\partial x}\right)^2 + 2A_{12}\frac{\partial u_0}{\partial x}\frac{\partial v_0}{\partial y}\right.$$

$$+ A_{22}\left(\frac{\partial v_0}{\partial y}\right)^2 + A_{66}\left(\frac{\partial u_0}{\partial y} + \frac{\partial v_0}{\partial x}\right)^2$$

$$+ D_{11}\left(\frac{\partial^2 w}{\partial x^2}\right)^2 + 2D_{12}\frac{\partial^2 w}{\partial x^2}\frac{\partial^2 w}{\partial y^2}$$

$$\left. + D_{22}\left(\frac{\partial^2 w}{\partial y^2}\right)^2 + 4D_{66}\left(\frac{\partial^2 w}{\partial x\, \partial y}\right)^2\right] dx\, dy$$

$$- \int_R qw\, dx\, dy - \int_{C_1} \hat{N}_n u_{n0}\, dS - \int_{C_2} \hat{N}_s u_{s0}\, dS$$

$$+ \int_{C_3} \hat{M}_n \frac{\partial w}{\partial n}\, dS + \int_{C_4} \hat{M}_s \frac{\partial w}{\partial s}\, dS - \int_{C_5} \hat{Q}_n w\, dS$$

3.
$$\frac{\partial^2 w}{\partial x^2} = \frac{\partial^2 w}{\partial r^2}\cos^2\theta + \frac{1}{r}\left(\frac{1}{r}\frac{\partial w}{\partial\theta} - \frac{\partial^2 w}{\partial r\, \partial\theta}\right)\sin 2\theta$$

$$+ \frac{1}{r}\left(\frac{1}{r}\frac{\partial^2 w}{\partial\theta^2} + \frac{\partial w}{\partial r}\right)\sin^2\theta$$

$$\frac{\partial^2 w}{\partial y^2} = \frac{\partial^2 w}{\partial r^2}\sin^2\theta + \frac{1}{r}\left(\frac{\partial^2 w}{\partial\theta\, \partial r} - \frac{1}{r}\frac{\partial w}{\partial\theta}\right)\sin 2\theta$$

$$+ \frac{1}{r}\left(\frac{1}{r}\frac{\partial^2 w}{\partial\theta^2} + \frac{\partial w}{\partial r}\right)\cos^2\theta$$

$$\frac{\partial^2 w}{\partial x\,\partial y} = \frac{1}{2}\left(\frac{\partial^2 w}{\partial r^2} - \frac{1}{r}\frac{\partial w}{\partial r} - \frac{1}{r^2}\frac{\partial^2 w}{\partial \theta^2}\right)\sin 2\theta$$

$$+\frac{1}{r}\left(\frac{\partial^2 w}{\partial r\,\partial\theta} - \frac{1}{r}\frac{\partial w}{\partial\theta}\right)\cos 2\theta$$

$$\Pi(w) = \frac{D}{2}\int_0^{2\pi}\int_0^a\left\{\left(\frac{\partial^2 w}{\partial r^2} + \frac{1}{r}\frac{\partial w}{\partial r} + \frac{1}{r^2}\frac{\partial^2 w}{\partial\theta^2}\right)^2\right.$$

$$+2(1-\nu)\left[\left(\frac{1}{r}\frac{\partial^2 w}{\partial r\,\partial\theta} - \frac{1}{r^2}\frac{\partial w}{\partial\theta}\right)^2\right.$$

$$\left.\left.-\frac{\partial^2 w}{\partial r^2}\left(\frac{1}{r}\frac{\partial w}{\partial r} + \frac{1}{r^2}\frac{\partial^2 w}{\partial\theta^2}\right)\right]\right\}r\,dr\,d\theta$$

$$-\int_R wq(r)r\,dr\,d\theta$$

4. $$M_r = -\left[D_{11}\frac{\partial^2 w}{\partial r^2} + D_{12}\left(\frac{1}{r}\frac{\partial w}{\partial r} + \frac{1}{r^2}\frac{\partial^2 w}{\partial\theta^2}\right)\right]$$

$$M_\theta = -\left[D_{12}\frac{\partial^2 w}{\partial r^2} + D_{22}\left(\frac{1}{r}\frac{\partial w}{\partial r} + \frac{1}{r^2}\frac{\partial^2 w}{\partial\theta^2}\right)\right]$$

$$M_{r\theta} = -D_{66}\left(\frac{1}{r}\frac{\partial^2 w}{\partial r\,\partial\theta} - \frac{1}{r^2}\frac{\partial w}{\partial\theta}\right)$$

6. $$w = \frac{Q_0 b r^2}{4D}(\ln r - 1) + \frac{c_1 r^2}{4} + c_2\ln r + c_3$$

$$w(a) = M_r(a) = 0, \qquad M_r(b) = 0$$

$$w = \frac{Q_0 a^2 b}{4D}\left\{\left(1 - \frac{r^2}{a^2}\right)\left[\frac{3+\nu}{2(1+\nu)} - \frac{b^2}{a^2 - b^2}\ln\left(\frac{b}{a}\right)\right] + \frac{r^2}{a^2}\ln\left(\frac{r}{a}\right)\right.$$

$$\left.+\frac{2b^2}{b^2 - a^2}\frac{1+\nu}{1-\nu}\ln\left(\frac{b}{a}\right)\ln\left(\frac{r}{a}\right)\right\}$$

7. $$c_1 = \frac{q_0 a^4}{64D}, \qquad c_2 = -\frac{q_0 a^2}{32D}, \qquad c_3 = \frac{q_1 a^3}{192D}, \qquad c_4 = -\frac{2q_1 a}{192D}$$

8. $$w = \frac{2q_0 a^4}{D\pi^6}\sum_{n=1,3,5\ldots}^{\infty}\frac{1}{n^6}\sin\frac{n\pi x}{a}\sin\frac{n\pi y}{a}$$

12. The result can be obtained by superposing the deflection w_1 of the simply supported, rectangular plate under uniformly distributed load and the deflection w_2 of a simply supported plate with edge moments M_y on the edges parallel to the x-axis. The expression for w_1 is given in Exercise 11. The expression for w_2 is given by

$$w_2 = \frac{a}{2\pi D} \sum_{m=1}^{\infty} \frac{\sin(m\pi x/a)}{m\cosh\alpha_m} M_m\left(\frac{b}{2}\tanh\alpha_m\cosh\frac{m\pi y}{a} - y\sinh\frac{m\pi y}{a}\right)$$

where M_m is the coefficient in the expression

$$M_y = \sum_{m=1}^{\infty} M_m\sin\frac{m\pi x}{a}$$

Determine M_m by requiring that the slopes of w_1 and w_2 are related by

$$\frac{\partial w_1}{\partial y} = -\frac{\partial w_2}{\partial y} \quad \text{at} \quad y = \pm\frac{b}{2}$$

We get

$$M_m = \frac{4q_0 a^2}{m^3\pi^3}\frac{\tanh\alpha_m(1 + \alpha_m\tanh\alpha_m) - \alpha_m}{\tanh\alpha_m(\alpha_m\tanh\alpha_m - 1) - \alpha_m}$$

14. $q = q_0\sin\dfrac{\pi x}{a}, \qquad w = w_p + w_h$

$$w_p = \frac{q_0 a^4}{D\pi^4}\sin\frac{\pi x}{a}$$

$$w_h = \frac{q_0 a^4}{D\pi^4}\left[\frac{\sinh\alpha_1}{\alpha_1 + \cosh\alpha_1\sinh\alpha_1}\left(\frac{\pi y}{a}\right)\sinh\frac{\pi y}{a}\right.$$
$$\left. - \frac{\alpha_1\cosh\alpha_1 + \sinh\alpha_1}{\alpha_1 + \cosh\alpha_1\sinh\alpha_1}\cosh\left(\frac{\pi y}{a}\right)\right]\sin\frac{\pi x}{a}$$

See Fig. E11 for the coordinate system.

15. $$q = \sum_{m=1}^{\infty} q_m\sin\frac{m\pi x}{a}$$

$$q_m = \begin{cases} 2q_0/m\pi, & m = 1,3,5,\ldots \\ 4q_0/m\pi, & m = 2,6,10\ldots \\ 0, & m = 4,8,12,\ldots \end{cases}$$

$w = w_p + w_h$

$$w_p = \frac{2q_0 a^4}{D\pi^5}\left[\sum_{m=1,3,5,\ldots}^{\infty}\frac{1}{m^5}\sin\frac{m\pi x}{a} + 2\sum_{m=2,6,10,\ldots}^{\infty}\frac{1}{m^5}\sin\frac{m\pi x}{a}\right]$$

$$w_h = -\frac{q_0 a^4}{D\pi^4} \sum_{m=1}^{\infty} \frac{W_m}{m^5} \left[\frac{\cosh\alpha_m}{2\cosh\beta_m}(2 + \beta_m\tanh\beta_m) \right.$$

$$\left. -\frac{\alpha_m}{2\cosh\beta_m}\sinh\alpha_m \right]\sin\frac{m\pi x}{a}$$

$$\alpha_m = \frac{m\pi y}{a}, \qquad \beta_m = \frac{m\pi b}{2a}$$

$$W_m = \begin{cases} 2, & m = 1,3,5,\ldots \\ 4, & m = 2,6,10,\ldots \\ 0, & m = 4,8,12,\ldots \end{cases}$$

For $a = b$, $w\left(\frac{a}{2},0\right) = 0.002028 q_0 a^4/D$.

16. Replace M_m by $4M_0/m\pi$ in w_2 of Exercise 12 to obtain the solution.

17. $w = c_1 + c_2 r^2 + \dfrac{q_0 r^5}{225aD}$;

$$c_1 = \frac{q_0 a^4}{45D}\left[\frac{4+\nu}{2(1+\nu)} - \frac{1}{5}\right], \qquad c_2 = -\frac{q_0 a^2}{90D}\left(\frac{4+\nu}{1+\nu}\right).$$

18. $w_c = \dfrac{Q_0 b^2}{16\pi D}\left[2\ln\left(\dfrac{b}{a}\right) + \left(\dfrac{a}{b}\right)^2 - 1\right].$

19. The Ritz approximations are given by

$$w_1 = c_1\left(1 - \frac{r^2}{a^2}\right), \qquad c_1 = \frac{q_0 a^4}{16(1+\nu)D}$$

$$w_2 = \left(1 - \frac{r^2}{a^2}\right)\left(c_1 + c_2\frac{r^2}{a^2}\right), \qquad c_1 = \frac{5+\nu}{1+\nu}\frac{q_0 a^4}{64D}, \qquad c_2 = -\frac{q_0 a^4}{64D}$$

The Galerkin approximation is given by

$$w_1 = c_1\left(1 - k_1\frac{r^2}{a^2} + k_2\frac{r^3}{a^3}\right), \qquad k_1 = \frac{3(2+\nu)}{4+\nu}, \qquad k_2 = \frac{2(1+\nu)}{4+\nu}$$

$$\int_0^a \left[\nabla^4 w - q_0\right] r\, dr = 0$$

where

$$\nabla^4 = \frac{1}{r}\frac{d}{dr}\left\{r\frac{d}{dr}\left[\frac{1}{r}\frac{d}{dr}\left(r\frac{d}{dr}\right)\right]\right\}$$

gives

$$c_1 = \frac{q_0 a^4}{15 D k_2} \left(\frac{10 - 5k_1 + 4k_2}{12 - 4k_1 + 3k_2} \right)$$

20.
$$\begin{bmatrix} k & \frac{k}{2}(a^2 - b^2) \\ \frac{k}{2}(a^2 - b^2) & 4(D_{11} + 2D_{12} + D_{22}) \\ & + \frac{k}{6}(a^4 + b^4 + a^2 b^2) \end{bmatrix} \begin{Bmatrix} c_1 \\ c_2 \end{Bmatrix} = \frac{q_0}{2} \begin{Bmatrix} 2 \\ (a^2 - b^2) \end{Bmatrix}$$

21. $w = c_1 \phi_1$, $\phi_1 = \left(1 - \frac{r^2}{a^2}\right)^2 \left(1 - \frac{rk_1}{a}\right)$, $k_1 = \frac{8a^2(2b^2 - a^2)}{17ab^4 + a^5 - 10a^3 b^2}$
and evaluate

$$\int_b^a \left\{ \frac{d}{dr}\left[\frac{1}{r}\frac{d}{dr}\left(r\frac{dw}{dr}\right)\right] - \frac{Q_0 b}{rD} \right\} r\, dr = 0.$$

22. $w = c_1 \left(\frac{x}{a}\right)^2 \left(\frac{y}{b}\right)^2 \left(1 - \frac{x}{a}\right)^2 \left(1 - \frac{y}{b}\right)^2$, $c_1 = \frac{6.125 q_0 a^4 b^4}{D[7(a^4 + b^4) + 4a^2 b^2]}$.

23. $c_{11} = \dfrac{P_0[1 - \cos(2\pi x_0/a)][1 - \cos(2\pi y_0/b)]}{4\pi^4[3D_{11}/a^4 + 2(2D_{66} + D_{12})/a^2 b^2 + 3D_{22}/b^4]}$.

24. $w = c_{11} \sin\dfrac{\pi x}{a}\sin\dfrac{\pi y}{b}$, where c_{11} is given by

$$\frac{4q_0(ab/\pi^2)}{(ab/4)\left[D_{11}(\pi/a)^4 + 2(D_{12} + 2D_{66})(\pi/a)^2(\pi/b)^2 + D_{22}(\pi/b)^4 + k\right]}$$

25. Same as in Exercise 24, except the numerator in c_{11} should be replaced by P_0.

26. $w = \displaystyle\sum_{m=1}^{\infty}\sum_{n=1}^{\infty} c_{mn}\sin\frac{m\pi x}{a}\sin\frac{n\pi y}{a}$, $m, n = 1, 3, 5, \ldots$

$$c_{mn} = \frac{32q_0 a^4 \sin(m\pi/2)}{m^2 n\pi^7 D(m^2 + n^2)^2}$$

$$w_{max} = w\left(\frac{a}{2}, \frac{a}{2}\right) = 0.002625\frac{q_0 a^4}{D}$$

Exercises 4.2

1. *Strains:*

$$e_r = \frac{du_0}{dr} + z\frac{d\psi}{dr}, \qquad e_\theta = \frac{u_0 + z\psi}{r}, \qquad e_z = 0$$

$$2e_{r\theta} = 0, \qquad 2e_{rz} = \psi + \frac{dw_0}{dr} \tag{a}$$

Equilibrium equations:

$$-\frac{d}{dr}(rN_r) + N_\theta = 0$$

$$-\frac{d}{dr}(rQ) = qr \tag{b}$$

$$-\frac{d}{dr}(rM_r) + M_\theta + Qr = 0$$

Constitutive equations:

$$N_r = A_{11}\frac{du_0}{dr} + A_{12}\frac{u_0}{r}$$

$$N_\theta = A_{12}\frac{du_0}{dr} + A_{22}\frac{u_0}{r}$$

$$M_r = D_{11}\frac{d\psi}{dr} + D_{12}\frac{\psi}{r} \tag{c}$$

$$M_\theta = D_{12}\frac{d\psi}{dr} + D_{22}\frac{\psi}{r}$$

$$Q = A_{55}\left(\psi + \frac{dw_0}{dr}\right)$$

2. The Euler equations are given by Eq. (b) of Exercise 1 [with N_r, N_θ and Q given by Eq. (c)], and

$$\frac{d\psi}{dr} + \bar{D}_{22}M_r - \bar{D}_{12}M_\theta = 0, \qquad \frac{\psi}{r} + \bar{D}_{11}M_\theta - \bar{D}_{12}M_r = 0$$

3. From Exercise 1, the governing equations are

$$-A_{55}\frac{d}{dr}\left[r\left(\psi + \frac{dw_0}{dr}\right)\right] = q_0 r$$

$$-\frac{d}{dr}\left[r\left(D_{11}\frac{d\psi}{dr} + D_{12}\frac{\psi}{r}\right)\right] + D_{12}\frac{d\psi}{dr} + D_{22}\frac{\psi}{r} + r\left(\psi + \frac{dw_0}{dr}\right)A_{55} = 0$$

These can be combined into a single equation ($A_{55} = K^2 G_{13} h$)

$$D_{11}\frac{d}{dr}\left(r\frac{d^2 w_0}{dr^2}\right) - D_{22}\left(\frac{1}{r}\frac{dw_0}{dr}\right) = \frac{q_0 r^2}{2} + (D_{22} - D_{11})\frac{q_0}{2A_{55}}$$

For isotropic case $[D_{11} = D_{22} = D = Eh^3/12(1 - \nu^2)]$, the solution is given by

$$w_0 = \frac{q_0 r^4}{64D} + \frac{c_1 r^2}{4} + c_2 \ln r + c_3$$

The constants c_1 and c_3 ($c_2 = 0$ because w_0 is finite at $r = 0$) are determined using the boundary conditions $(dw/dr)(a) = w(a) = 0$. We obtain

$$w_0 = \frac{q_0 a^4}{64D}\left[1 - \left(\frac{r^2}{a^2}\right)^2\right]^2 + \frac{q_0 a^2 h^2}{24KD(1 - \nu)}\left(1 - \frac{r^2}{a^2}\right)$$

4. $P_1 P_2 P_3(\omega^2)^3 + (K_{11}P_2 P_3 + K_{22}P_1 P_3 + K_{33}P_2 P_1)(\omega^2)^2$

$-[P_1(K_{23}^2 - K_{22}K_{33}) + P_2(K_{13}^2 - K_{11}K_{33}) + P_3(K_{12}^2 - K_{11}K_{22})]\omega^2$

$+ K_{11}K_{22}K_{33} + 2K_{12}K_{13}K_{23} - K_{11}K_{23}^2 - K_{22}K_{13}^2 - K_{33}K_{12}^2 = 0$

where K_{ij} are given in Eq. (4.2.26), P_i are given by $P_1 = \rho h$, $P_2 = P_3 = \rho h^3/12$, and ω is the frequency of natural vibration.

6. $$\begin{bmatrix} c_{11} & c_{12} & c_{13} \\ c_{12} & c_{22} & c_{23} \\ c_{13} & c_{23} & c_{33} \end{bmatrix}\begin{Bmatrix} W \\ X \\ Y \end{Bmatrix} = \begin{Bmatrix} q_0 \\ 0 \\ 0 \end{Bmatrix}$$

where

$c_{11} = (\alpha^2 K_2^2 G_{13} + \beta^2 K_1^2 G_{23})h$, $c_{12} = \alpha K_2^2 G_{13} h$,
$c_{13} = \beta K_1^2 G_{23} h$, $c_{22} = K_2^2 G_{13} h + \alpha^2 D_{11} + \beta^2 D_{66}$, $c_{23} = \alpha\beta(D_{12} + D_{66})$,
$c_{33} = K_1^2 G_{23} h + \alpha^2 D_{66} + \beta^2 D_{22}$.

Exercises 4.3

2. See Eqs. (1.5.29).

4. $$\begin{bmatrix} c_{11} & c_{12} & c_{13} \\ c_{12} & c_{22} & c_{23} \\ c_{13} & c_{23} & c_{33} \end{bmatrix}\begin{Bmatrix} u \\ v \\ w \end{Bmatrix} = \begin{Bmatrix} 0 \\ 0 \\ q_0 \end{Bmatrix}$$

where

$c_{11} = A_{11}\alpha^2 + A_{66}\beta^2$, $c_{12} = (A_{12} + A_{66})\alpha\beta$, $c_{13} = -B_{11}\alpha^3 - (B_{12} + 2B_{66})\alpha\beta^2$, $c_{22} = A_{66}\alpha^2 + A_{22}\beta^2$, $c_{23} = -(B_{12} + 2B_{66})\alpha^2\beta - B_{22}\beta^3$, $c_{33} = D_{11}\alpha^4 + 2(D_{12} + 2D_{66})\alpha^2\beta^2 + D_{22}\beta^4$.

5. Same as the expressions in Exercise 4, except for the following coefficients: $c_{13} = -3\alpha^2\beta B_{16} - \beta^3 B_{26}$, $c_{23} = -3\alpha\beta^2 B_{26} - \alpha^3 B_{16}$.

6. A special case of Exercise 7 below.

7. The same coefficient matrix as in Exercise 4 (with all $B_{ij} = 0$) but the right-hand column should be modified as follows:

$$\begin{Bmatrix} F_1 \\ F_2 \\ q_0 \end{Bmatrix}, \qquad \begin{aligned} F_1 &= \alpha\left(A_{11}\alpha_x + A_{12}\alpha_y\right)T_0 \\ F_2 &= \beta\left(A_{12}\alpha_x + A_{22}\alpha_y\right)T_0 \end{aligned}$$

where α_x and α_y coefficients of thermal expansion along the x and y-coordinates, respectively.

(a) For uniform temperature change, the matrix equation should be solved for each m and n (which are included in the definition of α and β) and should be summed as in Eq. (4.3.28).

(b) For sinusoidal temperature change, the solution is given by setting $m = n = 1$.

8. $P = [D_{11}\alpha^2 + 2(D_{12} + 2D_{66})\beta^2 + D_{22}(\beta^4/\alpha^2)]$

The smallest value of P occurs when $n = 1$ ($\beta = \pi/b$); the smallest value of P for various m ($\alpha = m\pi/a$) varies with the plate aspect ratio and stiffnesses, D_{ij}.

Exercises 4.4

1. $u_3 = -\dfrac{q_0 R^2}{2 Eh}[2 - e^{-\beta b}\cos\beta b - e^{-\beta c}\cos\beta c]$

where $b + c = L_0$, b is the distance from the origin of the coordinate system (See Fig. 4.19) to the left end of the load distribution, and c is the distance from the origin to the right end of the load distribution.

2. Take the origin of the coordinate system at the left-most point of the bottom, with x_1 along the vertical wall and x_3 to the right along the base of the tank (note that the front view of the tank has its height along the x_1-axis). The boundary conditions are

$$u_3(0, x_3) = \frac{du_3}{dx_1}(0, x_3) = 0$$

The load q in the first of Eq. (4.4.27) should be replaced by $q = \gamma(b - x_1)$. The solution is: $u_3 = u_{3p} + u_{3h}$; $u_{3p} = -\gamma(b - x_1)a^2/Eh$; $u_{3h} =$

$e_1^{-\beta x}[c_1\cos\beta x_1 + c_2\sin\beta x_1]$. The $e^{\beta x_1}$ term should be zero because u_3 is finite for all x_1. Thus, the radial deflection is

$$u_3 = -\frac{\gamma a^2 b}{Eh}\left\{1 - \frac{x_1}{b} - e^{\beta x_1}\left[\cos\beta x_1 + \left(1 - \frac{1}{\beta b}\right)\sin\beta x_1\right]\right\}$$

The second equation in Eq. (4.4.27) gives the vertical deflection

$$u_1 = \int_0^b v\frac{u_3}{a}\,dx_1 = u_0, \qquad u_1(0) = 0$$

The stresses are

$$N_\theta = -\frac{Eh}{a}u_3, \qquad M_1 = -D\frac{d^2u_3}{dx_1^2}, \qquad M_\theta = vM_1$$

3. $u_3 = -\alpha RT\left[e^{-\beta x_1}(\cos\beta x_1 + \sin\beta x_1) - 1\right]$

$M_1 = -2\alpha RT\beta^2 De^{-\beta x_1}\cos\beta x_1$

where T is the temperature change (see Fig. 4.20 for the coordinate system).

Exercises 4.5

1. $[K^{14}] = \dfrac{1}{4}\begin{bmatrix} 1 & 0 & -2 & 0 \\ 0 & -1 & 0 & 2 \\ -2 & 0 & 1 & 0 \\ 0 & 2 & 0 & -1 \end{bmatrix}$

$[K^{44}] = \dfrac{ab}{36D_{66}}\begin{bmatrix} 4 & 2 & 1 & 2 \\ 2 & 4 & 2 & 1 \\ 1 & 2 & 4 & 2 \\ 2 & 1 & 2 & 4 \end{bmatrix}$

3. $[\hat{K}^{11}]\{\hat{W}\} + [\hat{K}^{12}]\{\hat{M}\} = \lambda_v[\hat{S}]\{\hat{W}\} + \lambda_b[\hat{G}]\{\hat{W}\}$

$[\hat{K}^{12}]^T\{\hat{W}\} - [\hat{K}^{22}]\{\hat{M}\} = \{0\}$

Solve the second equation for $\{\hat{M}\}$ and substitute into the first equation to obtain the required result

$$[\bar{K}] = [\hat{K}^{11}] + [\hat{K}^{12}][\hat{K}^{22}]^{-1}[\hat{K}^{12}]^T$$

4.
$$\begin{bmatrix} [K^{11}] & [K^{12}] \\ [K^{12}]^T & [K^{22}] \end{bmatrix} \begin{Bmatrix} \{u\} \\ \{v\} \end{Bmatrix} = \begin{Bmatrix} \{F^1\} \\ \{F^2\} \end{Bmatrix}$$

$$\begin{bmatrix} [\hat{K}^{11}] & [\hat{K}^{12}] & [\hat{K}^{13}] \\ [\hat{K}^{12}]^T & [\hat{K}^{22}] & [\hat{K}^{23}] \\ [\hat{K}^{13}]^T & [\hat{K}^{23}]^T & [\hat{K}^{33}] \end{bmatrix} \begin{Bmatrix} \{w\} \\ \{\psi_x\} \\ \{\psi_y\} \end{Bmatrix} = \begin{Bmatrix} \{\hat{F}^1\} \\ \{\hat{F}^2\} \\ \{\hat{F}^3\} \end{Bmatrix}$$

$$K_{ij}^{11} = \int_{R^e} \left(A_{11} \frac{\partial \psi_i}{\partial x} \frac{\partial \psi_j}{\partial x} + A_{66} \frac{\partial \psi_i}{\partial y} \frac{\partial \psi_j}{\partial y} \right) dx\, dy$$

$$K_{ij}^{12} = \int_{R^e} \left(A_{12} \frac{\partial \psi_i}{\partial x} \frac{\partial \psi_j}{\partial y} + A_{66} \frac{\partial \psi_i}{\partial y} \frac{\partial \psi_j}{\partial x} \right) dx\, dy$$

$$K_{ij}^{22} = \int_{R^e} \left(A_{66} \frac{\partial \psi_i}{\partial x} \frac{\partial \psi_j}{\partial x} + A_{22} \frac{\partial \psi_i}{\partial y} \frac{\partial \psi_j}{\partial y} \right) dx\, dy$$

$$\hat{K}_{ij}^{11} = \int_{R^e} \left(A_{55} \frac{\partial \psi_i}{\partial x} \frac{\partial \psi_j}{\partial x} + A_{44} \frac{\partial \psi_i}{\partial y} \frac{\partial \psi_j}{\partial y} \right) dx\, dy$$

$$\hat{K}_{ij}^{12} = \int_{R^e} A_{55} \frac{\partial \psi_i}{\partial x} \psi_j \, dx\, dy$$

$$\hat{K}_{ij}^{13} = \int_{R^e} A_{44} \frac{\partial \psi_i}{\partial y} \psi_j \, dx\, dy$$

$$\hat{K}_{ij}^{22} = \int_{R^e} \left(D_{11} \frac{\partial \psi_i}{\partial x} \frac{\partial \psi_j}{\partial x} + D_{66} \frac{\partial \psi_i}{\partial y} \frac{\partial \psi_j}{\partial y} + A_{55} \psi_i \psi_j \right) dx\, dy$$

$$\hat{K}_{ij}^{23} = \int_{R^e} \left(D_{12} \frac{\partial \psi_i}{\partial x} \frac{\partial \psi_j}{\partial y} + D_{66} \frac{\partial \psi_i}{\partial y} \frac{\partial \psi_j}{\partial x} \right) dx\, dy$$

$$\hat{K}_{ij}^{33} = \int_{R^e} \left(D_{66} \frac{\partial \psi_i}{\partial x} \frac{\partial \psi_j}{\partial x} + D_{22} \frac{\partial \psi_i}{\partial y} \frac{\partial \psi_j}{\partial y} + A_{44} \psi_i \psi_j \right) dx\, dy$$

5.
$$K_{ij}^{11} = \int_{r_e}^{r_{e+1}} \left[A_{11} \frac{d\psi_i}{dr} \frac{d\psi_j}{dr} r + A_{12} \left(\psi_i \frac{d\psi_j}{dr} + \frac{d\psi_i}{dr} \psi_j \right) + A_{22} \frac{1}{r} \psi_i \psi_j \right] dr$$

$$K_{ij}^{12} = \int_{r_e}^{r_{e+1}} \frac{dw}{dr} \left[A_{11} r \frac{d\psi_i}{dr} \frac{d\psi_j}{dr} + A_{12} \psi_i \frac{d\psi_j}{dr} \right] dr$$

$$K_{ij}^{23} = \int_{r_e}^{r_{e+1}} A_{33} \frac{d\psi_i}{dr} \psi_j r \, dr$$

$$K_{ij}^{22} = \int_{r_e}^{r_{e+1}} \left[A_{11} \frac{2}{3} \left(\frac{dw}{dr} \right)^2 + A_{33} \right] r \frac{d\psi_i}{dr} \frac{d\psi_j}{dr} \, dr$$

$$K_{ij}^{33} = \int_{r_e}^{r_{e+1}} \left[A_{33} \psi_i \psi_j r + D_{11} \frac{d\psi_i}{dr} \frac{d\psi_j}{dr} r + D_{12} \left(\psi_i \frac{d\psi_j}{dr} + \frac{d\psi_i}{dr} \psi_j \right) \right.$$

$$\left. + D_{22} \frac{1}{r} \psi_i \psi_j \right] dr$$

6. The coefficients $K_{ij}^{\alpha\beta}$ ($\alpha, \beta = 1, 2, 3$) are as defined in Exercise 5, and

$$\overline{K}_{ij}^{33} = A_{33} \int_{r_e}^{r_{e+1}} r \psi_i \psi_j \, dr$$

$$K_{ij}^{34} = \int_{r_e}^{r_{e+1}} \frac{d\psi_i}{dr} \psi_j r \, dr$$

$$K_{ij}^{35} = \int_{r_e}^{r_{e+1}} \psi_i \psi_j \, dr$$

$$K_{ij}^{44} = \int_{r_e}^{r_{e+1}} \overline{D}_{22} r \psi_i \psi_j \, dr$$

$$K_{ij}^{45} = -\int_{r_e}^{r_{e+1}} \overline{D}_{12} \psi_i \psi_j r \, dr$$

$$K_{ij}^{55} = \int_{r_e}^{r_{e+1}} \overline{D}_{11} \psi_i \psi_j r \, dr$$

11. For simplicity, assume that the x-axis coincides with the interface ($y = 0$) of two triangular plate bending elements. From Eq. (4.5.12), we have

$$w^{(e)} = a_1^{(e)} + a_2^{(e)} x + a_4^{(e)} x^2 + a_6^{(e)} x^3$$

$$\left(\frac{\partial w}{\partial x} \right)^{(e)} = a_2^{(e)} + 2a_4^{(e)} x + 3a_6^{(e)} x^2$$

for $e = 1, 2$. Thus, $\partial w / \partial x$ varies quadratically along the interface. However, only two values of $\partial w / \partial x$ are known in each element on the interface. Since three constants are needed to uniquely define a quadratic function, it follows that $\partial w / \partial x$ from the two elements is not the same along the interface.

12. Yes. The mixed and shear deformable elements are Lagrange type elements (see Reddy, 1984) in which nodal variables are only the dependent variables, but not their derivatives.

13. The refined theory contains the fourth-order derivatives of the transverse deflection (like in the classical plate theory) and, therefore, the same interpolation functions as in the classical plate theory are required to interpolate the transverse deflection.

INDEX

ABOUT THE AUTHOR

J. N. Reddy is a Professor in the Engineering Science and Mechanics Department at Virginia Polytechnic Institute and State University. Previously, he served as Associate Professor, School of Aerospace, Mechanical and Nuclear Engineering, University of Oklahoma, and as a research scientist at Lockheed Missiles and Space Company. He earned his Ph.D. in Engineering Mechanics at the University of Alabama at Huntsville, his M.S. in Mechanical Engineering at Oklahoma State University, and his B.E. in Mechanical Engineering at Osmania University in India. He is co-author (with J. T. Oden) of two previous books: *An Introduction to the Mathematical Theory of Finite Elements* (Wiley, 1976) and *Variational Methods in Theoretical Mechanics*. Dr. Reddy is listed in *Who's Who in Technology Today*, *Who's Who in Computer Education and Research*, *Who's Who in South and Southwest*, and *American Men and Women of Science*. He received the award for Outstanding Faculty Achievement in Research from the College of Engineering, University of Oklahoma, and is a member of numerous professional organizations.